Advances in Phytomedicine Series, Volume Two

Editors: Mahmud T H Khan and Arjumand Ather

Lead Molecules from Natural Products: Discovery and New Trends

Advances in Phytomedicine 2

Lead Molecules from Natural Products: Discovery and New Trends

Mahmud T H Khan
Department of Pharmacology, Institute of Medical Biology, University of Tromso, Tromso, Norway

Arjumand Ather
H.E.J. Research Institute of Chemistry, International Center for Chemical Sciences, University of Karachi, Karachi, Pakistan

ELSEVIER

Amsterdam • Boston • Heidelberg • London • New York • Oxford • Paris
San Diego • San Francisco • Singapore • Sydney • Tokyo

Elsevier
Radarweg 29, PO Box 211, 1000 AE Amsterdam, The Netherlands
The Boulevard, Langford Lane, Kidlington, Oxford OX5 1GB, UK

First edition 2006

Library of Congress Cataloguing-in-Publication Data
A catalogue record for this book is available from the Library of Congress

British Library Cataloguing in Publication Data
A catalogue record for this book is available from the British Library

ISBN-13: 978-0-444-51619-0
ISBN-10: 0-444-51619-0
ISSN: 1572-557X

Transferred to Digital Printing 2007

Contents

This book is dedicated to our parents, and to our son Araz.

Preface to the Series

The systematic study of herbal medicinal products and the investigation of the biologically active principles of phytomedicines, including their clinical applications, standardization, quality control, mode of action and potential drug interactions have emerged as one of the most exciting developments in modern therapeutics and medicine. Studies in phytomedicine have moved from purely descriptive analytical studies to conceptual inquiries on the pharmacodynamic advantages and limitations of plant medicines for the treatment of moderate or moderately severe diseases and prevention. Healthcare practitioners and medical scientists have come to accept herbal medicinal products as drugs that are different from the pharmacologically active molecules that they may contain. Several comparative clinical studies have been published to show that these plant medicines could have full therapeutic equivalence with chemotherapeutic agents, while retaining the simultaneous advantage of being devoid of serious adverse effects. Developments in molecular biology and information technology have now made it possible for us to begin to understand the mechanism of action of many herbal drugs and the associated phytomedicines, which differ in many respects from that of synthetic drugs or single chemical entities. Herbal medicinal products are now generally available in both industrialized countries and traditional societies. With the current lack of standardization and regulation of herbal products, it is important to develop common criteria for judging safety and efficacy of phytotherapeutic agents.

This 'new' science demands different approaches to the classical methods of drug analysis, dosage formulations, manufacturing and claims substantiation. The therapeutic response observed with most herbs and phytomedicines are often not fully explainable using the currently available methods. Their activity usually characterized as polyvalent and interpreted as an aggregate or additive outcome of several constituents in the plant medicines are subjects of intense pharmacological studies. In most cases, a rationale does not even exist for the observed pharmacodynamic effects of very low doses of phytomedicines after prolonged or long-term application.

The public press is replete with lay information and claims about the use of herbal remedies, however, there is scarcity of scientifically accurate reviews and guides on the efficacy and safety of plant medicines. The time therefore seemed ripe to broaden the communication on the use and benefits of phytomedicines as safe and useful natural health care products aimed at the health professionals and scientists. It is also important to provide a broader dissemination of the extant scientific literature on phytomedicines, to enable the conventional medical community to fully appreciate the fact that plant extracts specifically, and natural products generally, offer valuable and needed benefits in the treatment and prevention of diseases, especially for conditions where there is no effective or generally acceptable drugs.

Although there are many papers published yearly on the use and analysis of plants as sources of biologically active molecules and some published materials on the use of plants as medicinal substances, volumes specifically addressing the needs of scientists and clinicians in the use of herbal products and standardized extracts as medicinal agents are far-in-between. We considered it therefore appropriate to fill this gap by producing an entirely new multi-volume series on the sourcing, selection, standardization, safety and clinical application of herbal medicinal products. As the name of the series implies, emphasis will be on those herbal medicinal products that are well characterized, standardized and substantiated as phytomedicines. The series will also provide timely reviews on the industrial production, regulatory and policy issues related to the use of phytomedicines. Although the literature in this field is evolving so rapidly that books on the subject become obsolete soon after they leave the press, the volumes in this series will aim at capturing the fundamental framework of each topic while remaining thoroughly up-to-date and comprehensive in scope.

The aim of this series is to present to the scientists and clinicians the state of current knowledge in various fields of phytomedicine research, development and use. The approach is to provide the historical background to each topic, discuss methodological issues and illustrate current trends with case studies and critical examples, and when ever possible, the authors will indicate those plant medicines that are available for immediate use in clinical settings. The series is not intended to serve or replace the many excellent journals in this field but it will rather attempt to distill information from primary references and introduce elements of medicinal plants research and development that are in transition from speculative knowledge to standard practice. *Advances in Phytomedicine* is also not meant to be a textbook of the various topics covered in the series. It will, however, provide a guide to specialized articles and books on the topics that are relevant to scientific research and development of plant medicines, as well as information on the regulatory issues, clinical trials and application of phytomedicines for healthcare.

Advances in Phytomedicine will therefore serve as a platform for reviewing recent developments in the use of herbal medicinal products. The coverage will include reviews of studies and use of all plant medicines, phytotherapeutic agents, nutraceuticals, plant cosmetics and therapeutically important molecules derived from these plant medicines. The first volume in this series has been devoted to exploring the ethnomedical approaches to drug discovery. It is indeed a very important starting point in addressing the relationship between plants and human health. Subsequent volumes will deal with other aspects of the use of herbal medicinal products. Selection of subjects will be through consultations with experts in the various fields of interest. We shall be guided by the principles of the so-called 6S, that is herb selection, sourcing, structure, standardization, safety and substantiation. Acknowledged experts and authorities in the various aspects of phytomedicine will be invited to assemble and edit specific volumes in the series.

The series is the outcome of extensive consultations among several biomedical scientists and clinicians who participated in the workshops and conferences organized by the Bioresources Development and Conservation Program (BDCP) on related subjects. I am immensely grateful to these colleagues for their support in developing the original concept. Many thanks go also to Ms. Kim Briggs and Ms. Joke Zwetsloot of Elsevier Science for their suggestions and help in producing this series. I am indebted to my colleagues at BDCP and the International Centre for Ethnomedicine and Drug Development for their contribution; and to my wife, Kate for her love and support. I acknowledge the International Cooperative Biodiversity

Group (ICBG) of the Fogarty International Centre, United States National Institutes of Health for providing financial support to my research group at the Walter Reed Army Institute of Research and BDCP.

Maurice M. Iwu
M.Pharm., Ph.D.
Séries Editor

Preface

Lead Molecules from Natural Products: Discovery and New Trends provides the reader with a thorough overview of current discoveries and trends in Natural Products research. This book consists of 22 chapters from well known scientists all over the world, with topics ranging from Natural Product Chemistry and Phytochemistry in their most basic form, to Molecular Biology and to *in silico* Drug design. The chapters, all of equally high quality, summarize years of extensive research in each area.

In Chapter 1, Do and Bernard, cite Michel de Montaigne, a French writer of the 16th century "Mother Nature can do everything and does everything", thus setting the tone of their chapter.

Nature provides the modern medical world with a number of highly potential drugs. The World Health Organization (WHO) states that countries in Africa, Asia and Latin America use traditional medicine (TM) to meet their primary health care needs, in Africa 80% of the population uses TM. This chapter gives special attention to Traditional Medicine and summarizes the discovery of potential drug molecules. Do and Bernard introduce a new approach named "*reverse pharmacognosy*". By coupling this approach to high-throughput screening (HTS), to *in silico* technologies (like virtual screening and computational ligand design), and with a knowledgebase from traditional medicine, this new approach will no doubt increase the speed of the drug discovery process.

In Chapter 2, Mischiati et al. explain how DNA hybridization arrays (macro-arrays and microarrays) are used for the analysis of gene expression profiles in human pathologies showing examples from their own research, using Bangladeshi plant extracts, like *Moringa oleifera*, *Emblica officinalis*, etc.

In Chapter 3, from the same research group, Lampronti et al. discuss the effects of medicinal plant extracts on the molecular interactions between DNA and different transcription factors. In an effort to determine whether inhibitory activities are present in extracts from medicinal plants, different plant extracts are tested for their ability in inhibiting the interactions between nuclear factors isolated from the human leukemic K562 cells and a GATA-1, AP-1, Sp1, NF-kB, NF-IL2A target double stranded oligonucleotide. They describe the discovery of several molecules responsible for these activities using GC coupled with MS technologies.

Lampronti et al. in chapter 4 discuss their experiments and results with the Bangladeshi plant extracts (e.g., *Emblica officinalis*, *Aegle marmelos*, etc.) as anti-tumor resources, and discuss the possibilities of several high-throughput methods, such as LC-MS, GC-MS, Surface Plasmon Resonance (SPR)-based Biospecific Interaction Analysis (BIA) employing biosensors, etc., to identify responsible and promising molecules as potential *leads*.

In chapter 5, Thompson discusses the impact of herbal medicines and plants extracts on viral infections, especially on the infections of herpes simplex virus (HSV).

Nosalova et al. discuss the cough reflex and its pathophysiology, as well as drugs modulating the reflex mechanisms in chapter 6. Franova et al. in chapter 7 discuss phytotherapy of this cough reflex. They also discuss different plant materials, their composition and their therapeutic uses.

Gali-Muhtasib and Schneider-Stock in chapter 8, discuss the composition and medicinal potentials of *Nigella sativa*, medicines derived from *Nigella sativa* are commonly used in TM, especially in the Middle East and Asia. In chapter 9, Gali-Muhtasib discusses recent advancements and future prospects of cyclin-dependent kinase (CDKs) inhibitors from natural resources. The same author, in chapter 10, discusses details of the anticancer potentials of the Eastern Mediterranean Sage (*Salvia triloba*).

The marine atmosphere contains an enormous resource of potent bioactive molecules. Costa-Lotufo et al. in chapter 11, provide an in-depth discussion of potential anti-cancer molecules from marine organisms. These bioactive molecules have been discovered during their years of laboratory experiences. Pessoa et al. from the same research group also explained in the chapter 12 about the anticancer potentials of the Brazilian terrestrial plants, with published and unpublished data from their own and other laboratories.

In chapter 13, Moraes et al. discuss different legal, ethical, safety and efficacy issues of phytomedicine and Traditional Medicine. These issues are important especially in the developing countries where, according to the WHO, a majority of the populations are depending on TM for their primary health care system.

Maciel et al. in their chapter 14, present a huge amount of data on biochemical profiling and pharmacological potentials about some prospective lead molecules from an important Amazonian medicinal plant – *Croton cajucara*.

In chapter 15, Matsui and Matsumoto talked about the potentials of natural antihypertensive peptides with large number of examples and their mechanism of actions.

Diderot et al. in their chapter (16) discuss the chemistry and pharmacology of xanthons, their discussions founded with a multitude of chemical data and results.

Gambari in chapter 17 systematically reviews nature as a potential resource for anti-HIV-1 molecules, with large tabular data for the readers. He carefully analyses the molecules according to their action(s) on different steps in the life cycle of human immunodeficiency virus type 1.

In chapter 18, Fernández discusses the anticancer potentials of saffron (*Crocus sativus*) from his own research facilities. Orhan and Şener in their chapter 19, discussed about their discovery of large number of molecules of Turkish origin as potential '*Lead*'. These are mainly of alkaloids, terpenoids and steroids, which possess significant anti-cholinesterase, anti-cholinergic, anti-hypertensive, anti-thrombocytes, anti-inflammatory, anti-malarial, antibacterial, activities.

Kubo et al,. in chapter 20, describe the molecular design of multifunctional anti-*Salmonella* agents based on natural products. They discuss a series of aliphatic (2*E*)-alkenals and alkanals from C_5 to C_{13}.

Céspedes et al. in chapter 21, discusses the plant growth inhibitors (PGIs) from secondary metabolites isolated and reported from the Latin American flora. In this chapter they review a large number of PGI compounds, monoterpenes, diterpenes, sesquiterpene lactones, limonoids, triterpenes, coumarins and flavonoids, etc., from plants belonging to the families Asteraceae, Celastraceae, Gomortegaceae, Lauraceae, Meliaceae, Monimiaceae, Poaceae and Winteraceae, etc.

In chapter 22, we, the editors Khan and Ather, discuss the relation between the different issues on the current hot theme '*Metabolomics*'.

The contents of the book are structured according to the central and recent hot topics for the natural product researcher. Most contributors discuss and describe their own laboratory experiences – their findings, including legal issues. So we believe that this book may be considered a *handbook* for researchers and for readers.

We are confident that this book will give new insight and shed light on new themes of natural product research, and thus will help us to predict promising leads, useful for physicians in the treatment of different diseases and disease manifestations.

The editors are grateful to the many well-known scientists, all experts in their fields, from all over the globe who contributed a chapter in this book.

Mahmud Tareq Hassan Khan, Arjumand Ather
Tromso, Norway
January, 2006

M.T.H. Khan and A. Ather (eds.)
Lead Molecules from Natural Products

Reverse pharmacognosy: a new concept for accelerating natural drug discovery

QUOC-TUAN DO, PHILIPPE BERNARD

Abstract

Combinatorial chemistry and high throughput screening (HTS) led to identification of numerous *in vitro* active and selective hits. Nevertheless, finding *in vivo* active drugs remains a difficult challenge. Traditional medicine cures based on natural materials have proven useful for many populations worldwide. These huge and disperse amounts of knowledge are sometimes neglected in Western research because of great differences in the concepts of illness. We introduce here a new approach named "reverse pharmacognosy" (from diverse molecules to plants) coupled with pharmacognosy (from biodiverse plants to molecules). Reverse pharmacognosy utilizes new techniques, such as HTS, virtual screening and knowledge database with traditional usage of plants. Integrating pharmacognosy and reverse pharmacognosy in the research process, provided an efficient and rapid tool for natural drug discovery.

Keywords: *pharmacognosy, reverse pharmacognosy, molecular diversity, biodiversity, HTS, virtual screening, database*

I. Introduction

The development of a new drug is a long and costly process. A survey in 2001 by Tollman et al. (2001) showed that the average time is 15 years and the cost is $880 million from the target validation to regulatory approval. And failures along the discovery process represent 75% of the whole cost. Effort to accelerate and rationalize the drug discovery lead to high throughput methods, such as combinatorial chemistry to generate potential hits and high throughput screening (HTS) for biological testing. Unfortunately, these techniques did not increase the discovery of new active entities (Myers, 1997; Lahana, 1999). Moreover, crucial questions appeared at different levels. From the point of view of chemistry: how to get diverse libraries? As they depend on corporate libraries of all sources and the chemistry know-how of research groups. The postulate is that the more diverse a chemical library is, the more likely it contains molecules that interact with biological targets. After finding hits, which one(s) will become lead(s) i.e. will be promising for chemistry modulation and optimization? Considering biology: there is the issue of target validation (which is not the scope of our article) and the drug-likeness of compounds being evaluated for their biological activity, i.e. absorption, distribution, metabolism, excretion and toxicity properties (ADMET), and their behavior *in vitro* and *in vivo*. For example, the well-known phosphodiesterase 4 inhibitors (IPDE4) case, most of IPDE4 are very active and selective *in vitro*, but most of them are inactive *in vivo*, and the active ones suffer from emetic side effects (Burnouf & Pruniaux, 2002). The patentability of a compound is also a crucial issue, and backup leads or the ability to jump to another class of molecules must be considered in early stages of the discovery process, in case of patent problems.

Owing to legislation and market pressures to have proven active products with an underlined biological mechanism, the cosmetic industry tends to adopt some pharmaceutical approaches, though the means are not comparable. Nevertheless, the goals of pharmaceutical and cosmetic industries are the same: the discovery of new bioactive entities. And they both have to address some, if not all the questions mentioned above.

"Mother Nature can do everything and does everything": this quote by Michel de Montaigne, a French writer of 16th century, appears particularly true concerning biodiversity and Nature chemistry that respond to the lack of the chemical diversity of synthetic compounds. Chemical "know-how" of Mother Nature has no comparison (Verdine, 1996). A study from Feher and Schmidt (2003) showed that for a natural compound database and a combinatorial chemistry database gathered from commercial providers, the mean number of chiral centers per molecule is, respectively 6.2 and 0.4. Biological wise, natural compounds can be extremely active, such is tetrodotoxin from Tetraodons, which has a dissociation constant to Na^+ channels of $K_d = 10^{-10}$ M (Kao, 1972). Nowadays, drugs derived from natural sources represent 25% of the prescription medicine market; this number would be higher considering the whole drug market (Duke, 1990). This renew interest for natural materials (Harvey, 1999, 2000) from plants, fungi, microorganisms or invertebrates led to the creation of several companies that work with natural substances e.g. Entomed for insects, Pharmamar for marine organisms. Traditional medicines also revive lots of enthusiasm. As a matter of fact, most of the traditional cures have been validated for their efficiency over a long time, so their *in vivo* activity is demonstrated. Well-established disciplines as pharmacognosy, particularly ethnopharmacology (Massiot, 1991; Anton, 1999) and phytochemistry (Tiew et al., 2003; Jossang et al., 2003) contributed and still do a lot for the discovery of new bioactive natural molecules e.g. taxol (Potier et al., 1994). But using the knowledge of a medicine man and traditional pharmacopoeia is not an easy task. Several works have been published in favor of integrating traditional knowledge into user-friendly database, and integrating them into early drug discovery stages (Yan et al., 1999; Bernard et al., 2001).

Greenpharma discovery platform was created to accelerate the discovery of new bioactive principles for cosmetic and pharmaceutical industries by answering the questions raised above. This unique platform allows us to set up two convergent parallel approaches: a well-established pharmacognosy approach and a new concept we name reverse pharmacognosy. As illustrated in Figure 1, pharmacognosy is from plants to molecules and reverse pharmacognosy from molecules to the plants.

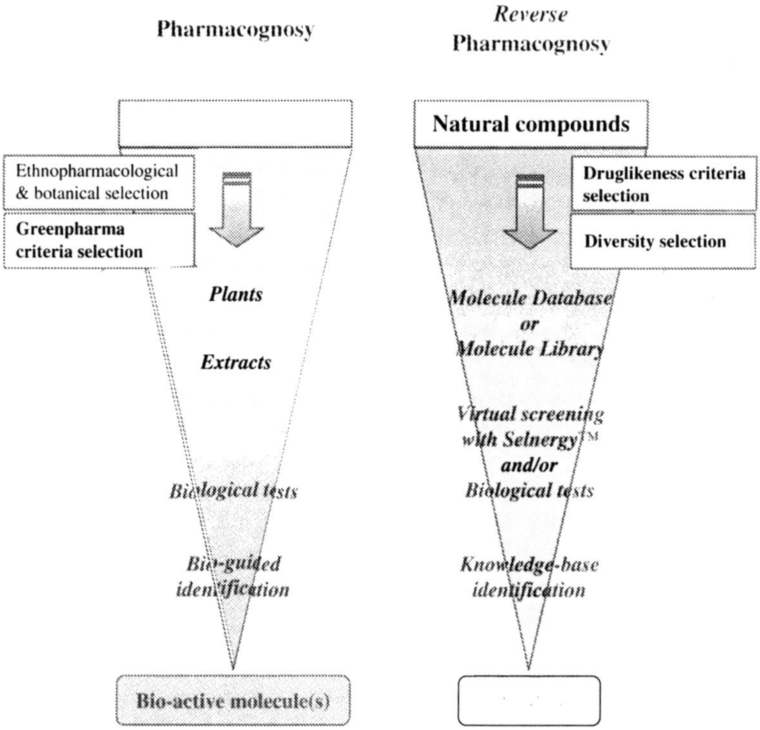

Fig. 1. Pharmacognosy and reverse pharmacognosy. *Pharmacognosy*: Plants are selected by means of ethnopharmacological knowledge, taxonomic biotope diversity coupled with in-house "patentability" criteria, then extracts are made and tested. Bio-guided characterization is undertaken for extracts of interest to isolate bio-active molecule(s). *Reverse pharmacognosy*: Natural compounds are selected by in-house druglikeness and chemical diversity criteria. They are submitted to Selnergy™ for prediction of activity then validated in biological tests, or the molecules are evaluated directly in biological assays. The last step consists in using our internal knowledge database to identify the source(s) of the validated compounds.

II. Pharmacognosy

Pharmacognosy, which is the study of drugs of natural origin focuses mainly in the pharmacochemistry of natural raw materials, but not exclusively from plants (Bruneton, 1993) for pharmaceutical, diet and cosmetic purposes. This discipline includes several fields of expertise: in botany (taxonomy, ethnobotany…), in chemistry (extraction, purification and characterization of compounds) and in pharmacology (*in vitro* and *in vivo* biological tests).

Ethnopharmacology is considered a part of pharmacognosy that focuses on the study of biological effects of raw materials from traditional medicine. Its goal is the validation or otherwise, of the use of plants, fungi, animal or mineral extracts for medical, diet or cosmetic purposes. It will try to establish a link between the traditional usage and an appropriate biological target, since it is the belief in Western pharmaceutical research that one drug needs to interact with a specific biological target. This is a complex task because it means to "decode" a know-how from a population who often employs plants in mixture or in a particular context i.e. rites, ceremonies, etc. These data can be sparse and transmitted orally, such as shaman knowledge or in sophisticated written forms such as in Indian Ayurvedic pharmacopoeia. Moreover, the vision of illness in traditional medicines can be very different from Western point of view. This vision is often associated with an entire philosophy of human being and his place in Universe. For instance, illness is considered as an imbalance or disturbance of ying and yang, the two "energies" that support the world in ancient Chinese philosophy (original texts of the Yellow Emperor's Classic of Medicine, 3000 BC). Mind and body are not dissociated and are interdependent. So the link between a traditional usage and the Western therapeutic classification is not always obvious to establish. For instance, a liana called ayahuasca, used by Amazonian shamans to free the mind from its body and to communicate with spirits has been evaluated in alcohol and drug addiction with encouraging results (Plotkin, 2000). However, one can encounter some surprising problem. In the same book, Plotkin cited an example of a cure by a shaman, which gave amazing results against severe diabetes. He could not find the antidiabetic activity when he tried to remake the mixture nor when he brought the mixture by the shaman back to his laboratory. So the crucial question is what kind of plants, animals to collect and how?

II.A. Sample collection

We employ two collection procedures: a therapy oriented one and a biodiversity one depending on the type of projects.

When a specific therapeutic area is desired, plants are collected on the field according to ethnopharmacological knowledge. In the example of central nervous system (CNS) applications, ethnobotanists and ethnopharmacologists will select a set of plants that gives hallucinations, enables to speak to spirits, etc. in a particular folk usage from different areas and traditions: Amazonian shamans, African tribes (knowledge that circulates orally) and Chinese and Ayurvedic pharmacopoeia (knowledge codified in a written form). Using different knowledge from different cultures will increase our probability and ability to find effective materials, which correspond to our therapeutic area. This approach will also avoid industrial property (IP) problems: if a plant is well-known for CNS in a particular traditional medicine, it is likely that its derivatives are already patented by others. So diversifying our sources will allow to jump into another less IP problematic plant.

When we want to augment our probability to get hits in a screening campaign, phytochemical diversity is sought. So we spot by diverse taxonomy (family, genus and species), geography (countries, continents, etc.) and/or biotope (rainforest, deserts, mountains, etc.). The postulate is that chemical diversity in the secondary metabolites will be reflected by taxonomic, biotope diversity. Though some chemicals are almost ubiquitous in vegetal realm, it is well-known that Rutaceae are rich in alkaloids. And one can reasonably think that a plant that grows in high mountains is more likely to contain sun-protecting compounds than a plant in underwoods. Crucial IP issue is also taken into account by in-house "patentability" criteria.

Of course, samples have to be stocked in sufficient quantities in order to allow resupply and retesting. And they are compliant to the international Convention on Biological Diversity signed in 1992 at the Earth Summit in Rio de Janeiro.

II.B. Extraction

Dry plants are used to extract the chemicals. It is done in solvents of different polarity such as methanol, ethyl acetate etc., in order to get as much chemicals as possible. Polymer filters are employed to discard high molecular weight molecules, such as tannins, polyphenols, which are sources of false positives in biological tests (Collins et al., 1998). Some specialists (Luu, 2003) estimate that we need to conserve fresh plant materials to maintain their biological activities for plants selected according to ethnobotanical criteria and more precisely if these plants are used in fresh form in traditional usage. This may explain sometimes the loss of activity. Nevertheless, this option implies drastic organization of supply network and of the extraction process.

II.C. Activity-guided fractionation

Extracts are tested on binding or cellular tests. Active ones will be fractioned furthermore then new fractions retested in an iterative process of testing and fractioning, until having the active isolated compound(s). Purification and identification steps involve liquid chromatography coupled with an evaporative light scattering (Stolyhwo et al., 1983) and photodiode array detectors and a mass spectrometer. The structure of compound(s) is elucidated by NMR.

III. Reverse pharmacognosy

Reverse pharmacognosy is a new concept introduced by Greenpharma that complements pharmacognosy in the research of new compounds or plants in pharmacy or cosmetics. It is organized into five components, which are a three–dimensional (3-D) structural database of natural compounds with its equivalent of natural compound library formatted for HTS, a target database, a virtual screening tools and a ethnopharmacological database. Figure 2 shows the relationship between these components, and a detailed description of each part as follows.

III.A. Virtual chemical database

The virtual chemical database (VCDB) is built by gathering natural compounds found in commercial database and in Greenpharma proprietary database. These compounds are available from commercial providers or from in-house extraction, purification and characterization platform. Their sources are known. Natural compounds from the literatures are also gathered. As their sources are available and most of the time the method applied for their extraction is described, such molecules can be obtained with our platform. This virtual compound library contains more than 100,000 natural compounds with their 3-D coordinates generated by Concord [Pearlman]. We are enriching VCDB constantly with new structures.

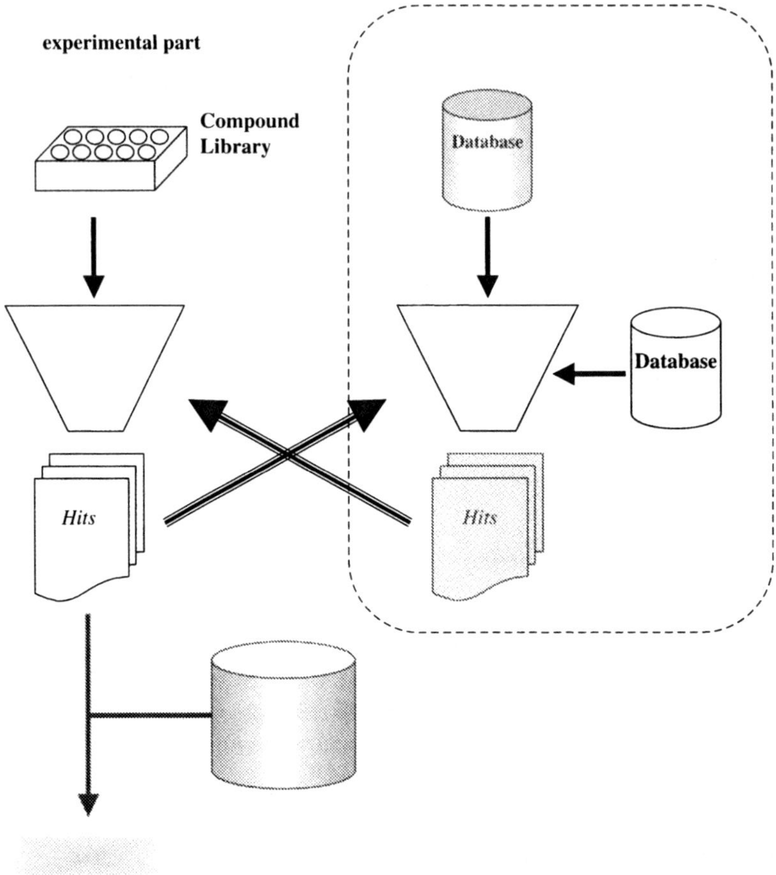

Fig. 2. Schema of reverse pharmacognosy. Line arrows represent information flow and 3-D arrows correspond to process flow. There are two parts in reverse pharmacognosy: one is experimental and the other relies on in silico tools in the initial steps. Data from ethnopharmacological database (ETPHDB) are utilized for "knowledge validation" in virtual to real hits and to retrieve the sources of compounds. Experimental validation process consists in internal biological tests and/or data gathered in scientific literature. Real hits or real inactive candidates are interesting for the refinement of our target models. VCDB and target DB are the input for Selnergy.

The database is created in Unity [Tripos] format which fingerprints can be used to evaluate diversity or extract a representative subset of the database. Unity fingerprints have been validated for these tasks (Matter, 1997). However, default parameters for generating unity fingerprints may not be optimal for the description of natural compounds, so we are examining a derived set of parameters to generate customized fingerprints.

For virtual screening purpose, we extract a subset from VCDB. Selected molecules share in-house drug-like criteria derived from Lipinski rules (Lipinski et al., 1997). Several authors raised the idea that the "rule of five" are too restrictive and exclude some drugs, especially natural compounds (Charifson and Walters, 2002). Feher and Schmidt (2003) have studied systematically the property distribution of drugs, combinatorial chemistry compounds and natural and derivative compounds. The most outstanding differences are in the number of chiral centers, prevalence of aromatic rings, saturation of molecules, complex ring systems and the number of heteroatoms. The authors suggested that mimicking natural compound property profile can increase structural diversity and biological relevance. At the sight of results cited above and the orientations of our internal projects, we have introduced derived criteria to filter natural compounds as listed below:

- Molecular weight in the range of 150–700 Da
- Maximum number of hydrogen bond donors: 10
- Maximum number of hydrogen bond acceptors: 5
- Maximum number of rotable bonds: 20
- Maximum number of sugar: 1
- No aliphatic chains longer than 6 atoms
- No peptides or nucleic acids
- No metals.

A last criteria to select compounds for virtual screening, is based on their chemical diversity. We use Optisim (Clark, 1997) from Sybyl package [Tripos] to extract a diverse subset of molecules from molecules, which passed the property profiling of VCDB.

VCDB allowed the analysis of chemical diversity and natural compound properties. A most diverse subset of VCDB was derived and compound available commercially or easily synthesized, formed a pure natural chemical library, suitable for experimental screening and HTS. This library built by considering chemical diversity and druglikeness is likely to give hits on a large variety of biological assays.

III.B. Target database

The Target Database contains an ensemble of protein 3-D structures determined by X-ray crystallography or by in-house homology models (currently more than 1500 structures). Each structure is indexed by its Protein Data Bank (PDB) code or an internal code, followed by the method used to obtain it, i.e. X-ray crystallography, NMR or homology modeling. Their biological source is described. Though most of them are from Human, the target database also contains proteins from other mammals, bacteria, virus, protozoa, etc. The proteins are classified by their biological mechanism e.g. protease, oxygenase, etc. This classification reflects the molecular properties of the targets. They are also classified by their biological activities e.g. pigmentation, lipolysis, inflammation, etc. Of course, a protein can be a member of several therapeutic families. This classification reflects biological properties and pathways. Structure of the co-crystallized ligand, if any, is recorded separately from the protein for calibration and validation of the virtual screening.

Data for our virtual screening tools Selnergy are recorded for the protein/ligand pairs, such as FlexX (Rarey et al., 1996) interaction energy, the root mean square deviation (RMSD) of the prediction placements and the co-crystallized ligand position. This RMSD is an estimation of the deviation in space of the calculated position of a ligand with its position found by X-ray crystallography, thus corresponding to the accuracy of a docking. Models of which RMSD is greater than 1.5 should be considered with care. Another validation step consists in retrieving known active ligands cited in scientific literature from a set of 100 non-active compounds. This step gives the discrimination power of our virtual screening process.

Recently, Bissantz et al. (2003) published interesting results from virtual screening on models derived from homology modeling of rhodopsin. Screening on G-protein coupled receptors (GPCR) for antagonists and agonists gave amazing hit rates, which can top at 70%. We have also included models in our target database, they are GPCR or protein with a crystallized homologue, e.g. protein subtypes, such as phosphodiesterases. These models are validated with data from literature such as site directed mutagenesis, labeling and tested for their screening power as described above.

We are also implementing other interaction energy functions in order to benefit from a consensus scoring method, to access more robust ligand/protein interaction solutions (Bissantz et al., 2000). Others (Paul and Rognan, 2002) have also employed consensus docking method for screening to strengthen virtual screening prediction.

III.C. Selnergy^{TM}

Virtual screening is a domain in molecular modeling, which is in fast development. Two approaches can be distinguished: screening based on ligand properties i.e. physico-chemical properties (1-D data), fragmental description (2-D), pharmacophores (3-D) (Martin et al., 1993). The techniques are quantitative structure–activity relationship (QSAR) (Herve et al., 1998; Baurin et al., 2000) comprising also, binary QSAR (Gao et al., 1999), ComFA (Cramer et al., 1988) or neural networks (Sadowski and Kubinyi, 1998; Kohonen and Somervuo, 2002) with fuzzy logic (Pintore et al., 2003); and screening based on target properties which necessitates the knowledge of the 3-D structure of the target and the ligand. The techniques are *de novo* design (Böhm, 1992), docking (Shoichet and Kuntz, 1993; Rarey et al., 1996; Jones et al., 1997), which consists in generating new ligands or adjusting ligands in the active site of the target, respectively. A scoring step allows classifying active/inactive ligands according to an interaction energy function. For a detailed review of virtual screening methods, refer to the book by Böhm and Schneider (2000).

Selnergy is a proprietary virtual high throughput screening (vHTS) tool that integrates the validated 3-D Target Database described in Section III.B. The docking engine is FlexX. Selnergy is implemented in Sybyl 6.9 of Tripos package. It allows the screening of VCDB to a set of proteins from the Target Database. A protein set can be built as a Boolean combination of sources, methods, protein families and/or therapeutic domains entries as shown in Figure 3. For instance, get all RX Human proteins involved in inflammation and pertaining to oxygenase.

The post-processing interface was also developed internally, and associate several filtering methods to discard irrelevant "hits". One filter is position based: in or out of the active site. Another filter is based on FlexX interaction energy. The user can

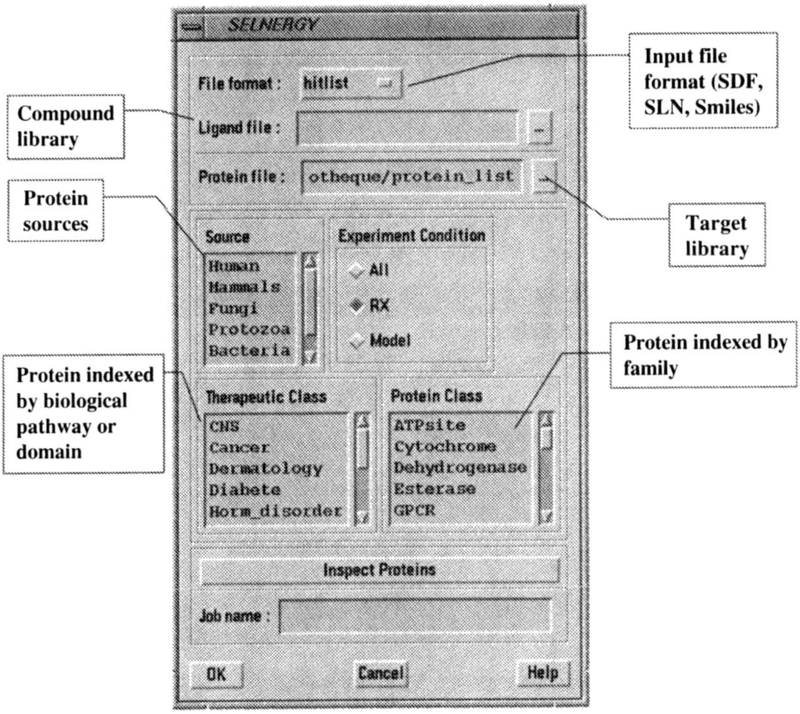

Fig. 3. Selnergy™ module. Selnergy™ is a proprietary tool developed by Greenpharma for biological profiling, able to analyze synergy and selectivity for a set of compounds vs. several biological targets. It couples an in-house target library, a worldwide standard docking program and an analysis module. The target library includes more than 1500 protein structures from crystallography and homology models. They are classified according to their experimental origin, their source, the therapeutic class and protein class. This module is adequate for bio-focused libraries. For example, the development of a reduced library of compounds with potential anti-inflammatory activity. All targets involved in inflammatory diseases will be screened.

adjust an energy cut-off in reference to the FlexX interaction energy registered in the Target Database.

We are implementing consensus scoring, clustering and statistical methods to ease the post-process step and to get more robust solutions.

III.D. Ethnopharmacological database (ETPHDB)

We have developed a proprietary database containing ethnobotanical data, botanical data, natural chemical structures and biological-testing results for extracts and compounds. The data are family, genus, species, the common names and synonyms, organs used in the traditional medicine and the ethnic groups. Molecular structures and families are also stored with their biological activities from the scientific literature or internal tests. Figure 4 corresponds to the user interface of the database. This database alone has shown its relevancy in accelerating the discovery process in

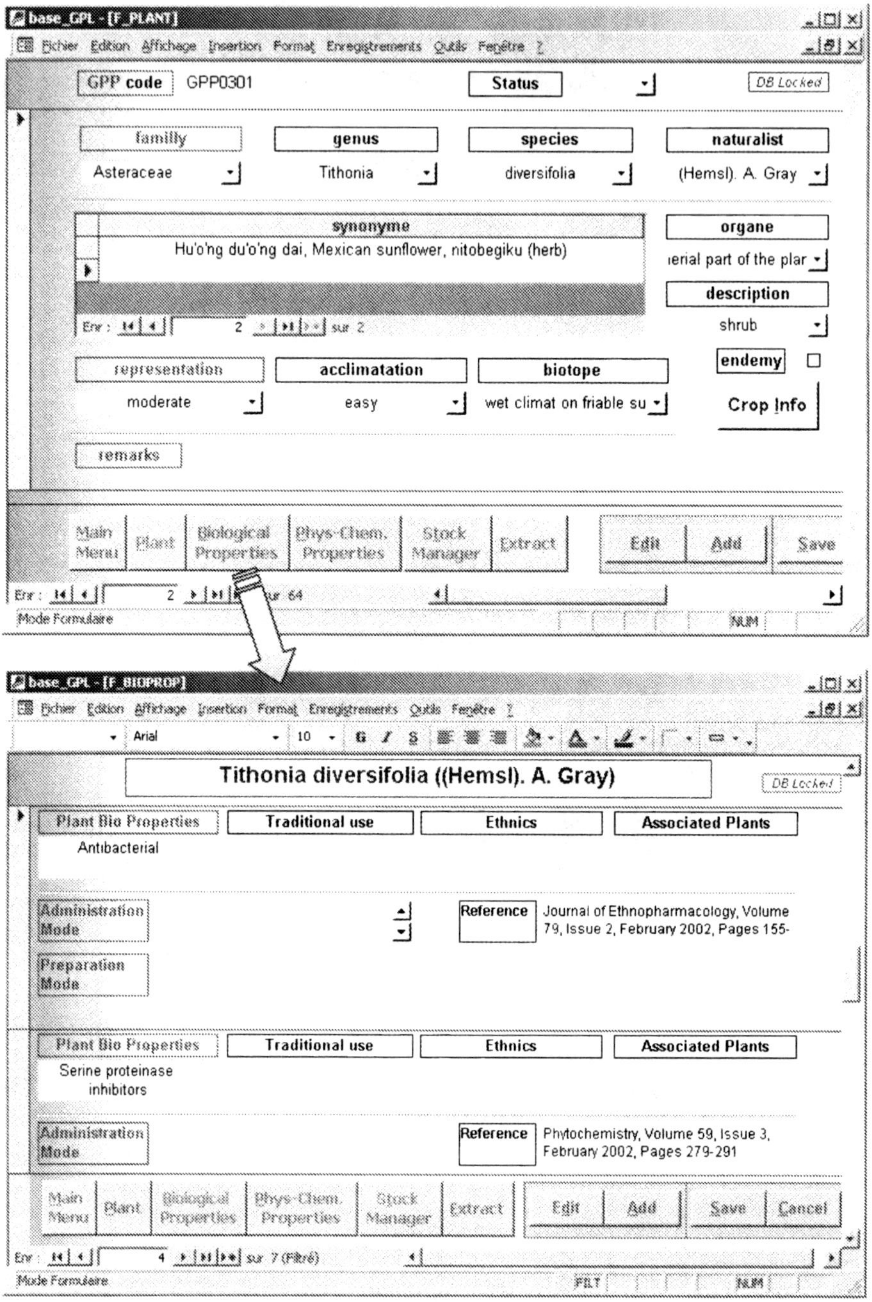

Fig. 4. Ethnopharmacological database (ETPHDB) contains botanical information on plants, traditional usage and phytochemistry data associated with biological activity information. So it is straightforward to link plants/molecules/activity.

bioactive entities as shown by Bernard et al. (2001), where it was used to find an anti-inflammatory substance.

IV. Applications of Selnergy^{TM}

IV.A. Building therapeutic focused libraries

The profits from applying a vHTS tool coupled with a knowledge database are an efficient and robust way to build therapeutic focused compound libraries.

An example of use is to select proteins from a same biological pathway, such as pigmentation in order to detect putative synergistic activities of one molecule and/or several molecules of a set of targets. In this case, the targets involved are tyrosinase, guanylate cyclase, melanocortin receptor 1 and endothelin receptors (Bertolotto and Ballotti, 1999). Moreover, we can estimate the affinity of these compounds to other pathway targets, proteins involved in ADMET, such as cytochrome P450, plasmatic proteins (Li, 2001). In this way, we can detect potential interactions with other proteins irrelevant of the desired activity, which can lead to putative side effects. So it gives us a fast answer as for selectivity and/or synergy properties or "Selnergy" properties of an ensemble of compounds to targets.

IV.B. Building family focused libraries

We can also screen a compound library on a particular protein family, such as kinases. Kinases dephosphorylate their substrate ATP and activate their target proteins by phosphorylating them. They are implied in different diseases, so in different biological pathways, which can be inflammation, cancer and so on (Levitzki, 1999). Nevertheless, they share similar structural motifs for the substrate interaction site.

IV.C. Estimating biological properties of a compound library

The use of docking tools has been evoked elsewhere for profiling a set of compounds or to detect molecules with similar biological activity (Lessel and Briem, 2000; Chen and Zhi, 2001). Selnergy can be used to estimate the biological properties of a compound library as one can screen compounds on the whole target library. By this mean we can generate a virtual bio-profile of a compound or an entire library enabling the comparison of compounds on the biological space. Such biological descriptor is similar to chemical descriptors to access chemical diversity. So we can use it to derive diverse biological activity structures from a database with same tools for chemical diversity (Optisim).

IV.D. Protein flexibility

Structures from crystallography are an instant "3-D picture" of a protein that means the intrinsic dynamics of the protein cannot be recorded. By co-crystallizing different ligands with the same protein and considering the protein alone, one can estimate the flexibility of an active site. Fibroblast growth factor receptor (fgfr) is a good illustration of a flexible and adjustable protein active site (Mohammadi et al., 1997). According to the shape of a ligand, an entire loop can flip over the active site to close it or remain in the open position as shown in Figure 5. We registered different conformations of fgfr in our target library to consider the flexibility of the protein at the ligand pocket.

IV.E. Lead leveraging

Another interesting use of Selnergy™ is the lead leveraging. We can use known active ligands to dock into a protein or a set of conformations of a protein to generate new compound ideas or to leverage the previous knowledge. The structure of phosphodiesterase 4D (PDE4D) was revealed by X-ray crystallography (Huai et al., 2003). We docked sophoflavescenol, a known inhibitor of PDE4 (Shin et al., 2002), which is a natural flavonoid into a model derived from the PDE4D structure (Lee et al., 2002). We found that there are 2 positions that show good fit with

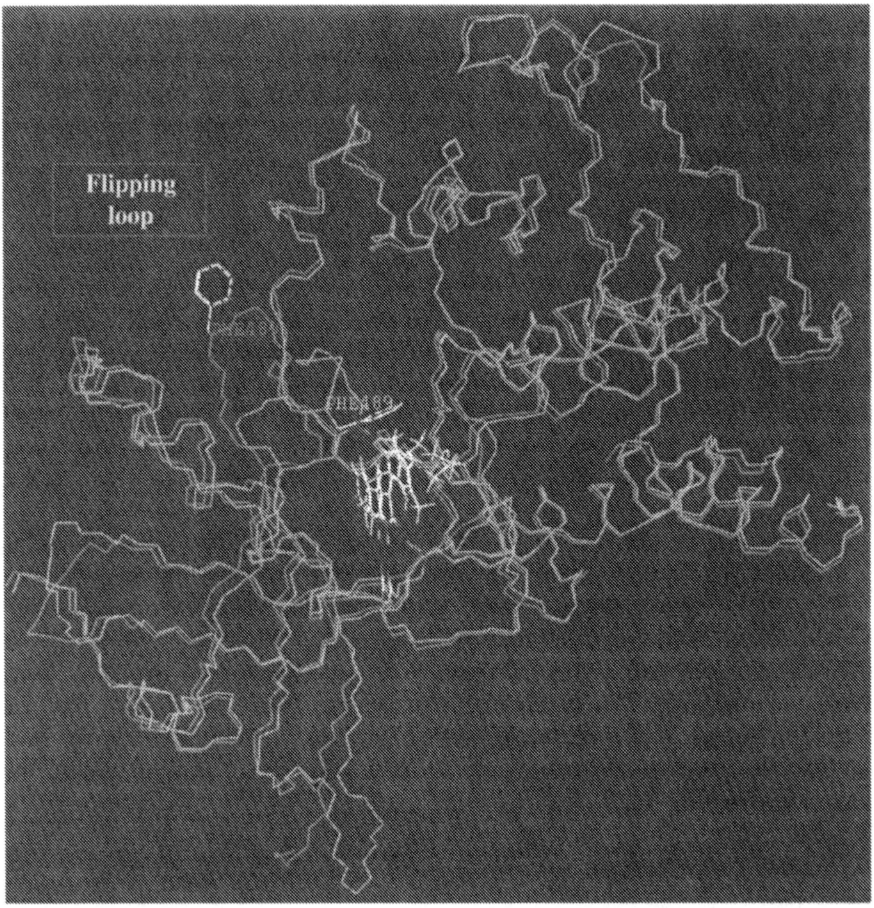

Fig. 5. Fibroblast growth factor receptor (fgfr) conformations. Two X-ray crystal structures of fgfr showing dynamic movements of the binding site of the receptor according to the type of ligands. A flipping loop can adjust to the size of binding ligands, with drastic change on the geometry of the active site.

important residues of the active site. Q369 is a hydrogen bond donor. F372 and F340 are involved in pi stacking or hydrophobic interactions (numbering as found in Huai et al., 2003 work). There are also possible chelations of the zinc ion by oxygen of the flavonoid. However, none of the solutions entirely occupied the active site. By overlaying the resulting placements (as illustrated in Figure 6A–C), the data

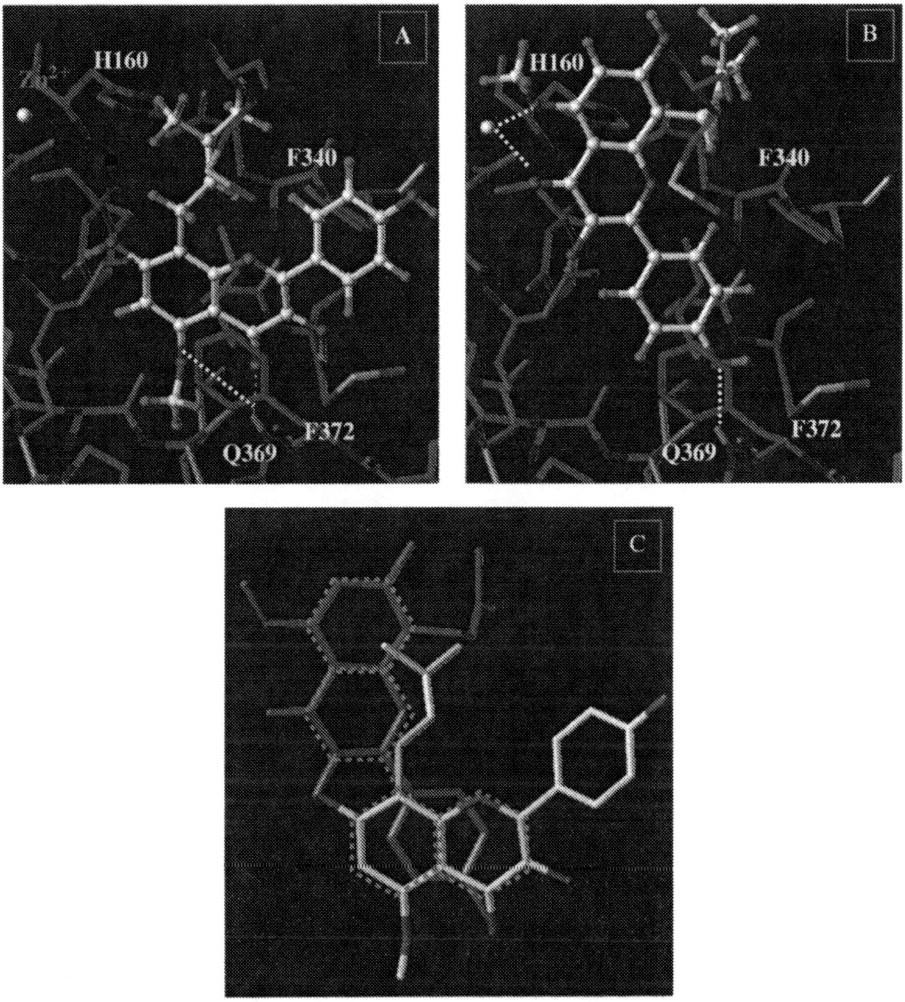

Fig. 6. Example of lead leveraging from a natural flavonoid docked into PDE4D active site. (A,B) Docking of sophoflavescenol in PDE4D active site. Colour codes: protein carbons are coloured in green while the ligand carbons are in white. The results showed two main placements of the compound in the active site, which do not occupy the entire site. In A, there are two hydrogen bond interactions with Q369 (dotted yellow lines). This interaction pattern is found in the crystal rolipram/PDE4D complex structure published by Huai et al. (2003). For B, only one H bond interaction can be seen. However, there seems to be a chelation of the zinc ion by the flavonoid oxygen atoms (yellow dotted lines). (C) By overlaying the two structures as there are docked (white placement is from A, green one is from B), we noticed that a biflavonoid template (highlighted by magenta dotted line) will fully occupied the active site, which can lead to more potent inhibitors.

suggested that biflavonoids are possible templates as phosphodiesterase inhibitors. Other works confirmed the hypothesis for biflavonoids (Beretz et al., 1986) and biflavones (Saponara and Bosisio, 1998).

Another example of lead leveraging comes of inhibitors of vascular endothelial growth factor receptor 2 (vegfr2) kinase sites from its X-ray structure (McTigue et al., 1999). It is known that ATP is the endogen ligand, so mimicking its structure can be a good starting point for designing inhibitors. SU5402 is a potent inhibitor of fgfr and vegfr2 at a lesser scale. Fgfr and vegfr2 kinase domains are very similar. As SU5402 was co-crystallized with fgfr (Mohammadi et al., 1997) and its interaction mode is known, one can reasonably assume that its mode of binding would be the same in vegfr2. In order to access more insight of the interaction mode of ATP (which was not co-crystallized with vegfr2) and shift to another class of compounds, we can overlay the 2 molecules. There is an intuitive but "wrong" way to align them as shown in Figure 7. A docking study revealed two families of placement for ATP. One privileges H bond interactions with N7 and N6 of the purine, respectively with

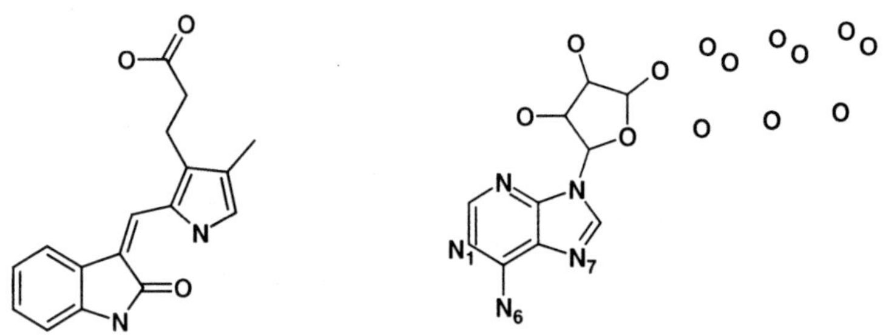

SU5402 ATP

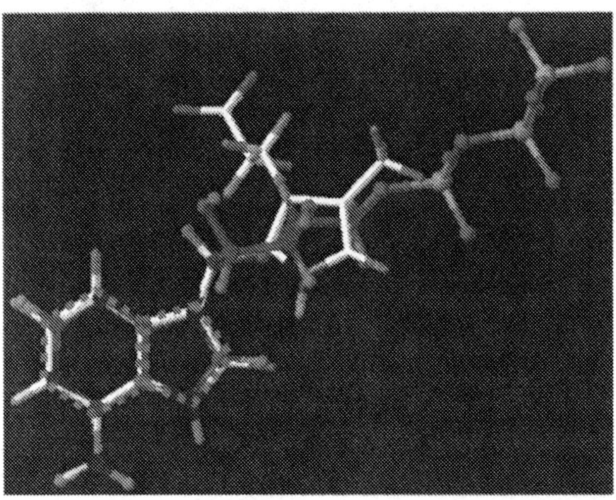

Fig. 7. SU5402 and ATP and their possible alignment. The endogen ligand ATP and SU5402, an antagonist inhibitor, bind to the kinase site of VEGFR2. An intuitive but "wrong" superposition of ATP and SU5402 would be overlying the two fused rings of the ligands (dotted lines in magenta).

backbone NH of C919 and backbone C=O of E917 (numbering as found in McTigue et al., 1999 work), in the other docking, N6 and N1 interact with the same receptor residues. We have chosen the latter one as in DNA; the H bond network in the double helix is made by N1 and N6. Superposing the 2 docked structures reveals a surprising alignment (see Figure 8A–C), which leads to a new scaffold: the α-carboline family. This class of compounds was patented for vegfr2 inhibition in angiogenesis process (Ator et al., 2001).

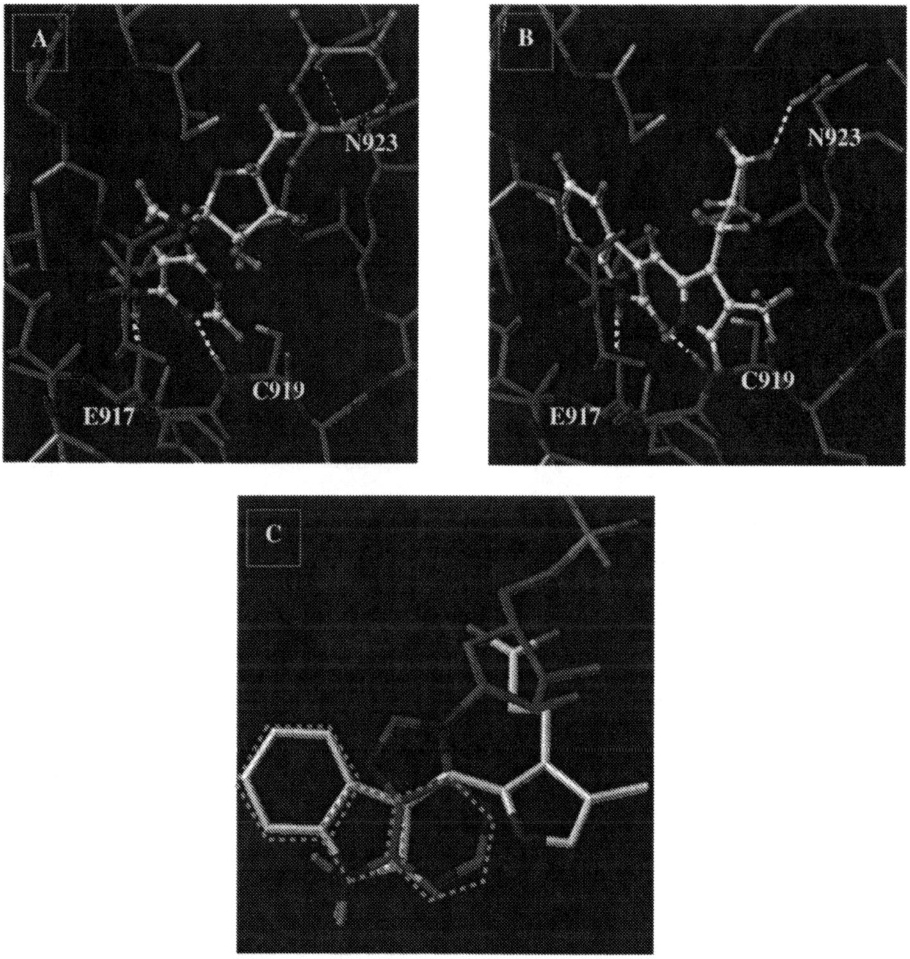

Fig. 8. Superposition of ATP and SU5402 leading to α-carboline class inhibitor of vegfr2. (A) Docking of ATP into the kinase site of vegfr2. The docking program found 2 families of placement, only the solution involving H bond interaction of N6 and N10 of the purine ring with backbone C=O of E917 and backbone NH of C919 is represented. (B) Docking of SU5402, we can notice the same interaction residues as in A. (C) Overlaying the two structures as they are docked into the kinase site, leads to a new class of molecules that can be potent inhibitors to VEGFR2: α-carbolines, highlighted with dotted magenta lines. This hypothesis was confirmed by others' work (Ator).

V. Pharmacognosy and reverse pharmacognosy in natural drug discovery process

An example of the implementation of pharmacognosy and reverse pharmacognosy (PrP) in our research process is given here. We have successfully used this approach for discovering new bioactive entities. In general, these entities are pure compounds in case of pharmaceutical oriented projects and plant extracts for cosmetics. PrP was applied in parallel in the discovery of new anti-inflammatory compounds with plants and the biological mechanism involved.

V.A. From pharmacognosy to lead

Inflammation was statistically correlated with plants used in case of snake bite. Among plants for this application, the ones which were the most cited by several folk sources were chosen for experimental evaluation in priority. Chemical information in our database indicated that one of the most interesting common compounds found in these plants is betulinic acid. Then a virtual screening study was undertaken with Selnergy^{TM}, on proteins implicated in inflammation pathway. Virtual hits were found for phospholipase A2 (PLA2). The pure compound was experimentally evaluated and showed potent activity against PLA2. The database also allows us to identify other more accessible natural sources where betulinic acid is in high concentration, for straightforward development (Bernard et al., 1999). After these *in vitro* binding assays, *in vivo* validations were undertaken. So a new anti-inflammatory compound was found with its natural sources and its biological target in inflammation pathway.

Another study on pigmentation/depigmentation process has also shown initial promising results (Baurin et al., 2002).

V.B. From reverse pharmacognosy to lead

Natural compounds from commercial sources were collected and recorded in a virtual natural compound library with their 3-D structure generated by Concord [Pearlman]. Molecule properties were calculated and Greenpharma drug-like criteria applied to get a subset of compounds. Then a structural diverse subset was selected from the latter set and submitted to Selnergy^{TM} for screening inflammation related proteins.

Most promising virtual hits were retrieved, and from the ETPHDB database, their sources were identified along with their traditional usage data. In order to respond to the pharmaceutical and cosmetic industries, molecules and plants containing these hits, which ethnopharmacological usage were intended for insect stings, animal bites, etc. were, in priority, submitted to biological tests for *in vitro* validation. Plant extracts were made with different polarity solvents with our extraction platform. Molecules are re-supplied or purified by us or re-synthesized. We are evaluating *in vitro* activities of several natural compounds with original scaffolds and extracts on cyclooxygenase 2. Similar molecules to chosen ones were gathered to build a bio-focused chemical library (synthetic derivatives can also be synthetized for the pharmaceutical industry), and "similar plants" to chosen ones (i.e. same family, biotope) were also extracted to make a bio-focused extract library. These bio-focused libraries will allow us to find more potent hits, jump to similar compounds or plants in case of supply problems or patent. Knowledge or virtual screening-based molecule selections are efficient tools for prioritizing which compounds must be tested and *in vitro* biological tests are always necessary for validating hits: passing from virtual and potential hits to real hits. These tests include binding and/or cellular assays.

V.C. Repositioning of compounds

Chemical compounds are usually studied and developed for a single application. However, it is well known that molecules can produce multiple biological effects. The obvious example is the therapeutic and side effects of drugs. This is generally due to the lack of selectivity against unwanted protein interactions. Selectivity is a relative concept for it depends on the concentration of the compound. It also depends on the number of side interactions one can afford to study. Usually, selectivity is investigated for subtype proteins. This one molecule on target paradigm has therefore occulted other possible "side" uses for a given molecule.

We undertake a study on ε-viniferin (EV) to find new biological mechanism, i.e. proteins not yet identified for interacting with EV and application for this molecule. Former studies have demonstrated its potency in numerous domains, such as antitumor (Mishima et al., 2003), antioxidant (Piver et al., 2003) and hepatoprotector (Oshima et al., 1995). EV was screened on the entire target library with Selnergy. Several targets were identified as putative proteins to interact with EV. Phosphodiesterase 4 (PDE4, the only subtype identified by Selnergy) was selected for further experimental studies. The activity of EV on PDE4 was confirmed by PDE4 binding assays ($IC50 = 4.6\,\mu M$). EV was also validated on cellular assays for the secretion of TNF-α and Interleukin-8. Our data demonstrated that EV possesses anti-inflammatory properties by inhibiting PDE4 subtype (Do et al., 2005).

VI. Conclusions and perspectives

Finding *in vivo* active compounds remains a real challenge either in the molecule research or in the biological target selection. Natural products neglected in last decades remain relevant tools for finding *in vivo* active compounds. In order to accelerate the natural hit discovery, we introduced the new concept of reverse pharmacognosy that complements pharmacognosy. Based on computation techniques, such as virtual screening tools, relational databases, Selnergy, our proprietary virtual screening tool is the key point of the reverse pharmacognosy. Moreover, these approaches of PrP combine many kinds of information available in literature, but sparse and not exploitable in original form; and a knowledge management strategy is needed to put all these data into an intelligible and usable form. Integrated into our workflow, PrP and a knowledge database have proved their abilities to accelerate the natural drug discovery process when used in parallel. Moreover, this global approach is an answer to the pharmaceutical and cosmetic specific needs in new *in vivo* active entities: the pharmaceutical industry requires pure natural or synthetic compounds, and cosmetic industries, natural compounds or extracts.

Though we have exemplified our approach with natural compounds, most of our methodologies, e.g. Selnergy™ can be applied to hemisynthetic and synthetic molecules, or shift results obtained from natural product studies to synthetic compounds.

References

Anton R. (1999) Plantes thérapeutiques. Cachan, France: Tec & Doc.

Ator MA, Bihovsky R, Chatterjee S, Dunn D, Hudkins RL. (2001) Preparation of novel multicyclic compounds and their amino acid derivatives as inhibitors of enzymes such as poly(ADP-ribose) polymerase. Patent No. WO2001085686.

Baurin N, Arnoult E, Scior T, Do QT, Bernard P. (2002) Preliminary screening of some tropical plants for anti-tyrosinase activity. J Ethnopharm 82:155–8.

Baurin N, Vangrevelinghe E, Morin-Allory L, Merour JY, Renard P, Payard M, Guillaumet G, Marot C. (2000) 3D-QSAR CoMFA study on imidazolinergic I(2) ligands: a significant model through a combined exploration of structural diversity and methodology. J Med Chem 43:1109–22.

Beretz A, Briancon-Scheid F, Stierle A, Corre G, Anton R, Cazenave JP. (1986) Inhibition of human platelet cyclic AMP phosphodiesterase and of platelet aggregation by a hemisynthetic flavonoid, amentoflavone hexaacetate. Biochem Pharmacol 35:257–62.

Bernard P, Berthon JY, Chrétien JR. (1999) Computer-aided molecular selection and design of natural bioactive molecules. Curr Opin Drug Disc Develop 2:213–23.

Bernard P, Scior T, Didier B, Hibert M, Berthon JY. (2001) Ethnopharmacology and bioinformatic combination for lead discovery: application to phospholipase A_2 inhibitors. Phytochemistry 58:865–74.

Bertolotto C, Ballotti R. (1999) La tyrosinase est l'enzyme clé de la mélanogenèse. Cosmétologie 24:20–2.

Bissantz C, Bernard P, Hibert M, Rognan D. (2003) Protein-based virtual screening of chemical databases. II. Are homology models of G-protein coupled receptors suitable targets? Proteins 50:5–25.

Bissantz C, Folkers G, Rognan D. (2000) Protein-based virtual screening of chemical databases. 1. Evaluation of different docking/scoring combinations. J Med Chem 43:4759–67.

Böhm HJ. (1992) The computer program LUDI: a new method for the *de novo* design of enzyme inhibitors. J Comput Aided Mol Des 6:61–78.

Böhm HJ, Schneider G. (2000) Virtual screening for bioactive molecules. Methods and principles in medicinal chemistry. Germany: Wiley.

Bruneton J. (1993) Pharmacognosie, phytochimie, plantes médicinales, 2nd edition. Paris: Technique et Documentation – Lavoisier, p. viii.

Burnouf C, Pruniaux MP. (2002) Recent advances in PDE4 inhibitors as immunoregulators and anti-inflammatory drugs. Curr Pharm Des 8:1255–96.

Charifson PS, Walters WP. (2002) Filtering databases and chemical libraries. J Comput-Aided Mol Des 16:311–23.

Chen YZ, Zhi DG. (2001) Ligand-protein inverse docking and its potential use in the computer search of protein targets of a small molecule. Proteins 43:217–26.

Clark RD. (1997) OptiSim: an extended dissimilarity selection method for finding diverse representative subsets. J Chem Inf Comput Sci 37:1181–8.

Collins RA, Ng TB, Fong WP, Wan CC, Yeung HW. (1998) Removal of polyphenolic compounds from aqueous plant extracts using polyamide minicolumns. Biochem Mol Biol Int 45:791–6.

CONCORD®. (1997) 4.0.7A R.S. Pearlman. Concord user's manual. St. Louis, MO: Tripos Inc.

Cramer III R, Patterson DE, Bunce JD. (1988) Comparative molecular field analysis (CoMFA). 1. Effect of shape on binding of steroids to carrier proteins. J Am Chem Soc 110:5959–67.
Do QT, Rénimel I, André P, Lugnier C, Muller CD, Bernard P. (2005) Reverse Pharmacognosy: application of Selnergy, a new tool for lead discovery. The example of ε-viniferin. Curr. Drug Disc. Techn. 2, 161–7.
Duke JA. (1990) Promising phytomedicinals. In: Janick J, Simon JE editors. Advances in new crops. Portland, OR: Timber Press, pp. 491–8.
Feher M, Schmidt J. (2003) Property distributions: differences between drugs, natural products and molecules from combinatorial chemistry. J Chem Inf Comput Sci 43:218–27.
Gao H, Williams C, Labute P, Bajorath J. (1999) Binary quantitative structure-activity relationship (QSAR) analysis of estrogen receptor ligands. J Chem Inf Comput Sci 39:164–8.
Harvey A. (1999) Medicines from nature: are natural products still relevant to drug discovery? TiPS 20:196–8.
Harvey A. (2000) Strategies for discovering drugs from previously unexplored natural products. Drug Discov. Today 5:294–300.
Herve F, Caron G, Duche JC, Gaillard P, Abd Rahman N, Tsantili-Kakoulidou A, Carrupt PA, d'Athis P, Tillement JP, Testa B. (1998) Ligand specificity of the genetic variants of human alpha1-acid glycoprotein: generation of a three-dimensional quantitative structure-activity relationship model for drug binding to the A variant. Mol Pharmacol 54:129–38.
Huai Q, Wang H, Sun Y, Kim HY, Liu Y, Ke H. (2003) Three-dimensional structures of PDE4D in complex with roliprams and implication on inhibitor selectivity. Structure 11:865–73.
Jones G, Willett P, Glen RC, Leach AR, Taylor R. (1997) Development and validation of a genetic algorithm for flexible docking. J Mol Biol 267:727–48.
Jossang A, Bitar HE, Pham VC, Sevenet T. (2003) Daphcalycine, a novel heptacycle fused ring system alkaloid from daphniphyllum calycinum. J Org Chem 68:300–4.
Kao CY. (1972) Pharmacology of tetrodotoxin and saxitoxin. Fed Proc 31:1117–23.
Kohonen T, Somervuo P. (2002) How to make large self-organizing maps for non vectorial data. Neural Netw 15:945–52.
Lahana R. (1999) How many leads from HTS? Drug Discov Today 4:447–8.
Lee ME, Markowitz J, Lee JO, Lee H. (2002) Crystal structure of phosphodiesterase 4D and inhibitor complex (1). FEBS Lett 530:53–8.
Lessel UF, Briem H. (2000) Flexssim-X: a method to detect molecules with similar biological activity. J Chem Inf Comput Sci 40:246–53.
Levitzki A. (1999) Protein tyrosine kinase inhibitors as novel therapeutic agents. Pharmacol Ther 82:231–9.
Li AP. (2001) Screening for human ADME/Tox drug properties in drug discovery. DDT 6:357–66.
Lipinski CA, Lombardo F, Dominy BW, Feeney PJ. (1997) Experimental and computational approaches to estimate solubility and permeability in drug discovery and development settings. Adv Drug Del Rev 23:3–25.
Luu DV. (2003) Quelle est l'importance de l'eau dans la plante et dans les produits à base de plantes ? Incidence du Bio Euroforum: Produits à base de plantes. France
Martin YC, Bures MG, Danaher EA, DeLazzer J, Lico I, Pavlik PA. (1993) A fast new approach to pharmacophore mapping and its application to dopaminergic and benzodiazepine agonists. J Comput Aided Mol 7:83–102.
Massiot G. (1991) How far can one go in the field of structural elucidation of natural products? J Ethnopharmacol 32:103110.
Matter H. (1997) Selecting optimally diverse compounds from structure databases: a validation study of two-dimensional and three-dimensional molecular descriptors. J Med Chem 40:1219–29.
McTigue MA, Wickersham JA, Pinko C, Showalter RE, Parast CV, Tempczyk-Russell A, Gehring MR, Mroczkowski B, Kan CC, Villafranca JE, Appelt K. (1999) Crystal structure of the kinase domain of human vascular endothelial growth factor receptor 2: a key enzyme in angiogenesis. Structure Fold Des 7:319–30.

Mishima S, Matsumoto K, Futamura Y, Araki Y, Ito T, Tanaka T, Iinuma M, Nozawa Y, Akao Y. (2003) Antitumor effect of stilbenoids from *Vateria indica* against allografted sarcoma S-180 in animal model. J Exp Ther Oncol 3:283–8.
Mohammadi M, McMahon G, Sun L, Tang C, Hirth P, Yeh BK, Hubbard SR, Schlessinger J. (1997) Structures of the tyrosine kinase domain of fibroblast growth factor receptor in complex with inhibitors. Science 276:955–60.
Myers PL. (1997) Will combinatorial chemistry deliver real medicines? Curr Opin Biotechnol 8:701–7.
Oshima Y, Namao K, Kamijou A, Matsuoka S, Nakano M, Terao K, Ohizumi Y. (1995) Powerful hepatoprotective and hepatotoxic plant oligostilbenes, isolated from the oriental medicinal plant *Vitis coignetiae* (Vitaceae). Experientia 51:63–6.
Paul N, Rognan D. (2002) ConsDock: a new program for the consensus analysis of protein–ligand interactions. Proteins 47:521–33.
Pintore M, Piclin N, Benfenati E, Gini G, Chretien JR. (2003) Database mining with adaptive fuzzy partition: application to the prediction of pesticide toxicity on rats. Environ Toxicol Chem 22:983–91.
Piver B, Berthou F, Dreano Y, Lucas D. (2003) Differential inhibition of human cytochrome P450 enzymes by epsilon-viniferin, the dimer of resveratrol: comparison with resveratrol and polyphenols from alcoholized beverages. Life Sci 73:1199–213.
Plotkin M. (2000) Les Médicaments du Futur sont dans la Nature, 1st edition. Paris: FIRST, p. 228.
Potier P, Gueritte-Voegelein F, Guenard D. (1994) Taxoids, a new class of antitumour agents of plant origin: recent results. Nouv Rev Fr Hematol 36(Suppl 1):S21–3.
Rarey M, Kramer B, Lengauer T, Klebe G. (1996) A fast flexible docking mehod using an incremental construction algorithm. J Mol Biol 261:470–89.
Sadowski J, Kubinyi H. (1998) A scoring scheme for discriminating between drugs and non drugs. J Med Chem 41:3325–9.
Saponara R, Bosisio E. (1998) Inhibition of cAMP-phosphodiesterase by biflavones of *Ginkgo biloba* in rat adipose tissue. J Nat Prod 61:1386–7.
Shin HJ, Kim HJ, Kwak JH, Chun HO, Kim JH, Park H, Kim DH, Lee YS. (2002) A prenylated flavonol, sophoflavescenol: a potent and selective inhibitor of cGMP phosphodiesterase 5. Bioorg Med Chem Lett 12:2313–6.
Shoichet BK, Kuntz ID. (1993) Matching chemistry and shape in molecular docking. Protein Eng 6:723–32.
Stolyhwo A, Colin H, Guiochon G. (1983) Use of light scattering as a detector principle in liquid chromatography. J Chromatogr A 265:1–18.
SYBYL®. (2004) The Tripos Bookshelf 6.9. 6.9 Tripos Inc., 1699 South Hanley Rd., St. Louis, MO: Tripos Inc.
Tiew P, Ioset JR, Kokpol U, Chavasiri W, Hostetfann K. (2003) Antifungal, antioxidant and larvicidal activities of compounds isolated from the heartwood of *Mansonia gagei*. Phytother Res 17:190–3.
Tollman P, Guy P, Altshuler J, Flanagan A, Steiner M. (2001) A revolution in R&D. How genomics and genetics are transforming the biopharmaceutical industry. The Boston Consulting Group. http://www.bcg.com/publications/files/eng_genomicsgenetics_rep_11_01.pdf
UNITY®. (2004) The Tripos Bookshelf 6.9 4.4 Tripos Inc., 1699 South Hanley Rd., St. Louis, MO: Tripos Inc.
Verdine GL. (1996) The combinatorial chemistry of nature. Nature 384:11–3.
Yan X, Zhou J, Xu Z. (1999) Concept design of computer-aided study on traditional Chinese drugs. J Chem Inf Comput Sci 39:86–9.

M.T.H. Khan and A. Ather (eds.)
Lead Molecules from Natural Products

Effects of plant extracts on gene expression profiling: from macroarrays to microarray technology

Carlo Mischiati, Alessia Sereni, Mahmud Tareq Hassan Khan, Ilaria Lampronti, Roberto Gambari

Abstract

DNA hybridization arrays (macro- and microarrays) are very useful tools for the analysis of gene expression profiles in human pathologies. In addition, macro- and microarrays can be used in pharmacogenomic and toxicogenomic experiments, aimed at extensive analyses of the effects of therapeutic drugs on overall gene expression of target cells. Despite the fact that extracts from medicinal plants have been described to retain interesting biological activity, including anti-inflammatory, antitumor and antimicrobial effects, few data are present in the literature on gene expression profile studies. Here, we review results on the effects of plant extracts from *Moringa oleifera* on the gene expression profile of a human tumor cell line, the K562 cell line, originally isolated from a patient with chronic myelogenous leukemia (CML) in blast crisis, very useful for the identification of antitumor compounds. The data obtained using macroarrays were compared with those obtained using reverse-transcription polymerase-chain reaction (RT-PCR). Effects of *M. oleifera* extracts were compared to those of *Emblica officinalis*. The results obtained suggest that a general strategy for the development of specific therapeutic approaches could be proposed starting from gene expression studies employing macro- or microarrays. Treatment of target cells with plant extracts will allow to identify genes, which are down- or up-regulated. For these genes, the molecular analysis of the promoter regions and coding sequences could allow to design decoy oligonucleiodes (ODN), antisense DNA or RNA, peptides and monoclonal antibodies expected to mimic the biological effects of the employed plant extracts.

Keywords: *Macroarray, RT-PCR, Medicinal plants, Moringa oleifera, Anti-tumor agents, Ethnopharmacology*

I. Introduction

A growing set of experimental data are present in the recent literature showing that DNA hybridization arrays are very useful tools for the analysis of gene expression profiles, sequence variation and amplification of genes in human pathologies as well as in animal and cellular model systems (De Backer et al., 2001; Debouck and Goodfellow, 1999; Fernandez et al., 2001; Freeman et al., 2000). These diagnostic tools, also known as macroarrays, microarrays and/or high-density oligonucleotide arrays (Gene Chips), allow the analysis of gene expression at genomic scale, since investigators are able to simultaneously examine changes in the expression of hundreds or even thousands of genes.

For hybridization arrays, the general approach is to immobilize gene-specific sequences (for instance oligonucleotide probes or cDNAs) on a solid-state matrix (nylon membranes, glass microscope slides, silicon/ceramic chips). These gene sequences are then hybridized with labeled copies of nucleic acids produced from biological samples (targets). It is assumed that a correlation exists between the expression of a gene (and therefore the amount of labeled target) and the output signal.

Accordingly, macro- and microarrays-based analysis of gene expression on a medium- or large scale is an increasingly recognized method for functional and clinical investigations. This approach is of course greatly facilitated by the now extensive catalog of known or partially sequenced genes.

For instance, Cooper (2001) has recently reviewed arrays-based approaches for the study of quantitative expression levels for hundreds of genes in six cell lines, including three mammary carcinoma cell lines. The extensive data assembled in this survey identified potential targets of carcinogenesis, for example the CRABP2 and GATA3 transcription factor genes.

In addition to this and similar approaches, DNA hybridization arrays can be used in pharmacogenomic and toxicogenomic experiments, aimed at extensive analyses of the effects of therapeutic drugs on overall gene expression of target cells.

With respect to the study of the effects on transcription of treatment of target cells with crude extracts or purified compounds from medicinal plants, very few data employing macro- or microarrays are available in the recent literature. An example of the application of the microarray technology to this specific field is the recent paper by Chen et al. (2001), on the microarray profiling of gene expression patterns in bladder tumor cells treated with the soy isoflavone genistein. These authors employed a susceptible bladder tumor line, treated with the inhibitory dose (50 μM) of genistein for various periods of time, followed by mRNA isolations, cDNA probe preparations and hybridization individually to cDNA chips containing 884 sequence-verified known human genes. After analyzing the hybridization signals with a simple quantitative method developed by this study, they detected changes in gene expression profile, as for example the expression of EGR-1, associated with proliferation and differentiation. This gene was induced by the treatment with genistein, and its expression changes were later confirmed by RT-PCR. Other reports on the effects of bioactive compounds on target cell lines or primary cells were recently published by several research groups (Cunningham et al., 2000; Jain, 2000; Reilly et al., 2001).

Taken together, the available data show that the microarray technology is a reliable and powerful tool for profiling gene expression patterns in many biological systems related to cancer. The objective of these studies is the identification of the group of genes with distinct expression profiles and selectively up- or down-regulated by treatments with bioactive compounds.

Here we present preliminary results on the effects of plant extracts from *M. oleifera* on the gene expression profile of a human tumor cell line, the K562 cell line, originally isolated from a patient with CML in blast crisis (Lozzio and Lozzio, 1977).

This philadelphia-positive cell line has been employed, over the past decade, for the identification of antitumor compounds. An example is the recent characterization of the tyrosine-kinase inhibitor STI-571, a drug able to selectively inhibit the growth of CML cell lines by modulating the Bcr-Abl activity (reviewed in Nimmanapalli and Bhalla, 2002).

As far as the choice of *M. oleifera* is concerned, we would like to point out that this medicinal plant exhibits very interesting biological properties (Faizi et al., 1998; Murakami et al., 1998; Ghasi et al., 2000; Tahiliani and Kar, 2000; Rao et al., 2001; Pari and Kumar, 2002; Vlahov et al., 2002). The pharmacological properties of *M. oleifera* have been described, with respect to anti-microbial activity, by Caceres et al. (1991, 1992) who have investigated the effects of extracts from leaves, roots, bark and seeds on the *in vitro* activity against bacteria, yeast, dermatophytes and helminths pathogenic to man. By a disk-diffusion method, it was demonstrated that the fresh leaf juice and aqueous extracts from the seeds inhibit the growth of *Pseudomonas aeruginosa* and *Staphylococcus aureus*. Interestingly, no activity was demonstrated against four other pathogenic Gram-positive and Gram-negative bacteria and *Candida albicans*. Since this report, other papers were published demonstrating that *M. oleifera* retains antispasmodic, hypotensive, anti-inflammatory, diuretic and hypocholesterolemic activities.

In addition, in a recent paper, Guevara et al. (1999) examined ethanol extract from the seeds of *M. oleifera*. These authors have fractionated compounds, obtained in relatively good yield, exhibiting potential anti-tumor activity, as assessed by an *in vitro* assay based on inhibitory effects on Epstein–Barr virus-early antigen (EBV-EA) activation in Raji cells induced by the tumor promoter 12-O-tetradecanoyl-phorbol-13-acetate (TPA). All the tested compounds showed inhibitory activity against EBV-EA activation. Among them, niazimicin was subjected to further *in vivo* investigation and found to be a potent antitumor drug able to inhibit the two-stage carcinogenesis provoked in mouse skin by using 7,12-dimethylbenz(a)anthracene (DMBA), as initiator, and TPA, as tumor promoter.

All these studies encourage further investigation at molecular level for the identification of genes whose expression changes following treatment of cells with *M. oleifera* extracts. To this aim, the approaches involving macro- and/or microarrays are much promising.

In this paper, we review recent data obtained in our laboratory, focused to determine whether extracts from different medicinal plants exhibit different effects on the gene expression profiling (also called transcriptome). To this aim, the gene expression profiling studies were performed by the macroarray approach. Two plant extracts obtained from *M. oleifera* and from *E. officinalis* were compared. The dried fruits of *E. officinalis* were extracted with absolute ethanol and the yield was 9.33%. A similar procedure was followed to prepare extracts from *M. oleifera*. In this case, the obtained yield was 15.31%.

In the gene profiling experiments, in order to compare the effects of different plant extracts on gene expression, we evaluate the IC_{50} concentration of each one. This value represents the plant extracts concentration causing 50% inhibition of growth of the human erythroleukemia K562 cells after three days of culture. After this time, 5 μg of total RNA isolated by the TRIZOL reagent (Life Technologies, Inc.) were reverse-transcribed using gene-specific primers and reverse transcriptase (Clontech) in the presence of [α-^{32}P] dATP. Radiolabeled cDNAs were purified, denatured with 0.25 M NaOH to remove RNA, and hybridized to identical filters (Atlas Human Cancer 1.2 Arrays, purchased from Clontech) that contain 1200 non-redundant human cDNA clones. Filters were exposed to a phosphor-imager, and spots were detected and quantified using the AtlasImage 2.0 software (Clontech). The intensity values for filters containing cDNA derived from control and plant extracts-treated cells were compared in order to determine the fold induction or reduction. A twofold change was assumed as cut-off in order to identify differentially expressed genes.

II. Treatment of K562 cells with *M. oleifera* extracts leads to changes in gene expression profile

In Figure 1 we show the dose-dependent effects of *M. oleifera* extracts on the proliferation of the human erythroleukemia K562 cells. This preliminary experiment was relevant to determine the IC_{50} value of these plant extracts, that was found be 10.7 μg/ml.

Afterwards, we have evaluated the effects of *M. oleifera* extracts at molecular level. The K562 cells were cultured for three days in the absence or in the presence of

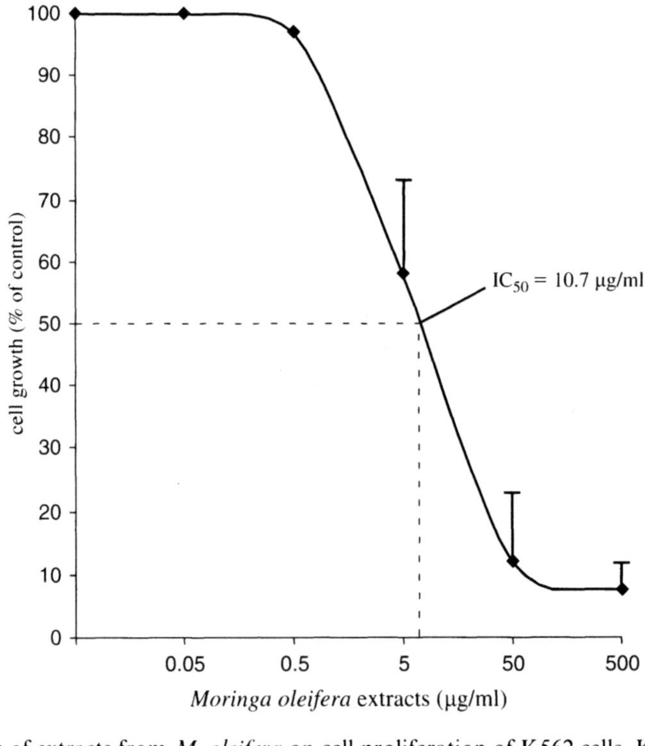

Fig. 1. Effects of extracts from *M. oleifera* on cell proliferation of K562 cells. K562 cells were seeded at the initial cell concentration of 30,000 cells/ml and then cultured for three days in the presence of the indicated amounts of plant extracts. Cell number/ml was determined and compared to the values of control untreated cells.

IC_{50} concentration of plant extracts and purified RNAs were used in cDNA arrays experiments. In Figure 2A we show the comparison of gene expression profile of untreated K562 cells (upper panel) and cells treated with *M. oleifera* extracts (lower panel). In this experiment we obtained genes that were up-, down- or not-modulated. In Figure 2 (panels B–D) we show three typical examples of (a) an up-regulated gene (cytosolic superoxide dismutase 1, see Figure 2B), (b) a down-regulated gene (microtubule-associated protein 1B, see Figure 2C) and (c) a gene whose expression is not altered following the treatment with *M. oleifera* extracts (heterogeneous nuclear ribonucleoprotein K, see Figure 2D). It should be noted that, while the macroarray technology should be considered informative for differential gene expression pathways, the results should be routinely confirmed by molecular approaches such as, for instance, the reverse transcription (RT)-PCR. In Table 1 we have summarized interesting genes that were found differentially expressed during the treatment. Among these genes, *M. oleifera* extracts up-regulated the expression of genes involved in the survival mechanism, such as the checkpoint suppressor 1 (CHES1) (Pati et al., 1997) and E4BP4 (Altura et al., 1998; Nishimura and Tanaka, 2001), or in the detoxification mechanism, such as cytosolic superoxide dismutase (SOD1) (Schwartz et al., 1998). Furthermore, the increased expression of cadherin 5 (CDH5) gene observed after the treatment of K562 cells with *M. oleifera* extracts should be considered as a molecular feature linked to cell-clustering produced by the treatment, also associated

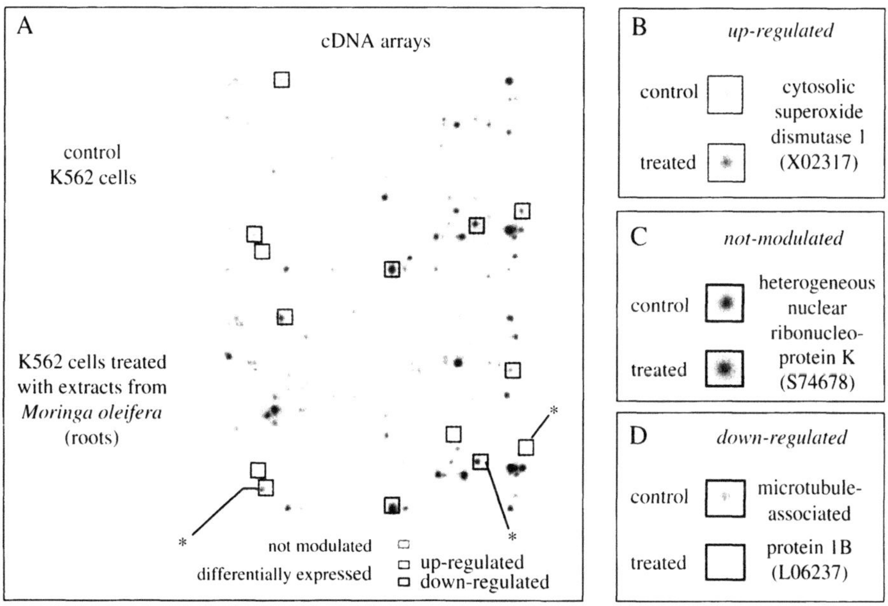

Fig. 2. Effects of *M. oleifera* extracts on gene expression profile. The K562 cells were cultured for three days in the absence or in the presence of IC_{50} concentration of plant extracts and purified RNAs were used in cDNA arrays experiments. (A) Gene expression profiles of untreated K562 cells (upper panel) and cells treated with *M. oleifera* extracts (lower panel). (B–D) Magnification of three example spots, representing an up- (B), a down- (D) and a not-modulated (C) gene, from the experiment depicted in panel (A). Brackets contain the GenBank entry of each gene considered.

Table 1
List of genes differentially expressed during the treatment of human erythroleukemic K562 cells with *M. oleifera* extracts (partial list)

Gene name	GenBank accession	Modulation
Cell cycle protein P38-2G4 homolog	U59435	Up
Nucleoside diphosphate kinase B (NDP kinase B; NDKB)	L16785	Up
Ras homolog gene family member A (RHOA; ARHA)	L25080	Up
Growth factor receptor-bound protein 2 (GRB2)	L29511	Down
e4bp4	X64318	Up
Growth arrest & DNA damage-inducible protein 153 (GADD153)	S40706	Down
Growth arrest & DNA damage-inducible protein (GADD45)	M60974	Down
Caspase 10 (CASP10)	U60519	Down
Checkpoint suppressor 1	U68723	Up
Cadherin 5 (CDH5)	X79981	Up
Cytosolic superoxide dismutase 1 (SOD1)	K00065	Up
Macrophage inhibitory cytokine 1 (MIC1)	AF019770	Down
Vimentin (VIM)	X56134	Down
Heterogeneous nuclear ribonucleoprotein K (HNRNPK)	S74678	Down
Microtubule-associated protein 1B	L06237	Down

to the cytotoxic effects of this plant extracts (Caveda et al., 1996; Carmeliet et al., 1999). All these molecular features highlight the cytotoxic effects of *M. oleifera* extracts on human cancer cells. In addition, the up-regulation of both the nucleoside diphosphate kinase B gene, a potential negative regulator of cancer metastasis involved in the regulation of c-myc gene expression (Lee et al., 1997), and the cell cycle protein P38-2G4 homolog (Lamartine et al., 1997), suggests that *M. oleifera* extracts could modulate the cell cycle.

III. Modifications of the gene expression profile in human K562 cells treated with *M. oleifera* and *E. officinalis* plant extracts

Is the macroarray approach suitable to identify genes direct or indirect targets of bioactive compounds present in medicinal plant extracts? To verify this hypothesis, we have compared the gene expression profiles obtained in human K562 cells treated with plant extracts from *E. officinalis* or *M. oleifera*, two plant extracts chosen because GC/MS analyses demonstrated the presence of sharply different compounds (data not shown). This analysis has been summarized in Figure 3. Despite the different number of modulated genes found in K562 cells treated with *M. oleifera* and *E. officinalis* extracts (41 and 29, respectively), the ratio between up- and down-regulated was found to be quite similar. Among the 11 genes, which were found up-regulated by the treatment of K562 cells with *M. oleifera* extracts, only two genes were found up-regulated also following the treatment with *E. officinalis* extracts.

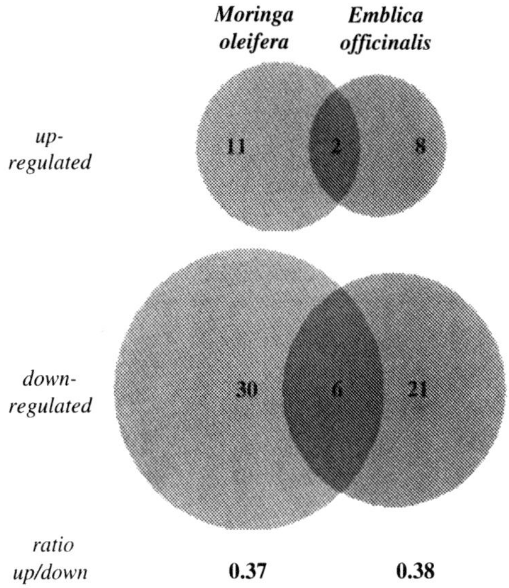

Fig. 3. Comparison of the number of genes found modulated in human K562 cells treated with plant extracts from *E. officinalis* and *M. oleifera*. The number of genes modulated by both plant extracts has been reported in the pink-overlapping zone. The ratio among up- and down-regulated genes found has also been calculated.

Similarly, only six genes were found down-regulated by the treatments with both *M. oleifera* and *E. officinalis* extracts. The results summarized in Figure 3 indicate that the macroarray approach is suitable to identify genes, which could be the direct or indirect target of bioactive compounds present in extracts from medicinal plants. The fact that the majority of the up-regulated and down-regulated genes found specifically modulated by the plant extracts used are different, is in agreement with the different chemical nature of *M. oleifera* and *E. officinalis* extracts.

IV. Validation of the macroarray data by RT-PCR

In order to confirm the macroarray data, RT-PCR analysis was performed on three representative genes. The comparison between the RT-PCR data (right side on each panel) and the macroarray data (left side of each panel) has been reported in Figure 4. As evident, the expression of both the E4BP4 gene (see Figure 4A) and the macrophage inhibitor cytokine 1 gene (see Figure 4C) were found respectively up- and down-modulated in both experimental approaches. Similarly, the expression of the vimentin gene (see Figure 4B) was found not modulated both in the macroarray technology and in the RT-PCR assay.

V. Conclusions and future perspectives

Membrane arrays are now available from a number of biotechnology companies to study gene expression profiles, including GenoSystem (Sigma) (Panorama Arrays),

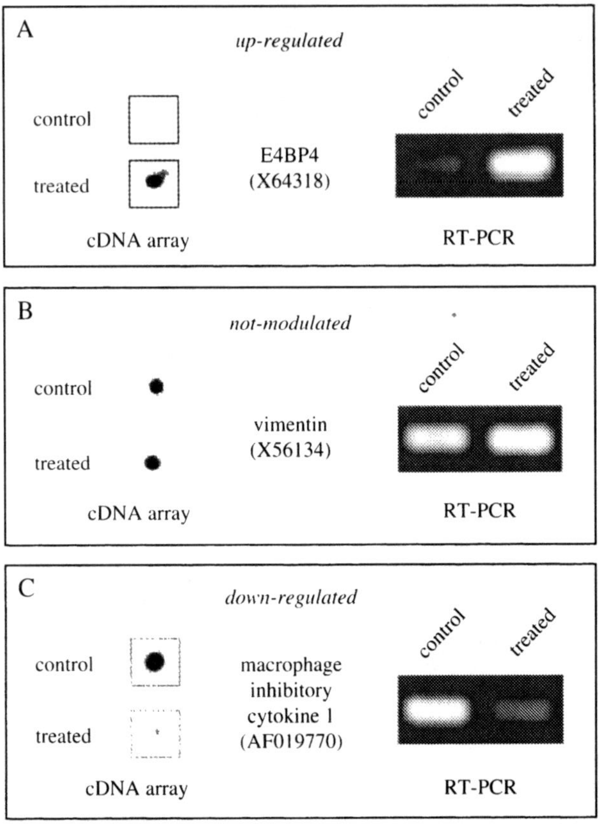

Fig. 4. RT-PCR analysis of the RNA messenger levels for three selected genes in human K562 cells treated or not with the plant extracts from *M. oleifera*. The results are relative to an up- (A), a down- (C) and a not-modulated (B) gene. In the left side of each panel it was reported the hybridization spot obtained for that gene in the cDNA array experiment. Brackets contain the GenBank entry of each gene considered.

Research Genetics (GeneFilters microarrays), R&D systems (DNA Expression Arrays), Clontech (Atlas Nylon Arrays), GenoTechnology, Inc., SuperArray Inc. (GEArray system), Ambion, Inc. (ULTRArray gene expression membranes).

In addition, microarrayer systems, microarray scanners and microarray chips have been recently introduced and commercially available from several biotechnology companies, including Affymetrix (Santa Clara, CA), Agilent Technologies (Palo Alto, CA), Incyte (Palo Alto, CA), ResGen (Huntsville, AL), Axon Instruments (Union City, CA), Amersham Pharmacia Biotech (Piscataway, NJ), Packard Bio-Chip (a wholly owned subsidiary of Packard BioScience Company), BioDiscovery (Los Angeles, CA), BioSieve (San Jose, CA), GeneData AG (Basel, Switzerland), Imaging Research Inc., St. Catharines, Ontario, Canada), LION bioscience AG (Heidelberg, Germany), Partek (St. Charles, MO), Rosetta Inpharmatics (Kirkland, WA), Silicon Genetics (Redwood City, CA), Gene Logic (Gaithersburg, MD), Tele-Chem International, Inc. (Sunnyvale, CA).

Both macroarray and microarray technologies are expected to be employed in the post-genomic research in several applied research fields. In addition to experiments aimed at the definition of gene expression (a) in differentiated cells or tissues (Fernandez et al., 2001; Watakabe et al., 2001; Bertucci et al., 1999; Dales et al., 2001; Hu et al., 2001; Kallioniemi et al., 2001; Kan et al., 2001; Rew, 2001; Schuldiner and Benvenisty, 2001; Takahashi et al., 2001; Takemasa et al., 2001), (b) during neoplastic transformation, (c) during development (Freeman et al., 2000); therefore, gene expression profile studies are expected to be in the near future largely employed for determining the effects of drugs on genome expression. These pharmacogenomic approaches will permit to obtain important information on one hand on the genotoxicity of drug treatments (Afshari et al., 1999; Kan et al., 2001) and on the other hand on the identification of genes whose expression is altered by the treatment (Nuwaysir et al., 1999; Loguinov et al., 2001; Ueda, 2001). This second issue is of great interest for the design of more selective molecular approaches (antisense therapy, transcription factor decoy treatment, treatment with triple-helix forming oligonucleotides) for the development of non-viral gene therapy of human diseases (Baba, 2001).

In this respect, extracts from medicinal plants are of great interest. Medicinal plants have been described to exhibit a variety of therapeutic properties (Ahmad et al., 1998; Datta et al., 1998; Ankli et al., 2002; Neto et al., 2002). For instance, medicinal plants are used in prenatal care (Velazco, 1980; Abo et al., 2000), in obstetrics (Pinn, 2001), in gynecology (Abo et al., 2000), in respiratory disorders (Ankli et al., 2002; Neto et al., 2002), in skin disorders (Graf, 2000), in cardiac diseases (Ankli et al., 2002), in nervous and muscular disorders (Datta et al., 1998) and in mental health (Ahmad et al., 1998). Examples are a variety of herbal drugs lowering cholesterol (useful in chronic angina and congenital heart diseases) (Mishra et al., 1981; Mathur et al., 1996; Ram et al., 1997), herbal treatments for a wide range of skin conditions (Graf, 2000), including diseases like Psoriasis and Erysipelas (Ahmad et al., 1998; Datta et al., 1998; Ankli et al., 2002; Neto et al., 2002). For instance, *Terminalia arjuna* is used in angina pectoris (Dwivedi et al., 1994), congestive heart failure (Bharani et al., 1995), hyperlipidemia (Ram et al., 1997); *Paederia foetida* exhibits anti-inflammatory activity (De et al., 1994); *Hygrophila auriculata* and *Hemidesmus indicus* have been described to retain hepatoprotective activities for the treatment of liver ailments (Singh and Handa, 1995); *E. officinalis* exhibits anti-inflammatory (Asmawi et al., 1993), anti-fungal (Datta et al., 1998), anti-carcinogenetic (Sharma et al., 2000; Jose et al., 2001) and hepatoprotective activities (Jose and Kuttan, 2000). Recently, it was shown that methanolic extracts of *Oroxylum indicum* strongly inhibit the mutagenicity of Trp-P-1 in an Ames test (Nakahara et al., 2001). Anti-diarrhoeal activity and anti-fungal activity of *Aegle marmelos* were also reported (Shoba and Thomas, 2001).

In conclusion, the analysis of gene expression profiles by the macro- or the micro-array approaches performed in cells treated with medicinal plant extracts, could be a starting point useful in order to develop innovative therapies (Du et al., 2003). In fact, as reported in the procedural scheme depicted in Figure 5, the treatment with plant extracts will allow the identification of genes that are down- or up-regulated. In prospective, the expression of these genes could be modulated by gene-specific decoy oligonucliotides (ODNs), antisense DNA or RNA, peptides or monoclonal

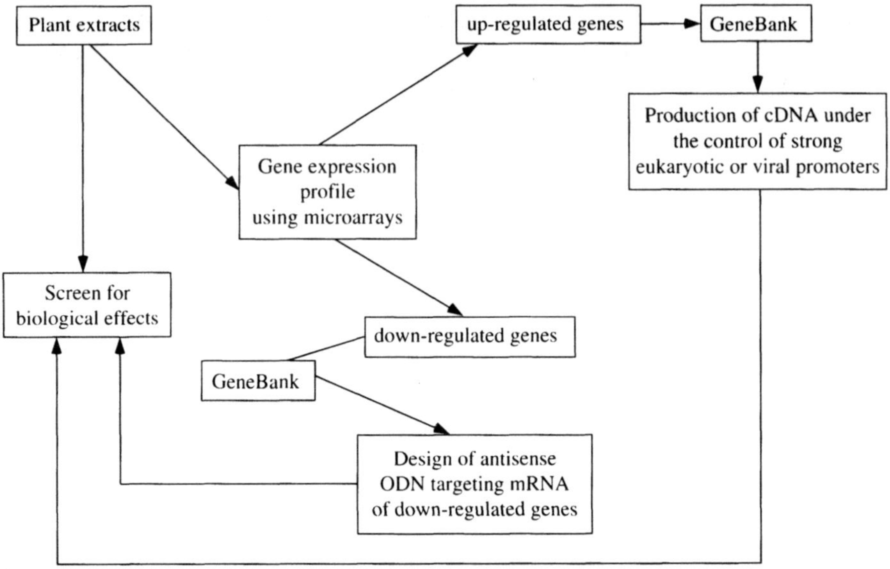

Fig. 5. Flow chart indicating the interplay between different technologies in drug design based on the informations derived from studies on gene expression profile using micro- and macroarrays.

antibodies, thus mimicking the biological effects of the employed plant extracts and limiting the possible side effects.

Acknowledgments

R.G. is granted by AIRC, Italian Cystic Fibrosis Foundation, AVLT and Fondazione CARIPARO. One of us (MTHK) gratefully acknowledges the travel support from Third World Academy of Sciences (TWAS), Italy, under the "South–South Fellowship" scheme. He was also the recipient of the fellowships from the UNESCO-MCBN (Grant no. 1056), the CIB, Italy, and the AVTL, Italy.

References

Abo KA, Adeyemi AA, Adeite DA. (2000) Ethnobotanical survey of plants used in the treatment of infertility and sexually transmitted diseases in southwest Nigeria. Afr J Med Med Sci 29:325–7.

Afshari CA, Nuwaysir EF, Barrett JC. (1999) Application of complementary DNA microarray technology to carcinogen identification, toxicology, and drug safety evaluation. Cancer Res 59:4759–60.

Ahmad I, Mehmood Z, Mohammad F. (1998) Screening of some Indian medicinal plants for their antimicrobial properties. J Ethnopharmacol 62:183–93.

Altura RA, Inukai T, Ashmun RA, Zambetti GP, Roussel MF, Look AT. (1998) The chimeric E2A-HLF transcription factor abrogates p53-induced apoptosis in myeloid leukemia cells. Blood 92:1397–405.

Ankli A, Heinrich M, Bork P, Wolfram L, Bauerfeind P, Brun R, Schmid C, Weiss C, Bruggisser R, Gertsch J, Wasescha M, Sticher O. (2002) Yucatec Mayan medicinal plants: evaluation based on indigenous uses. J Ethnopharmacol 79:43–52.

Asmawi MZ, Kankaanranta H, Moilanen E, Vapaatalo H. (1993) Anti-inflammatory activities of *Emblica officinalis* Gaertn leaf extracts. J Pharm Pharmacol 45:581–4.

Baba Y. (2001) Development of novel biomedicine based on genome science. Eur J Pharm Sci 13:3–4.

Bertucci F, Van Hulst S, Bernard K, Loriod B, Granjeaud S, Tagett R, Starkey M, Nguyen C, Jordan B, Birnbaum D. (1999) Expression scanning of an array of growth control genes in human tumor cell lines. Oncogene 18:3905–12.

Bharani A, Ganguly A, Bhargave KD. (1995) Salutare effects of *Terminalia arjuna* in patients with severe refractory heart failure. Int J Cardiol 49:191–9.

Caceres A, Cabrera O, Morales O, Mollinedo P, Mendia P. (1991) Pharmacological properties of *Moringa oleifera*. Preliminary screening for antimicrobial activity. J Ethnopharmacol 33:213–6.

Caceres A, Saravia A, Rizzo S, Zabala L, De Leon E, Nave F. (1992) Pharmacologic properties of *Moringa oleifera*. Screening for antispasmodic, anti-inflammatory and diuretic activity. J Ethnopharmacol 36:233–7.

Carmeliet P, Lampugnani MG, Moons L, Breviario F, Compernolle V, Bono F, Balconi G, Spagnuolo R, Oostuyse B, Dewerchin M, Zanetti A, Angellilo A, Mattot V, Nuyens D, Lutgens E, Clotman F, de Ruiter MC, Gittenberger-de Groot A, Poelmann R, Lupu F, Herbert JM, Collen D, Dejana E. (1999) Targeted deficiency or cytosolic truncation of the VE-cadherin gene in mice impairs VEGF-mediated endothelial survival and angiogenesis. Cell 98:147–57.

Caveda L, Martin-Padura I, Navarro P, Breviario F, Corada M, Gulino D, Lampugnani MG, Dejana E. (1996) Inhibition of cultured cell growth by vascular endothelial cadherin (cadherin-5/VE-cadherin). J. Clin. Invest. 98:886–93.

Chen CC, Shieh B, Jin YT, Liau YE, Huang CH, Liou JT, Wu LW, Huang W, Young KC, Lai MD, Liu HS, Li C. (2001) Microarray profiling of gene expression patterns in bladder tumor cells treated with genistein. J. Biomed. Sci. 8:214–22.

Cooper CS. (2001) Applications of microarray technology in breast cancer research. Breast Cancer Res. 3:158–75.

Cunningham MJ, Liang S, Fuhrman S, Seilhamer JJ, Somogyi R. (2000) Gene expression microarray data analysis for toxicology profiling. Ann. N. Y. Acad. Sci. 919:52–67.

Dales JP, Plumas J, Palmerini F, Devilard E, Defrance T, Lajmanovich A, Pradel V, Birg F, Xerri L. (2001) Correlation between apoptosis microarray gene expression profiling and histopathological lymph node lesions. Mol. Pathol. 54:17–23.

Datta BK, Rahman I, Das TK. (1998) Antifungal activity of Indian plant extracts. Mycoses 41:535–6.

De Backer MD, Ilyina T, Ma XJ, Vandoninck S, Luyten WH, Vanden Bossche H. (2001) Genomic profiling of the response of *Candida albicans* to itraconazole treatment using a DNA microarray. Antimicrob Agents Chemother 45:1660–70.

De S, Ravishankar B, Bhavsar GC. (1994) Investigation of the anti-inflammatory effects of *Paederia foetida*. J. Ethnopharmacol, 43:31–8.

Debouck C, Goodfellow PN. (1999) DNA microarrays in drug discovery and development. Nat Genet 21:48–50.

Du ZY, Li XY, Li YC, Wang SY. (2003) Analysis of triptolide-regulated gene expression in Jurkat cells by complementary DNA microarray. Acta Pharmacol Sin 24:864–72.

Dwivedi S, Agarwal MP. (1994) Antianginal and cardioprotective effects of *Terminalia arjuna*, an indigenous drug, in coronary artery disease. JAPI 42:287–9.

Faizi S, Siddiqui BS, Saleem R, Aftab K, Shaheen F, Gilani AH. (1998) Hypotensive constituents from the pods of *Moringa oleifera*. Planta Med. 64(2):25–8.

Fernandez PL, Nayach I, Fernandez E, Fresno L, Palacin A, Farre X, Campo E, Cardesa A. (2001) Tissue macroarrays ("microchops") for gene expression analysis. Virchows Arch. 438:591–4.

Freeman WM, Robertson DJ, Vrana KE. (2000) Fundamentals of DNA hybridization arrays for gene expression analysis. Biotechniques. 29:1042–6.

Ghasi S, Nwobodo E, Ofili JO. (2000) Hypocholesterolemic effects of crude extract of leaf of *Moringa oleifera* Lam in high-fat diet fed wistar rats. J. Ethnopharmacol 69:21–5.

Graf J. (2000) Herbal anti-inflammatory agents for skin disease. Skin Therapy Lett 5:3–5.

Guevara AP, Vargas C, Sakurai H, Fujiwara Y, Hashimoto K, Maoka T, Kozuka M, Ito Y, Tokuda H, Nishino H. (1999) An antitumor promoter from *Moringa oleifera* Lam. Mutat Res. 440:181–8.

Hu YC, Lam KY, Law S, Wong J, Srivastava G. (2001) Identification of differentially expressed genes in esophageal squamous cell carcinoma (ESCC) by cDNA expression array: overexpression of Fra-1, Neogenin, Id-1, and CDC25B genes in ESCC. Clin. Cancer Res. 7:2213–21.

Jain KK. (2000) Applications of biochip and microarray systems in pharmacogenomics. Pharmacogenomics 1:289–307.

Jose JK, Kuttan G, Kuttan R. (2001) Antitumour activity of *Emblica officinalis*. J. Ethnopharmacol, 75:65–9.

Jose JK, Kuttan R. (2000) Hepatoprotective activity of *Emblica officinalis* and Chyavanaprash. J. Ethnopharmacol, 72:135–40.

Kallioniemi OP, Wagner U, Kononen J, Sauter G. (2001) Tissue microarray technology for high-throughput molecular profiling of cancer. Hum Mol Genet 10:657–62.

Kan T, Shimada Y, Sato F, Maeda M, Kawabe A, Kaganoi J, Itami A, Yamasaki S, Imamura M. (2001) Gene expression profiling in human esophageal cancers using cDNA microarray. Biochem Biophys Res Commun 286:792–801.

Kan T, Shimada Y, Sato F, Maeda M, Kawabe A, Kaganoi J, Itami A, Yamasaki S, Imamura M, Kudoh K, Ramanna M, Ravatn R, Elkahloun AG, Bittner ML, Meltzer PS, Trent JM, Dalton WS, Chin KV. (2000) Monitoring the expression profiles of doxorubicin-induced and doxorubicin-resistant cancer cells by cDNA microarray. Cancer Res 60:4161–6.

Lamartine J, Seri M, Cinti R, Heitzmann F, Creaven M, Radomski N, Jost E, Lenoir GM, Romeo G, Sylla BS. (1997) Molecular cloning and mapping of a human cDNA (PA2G4) that encodes a protein highly homologous to the mouse cell cycle protein p38-2G4. Cytogenet Cell Genet 78:31–7.

Lee IH, Chang SI, Okada K, Baba H, Shiku H. (1997) Transcription effect of nm23-M2/NDP kinase on c-myc oncogene. Mol Cells 7:589–93.

Loguinov AV, Anderson LM, Crosby GJ, Yukhananov RY. (2001) Gene expression following acute morphine administration. Physiol Genomics 6:169–81.

Lozzio BB, Lozzio CB. (1977) Properties of the K562 cell line derived from a patient with chronic myeloid leukemia. Int. J. Cancer, 19:136–41.

Mathur R, Sharma A, Dixit VP, Varma M. (1996) Hypolipidaemic effect of fruit juice of *Emblica officinalis* in cholesterol-fed rabbits. J. Ethnopharmacol 50:61–8.

Mishra M, Pathak UN, Khan AB. (1981) *Emblica officinalis* Gaertn and serum cholesterol level in experimental rabbits. Br J. Exp. Pathol, 62:526–8.

Murakami A, Kitazono Y, Jiwajinda S, Koshimizu K, Ohigashi H. (1998) Niaziminin, a thiocarbamate from the leaves of *Moringa oleifera*, holds a strict structural requirement for inhibition of tumor-promoter-induced Epstein–Barr virus activation. Planta Med 64:319–23.

Nakahara K, Onishi-Kameyama M, Ono H, Yoshida M, Trakoontivakorn G. (2001) Antimutagenic activity against trp-P-1 of the edible Thai plant, *Oroxylum indicum* vent. Biosci Biotechnol Biochem 65:2358–60.

Neto CC, Owens CW, Langfield RD, Comeau AB, Onge JS, Vaisberg AJ, Hammond GB. (2002) Antibacterial activity of some Peruvian medicinal plants from the Callejon de Huaylas. J Ethnopharmacol 79:133–8.

Nimmanapalli R, Bhalla K. (2002) Novel targeted therapies for Bcr-Abl positive acute leukemias: beyond STI571. Oncogene 21:8584–90.

Nishimura Y, Tanaka T. (2001) Calcium-dependent activation of nuclear factor regulated by interleukin 3/adenovirus E4 promoter-binding protein gene expression by calcineurin/

nuclear factor of activated T cells and calcium/calmodulin-dependent protein kinase signaling. J. Biol. Chem. 276:19921–8.

Nuwaysir EF, Bittner M, Trent J, Barrett JC, Afshari CA. (1999) Microarrays and toxicology: the advent of toxicogenomics. Mol Carcinog 24:153–9.

Pari L, Kumar NA. (2002) Hepatoprotective Activity of *Moringa oleifera* on antitubercular drug-induced liver damage in rats. J. Med Food 5:171–7.

Pati D, Keller C, Groudine M, Plon SE. (1997) Reconstitution of a MEC1-independent checkpoint in yeast by expression of a novel human fork head cDNA. Mol Cell Biol 17:3037–4306.

Pinn G. (2001) Herbs used in obstetrics and gynaecology. Aust Fam Physician 30:351–6.

Ram A, Lauria P, Gupta R, Kumar P, Sharma VN. (1997) Hypocholesterolaemic effects of *Terminalia arjuna* tree bark. J Ethnopharmacol 55:165–9.

Rao AV, Devi PU, Kamath R. (2001) In vivo radioprotective effect of *Moringa oleifera* leaves. Indian J Exp Biol 39:858–63.

Reilly TP, Bourdi M, Brady JN, Pise-Masison CA, Radonovich MF, George JW, Pohl LR. (2001) Expression profiling of acetaminophen liver toxicity in mice using microarray technology. Biochem Biophys Res Commun 282:321–8.

Rew DA. (2001) DNA microarray technology in cancer research. Eur J Surg Oncol 27:504–8.

Schuldiner O, Benvenisty N. (2001) A DNA microarray screen for genes involved in c-MYC and N-MYC oncogenesis in human tumors. Oncogene 20:4984–94.

Schwartz PJ, Coyle JT. (1998) Effects of overexpression of the cytoplasmic copper-zinc superoxide dismutase on the survival of neurons in vitro. Synapse 29:206–12.

Sharma N, Trikha P, Athar M, Raisuddin S. (2000) In vitro inhibition of carcinogen-induced mutagenicity by *Cassia occidentalis* and *Emblica officinalis*. Drug Chem Toxicol 23:477–84.

Shoba FG, Thomas M. (2001) Study of antidiarrhoeal activity of four medicinal plants in castor-oil induced diarrhoea. J Ethnopharmacol 76:73–6.

Singh A, Handa SS. (1995) Hepatoprotective activity of *Apium graveolens* and *Hygrophila auriculata* against paracetamol and thioacetamide intoxication in rats. J Ethnopharmacol 49:119–26.

Tahiliani P, Kar A. (2000) Role of *Moringa oleifera* leaf extract in the regulation of thyroid hormone status in adult male and female rats. Pharmacol Res 41:319–23.

Takahashi M, Rhodes DR, Furge KA, Kanayama Ho, Kagawa S, Haab BB, The BT. (2001) Gene expression profiling of clear cell renal cell carcinoma: gene identification and prognostic classification. Proc Natl Acad Sci 98:9754–9.

Takemasa I, Higuchi H, Yamamoto H, Sekimoto M, Tomita N, Nakamori S, Matoba R, Monden M, Matsubara K. (2001) Construction of preferential cDNA microarray specialized for human colorectal carcinoma: molecular sketch of colorectal cancer. Biochem Biophys Res Commun 285:1244–9.

Ueda K. (2001) Detection of the retinoic acid-regulated genes in a RTBM1 neuroblastoma cell line using cDNA microarray. Kurume Med J 48:159–64.

Velazco BA. (1980) Traditional herbal practices and motherhood. Philipp J Nurs 50:95–9.

Vlahov G, Chepkwony PK, Ndalut PK. (2002) (13)C NMR characterization of triacylglycerols of *Moringa oleifera* seed oil: an "oleic-vaccenic acid" oil. J Agric Food Chem. 50:970–5.

Watakabe A, Sugai T, Nakaya N, Wakabayashi K, Takahashi H, Yamamori T, Nawa H. (2001) Similarity and variation in gene expression among human cerebral cortical subregions revealed by DNA macroarrays: technical consideration of RNA expression profiling from postmortem samples. Brain Res Mol Brain Res. 88:74–82.

M.T.H. Khan and A. Ather (eds.)
Lead Molecules from Natural Products

Effects of medicinal plant extracts on molecular interactions between DNA and transcription factors

Ilaria Lampronti, Mahmud Tareq Hassan Khan, Nicoletta Bianchi, Giordana Feriotto, Carlo Mischiati, Monica Borgatti, Roberto Gambari

Abstract

Alteration of gene transcription is one of the approaches to control the expression of selected genes and could be achieved by molecules interfering with the interactions between transcription factors and DNA. In this respect, DNA-binding drugs are of great interest and the object of a large number of research articles. The most suitable techniques for a fast screening of these compounds are the electrophoretic mobility shift assay and the filter-binding approach. In this communication, we present evidence showing that extracts from medicinal plants could inhibit the interactions between nuclear factors and specific target DNA sequences. Among the results obtained, we found that low concentrations (6–12 μg/binding reaction) of *Aegle marmelos* extracts, and very low concentrations (0.5 μg/binding reaction) of *Emblica officinalis* extracts inhibit GATA-1/DNA interactions. On the contrary, up to 400 μg/binding reaction of extracts from *Argemone mexicana* were still unable to inhibit GATA-1/DNA interactions. The employment of several analytical and preparative procedures, among which high-performance liquid chromatography (HPLC) and gas chromatography/mass spectrometry (GC/MS), will help in identifying the bioactive compounds responsible for this very interesting feature.

Keywords: *medicinal plants, anti-tumor agents, ethnopharmacology, transcription*

I. Introduction

Alteration of gene transcription is one of the approaches to control the expression of selected genes (Agarwal and Gewirtz, 1999; Praseuth et al., 1999; Gorman and Glazer, 2001). One of the most important strategies for the modulation of gene transcription is to interfere with the interaction between transcription factors (TFs) and target DNA elements present within the promoters of the genes whose expression must be modified (Izumiya et al., 2003; El-Tanani et al., 2001). Among the different approaches used in non-viral gene therapy, the left side of Figure 1 outlines the TF decoy approach (Cho-Chung et al., 1999; Mann and Dzau, 2000; Borgatti et al., 2003; Yamasaki et al., 2003), the use of triple-helix forming oligonucleotides

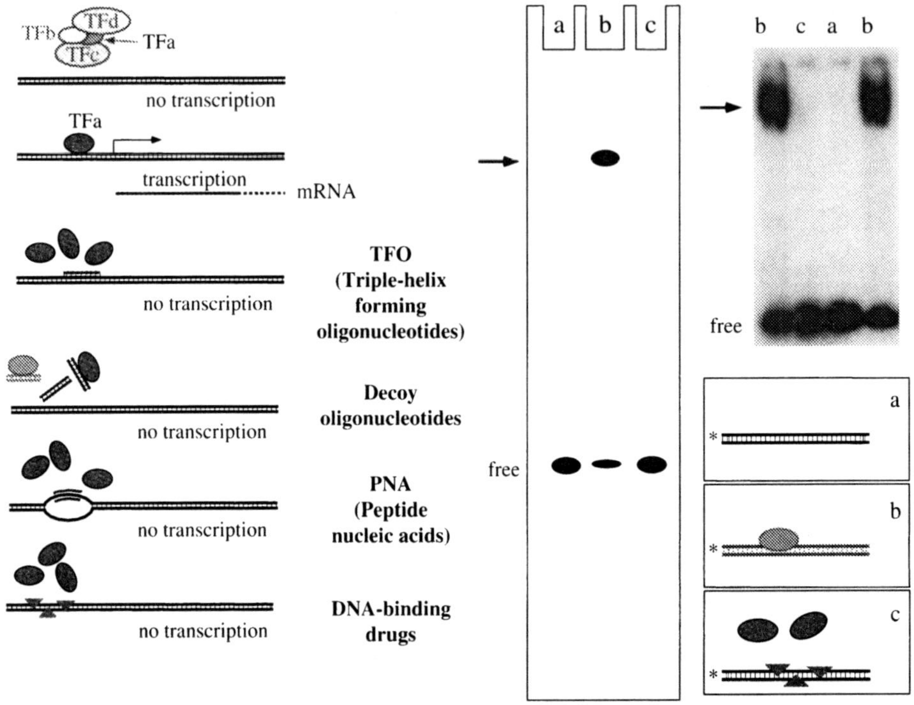

Fig. 1. Left side of the panel shows the strategy causing alteration of gene expression. In the middle and right side of the panel, EMSA assay is shown. ^{32}P-labeled Sp1 oligonucleotide migrates as a free molecule (as shown in "a" lanes and insert "a") or as a bound molecule associated with the specific TF (as shown in "b" lanes and insert "b"). This interaction is inhibited by chromomycin as shown in "c" lanes and in the "c" insert.

(Aggarwal et al., 1996; Rutigliano et al., 1998; Chan and Glazer, 1997), strand invasion by peptide nucleic acids (PNAs) (Mischiati et al., 2002) and the use of DNA-binding drugs (Gambari, 2000).

These approaches have been applied to the development of antitumor and antiviral therapeutics as well as intervention in several human pathologies (Denny, 1989; Bloemink and Reedijk, 1996; Chan et al., 1997; Dean, 2000; Jo et al., 2002; McMahon et al., 2002; Tomita et al., 2002; Trojan et al., 2002).

As far as development of antitumor agents, DNA-binding drugs are of great interest and are the object of a large number of research articles. For instance, the DNA-binding drugs distamycin, tallimustine, mithramycin, cisplatin and cisplatin analogs are all used in antitumor therapy (Bianchi et al., 2001). In addition, some of these molecules retain the very interesting property of inducing differentiation and apoptosis. As an example, we have recently shown that tallimustine (Bianchi et al., 2001), mithramycin (Bianchi et al., 1999), cisplatin and cisplatin analogs (Bianchi et al., 2000) are potent inducers of erythroid differentiation of the human chronic myelogenous leukemia K562 cell line, as well as apoptosis (Bianchi et al., 1999).

In respect to the mechanism of action of DNA-binding drugs, our research group and other research groups demonstrated that DNA-binding drugs inhibit the molecular interactions between DNA and TFs, leading to inhibition of transcription. For instance, Feriotto et al. (1994) studied the pharmacological modulation of the interaction between TFs and target DNA sequences using the A/T selective distamycin as the powerful inhibitor of the interaction between nuclear factors and the long terminal repeat (LTR) of the human immunodeficiency type 1 (HIV-1) retrovirus. These authors showed that distamycin binds to different regions of the HIV-1 LTR depending on the DNA sequence. In another research paper, Welch et al. (1994) demonstrated that several intercalating, minor groove binding, and covalently binding drugs, evaluated by mobility shift assays, were able to interfere with TFs binding to their respective DNA recognition sequences. The Cys2His2 zinc finger proteins EGR1, WT1, and NFIL2A, the basic leucine-zipper protein Jun/Fos, and the minor groove binding protein hTBP were chosen as representative TFs. The intercalators nogalamycin and hedamycin, and the G/C-specific minor groove binding drug chromomycin A3 were the most potent drugs, preventing TF–DNA complex formation at concentrations less than 1 μM. Similar concentrations of chromomycin A3 disrupted preformed complexes, while nogalamycin and hedamycin were 50-fold less potent if proteins were allowed to bind to DNA before drug treatment. Echinomycin inhibited EGR1–DNA complex formation 50% at 5 μM, but had little effect on the formation of NIL2A–DNA complexes. Conversely, doxorubicin was found to inhibit NFIL2A complex formation 50% at less than 1 μM, but did not achieve this level of inhibition of EGR1/DNA complex formation even at 50 μM. The A/T-directed minor groove binding drugs, while inhibiting hTBP at submicromolar concentrations, had no effect on either EGR1 or NFIL2A. As an additional example, Bianchi et al. (1996) studied the biochemical effects of the sequence-selective binding of chromomycin to the Sp1 binding sites of the HIV-1 LTR, showing that the molecular interactions between nuclear proteins and Sp1 binding sites are inhibited by chromomycin, and this effect leads to a sharp inhibition of *in vitro* transcription.

As reported in several papers (Cragg and Newman, 1999; Couladis et al., 2002; Khan et al., 2002; Stevigny et al., 2002; Katsube et al., 2003), an attractive property of plant extracts is related to their very interesting effects on tumor cells. In this respect, antitumor activity of herbal medicines has been described both *in vitro* and *in vivo* (Mukherjee et al., 2001; Popov et al., 2001; Richardson, 2001; Steenkamp et al., 2001; Wargovich et al., 2001; Yu et al., 2001; Chang et al., 2002; Ruffa et al., 2002; Tatman and Mo, 2002). For instance, in recent reports antitumor activity of *Emblica officinalis* has been shown (Jose et al., 2001; Khan et al., 2002). Aqueous extracts of *E. officinalis* were found to display cytotoxic effects on L929 cells in culture in a dose-dependent manner. Accordingly, *E. officinalis* extracts were found to reduce ascites and solid tumors in tumor-bearing mice (Jeena et al., 1999; Sharma et al., 2000; Jose et al., 2001).

However, in the field of ethnopharmacology, it should be underlined that, in addition to antitumor activity, several plant extracts as well as isolated compounds from plants are used in medicine for other properties, most of which have been well established since a long time in developing countries and later described following detailed *in vitro* and *in vivo* studies. Examples of known medicinal plants are *Terminalia arjuna*, used in angina pectoris (Bharani et al., 2002), congestive heart failure (Bharani et al., 1995), and hyperlipidemia (Ram et al., 1997); *Paederia foetida* and *Emblica officinalis*, exhibiting anti-inflammatory activity (Asmawi et al., 1993; De et al., 1994); *Hygrophila auriculata* and *Hemidesmus indicus*, described to retain hepatoprotective activities for the treatment of liver ailments (Singh and Handa, 1995; Prabakan et al., 2000). Several reports and reviews on the use of medicinal plants in ethnopharmacology have been recently published (Elisabetsky, 1991; Hedberg, 1993; Wickberg, 1993; Elisabetsky and Shanley, 1994; Borchers et al., 2000; Heinrich and Gibbons, 2001).

II. The electrophoretic mobility shift assay (EMSA) as a tool for identification of inhibitors of protein–DNA interactions

The most frequently used experimental approach to study effects of DNA-binding drugs on protein–DNA interactions is the electrophoretic mobility shift assay (EMSA). This assay is performed using double-stranded ^{32}P-labeled oligonucleotides as target DNA. Binding reactions are set up as described elsewhere (Feriotto et al., 1994) in binding buffer (10% glycerol, 0.05% NP-40, 10 mM Tris-HCl pH 7.5, 50 mM NaCl, 0.5 mM DTT, 10 mM $MgCl_2$), in the presence of poly(dI:dC).poly (dI:dC) (Pharmacia, Uppsala, Sweden), 50 ng of purified TFs (or 2–5 μg of crude nuclear extracts), and 0.25 ng of labeled oligonucleotides, in a total volume of 25 μl. After 30 min binding of protein factors to synthetic oligonucleotides at room temperature, samples are electrophoresed at constant voltage (200 V for 1.5 h) through a low ionic strength (0.25 × TBE buffer) (1 × TBE = 0.089 M Tris-borate, 0.002 M EDTA) on 6% polyacrylamide gels until tracking dye (bromophenol blue) reached the end of a 16 cm slab. Gels are dried and exposed for autoradiography with intensifying screens at -80 C. In these experiments, DNA/protein complexes migrate through the gel with slower efficiency. Figure 1 (middle panel) shows the EMSA approach, together with the expected effects on protein/DNA interactions of inhibitors, such as DNA-binding drugs. In order to test inhibitors of protein/DNA interactions, addition of the reagents is as follows: (a) poly(dI:dC).poly(dI:dC); (b) labeled target DNA; (c) inhibitors of protein/DNA interactions (as DNA-binding drugs); (d) binding buffer; (e) nuclear factors.

In Figure 1 (right side of the panel) an example is shown displaying the inhibitory effects of chromomycin on Sp1/DNA interactions. In other related papers (Ciucci et al., 1994, 1996; Feriotto et al., 1995; Passadore et al., 1995), the effects of distamycin and distamycin analogs were determined. Taken together, these reports support the hypothesis that EMSA is a fast analysis that allows a rapid screening of DNA-binding drugs able to interfere with DNA–protein interactions.

III. Filter-binding assay for the screening of inhibitors of TF/DNA interactions

The filter binding assay (Nozaki et al., 1994; Mischiati et al., 2001) is performed using double-stranded ^{32}P-labeled target oligonucleotides. In this assay, nuclear factors are isolated and separated by electrophoresis on a 10% SDS-polyacrylamide gel. After blotting to nitrocellulose filters, each lane is cut and the strips are incubated in the presence of ^{32}P-labeled target oligonucleotides. Binding reactions are set up as described elsewhere (Baraldi et al., 2000; Mischiati et al., 2001) in the presence of poly(dI:dC).poly(dI:dC) (Pharmacia, Uppsala, Sweden), 0 25 ng of labeled oligonucleotide, in a total volume of 50 μl. After 30 min binding at room temperature, the strips are washed, dried and autoradiography is performed using a Kodak XOmat AR film.

IV. Plant extracts and inhibition of protein–DNA interactions

In an effort to determine whether inhibitory activities are present in extracts from medicinal plants, different plant extracts were tested for their ability to inhibit the interactions between nuclear factors isolated from the human leukemic K562 cells and GATA-1, AP-1, Sp1, NF-kB, NF-IL2A target double-stranded oligonucleotides. Nuclear extracts from several medicinal plants were considered, including *Terminalia arjuna*, *Vernonia anthelmintica*, *Oroxylum indicum*, *Saraca asoka*, *Rumex maritimus*, *Cuscuta reflexa*, *Argemone mexicana*, *Emblica officinalis*, and *Aegle marmelos*.

Representative results obtained are shown in Figure 2. The EMSA was performed by using the double-stranded synthetic oligonucleotides mimicking the binding sites of the GATA-1 TF. The synthetic oligonucleotide was 5′-end-labeled using [γ- ^{32}P] ATP and T4 polynucleotide kinase (MBI Fermentas, Italy). Nuclear extracts were prepared from human leukemic K562 cells. The 5′-labeled oligonucleotide was incubated with plant extracts for 15 min and then nuclear extracts were added. After 30 min binding at room temperature, the samples were electrophoresed at constant voltage (200 V) under low ionic strength conditions (0.25 × TBE buffer = 22 mM Tris-borate, 0.4 mM EDTA) on 6% polyacrylamide gels. Gels were dried and subjected to standard autoradiographic procedures. The results obtained show that low concentrations (6–12 μg/binding reaction) of *Aegle marmelos* extracts, and very low concentrations (0.5 μg/binding reaction) of *Emblica officinalis* extracts inhibit GATA-1/DNA interactions. On the contrary, 400 μg/binding reaction of extracts from *Argemone mexicana* were still unable to inhibit GATA-1/DNA interactions.

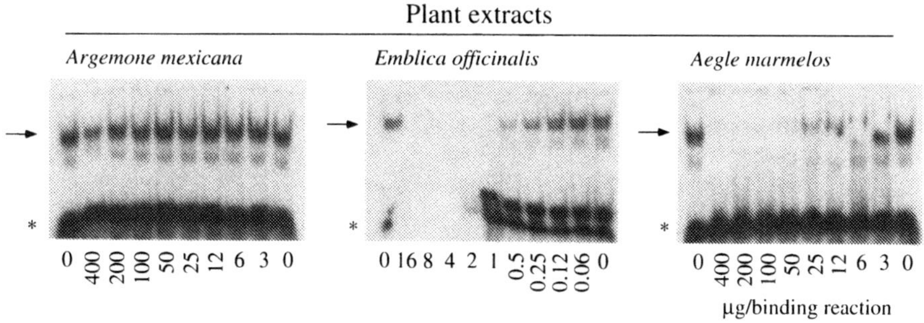

Fig. 2. Effects of increasing amounts of extracts from *Argemone mexicana* (left side of the panel), *Emblica officinalis* (middle of the panel) and *Aegle marmelos* (right side of the panel) on GATA-1/DNA interactions. * = free probe; arrows indicate GATA-1/DNA complexes.

V. Future perspectives

The results obtained demonstrate that plant extracts could retain the very interesting possibility of inhibiting molecular interactions between TFs and DNA-binding drugs. Of course, the identification of the bioactive compounds from plant extracts exhibiting inhibitory activities on DNA/protein interactions is of great interest (Marquez et al., 2004; Lampronti et al., 2005). Methods involving the differential extraction of bioactive compounds could be of interest. Furthermore, the research in this field will take advantages from several procedures, among which are preparative and analytic high-performance liquid chromatography (HPLC) and gas chromatography/mass spectrometry (GC/MS).

Acknowledgments

RG is supported by AIRC, Italian Cystic Fibrosis Foundation, AVLT and Fondazione CARIPARO. MTHK is supported by the "Third World Academy of Sciences (TWAS)" for his visit to Pakistan.

References

Agarwal N, Gewirtz AM. (1999) Oligonucleotide therapeutics for hematologic disorders. Biochim Biophys Acta 1489:85–96.

Aggarwal BB, Schwarz L, Hogan ME, Rando RF. (1996) Triple helix-forming oligodeoxyribonucleotides targeted to the human tumor necrosis factor (TNF) gene inhibit TNF production and block the TNF-dependent growth of human glioblastoma tumor cells. Cancer Res 56:5156–64.

Asmawi MZ, Kankaanranta H, Moilanen E, Vapaatalo H. (1993) Anti-inflammatory activities of *Emblica officinalis* Gaertn leaf extracts. J Pharm Pharmacol 45:581–4.

Baraldi PG, Cacciari B, Guiotto A, Romagnoli R, Spalluto G, Leoni A, Bianchi N, Feriotto G, Rutigliano C, Mischiati C, Gambari R. (2000) [2,1-c][1,4]benzodiazepine (PBD)-distamycin hybrid inhibits DNA binding to transcription factor Sp1. Nucleosides Nucleotides Nucleic Acids 19:1219–29.

Bharani A, Ganguly A, Mathur LK, Jamra Y, Raman PG. (2002) Efficacy of *Terminalia arjuna* in chronic stable angina: a double-blind, placebo-controlled, crossover study comparing *Terminalia arjuna* with isosorbide mononitrate. Indian Heart J 54:170–5.

Bharani A, Ganguly A, Bhargava KD. (1995) Salutary effect of *Terminalia arjuna* in patients with severe refractory heart failure. Int J Cardiol 49:191–9.

Bianchi N, Chiarabelli C, Borgatti M, Mischiati C, Fibach E, Gambari R. (2001) Accumulation of gamma-globin mRNA and induction of erythroid differentiation after treatment of human leukaemic K562 cells with tallimustine. Br J Haematol 113:951–61.

Bianchi N, Ongaro F, Chiarabelli C, Gualandi L, Mischiati C, Bergamini P, Gambari R. (2000) Induction of erythroid differentiation of human K562 cells by cisplatin analogs. Biochem Pharmacol 60:31–40.

Bianchi N, Osti F, Rutigliano C, Corradini FG, Borsetti E, Tomassetti M, Mischiati C, Feriotto G, Gambari R. (1999) The DNA-binding drugs mithramycin and chromomycin are powerful inducers of erythroid differentiation of human K562 cells. Br J Haematol 104:258–65.

Bianchi N, Passadore M, Rutigliano C, Feriotto G, Mischiati C, Gambari R. (1996) Targeting of the Sp1 binding sites of HIV-1 long terminal repeat with chromomycin. Biochem Pharmacol 52:1489–98.

Bloemink MJ, Reedijk J. (1996) Cisplatin and derived anticancer drugs: mechanism and current status of DNA binding. Met Ions Biol Syst 32:641–85.
Borchers AT, Keen CL, Stern JS, Gershwin ME. (2000) Inflammation and Native American medicine: the role of botanicals. Am J Clin Nutr 72:339–47.
Borgatti M, Lampronti I, Romanelli A, Pedone C, Saviano M, Bianchi N, Mischiati C, Gambari R. (2003) Transcription factor decoy molecules based on a peptide nucleic acid (PNA)-DNA chimera mimicking Sp1 binding sites. J Biol Chem 278:7500–9.
Chan PP, Glazer PM. (1997) Triplex DNA: fundamentals, advances, and potential applications for gene therapy. J Mol Med 75:267–82.
Chang FR, Hsieh TJ, Huang TL, Chen CY, Kuo RY, Chang YC, Chiu HF, Wu YC. (2002) Cytotoxic constituents of the stem bark of *Neolitsea acuminatissima*. J Nat Prod 65:255–8.
Cho-Chung YS, Park YG, Lee YN. (1999) Oligonucleotides as transcription factor decoys. Curr Opin Mol Ther 1:386–92.
Ciucci A, Feriotto G, Mischiati C, Gambari R, Animati F, Lombardi P, Natali PG, Arcamone F, Giacomini P. (1994) Distamycin analogues with improved sequence-specific DNA binding activities. Biochem Pharmacol 48:1583–91.
Ciucci A, Manzini S, Lombardi P, Arcamone F. (1996) Backbone and benzoyl mustard carrying moiety modifies DNA interactions of distamycin analogues. Nucleic Acids Res 24:311–5.
Couladis M, Badisa RB, Baziou P, Chaudhuri SK, Pilarinou E, Verykokidou E, Harvala C. (2002) Antioxidant and cytotoxic activities of *Hypericum* sp. on brine shrimps and human cancer cell lines. Phytother Res 16:719–22.
Cragg GM, Newman DJ. (1999) Discovery and development of antineoplastic agents from natural sources. Cancer Invest 17:153–63.
De S, Ravishankar B, Bhavsar GC. (1994) Investigation of the anti-inflammatory effects of *Paederia foetida*. J Ethnopharmacol 43:31–8.
Dean DA. (2000) Peptide nucleic acids: versatile tools for gene therapy strategies. Adv Drug Deliv Rev 44:81–95.
Denny WA. (1989) DNA-intercalating ligands as anti-cancer drugs: prospects for future design. Anticancer Drug Des 4:241–63.
Elisabetsky E, Shanley P. (1994) Ethnopharmacology in the Brazilian Amazon. Pharmacol Ther 64:201–14.
Elisabetsky E. (1991) Sociopolitical, economical and ethical issues in medicinal plant research. J Ethnopharmacol 32:235–9.
El-Tanani M, Fernig DG, Barraclough R, Green C, Rudland P. (2001) Differential modulation of transcriptional activity of estrogen receptors by direct protein–protein interactions with the T cell factor family of transcription factors. J Biol Chem 276:41675–82.
Feriotto G, Ciucci A, Mischiati C, Animati F, Lombardi P, Giacomini P, Arcamone F, Gambari R. (1994) Binding of Epstein–Barr virus nuclear antigen 1 to DNA: inhibition by distamycin and two novel distamycin analogues. Eur J Pharmacol 267:143–9.
Feriotto G, Mischiati C, Bianchi N, Passadore M, Gambari R. (1995) Binding of distamycin and chromomycin to human immunodeficiency type 1 virus DNA: a non-radioactive automated footprinting study. Eur J Pharmacol 290:85–93.
Feriotto G, Mischiati C, Gambari R. (1994) Sequence-specific recognition of the HIV-1 long terminal repeat by distamycin: a DNAase I footprinting study. Biochem J 299:451–8.
Gambari R. (2000) A guided experimental approach to practical molecular pharmacology teaching: demonstration of sequence selectivity of DNA-binding drugs by arrested polymerase chain reaction. Biochem Educ 28:100–3.
Gorman L, Glazer PM. (2001) Directed gene modification via triple helix formation. Curr Mol Med 1:391–9.
Hedberg I. (1993) Botanical methods in ethnopharmacology and the need for conservation of medicinal plants. J Ethnopharmacol 38:121–8.
Heinrich M, Gibbons S. (2001) Ethnopharmacology in drug discovery: an analysis of its role and potential contribution. J Pharm Pharmacol 53:425–32.

Izumiya Y, Lin SF, Ellison T, Chen LY, Izumiya C, Luciw P, Kung HJ. (2003) Kaposi's sarcoma-associated herpesvirus K-bZIP is a coregulator of K-Rta: physical association and promoter-dependent transcriptional repression. J Virol 77:1441–1451.

Jeena KJ, Joy KL, Kuttan R. (1999) Effect of *Emblica officinalis*, *Phyllanthus amarus* and *Picrorrhiza kurroa* on N-nitrosodiethylamine-induced hepatocarcinogenesis. Cancer Lett 136:11–6.

Jo N, Ogata N, Aoki M, Otsuji T, Morishita R, Kaneda Y, Matsumura M. (2002) Effective transfection of a cis element "decoy" of the nuclear factor-kappaB binding site into the experimental choroidal neovascularization. Curr Eye Res 24:465–73.

Jose JK, Kuttan G, Kuttan R. (2001) Antitumour activity of *Emblica officinalis*. J Ethnopharmacol 75:65–9.

Katsube N, Iwashita K, Tsushida T, Yamaki K, Kobori M. (2003) Induction of apoptosis in cancer cells by Bilberry (*Vaccinium myrtillus*) and the anthocyanins. J Agric Food Chem 51:68–75.

Khan MTH, Lampronti I, Martello D, Bianchi N, Jabbar S, Choudhuri MSK, Datta BK, Gambari R. (2002) Identification of pyrogallol as an antiproliferative compound present in extracts from the medicinal plant *Emblica officinalis*: effects on *in vitro* cell growth of human tumor cell lines. Int J Oncol 21:187–92.

Lampronti I, Khan MTH, Bianchi N, Ather A, Borgatti M, Vizziello L, Fabbri E, Gambari R. (2005) Bangladeshi medicinal plant extracts inhibiting molecular interactions between nuclear factors and target DNA sequences mimicking NF-kB binding sites. Med Chem 1:327–33.

Mann MJ, Dzau VJ. (2000) Therapeutic applications of transcription factor decoy oligonucleotides. J Clin Invest 106:1071–5.

Marquez N, Sancho R, Ballero M, Bremner P, Appendino G, Fiebich BL, Heinrich M, Munoz E. (2004) Imperatorin inhibits T-cell proliferation by targeting the transcription factor NFAT. Planta Med 70:1016–21.

McMahon BM, Stewart J, Fauq A, Younkin S, Younkin L, Richelson E. (2002) Using peptide nucleic acids as gene-expression modifiers to reduce beta-amyloid levels. J Mol Neurosci 19:71–6.

Mischiati C, Jeang KT, Feriotto G, Breda L, Borgatti M, Bianchi N, Gambari R. (2001) Aromatic polyamidines inhibiting the Tat-induced HIV-1 transcription recognize structured TAR-RNA. Antisense Nucleic Acid Drug Dev 11:209–17.

Mischiati C, Borgatti M, Feriotto G, Rutigliano C, Breda L, Bianchi N, Gambari R. (2002) Inhibition of HIV-1 LTR-driven *in vitro* transcription by molecular hybrids based on peptide nucleic acids mimicking the NF-kappaB binding site. Int J Mol Med 9:633–9.

Mukherjee AK, Basu S, Sarkar N, Ghosh AC. (2001) Advances in cancer therapy with plant-based natural products. Curr Med Chem 8:1467–86.

Nozaki T, Masutani M, Akagawa T, Sugimura T, Esumi H. (1994) Non-covalent interaction between poly(ADP-ribose) and cellular proteins: an application of a poly(ADP-ribose)-western blotting method to detect poly(ADP-ribose) binding on protein-blotted filter. Biochem Biophys Res Commun 198:45–51.

Passadore M, Bianchi N, Feriotto G, Mischiati C, Giacomini P, Piva R, Gambari R. (1995) Differential effects of distamycin analogues on amplification of human gene sequences by polymerase-chain reaction. Biochem J 308:513–9.

Popov AM, Atopkina LN, Uvarova NI, Elyakov GB. (2001) The antimetastatic and immunomodulating activities of ginseng minor glycosides. Dokl Biochem Biophys 380:309–12.

Prabakan M, Anandan R, Devaki T. (2000) Protective effect of *Hemidesmus indicus* against rifampicin and isoniazid-induced hepatotoxicity in rats. Fitoterapia 71:55–9.

Praseuth D, Guieysse AL, Helene C. (1999) Triple helix formation and the antigene strategy for sequence-specific control of gene expression. Biochim Biophys Acta 1489:181–206.

Ram A, Lauria P, Gupta R, Kumar P, Sharma VN. (1997) Hypocholesterolaemic effects of *Terminalia arjuna* tree bark. J Ethnopharmacol 55:165–9.

Richardson MA. (2001) Biopharmacologic and herbal therapies for cancer: research update from NCCAM. J Nutr 131:3037S–40S.

Ruffa MJ, Ferraro G, Wagner ML, Calcagno ML, Campos RH, Cavallaro L. (2002) Cytotoxic effect of Argentine medicinal plant extracts on human hepatocellular carcinoma cell line. J Ethnopharmacol 79:335–9.

Rutigliano C, Bianchi N, Tomassetti M, Pippo L, Mischiati C, Feriotto G, Gambari R. (1998) Surface plasmon resonance for real-time monitoring of molecular interactions between a triple helix-forming oligonucleotide and the Sp1 binding sites of human *Ha-ras* promoter: effects of the DNA-binding drug chromomycin. Int J Oncol 12:337–43.

Sharma N, Trikha P, Athar M, Raisuddin S. (2000) In vitro inhibition of carcinogen-induced mutagenicity by *Cassia occidentalis* and *Emblica officinalis*. Drug Chem Toxicol 23:477–84.

Singh A, Handa SS. (1995) Hepatoprotective activity of *Apium graveolens* and *Hygrophila auriculata* against paracetamol and thioacetamide intoxication in rats. J Ethnopharmacol 49:119–26.

Steenkamp V, Stewart MJ, van der Merwe S, Zuckerman M, Crowther NJ. (2001) The effect of *Senecio latifolius* a plant used as a South African traditional medicine on a human hepatoma cell line. J Ethnopharmacol 78:51–8.

Stevigny C, Block S, De Pauw-Gillet MC, de Hoffmann E, Llabres G, Adjakidje V, Quetin-Leclercq J. (2002) Cytotoxic aporphine alkaloids from *Cassytha filiformis*. Planta Med 68:1042–4.

Tatman D, Mo H. (2002) Volatile isoprenoid constituents of fruits, vegetables and herbs cumulatively suppress the proliferation of murine B16 melanoma and human HL-60 leukemia cells. Cancer Lett 175:129–39.

Tomita N, Morishita R, Tomita T, Ogihara T. (2002) Potential therapeutic applications of decoy oligonucleotides. Curr Opin Mol Ther 4:166–70.

Trojan LA, Kopinski P, Wei MX, Ly A, Glogowska A, Czarny J, Shevelev A, Przewlocki R, Henin D, Trojan J. (2002) IGF-I: from diagnostic to triple-helix gene therapy of solid tumors. Acta Biochim Pol 49:979–90.

Wargovich MJ, Woods C, Hollis DM, Zander ME. (2001) Herbals, cancer prevention and health. J Nutr 131:3034–6.

Welch JJ, Rauscher 3rd FJ, Beerman TA. (1994) Targeting DNA-binding drugs to sequence-specific transcription factor–DNA complexes. Differential effects of intercalating and minor groove binding drugs. J Biol Chem 269:31051–8.

Wickberg B. (1993) Chemical methods in ethnopharmacology. J Ethnopharmacol 38:159–65.

Yamasaki K, Asai T, Shimizu M, Aoki M, Hashiya N, Sakonjo H, Makino H, Kaneda Y, Ogihara T, Morishita R. (2003) Inhibition of NFkappaB activation using cis-element 'decoy' of NFkappaB binding site reduces neointimal formation in porcine balloon-injured coronary artery model. Gene Ther 10:356–64.

Yu TX, Ma RD, Yu LJ. (2001) Structure–activity relationship of tubeimosides in anti-inflammatory, antitumor, and antitumor-promoting effects. Acta Pharmacol Sin 22:463–8.

M.T.H. Khan and A. Ather (eds.)
Lead Molecules from Natural Products

Plants with antitumor properties: from biologically active molecules to drugs

Ilaria Lampronti, Mahmud Tareq Hassan Khan, Nicoletta Bianchi, Elisabetta Lambertini, Roberta Piva, Monica Borgatti, Roberto Gambari

Abstract

Medicinal plants are of great interest as starting material for identification of new biologically active compounds. A large number of low-molecular-weight compounds isolated from plants or microorganisms have already been identified as effective in diseases as diverse as HIV infection, herpes simplex, neuroblastoma, and breast cancer. Some drugs that emerged from this process are already in the late-phase clinical trials. The first step for the identification of bioactive compounds in the biomedical field is the screening of extracts from different tissues of several medicinal plants for a given activity (for instance, *in vitro* antiproliferative activity). To this aim, the method of extraction of bioactive compounds is crucial. The second step is the fractionation and the characterization of the plant extracts exhibiting the desired biological activity. This step takes advantage from several analytical and preparative procedures, among which preparative and analytic high-performance liquid chromatography (HPLC), several HPLC-based methods, such as HPLC/MS, gas chromatography/mass spectrometry (GC/MS), surface plasmon resonance (SPR)-based biospecific interaction analysis (BIA) employing biosensors. Applications of these methodologies to the screening, identification, purification, and characterization of bioactive compounds from medicinal plants are described in this review.

Keywords: *medicinal plants, Emblica officinalis, Aegle marmelos, pyrogallol, antitumor agents, ethnopharmacology*

Abbreviations: HPLC, high-performance liquid chromatography; RP-HPLC, reverse-phase HPLC; APCI, atmospheric pressure chemical ionization; NMR, nuclear magnetic resonance; LC, liquid chromatography; DAD, diode array detection; UVD, ultraviolet detection; FTIR, Fourier-transformed infrared; ESI, electrospray ionization; GC/MS, gas chromatography/mass spectrometry.

I. Introduction

According to WHO estimates, more than 80% of people in developing countries depend on traditional medicines for their primary health needs (Datta et al., 1998; Ahmad et al., 1998; Ankli et al., 2002; Neto et al., 2002). Plants are of great interest as starting material for identification of new biologically active compounds. In fact, medicinal plants have been described as exhibiting a variety of therapeutic properties (Ahmad et al., 1998; Datta et al., 1998; Ankli et al., 2002; Neto et al., 2002). For instance, medicinal plants are used in prenatal care (Velazco, 1980; Abo et al., 2000), in obstetrics (Pinn, 2001), in gynecology (Abo et al., 2000), in respiratory disorders (Ankli et al., 2002; Neto et al., 2002), in skin disorders (Graf, 2000), in cardiac diseases (Ankli et al., 2002), in nervous and muscular disorders (Datta et al., 1998), and in mental health (Ahmad et al., 1998).

Examples of specific applications are a variety of herbal drugs that lower cholesterol (useful in chronic angina and congenital heart diseases) (Mishra et al., 1981; Mathur et al., 1996; Ram et al., 1997) and herbal treatments for a wide range of skin conditions (Graf, 1981), including diseases like Psoriasis and Erysipelas (Ahmad et al., 1998; Datta et al., 1998; Ankli et al., 2002; Neto et al., 2002). Therefore, medicinal plants should be considered as of great interest, since they could provide health security to rural people in primary health care.

A large number of low-molecular-weight compounds from plants or microorganisms have been already isolated that could be effective in diseases as diverse as HIV infection, herpes simplex, neuroblastoma, and breast cancer. It should be pointed out that some drugs that emerged from this process are already in the late-phase clinical trials (Ankli et al., 2002; Neto et al., 2002).

In addition, anti-inflammatory drugs have been isolated from a variety of extracts from medicinal plants, including *Salvia miltiorrhiza* (Kang et al., 2000), *Scutellaria baicalensis* Georgi (Li et al., 2000), *Curcuma longa* (Ramsewak et al., 2000), *Elaeagnus angustifolia* fruit (Ahmadiani et al., 2000), *Sida cordifolia* L. (Malva-branca) (Franzotti et al., 2000), *Tragia involucrata* Linn. (Dhara et al., 2000), *Ficus racemosa* Linn. (Mandal et al., 2000). The biological effects produced by most of the indicated extracts were comparable to those of phenylbutazone, a prototype of a nonsteroidal anti-inflammatory agent.

As already mentioned, an interesting property of plant extracts is their effects on tumor cells. In this respect, antitumor activity of herbal medicines has been described in several reports, both *in vitro* and *in vivo* (Mukherjee et al., 2001; Popov et al., 2001; Richardson, 2001; Steenkamp et al., 2001; Wargovich et al., 2001; Yu et al., 2001; Chang et al., 2002; Ruffa et al., 2002; Tatman and Mo, 2002). For instance, in a recent study antitumor activity of *Emblica officinalis* has been reported (Jose et al., 2001; Khan et al., 2002). Aqueous extracts of *E. officinalis* were found to display cytotoxic effects on L929 cells in culture in a dose-dependent manner. Interestingly, *E. officinalis* extracts were found to reduce ascites and solid tumors in tumor-bearing mice (Jeena et al., 1999; Sharma et al., 2000; Jose et al., 2001).

A general outline of approaches leading to the identification of bioactive compounds from plant extracts exhibiting biological activities of interest in the biomedical field is reported in Figure 1. The first step is the screening of extracts from different tissues of several medicinal plants for a given activity (for instance, *in vitro* antiproliferative activity). In this step, the method of extraction of bioactive compounds is crucial. A second step is the fractionation and the characterization of the plant extracts exhibiting the desired biological activity. This step takes advantage from several analytical and preparative procedures, among which preparative and analytic high-performance liquid chromatography (HPLC) and gas chromatography/mass spectrometry (GC/MS).

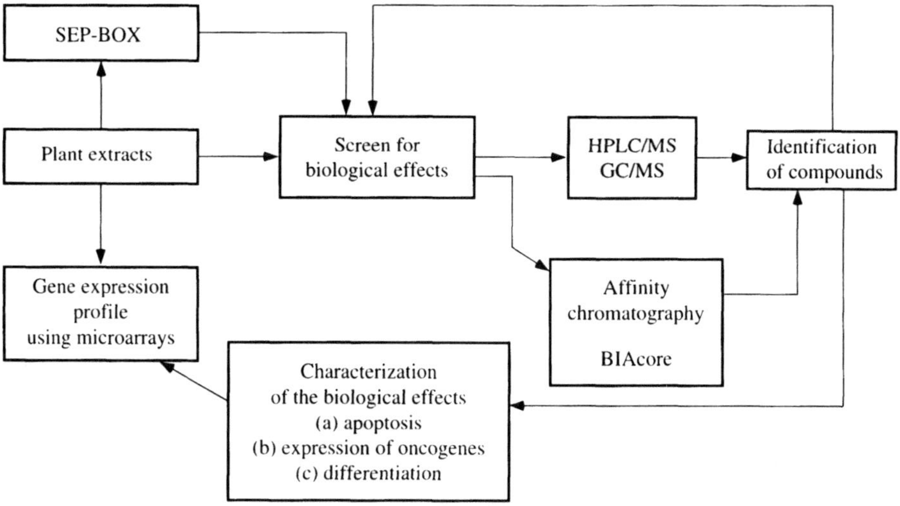

Fig. 1. Experimental strategies to identify bioactive compounds from plant extracts exhibiting biological activities of interest in the biomedical field.

II. High-performance liquid chromatography (HPLC)

This technique has been employed by several groups for isolation of bioactive compounds from medicinal plant extracts. A few examples are reported here. For instance, Ye et al. (2002) studied flavonoids contents in 40 samples of *Semen cuscutae* collected from areas all around China, by using reverse-phase HPLC (RP-HPLC). Five principal flavonoids, quercetin 3-*O*-*β*-D-galactoside-7-*O*-*β*-D-glucoside, quercetin 3-*O*-*β*-D-apiofuranosyl-(1→2)-*β*-D-galactoside, hyperoside, quercetin, and kaempferol, were analyzed simultaneously by using a RP-HPLC system with 0.025 M phosphoric acid–methanol as mobile phase. The recovery of the method was 97.0–102.9%, and all the flavonoids showed good linearity in a relatively wide concentration range. A rapid sensitive and reproductive RP-HPLC method with photo diode array detection was described by Singh et al. (2002) for the simultaneous quantification of major oleane derivatives: arjunic acid, arjunolic acid, arjungenin (Ankli et al., 2002), and arjunetin (Neto et al., 2002) in *Terminalia arjuna* extract. The method involves the use of a Waters Spherisorb S10 ODS2 column (250 mm × 4.6 mm I.D., 10 μm) and binary gradient mobile phase profile. Extraction efficiency, peak purity, and similarity were validated using a photo diode array detector.

A further example is the HPLC determination, employing a silica gel C-18 reverse-phase column, of aristolochic acid in medicinal plants and slimming products, published by Lee et al. (2001). The mobile system, 0.3% ammonium carbonate solution–acetonitrile (75:25, v/v) with pH 7.5, was the optimal buffer to clearly separate aristolochic acids I and II within 20 min. The recovery of aristolochic acids I and II in medicinal plants and slimming products was higher than 90%. The major component was aristolochic acid I in *Aristolochia fangchi* and aristolochic acid II was the major component for *Aristolochia contorta*. Twelve out of 16 samples of slimming pills and powders contained aristolochic acids I and/or II. The finding that the major component in most slimming products was aristolochic acid II allowed Lee and co-workers to conclude that slimming products were not mainly made of *A. fangchi.*

Two HPLC-based technological improvements have been recently proposed, one coupling high-efficiency HPLC with mass spectrometry (MS), the other using high-performance systems (such as the SEP-Box) useful for handling high amounts of crude materials and separating it in hundreds/thousands of fractions.

III. Recently developed HPLC-based methods

In a first example, HPLC and MS are coupled to purify and identify constituents of biological or pharmaceutical interest from crude plant extracts (Chan et al., 2000). Liquid chromatography combined with mass spectrometry (LC/MS) is also utilized in different fields of research because of being a rapid, sensitive, and selective method of analysis. For instance, LC/MS was utilized by Jensen et al. (2002) for the determination of the active terpene (ginkolides and bilobalide) present in *Ginko biloba*, using atmospheric pressure chemical ionization (APCI) in the negative ion mode, in order to evaluate the composition of active constituents in phytopharmaceuticals preparation or in extracts of medicinal plants. LC–APCI/MS detection allowed a considerable reduction in time analysis when compared to LC–UV. All compounds were selectively detected by single-ion monitoring of their specific deprotonated molecules $[M-H]^-$. With this method, the ginco terpene trilactones were detected on-line in the picogram range (Jensen et al., 2002).

Fabre et al. (2000) proposed the ion pair HPLC–ESI/MS/MS (HPLC coupled with electrospray ionization mass spectrometry/mass spectrometry) as a direct and quick method of characterization of 14 isoquinoline alkaloids in an extract of the aerial part of *Escholtzia californica*, a Papaveracea interesting for its increasing use in medicinal chemistry. In this study, the resulted tandem mass spectrometry (MS/MS) is used as a very efficient technique to identify these alkaloid derivatives with a specific and detailed fragmentation pattern (Fabre et al., 2000).

Silva et al. (2000) coupled HPLC to ultraviolet spectroscopy and electrospray ionization mass spectrometry (LC/UV/ESI/MS) to isolate and identify ellagitannins from *Terminalia macroptera* roots. *Terminalia macroptera* is a medicinal plant used in Guinea-Bissau and other West African countries to treat infectious diseases like gonorrhea. By using LC/UV/ESI/MS, four major compounds (ellagic acid, gallic acid, punicalagin, and terchebulin) were found in ethanol extracts, and three derivatives (3,3′-di-*O*-methylellagic acid, 3,4,3′,4′-tetra-*O*-methylellagic acid, and terflavin A) were separated and identified (Silva et al., 2000).

Starting in 1999, the *Ginseng* was also studied with LC/MS. An HPLC/MS/MS method has been developed for the characterization and quantification of ginsenosides (Rb1, Rb2, Rc, Rd, Re, Rf, Rg1) contained in the root of *Panax ginseng* (oriental ginseng) and *Panax quinquefolium* (American ginseng). Differentiation of ginsenosides was achieved through simultaneous detection of intact ginsenoside molecular ions and their characteristic thermal degradation products. An important parameter used to differentiate *P. ginseng* and *P. quinquefolium* was the presence of ginsenoside Rf and 24-(*R*)-preudoginsenoside F11 in oriental and American ginseng, respectively (Wang et al., 1999).

A third research on ginseng was conducted by Tran et al. (2001). In this study, extracts of *Panax vietnamensis* (Vietnamese ginseng) were analyzed with LC–ESI/MS, since the methanol fraction of this medicinal plant was found to possess hepatoprotective effects on D-galactosamine (D-GalN)/tumor necrosis factor-alpha (TNF-α)-induced cell death in primary cultured mouse hepatocytes. The LC–ESI/MS allowed to identify five known saponins: ginsenosides Rb1, Rb2, Rc, Rd, and Re; eight known dammarane-type triterpene saponins: majonoside R2, pseudoginsenoside RT4, vina ginsenosides R1, R2, and R10, ginsenosides Rg1, Rh1, and Rh4; and the known sapogenin protopanaxatriol oxide II. Further chemical investigations of the methanol extract resulted in two new dammarane-type triterpene saponins, ginsenoside Rh5, and vina ginsenoside R25. All the chemical structures were confirmed on the basis of the spectral analysis (Tran et al., 2001).

A well-known medicinal plant used frequently in Europe is *Hypericum perforatum* L. to treat mild to moderately severe depressive disorders. Recent pharmacological and clinical researches by Orth et al. (1999) demonstrated that hyperforin is the main active component of the drug. The identity and the purity of this isolated substance were determined by the authors by high-performance thin-layer chromatography (HPTLC), HPLC, with diode array and ultraviolet detection (DAD and UV), Fourier-transformed infrared (FT-IR), proton nuclear magnetic resonance (^{1}H-NMR) spectroscopy, and liquid chromatography coupled with positive ion electrospray ionization tandem mass spectrometry (LC–ESI(+)/MS/MS) (Orth et al., 1999). Moreover, the LC/MS/MS was used to determine hyperforin in plasma after oral administration of 300 mg/kg of alcoholic *H. perforatum* extracts in rats and human volunteers (Biber et al., 1998).

Another interesting example of isolation of new bioactive annonaceus acetogenins from *Rollinia mucosa* with LC/MS was reported by Gu et al. (1997). In this research, the structures of rollidecin C and rollidecin D were confirmed by analyses of the ^{1}H and ^{13}C NMR. The first compound exhibited selective cytotoxicity toward the colon tumor cell line (HT-29), while the second isolated derivative showed only borderline cytotoxicity in a panel of six human tumor cell lines.

The SEPBOX for isolation of high amounts of pure compounds from plant extracts

In this case of preparative approach for isolation of pure compounds from plant extracts, a skillful combination of HPLC and solid-phase extraction (SPE) technologically realized as an HPLC/SPE/HPLC/SPE coupling was developed (the SEPBOX-system by AnalityCon) (Bindseil, 1997). By this technique, fractionation of an extract (1–5 g) into 100–300 fairly pure compounds is possible within less than 24 h. In this way, SEPBOX guarantees an efficient provision of natural compounds for high-throughput screening (Mellor and Schulte, 1997). What's new about this technology is that it is possible to apply modern drug discovery methodologies to characterize natural products. For example, the German company AnalytiCon is isolating, purifying, and characterizing thousands of novel natural products from Bayer's proprietary database. In combination with SEPBOX, AnalytiCon uses HITSE (high-throughput structure elucidation) to determine the structures of the isolated compounds.

IV. Gas chromatography/mass spectrometry (GC/MS)

Several equipments are available for performing GC/MS. In recent studies, we employed a Fison (Thermo Finnigan, San Jose, CA) model GC 8000 gas chromatograph interfaced to a Fison model MD 800 quadrupole mass spectrometer. The fused-silica gas chromatographic capillary column was a MEGA SE 54 (methyl phenyl polysiloxanes), 25 m × 0.25 mm I.D., and 0.25 μm film thickness. The head pressure of the carrier gas (helium, 99.99% purity) was 50 kPa (7.2 psi). In these analyses, 1 μl of sample was dissolved into appropriate solvents and injected into the gas chromatograph. The injector and detector temperatures for the GC were 250 °C and 300 °C, respectively. The column oven temperature was increased linearly from 40 °C (held for 4 min) to 200 °C (held for 10 min) at 10 °C/min. The mass spectrometer operated at source and interface temperatures of 250 °C. The ionization mode was electron impact (EI) (70 eV), with an electron multiplier voltage of 50 V above the "tune" voltage. The "solvent delay," the time gap of a given analysis in which the mass spectrometer is turned off, was 4 min. The GC/MS system was operated in "full scan" mode. The software utilized was the MS data-handling system (version 1.12, Fisons, Thermo Finnigan, San Jose, CA) with NIST library to identify all the derivatives found in plant extracts. Figure 2 shows an example of chromatogram and related mass spectra of the petroleum ether fraction of *Aegle marmelos*.

Example 1. Identification of pyrogallol as an antiproliferative compound present in extracts from the medicinal plant *Emblica officinalis*: effects on *in vitro* cell growth of human tumor cell lines.

Terminalia arjuna, *Aphanamixis polystachya*, *Oroxylum indicum*, *Emblica officinalis*, *Cuscuta reflexa*, *A. marmelos*, *Saraca asoka*, *Rumex maritimus*, *Lagerstroemia speciosa*, and *Red Sandalwood* were used in this study. The production of plant extracts is described in detail elsewhere. The dried fruits of *E. officinalis* were extracted with absolute ethanol and the yield was 9.33%. This ethanolic extract of *E. officinalis* was defatted with petroleum ether, and the defatted extract was successively fractionated with different solvent systems on the polarity basis. The solvents were 100% dichloromethane, 25% ethylacetate–dichloromethane, 50% ethylacetate–dichloromethane, 75% ethylacetate–dichloromethane, 100% ethylacetate, butanol, and acetone and the remaining aqueous portions were separated. The *in vitro* antiproliferative activity of the studied plant extracts was assayed as follows. Cell number/ml was determined by using a model ZBI Coulter Counter (Coulter Electronics, Hialeah, FL). Usually, cells were seeded at the initial cell concentration of 30×10^3 cells/ml, and the cell number/ml was determined after 2, 3, 4, 5 days of cell culture. IC_{50} was determined usually after 4 days, when untreated cells are in the log phase of cell growth. In these studies, the kinetics of cell growth of untreated cells should be carefully determined, since all the assays should be made when the cell growth kinetic is in the log phase. Misleading results are on the contrary obtained when comparison is made on cell populations having reached plateau levels of cell growth. In the study published by Khan et al. (2002), the effects of the extracts of *E. officinalis* on *in vitro* proliferation of different human tumor cell lines, such as the leukemic K562, the B-lymphoid Raji, the T-lymphoid Jurkat, and the erythroleukemic HEL human cells lines were analyzed. The data obtained show that

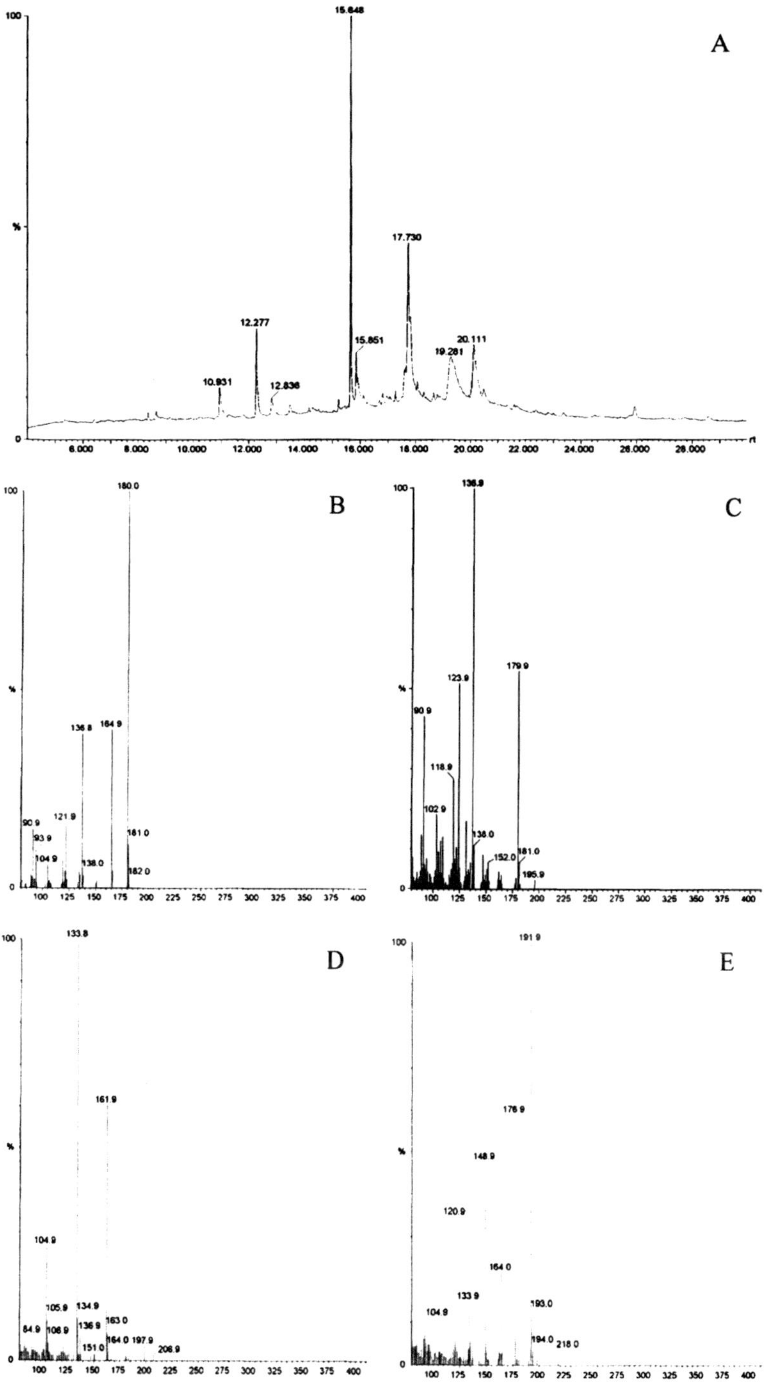

Fig. 2. (A) Chromatogram of the petroleum ether fraction of *Aegle marmelos* extracts, obtained with a Fison model GC 8000 gas chromatograph. (B), (C), (D), and (E) Mass spectra (Fison model MD 800 quadrupole mass spectrometer) related to the four major peaks of chromatogram (A) at the retention times of 15.64 min (peak 1, (B)), 17.73 min (peak 2, (C)), 19.28 min (peak 3, (D)), and 20.11 min (peak 4, (E)).

the IC_{50} of unfractionated extracts from *E. officinalis* was 0.4 ng/ml in the case of K562 cells, 8 ng/ml in the case of Raji cells, 2.5 ng/ml in the case of Jurkat cells, and 10 ng/ml in the case of erythroleukemic HEL cells. When GC/MS was performed on *E. officinalis* extracts to identify putative antiproliferative compounds, we were able to show that in *E. officinalis* extracts three peaks with different retention times (r.t. 13.4, 20.9, 22.3 min) are present. The mass spectrometric analysis demonstrated that these three peaks correspond to three derivatives: (a) pyrogallol (or 1,2,3-benzene-triol); (b) tetradecahydro-1,4α-dimethyl-7-(1-methylethyl)-1-phenanthrene methyl ester; and (c) decahydro-4*H*-cyclopentacycloocten-4-one, respectively (Figure 3), identified by using the NIST library (MS data-handling system, version 1.12, Fisons, Thermo Finnigan, San Jose, CA). As control, we analyzed commercial pyrogallol for comparison with the peaks found in *E. officinalis* extracts. The r.t. and the fragmentation pathways resulted to correspond perfectly. Accordingly, we hypothesized pyrogallol (Figure 3) to be one of those responsible for the antiproliferative activity of *E. officinalis* extracts. The effects of pyrogallol on *in vitro* cell growth of human tumor cell lines were then analyzed; the results firmly demonstrated that pyrogallol was one of the bioactive compounds against tumor cell proliferation. IC_{50} of pyrogallol on K562, Jurkat, HEL, and Raji cell lines was found to be in a range between 10 and 30 μM.

Example 2. *In vitro* antiproliferative effects on human tumor cell lines of extracts from the Bangladeshi medicinal plant *A. marmelos* Correa: identification of bioactive compounds.

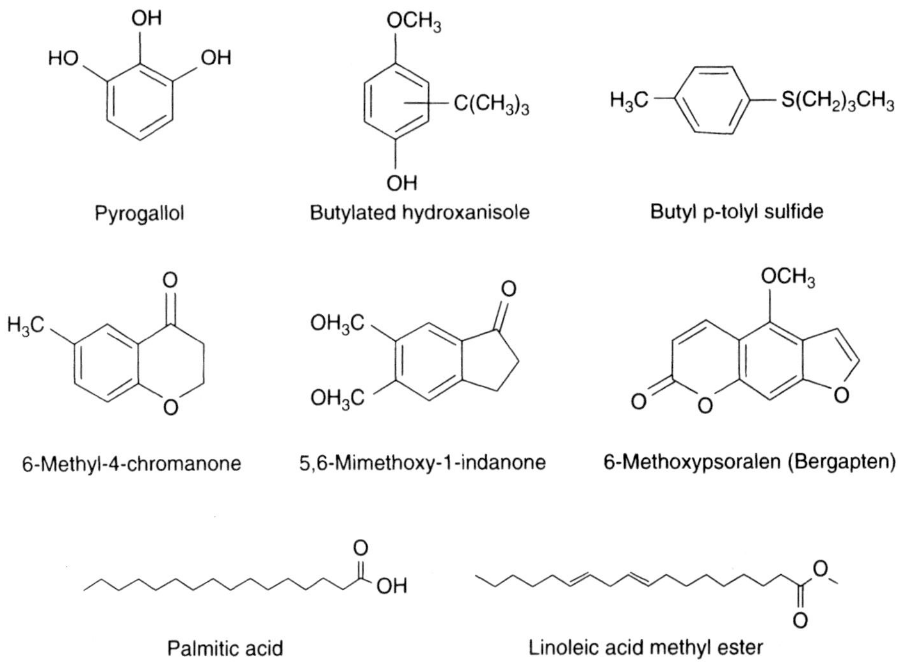

Fig. 3. Chemical structures of the identified derivatives of *Emblica officinalis* and *Aegle marmelos*.

In a second paper from our group (Lampronti et al., 2003), the biological activities of *A. marmelos* fractions, as shown in Figure 4A, were compared to those of *E. officinalis* and with those of other plants, including *Paederia foetida*, *Saraka asoka*, *Hemidemus indicus*, and *Cassia sophera*. The data obtained show that the IC_{50} of the petroleum ether fraction of *A. marmelos* extracts was about 25 μg/ml, a value higher than that obtained when *E. officinalis* extracts were used (0.4 μg/ml) (Figure 4B), but significantly lower than the IC_{50} values obtained with *P. foetida*, *H. indicus*, and *C. sophera*. The extracts from *S. asoka* were unable to inhibit K562 cell growth at the concentrations used. The GC/MS analysis (Figure 2) indicated that one peak (r.t. = 19.28 min), corresponding to 6-methyl-4-chromanone, was present in two fractions (ethyl acetate and carbontetrachloride). The other identified molecules differ between the analyzed fractions, but several identified compounds (butylated hydroxyanisole, butyl-*p*-tolyl sulfide, 6-methyl-4-chromanone, 5,6-dimethoxy-1-indanone, palmitic acid, methyl linoleate, 5-methoxypsoralen) (Figure 3) were commercially available and were tested for their antiproliferative activity. Lampronti and co-workers (2003) analyzed seven commercially pure compounds identified within *A. marmelos* extracts on K562 cell proliferation in order to identify active components (Table 1). Interestingly, all the molecules exhibited antiproliferative activity, even if at different concentrations. The most active compounds were butyl-*p*-tolyl sulfide (7 μM), 6-methyl-4-chromanone (15 μM), and butylated hydroxyanisole (BHA, 35 μM) present in the ethyl acetate fraction. The antiproliferative effects of the most active compounds were comparable to some of the most commonly used antitumor agents such as cisplatin (Bianchi et al., 2000), 5-fluorouracil (Roobol et al., 1984; Ozaki, 1996; Nagy et al., 2002), chromomycin (Baguley, 1982; Ono et al., 1982; Inoue et al., 1983; Bianchi et al., 2001), and cytosine arabinoside (Winter et al., 1985; Grant, 1998; Bianchi et al., 2001). Another interesting point of this paper is that we tested the ability of both *A. marmelos* extracts and identified compounds in inducing differentiation of K562 cells (Figure 5). After treatment with the analyzed compounds, K562 differentiation was analyzed as reported elsewhere by benzidine-staining procedure (Gambari et al., 1984). Table 2 shows that two compounds (butyl-*p*-tolyl sulfide and 6-methyl-4-chromanone) induced erythroid differentiation at concentrations higher than those able to give a 50% inhibition of K562 cell growth; one compound (5-methoxypsoralen) was able to induce K562 differentiation when used at concentrations lower than that causing 50% inhibition of K562 cell growth. The other compounds were unable to induce differentiation. These results suggest that the antiproliferative activity of butylated hydroxyanisole, 5,6-dimethoxy-1-indanone, palmitic acid, and methyl linoleate is not associated to induction of differentiation; on the contrary, the antiproliferative activity of butyl-*p*-tolyl sulfide, 6-methyl-4-chromanone, and 5-methoxypsoralen is associated to activation of the differentiation pattern of K562 cells.

Example 3. Effects of extracts from Bangladeshi medicinal plants on expression of tumor-associated genes.

In this specific field, the activity of plant extracts could be analyzed at the level of the entire transcriptome (Mischiati et al., 2003), or at the level of single tumor-associated genes. In the case of breast cancer, the identification of plant extracts able to alter the expression of human estrogen receptor alpha (ERα) gene appears, for

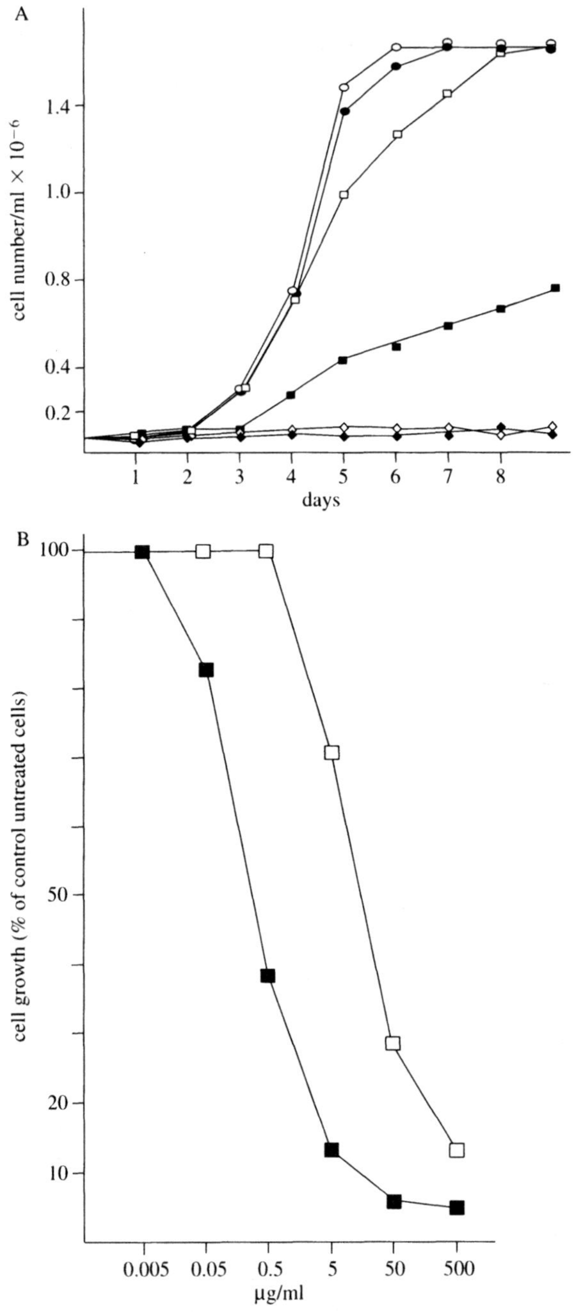

Fig. 4. (A) Extracts from *Aegle marmelos* inhibit *in vitro* proliferation of human leukemia K562 cells. K562 cells were seeded at the initial cell concentration of 30,000 cells/ml and then cultured for the indicated length of time in the absence (open circles) or in the presence of 0.001 μg/ml (closed circles), 0.01 μg/ml (open squares), 0.1 μg/ml (closed squares), 1 μg/ml (open rhombs), and 10 μg/ml (closed rhombs) of the petroleum ether fraction of *Aegle marmelos* extracts. (B) Effects of extracts from *Aegle marmelos* (open squares) and *Emblica officinalis* (closed squares) on *in vitro* cell proliferation of K562 cells. K562 cells were cultured for 4 days in the presence of the indicated amounts of extracts.

Table 1
Effects of purified compounds from *Aegle marmelos* on K562 cell growth (IC_{50})

Compounds	IC_{50} (μM)
Butylated hydroxyanisole	35
Butyl-*p*-tolyl sulfide	7
6-Methyl-4-chromanone	15
5,6-Dimethoxy-1-indanone	70
Palmitic acid	85
Methyl linoleate	250
5-Methoxypsoralen	100
Cisplatin	5
5-Fluorouracil	50
Chromomycin	5
Cytosine arabinoside	0.25

instance, of great interest. The usefulness of estrogen receptor measurements in primary breast tumors for the prediction of early recurrence is well known, since the absence of ERα in breast cancer is associated with early recurrence (Knight et al., 1977). Therefore, up-regulation of the estrogen receptor levels could have important implications in therapy (Nass and Davidson, 1999; Murphy and Watson, 2002). In a recent study, we determined the activity of extracts from Bangladeshi medicinal plants on human breast tumor cell lines, looking at their effects on the expression of the ERα gene. In these experiments, analysing the accumulation of ERα mRNA was performed by quantitative RT-PCR. Total RNA was extracted by using the SV total RNA isolation system (Promega, Madison, WI, USA), and cDNA was synthesized from 1 μg of RNA using the superscript preamplification and random primer system (Gibco BRL, Milan, Italy). Human ERα transcript was determined by RT followed by real-time *TaqMan* PCR analysis. The ERα probe was 5′-labeled with a reporter dye (FAM) and 3′-labeled with a quencher dye (TAMRA). The probe for glyceraldehyde phosphate dehydrogenase (GAPDH) reference was 5′-labeled with a different reporter dye (JOE). Real-time PCR was performed on an ABI-PRISM 7700 sequence detector using the software SDS 1.6 (PE Applied Biosystems, Foster City, CA, USA). *TaqMan* PCR reactions were carried out in a total volume of 25 μl in 1X *TaqMan* Universal PCR Master Mix containing: dATP, dCTP, dGTP (200 μM each); dUTP (400 μM); $MgCl_2$ 5.5 mM; AmpErase UNG 0.01/μl; and AmpliTaq Gold (0.05 U/μl). For ERα gene 200 nM probe and 300 nM of each primer were used, while for GAPDH gene 100 nM probe and 40 nM of each primer were used. The amplifications were performed in duplicate for each sample and the PCR optimal conditions were 50°C for 2 min, 95°C for 10 min followed by 40 cycles of 95°C for 15 s, and 60°C for 1 min. All signals were normalized to GAPDH signal in the same reaction. To normalize the content of cDNA samples, the comparative Ct (threshold cycle) method, consisting in the normalization of the number of target gene copies versus an endogenous reference gene such as GAPDH, was used. For comparative analysis of gene expression, data were obtained by using the ΔCt method derived from a mathematical elaboration

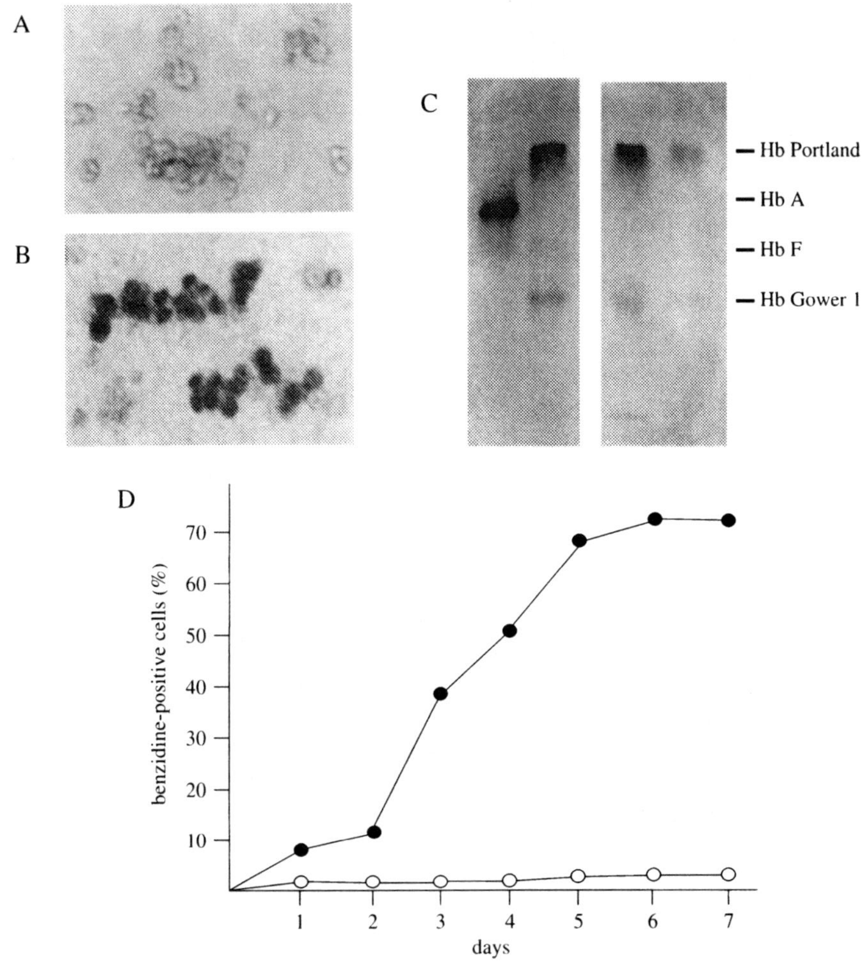

Fig. 5. (A) Untreated K562 cells, most of which are benzidine-negative. (B) Erythroid-induced K562 cells positive to the benzidine-stain (hemoglobin-containing cells). (C) Characterization of the hemoglobins produced by K562 cells after erythroid induction and analyzed by cellulose–acetate gel electrophoresis of postmitochondrial cell lysates; the major hemoglobin produced is Hb Portland ($\zeta_2\gamma_2$). (D) Effects of 5-methoxypsoralen (Bergapten) (closed circles) on K562 differentiation. K562 cells were seeded at the initial cell concentration of 30.000 cells/ml and then cultured for 6 days in the absence (open circles) and in presence of 50 μM 5-methoxypsoralen (closed circles).

previously described (Penolazzi et al., 2000; Bianchi et al., 2001; Holland et al., 1991; Lambertini et al., 2002).

The results obtained showed that in MCF7 and MDA-MB-231 cells treated with *E. officinalis* extracts a sharp increase of accumulation of ERα mRNA was detectable. This effect was found to be similar to that usually obtained using decoy oligonucleotides activating ERα gene expression (Penolazzi et al., 2000). Therefore, the antiproliferative effects of plant extracts on the breast cancer cell lines analyzed are consistent, for some of them, with their ability to induce a more

Table 2
Effects of purified compounds of *Aegle marmelos* on differentiation (% of benzidine-positive cells after 6 days culture at the indicated concentrations)

Compounds	Differentiation % (concentration)
Butylated hydroxyanisole	6 (1 μM)
Butyl-*p*-tolyl sulfide	25 (50 μM)
6-Methyl-4-chromanone	35 (50 μM)
5,6-Dimethoxy-1-indanone	10 (100 μM)
Palmitic acid	4 (100 μM)
Methyl linoleate	4 (200 μM)
5-Methoxypsoralen	60 (50 μM)
Cisplatin	75 (8 μM)
5-Fluorouracil	3 (60 μM)
Chromomycin	80 (0.2 μM)
Cytosine arabinoside	92 (0.5 μM)

differentiated phenotype associated with an increase in ERα gene expression. On the other hand, our data do not allow to identify the mechanism of action for several reasons, including the fact that plant extracts are likely to contain different molecules possibly involved in regulation of the breast phenotype through completely different pathways. In any case, this is, to our knowledge, the first report describing an effect of extracts from medicinal plants on the expression of the human ERα gene. This observation should encourage the identification and study of bioactive compounds from the same extracts (or chemical analogs) on ERα gene expression.

V. Plant extracts and plant-derived compounds in clinical trials

As far as plant extracts are concerned, several reports are available describing their use in clinical trials. For instance, Sastravaha et al. (2005) reported recently adjunctive periodontal treatment with *Centella asiatica* and *Punica granatum* extracts in supportive periodontal therapy. They found significant improvement of pocket depth, attachment level, bleeding index, and gingival index, associated to a reduction of IL-1 beta and IL-6 concentration, in treated versus control patients. For singly isolate molecules, compounds isolated from medicinal plants have been already studied in clinical trials as anti-HIV (Houghton, 1996) and anti-cancer (da Rocha et al., 2001) agents. For instance, vincristine, irinotecan, etoposide, and paclitaxel are examples of plant-derived compounds demonstrated to exhibit antitumor properties (da Rocha et al., 2001). Moreover, it was recently reported and clinically tested that arginine, yohimbine, *P. ginseng*, Maca, and *G. biloba* all have some degree of evidence that they may be helpful for erectile dysfunction (McKay, 2004). Furthermore, the pharmacological activities of Genistein in various types of diseases, such as osteoporosis, cardiovascular diseases, menopausal symptoms, were verified in human clinical trials (Suthar et al., 2001).

VI. Future perspectives: biospecific interaction analysis (BIA) using surface plasmon resonance (SPR) and biosensor technologies for identification, isolation and characterization of bioactive compounds

The recent development of surface plasmon resonance (SPR)-based biosensor technologies for biospecific interaction analysis (BIA) enables to monitor a variety of molecular reactions in real time (Johnsson et al., 1991; Vadgama and Crump, 1992; Malmqvist, 1993; Wood, 1993; Nilsson et al., 1995). In the BIAcore biosensor systems, plane polarized light is totally internally reflected from the gold-coated sensor chip, where the molecular interactions take place (SPR angle). Surface plasmon resonance in the gold layer results in extinction of the reflected light at a specific angle (SPR angle), which varies with the refractive index of the solution close to the other side of the sensor chip. When molecules bind to the chip, the refractive index changes and the change in SPR angle is monitored, generating an increase of response, measured in resonance units (RU). After ligand immobilization, the injection of analyte(s) results in a further increase of RU only when molecular interactions between the ligand and the analyte occur. A further injection of binding/running buffer allows to determine whether this interaction is stable or not. In the case of a stable complex, no significant decrease of RU will be detected; by contrast, in the case of unstable complexes, sharp decreases in RU are obtained. The sensor chip can be regenerated by removing all the bound analyte by short pulses with suitable buffers (e.g., 50 mM NaOH or 0.1% SDS) (Johnsson et al., 1991; Malmqvist, 1993; Nilsson et al., 1995).

SPR-based BIA offers many advantages with respect to most of the other available methodologies to study biomolecular interactions:

(a) most of the commercially available biosensors are fully automated instruments;
(b) no labeling is required, thus allowing the study of an extremely large variety of biomolecules;
(c) a large variety of activated sensor chips are commercially available, allowing to immobilize DNA, RNA, proteins, peptides, cells;
(d) the binding between ligands and analytes could be performed in the presence of low concentrations of organic solvents; therefore, enabling to work with biomolecules exhibiting low solubility in aqueous buffers;
(e) the amount of both ligand and analyte needed to obtain informative results is low (less than 0.5 μg of DNA and 1 μg of peptide/protein are necessary to generate suitable surfaces);
(f) most of the commercially available biosensors offer a wide temperature control range, usually from 4°C to 95°C;
(g) the assay is rapid, usually requiring 20–25 min for ligand immobilization and 4–5 min for a complete characterization of ligand–analyte interactions; it should be noted that the binding is a real-time interaction analysis, leading to informative results at the same time the binding occurs;
(h) the sensor chip could be re-used many times, leading to low running costs, with the only limitation of verifying the stability of the immobilized ligand;
(i) this technology allows to isolate the analyte(s) for further characterization.

While SPR-based BIA is a very useful approach to identify bioactive compounds of great interest in biomedical applications, very few examples are available on this specific field (Leckie et al., 1999; Nam et al., 2002; Xiao & Parkin, 2002). For instance, Leckie et al. (1999) studied two members of the pgip gene family (pgip-1 and pgip-2) of *Phaseolus vulgaris* L. expressed separately in *Nicotiana benthamiana*. In particular, they employed SPR to study the ligand specificity of pgip-1 and pgip-2. Polygalacturonase-inhibiting protein-1 (PGIP-1) was unable to interact with PG from *Fusarium moniliforme* and interacted with PG from *Aspergillus niger*. PGIP-2 interacted with both PGs. Interestingly, only eight amino acid variations distinguish the two proteins: five of them are confined within the beta-sheet/beta-turn structure and two of them are contiguous to this region. By site-directed mutagenesis, each of the variant amino acids of PGIP-2 was replaced with the corresponding amino acid of PGIP-1, in a loss-of-function approach. The mutated PGIP-2 s were expressed individually in *N. benthamiana*, purified and subjected to SPR analysis. Each single mutation caused a decrease in affinity for PG from *F. moniliforme*; residue Q253 made a major contribution, and its replacement with a lysine led to a dramatic reduction in the binding energy of the complex. Conversely, in a gain-of-function approach, amino acid K253 of PGIP-1 was mutated into the corresponding amino acid of PGIP-2, a glutamine. With this single mutation, PGIP-1 acquired the ability to interact with *F. moniliforme* PG.

Few reports are available on the use of SPR-based BIA to study the screening of low-molecular-weight compounds interacting with target molecules (either proteins, DNA, or RNA). In this specific case, SPR-based BIA could be performed in different complementary ways. The target protein could be immobilized onto sensor chips and the compounds (for instance, plant extracts) injected into suitable binding buffers. These direct bindings of the compounds give real-time indications on the affinity for the target proteins, as well as stability of the generated complexes. For instance, Karlsson et al. (2000) demonstrated that the sensitivity of BIAcore technology is sufficient for the detection and the characterization of binding events involving low-molecular-weight compounds and their immobilized protein targets. Eleven compounds with known binding specificity to thrombin and 159 additional compounds were investigated. All compounds with known binding specificity were identified at 1 and 10 μM concentration. In addition to direct binding, competitive assays can be carried on by SPR-based BIA. In this experimental approach, the target is immobilized to the sensor surface and specifically recognized protein injected together with the molecules under investigation.

It should be pointed out that SPR-based BIA is not only an analytical method. In a very important routine, recovery of the bound analyte could be easily obtained. Therefore, coupling of this procedure with other analytical methods could be performed. In this respect SPR-MS technology is of great interest. A development of SPR-based BIA has been recently described by Nedelkov & Nelson (2003) who reported the design and use of multi-affinity surfaces in biomolecular interaction analysis/mass spectrometry (BIA/MS). These results are important for the development of SPR/MS arrays. The feasibility of multi-affinity ligand surfaces in BIA/MS was explored by constructing multi-protein affinity surfaces using antibodies to beta-2-microglobulin, cystatin C, retinol-binding protein, transthyretin, serum amyloid P, and C-reactive protein. After an injection of diluted human plasma aliquots over the antibodies-derivatized surfaces, and subsequent MALDI-TOF MS analysis, signals representing targeted proteins were observed in the mass spectra. The ability to create such multi-affinity surfaces indicates that smaller-size ligand areas/spots can be employed in the BIA/MS protein interaction screening experiments; opens up the possibilities for construction of novel multi-arrayed SPR-MS platforms and methods for high-throughput parallel protein interaction investigations.

Acknowledgements

RG is supported by AIRC, Italian Cystic Fibrosis Foundation, AVLT and Fondazione CARIPARO.

References

Abo KA, Adeyemi AA, Adeite DA. (2000) Ethnobotanical survey of plants used in the treatment of infertility and sexually transmitted diseases in southwest Nigeria. Afr J Med Med Sci 29:325–7.

Ahmad I, Mehmood Z, Mohammad F. (1998) Screening of some Indian medicinal plants for their antimicrobial properties. J Ethnopharmacol 62:183–93.

Ahmadiani A, Hosseiny J, Semnanian S, Javan M, Saeedi F, Kamalinejad M, Saremi S. (2000) Antinociceptive and anti-inflammatory effects of *Elaeagnus angustifolia* fruit extract. J Ethnopharmacol 72:287–92.

Ankli A, Heinrich M, Bork P, Wolfram L, Bauerfeind P, Brun R, Schmid C, Weiss C, Bruggisser R, Gertsch J, Wasescha M, Sticher O. (2002) Yucatec Mayan medicinal plants. Evaluation based on indigenous uses. J Ethnopharmacol 79:43–52.

Baguley BC. (1982) Nonintercalative DNA-binding antitumour compounds. Mol Cell Biochem 43:167–81.

Bianchi N, Chiarabelli C, Borgatti M, Mischiati C, Fibach E, Gambari R. (2001) Accumulation of gamma-globin mRNA and induction of erythroid differentiation after treatment of human leukaemic K562 cells with tallimustine. Br J Haematol 113:951–61.

Bianchi N, Ongaro F, Chiarabelli C, Gualandi L, Mischiati C, Bergamini P, Gambari R. (2000) Induction of erythroid differentiation of human K562 cells by cisplatin analogs. Biochem Pharmacol 60:31–40.

Biber A, Fischer H, Romer A, Chatterjee SS. (1998) Oral bioavailability of hyperforin from hypericum extracts in rats and human volunteers. Pharmacopsychiatry 31:36–43.

Bindseil N. (1997) SEPBOX: automated isolation of natural products. In: International Symposium on Laboratory Automation & Robotics.

Chan TW, But PP, Cheng SW, Kwok IM, Lau FW, Xu HX. (2000) Differentiation and authentication of *Panax ginseng*, *Panax quinquefolius*, and ginseng products by using HPLC/MS. Anal Chem 72:1281–7.

Chang FR, Hsieh TJ, Huang TL, Chen CY, Kuo RY, Chang YC, Chiu HF, Wu YC. (2002) Cytotoxic constituents of the stem bark of *Neolitsea acuminatissima*. J Nat Prod 65:255–8.

da Rocha AB, Lopes RM, Schwartsmann G. (2001) Natural products in anticancer therapy. Curr Opin Pharmacol 1:364–9.

Datta BK, Rahman I, Das TK. (1998) Antifungal activity of Indian plant extracts. Mycoses 41:535–6.

Dhara AK, Suba V, Sen T, Pal S, Chaudhuri AK. (2000) Preliminary studies on the anti-inflammatory and analgesic activity of the methanolic fraction of the root extract of *Tragia involucrata* Linn. J Ethnopharmacol 72:265–8.

Fabre N, Claparols C, Richelme S, Angelin ML, Fouraste I, Moulis C. (2000) Direct characterization of isoquinoline alkaloids in a crude plant extract by ion-pair liquid

chromatography-electrospray ionization tandem mass spectrometry: example of Eschscholtzia californica. J Chromatogr A 904:35–46.
Franzotti EM, Santos CV, Rodrigues HM, Mourao RH, Andrade MR, Antoniolli AR. (2000) Anti-inflammatory, analgesic activity and acute toxicity of *Sida cordifolia* L. (Malva-branca). J Ethnopharmacol 72:273–7.
Gambari R, Del Senno L, Barbieri R, Viola L, Tripodi M, Raschellà G, Fantoni A. (1984) Human leukemia K562 cells: induction of erythroid differentiation by 5-azacytidine. Cell Differentiation 14:87–97.
Graf J. (2000) Herbal anti-inflammatory agents for skin disease. Skin Therapy Lett 5:3–5.
Grant S. (1998) Ara-C: cellular and molecular pharmacology. Adv Cancer Res 72:197–233.
Gu ZM, Zhou D, Lewis NJ, Wu J, Shi G, McLaughlin JL. (1997) Isolation of new bioactive annonaceous acetogenins from *Rollinia mucosa* guided by liquid chromatography/mass spectrometry. Bioorg Med Chem 5:1911–6.
Holland PM, Abramson RD, Watson R, Gelfand DH. (1991) Detection of specific polymerase chain reaction product by utilizing the 5′-3′ exonuclease activity of Thermus aquaticus DNA polymerase. Proc Natl Acad Sci USA 88:7276–81.
Houghton PJ. (1996) Compounds with anti-HIV activity from plants. Trans R Soc Trop Med Hyg 90:601–4.
Inoue K, Fujimoto S, Ogawa M. (1983) Antitumor efficacy of seventeen anticancer drugs in human breast cancer xenograft (MX-1) transplanted in nude mice. Cancer Chemother Pharmacol 10:182–6.
Jeena KJ, Joy KL, Kuttan R. (1999) Effect of *Emblica officinalis*, *Phyllanthus amarus* and *Picrorrhiza kurroa* on *N*-nitrosodiethylamine induced hepatocarcinogenesis. Cancer Lett 136:11–6.
Jensen AG, Ndjoko K, Wolfender JL, Hostettmann K, Camponovo F, Soldati F. (2002) Liquid chromatography–atmospheric pressure chemical ionisation/mass spectrometry: a rapid and selective method for the quantitative determination of ginkgolides and bilobalide in ginkgo leaf extracts and phytopharmaceuticals. Phytochem Anal 13:31–8.
Johnsson U, Fagerstam L, Ivarsson B, Johnsson B, Karlsson R. (1991) Real-time biospecific interaction analysis using surface plasmon resonance and a sensor chip technology. Biotechniques 11:620–7.
Jose JK, Kuttan G, Kuttan R. (2001) Antitumour activity of *Emblica officinalis*. J Ethnopharmacol 75:65–9.
Kang BY, Chung SW, Kim SH, Ryu SY, Kim TS. (2000) Inhibition of interleukin-12 and interferon-gamma production in immune cells bytanshinones from *Salvia miltiorrhiza*. Immunopharmacology 49:355–61.
Karlsson R, Kullman-Magnusson M, Hamalainen MD, Remaeus A, Andersson K, Borg P, Gyzander E, Deinum J. (2000) Biosensor analysis of drug-target interactions: direct and competitive binding assays for investigation of interactions between thrombin and thrombin inhibitors. Anal Biochem 278:1–13.
Khan MTH, Lampronti I, Martello D, Bianchi N, Jabbar S, Choudhuri MSK, Datta BK, Gambari R. (2002) Identification of pyrogallol as an antiproliferative compound present in extracts from the medicinal plant *Emblica Officinalis*: effects on *in vitro* cell growth human tumor cell lines. Int J Oncol 21:187–92.
Knight WA, Livingston RB, Gregory EJ, McGuire WL. (1977) Estrogen receptor as an independent prognostic factor for early recurrence in breast cancer. Cancer Res 37:4669–71.
Lambertini E, Penolazzi L, Aguiari G, del Senno L, Pezzetti F, Sollazzo V, Piva R. (2002) Osteoblastic differentiation induced by transcription factor decoy against human estrogen receptor a gene. Biochem Biophys Res Commun 292:761–70.
Lampronti I, Martello D, Bianchi N, Borgatti M, Lambertini E, Piva R, Jabbar S, Choudhuri MSK, Khan MTH, Gambari R. (2003) *In vitro* antiproliferative effects on human tumor cell lines of extracts from the Bangladeshi medicinal plant *Aegle marmelos* Correa. Phytomedicine, in press.
Leckie F, Mattei B, Capodicasa C, Hemmings A, Nuss L, Aracri B, De Lorenzo G, Cervone F. (1999) The specificity of polygalacturonase-inhibiting protein (PGIP): a single amino

acid substitution in the solvent-exposed beta-strand/beta-turn region of the leucine-rich repeats (LRRs) confers a new recognition capability. EMBO J 18(9):2352–63.

Lee KT, Sohn IC, Kim YK, Choi JH, Choi JW, Park HJ, Itoh Y, Miyamoto K. (2001) Tectorigenin, an isoflavone of *Pueraria thunbergiana* Benth., induces differentiation and apoptosis in human promyelocytic leukemia HL-60 cells. Biol Pharm Bull 24:1117–21.

Li BQ, Fu T, Gong WH, Dunlop N, Kung H, Yan Y, Kang J, Wang JM. (2000) The flavonoid baicalin exhibits anti-inflammatory activity by binding to chemokines. Immunopharmacology 49:295–306.

Malmqvist M. (1993) Biospecific interactions analysis using biosensor technology. Nature 361:186–7.

Mandal SC, Maity TK, Das J, Saba BP, Pal M. (2000) Anti-inflammatory evaluation of *Ficus racemosa* Linn. leaf extract. J Ethnopharmacol 72:87–92.

Mathur R, Sharma A, Dixit VP, Varma M. (1996) Hypolipidaemic effect of fruit juice of *Emblica officinalis* in cholesterol-fed rabbits. J Ethnopharmacol 50:61–8.

McKay D. (2004) Nutrients and botanicals for erectile dysfunction: examining the evidence. Altern Med Rev 9:4–16.

Mellor F, Schulte M. (1997) The chromatography of natural substances – active ingredients from natural sources. GIT Lab J 1/97:42–6.

Mischiati C, Sereni A, Gambari R. (2003) Use of macroarray technology to study the effects of DNA-binding drugs on gene expression profile of erythroid-induced human leukemic K562 cells. Minerva Biotecnologica 15:153–60.

Mishra M, Pathak UN, Khan AB. (1981) *Emblica officinalis* Gaertn and serum cholesterol level in experimental rabbits. Br J Exp Pathol 62:526–8.

Mukherjee AK, Basu S, Sarkar N, Ghosh AC. (2001) Advances in cancer therapy with plant based natural products. Curr Med Chem 8:1467–86.

Murphy LC, Watson P. (2002) Steroid receptors in human breast tumorigenesis and breast cancer progression. Biomed Pharmacother 56:65–77.

Nagy B, Mucsi I, Molnar J, Thurzo L. (2002) Combined effect of cisplatin and 5-fluorouracil with irradiation on tumor cells *in vitro*. Anticancer Res 22:135–8.

Nam SY, Yi HK, Lee JC, Kim JC, Song CH, Park JW, Lee DY, Kim JS, Hwang PH. (2002) Cortex mori extract induces cancer cell apoptosis through inhibition of microtubule assembly. Arch Pharm Res 25:191–6.

Nass SJ, Davidson NE. (1999) The biology of breast cancer. Hematology/Oncol Clin N Am 13:311–32.

Nedelkov D, Nelson RW. (2003) Design and use of multi-affinity surfaces in biomolecular interaction analysis-mass spectrometry (BIA/MS): a step toward the design of SPR/MS arrays. J Mol Recognit 16:15–9.

Neto CC, Owens CW, Langfield RD, Comeau AB, Onge JS, Vaisberg AJ, Hammond GB. (2002) Antibacterial activity of some Peruvian medicinal plants from the Callejon de Huaylas. J Ethnopharmacol 79:133–8.

Nilsson P, Persson B, Uhlén M, Nygren PA. (1995) Real-time monitoring of DNA manipulations using biosensor technology. Anal Biochem 224:400–8.

Ono Y, Kozai Y, Ootsu K. (1982) Antitumor and cytocidal activities of a newly isolated benzenoid ansamycin, macbecin I. Gann 73:938–44.

Orth HC, Rentel C, Schmidt PC. (1999) Isolation, purity analysis and stability of hyperforin as a standard material from *Hypericum perforatum* L. J Pharm Pharmacol 51:193–200.

Ozaki S. (1996) Synthesis and antitumor activity of 5-fluoruracil and derivatives. Med Res Rev 16:51–86.

Penolazzi L, Lambertini E, Aguiari G, del Senno L, Piva R. (2000) Cis element 'decoy' against the upstream promoter of the human estrogen receptor gene. Biochim Biophys Acta 1492:560–7.

Pinn G. (2001) Herbs used in obstetrics and gynaecology. Aust Fam Physician 30:351–6.

Popov AM, Atopkina LN, Uvarova NI, Elyakov GB. (2001) The antimetastatic and immunomodulating activities of ginseng minor glycosides. Dokl Biochem Biophys 380:309–12.

Ram A, Lauria P, Gupta R, Kumar P, Sharma VN. (1997) Hypocholesterolaemic effects of *Terminalia arjuna* tree bark. J Ethnopharmacol 55:165–9.

Ramsewak RS, DeWitt DL, Nair MG. (2000) Cytotoxicity, antioxidant and anti-inflammatory activities of curcumins I–III from *Curcuma longa*. Phytomedicine 7:303–8.

Richardson MA. (2001) Biopharmacologic and herbal therapies for cancer: research update from NCCAM. J Nutr 131:3037–40.

Roobol C, Sips HC, Theunissen J, Atassi G, Berheim JL. (1984) *In vitro* assessment of cytotossic agents in murine cancers: comparison between antiproliferative and antimetabolic assays. J Natl Caner Inst 72:661–6.

Ruffa MJ, Ferraro G, Wagner ML, Calcagno ML, Campos RH, Cavallaro L. (2002) Cytotoxic effect of Argentine medicinal plant extracts on human hepatocellular carcinoma cell line. J Ethnopharmacol 79:335–9.

Sastravaha G, Gassmann G, Sangtherapitikul P, Grimm WD. (2005) Adjunctive periodontal treatment with *Centella asiatica* and *Punica granatum* extracts in supportive periodontal therapy. J Int Acad Periodontol 7:70–9.

Sharma N, Trikha P, Athar M, Raisuddin S. (2000) *In vitro* inhibition of carcinogen-induced mutagenicity by *Cassia occidentalis* and *Emblica officinalis*. Drug Chem Toxicol 23:477–84.

Silva O, Gomes ET, Wolfender JL, Marston A, Hostettmann K. (2000) Application of high performance liquid chromatography coupled with ultraviolet spectroscopy and electrospray mass spectrometry to the characterisation of ellagitannins from *Terminalia macroptera* roots. Pharm Res 17:1396–401.

Singh DV, Verma RK, Singh SC, Gupta MM. (2002) RP-LC determination of oleane derivatives in *Terminalia arjuna*. J Pharm Biomed Anal 28:447–52.

Steenkamp V, Steward MJ, van der Merwe S, Zuckerman M, Crowther NJ. (2001) The effect of *Senecio latifolius* a plant used as a South African traditional medicine on a human hepatoma cell line. J Ethnopharmacol 78:51–8.

Suthar AC, Banavalikar MM, Biyani MK. (2001) Pharmacological activities of Genistein, an isoflavone from soy (Glycine max): part II – anti-cholesterol activity, effects on osteoporosis & menopausal symptoms. Indian J Exp Biol 39:520–5.

Tatman D, Mo H. (2002) Volatile isoprenoid constituents of fruits, vegetables and herbs cumulatively suppress the proliferation of murine B16 melanoma and human HL-60 leukemia. Cancer Lett 175:129–39.

Tran QL, Adnyana IK, Tezuka Y, Nagaoka T, Tran QK, Kadota S. (2001) Triterpene saponins from Vietnamese ginseng (*Panax vietnamensis*) and their hepatocytoprotective activity. J Nat Prod 64:456–61.

Velazco BA. (1980) Traditional herbal practices and motherhood. Philipp J Nurs 50:95–9.

Vadgama P, Crump PW. (1992) Biosensors: recent trends, a review. Analyst 117:1657–70.

Wang X, Sakuma T, Asafu-Adjaye E, Shiu GK. (1999) Determination of ginsenosides in plant extracts from *Panax ginseng* and *Panax quinquefolius* L. by LC/MS/MS. Anal Chem 71:1579–84.

Wargovich MJ, Woods C, Hollis DM, Zander ME. (2001) Herbals, cancer prevention and health. J Nutr 131:3034–6.

Winter JN, Variakojis D, Gaynor ER, Larson RA, Miller KB. (1985) Low-dose cytosine arabinoside (Ara-C) therapy in the myelodysplastic syndromes and acute leukemia. Cancer 56:443–9.

Wood SJ. (1993) DNA–DNA hybridization in real time using BIAcore. Microchem J 47: 330–7.

Xiao H, Parkin KL. (2002) Antioxidant functions of selected allium thiosulfinates and S-alk(en)yl-L-cysteine sulfoxides. J Agric Food Chem 50:2488–93.

Ye M, Li Y, Yan Y, Liu H, Ji X. (2002) Determination of flavonoids in *Semen Cuscutae* by RP-HPLC. J Pharm Biomed Anal. 28:621–8.

Yu TX, Ma RD, Yu LJ. (2001) Structure–activity relationship of tubeimosides in anti-inflammatory, antitumor, and antitumor-promoting effects. Acta Pharmacol Sin 22:463–8.

M.T.H. Khan and A. Ather (eds.)
Lead Molecules from Natural Products

Herbal extracts and compounds active against herpes simplex virus

Kenneth D. Thompson

Abstract

Herbal compounds are being investigated for antiviral activity against a variety of infectious agents including herpes simplex virus (HSV). Purification and identification of the active compound are very important in the development of these compounds as therapeutic agents. Much of the information presented here deals with testing of crude extracts, purifying active compounds, identifying, and characterizing plant molecules that have antiviral activity against HSV. In the development of antiviral agents from plants, it is also important to understand the targets and modes of action of these potentially useful new agents. Topical antiviral agents would be of benefit to many patients with recurrent HSV infection for several reasons including the ease of use, avoiding systemic exposure to drug, and providing greater levels of antiviral compound at the site of virus replication. Another application of these compounds would be their use as topical microbicides to prevent the transmission of sexually transmitted viruses. Herbal compounds may be ideally suited for their ability to prevent the transmission of viruses such as HSV.

Keywords: *herpes simplex virus, herbal extracts, herbal compounds, antiviral compounds, topical microbicides, genital herpes, sexually transmitted viruses*

Abbreviations: HSV, herpes simplex virus; HIV, human immunodeficiency virus; HCMV, human cytomegalovirus; PRV, pseudorabies virus; ACV, acyclovir; TK, thymidine kinase; IC_{50}, inhibitory concentration 50; STD, sexually transmitted disease.

I. Introduction

I.A. Epidemiology of herpes simplex virus

The most common disease-associated infections caused by herpes simplex viruses (HSV-1 and HSV-2) are oral and genital herpes although HSV can cause disease at many anatomical sites, including encephalitis and other systemic infections. During primary infections, HSV enters nerve cells to establish latency. After establishing latency, HSV can become reactive to cause recurrent disease at mucosal surfaces. In immunocompromised patients and neonates, HSV can cause serious systemic infections. In addition,

HSV-2 infection may be a risk factor for the transmission of human immunodeficiency virus (HIV) (Holmberg et al., 1988; Severson and Tyring, 1999).

Genital herpes is an important sexually transmitted disease (STD) most commonly caused by HSV-2 but a significant number of infections are also caused by HSV-1 (Johnson et al., 1989). Fleming et al. (1997) reported that the seroprevalence of HSV-2 in persons 12 years of age or older in the U.S. was 21.9%, a 30% increase from the previous decade. In a recent report on the seroprevalence of HSV, Xu et al. (2002) found that 51% of persons 12 years or older in the U.S. had antibody to HSV-1 only, 5.3% had antibody to HSV-2 only, and 16.6% had antibody to both viruses. Asymptomatic shedding of HSV-2 appears to be common and can play a role in the transmission of this virus (Wald et al., 1997). Independent predictors of HSV-2 seropositivity included female sex, black race, older age, less education, more lifetime sex partners, prior diagnosis of syphilis or gonorrhea, and lack of HSV-1 antibody. The majority of HSV-2-seropositive persons (84.7%) had never received a diagnosis of genital herpes (Gottlieb et al., 2002).

I.A.1. HSV structure and entry

Herpes simplex viruses (HSV) types 1 and 2 are members of the alphaherpesvirus subfamily of herpesviruses. HSV is a large enveloped-DNA virus that has a genome consisting of a double-stranded molecule and contains about 90 genes. The envelope surrounding the capsid and tegument contains at least 12 surface glycoproteins and some of the glycoproteins are essential for attachment and entry of the virus into the host cell.

Binding and entry of HSV into cells is mediated by viral glycoproteins. Initially, glycoprotein C (gC) and glycoprotein B (gB) bind to heparan sulfate. Next, gD binds to a cellular mediator of viral entry HveA (a TNF receptor), nectin 1 or nectin 2. Nectin 2 is an entry mediator for all herpesviruses tested so far and is also expressed on neurons.

After the viral and cellular membranes fuse, viral uncoating occurs and the viral genome enters the nucleus where viral gene expression begins. HSV gene expression is regulated in three distinct phases: immediate-early (IE), early (E), and late (L). IE RNA expression does not require protein synthesis. HSV-1 and HSV-2 are closely related viruses and after a primary infection, these viruses establish latency in neurons. HSV can reactivate from latency to cause recurrent infections.

I.A.2. Current anti-HSV therapy

Acyclovir (ACV) was the first effective anti-HSV therapy that was generally used to treat serious HSV infections. ACV is phosphorylated by the HSV thymidine kinase (TK) and ACV functions to inhibit DNA synthesis. ACV is generally nontoxic to patients. Other anti-HSV nucleoside analogs are penciclovir, valacyclovir, and famciclovir. All these compounds are activated by the viral-encoded TK and their mode of action is the inhibition of DNA synthesis. HSV resistance to ACV develops rather easily when mutations in the TK result in a nonfunctional or altered TK. ACV-resistant HSV can be treated with foscarnet, a compound that inhibits the viral DNA polymerase.

Although the current anti-HSV compounds can be used as prophylactic therapy for recurrent genital herpes, it must be used as daily systemic therapy. Some plant extracts that have exhibited antiviral activity *in vitro* may be good sources of topical microbicides and may be useful in preventing the transmission of some viral STDs (Vermani and Garg, 2002). Topical antiviral agents may also benefit many patients with recurrent HSV infection for several reasons including the ease of use, avoiding systemic exposure to drugs, and cost.

I.A.3. Herpes simplex virus vaccine

An important approach to the prevention of genital herpes would be an effective vaccine. Currently, there are no vaccines that are approved for preventing genital herpes. There are, however, clinical trials ongoing to determine if a vaccine can prevent genital herpes in women. A vaccine trial was conducted using HSV-2 gD with an adjuvant in a small number of men and women who did not have genital herpes but whose partner was infected with HSV (Stanberry et al., 2002). The vaccine appeared to prevent HSV disease in more than 70% of the HSV-negative women. The vaccine was not efficacious in women who were seropositive for HSV-1 and seronegative for HSV-2. The vaccine was not effective in males regardless of their HSV serological status.

I.B. Screening plants for antiviral activity

Historically, medicinal plants have been used by many societies for the treatment of human diseases. Although herbal medicines are currently utilized by the general public to treat a variety of illnesses, scientific data supporting their efficacies are often lacking. Nonetheless, laboratory investigations of traditional herbal medicines has shown that natural products may contain compounds with antiviral activity (Beuscher et al., 1994; Vlietinck et al., 1995; Taylor et al., 1996). A number of reports show that screening plants from around the world has yielded many potentially useful antiviral materials that have the possibility of being developed into useful antiviral compounds. The following reports were studies in which screening of a number of plant extracts were investigated as antiviral compounds against HSV and sometimes other viruses as well.

One hundred and forty-two traditional medicines used in China, Indonesia, and Japan were screened for *in vitro* antiviral activity against HSV-1, poliovirus type 1, and measles virus. Among the 32 extracts with anti-HSV-1 activity, three had anti-HSV-1 activity alone and the others showed anti-HSV-1 activity with antipoliovirus and/or antimeasles virus activities. The 32 extracts were further examined for their therapeutic efficacies of HSV-1 infection in mice. Twelve extracts were found to be effective in limiting the development of skin lesions and/or in prolonging the mean survival times of HSV-1-infected mice (Kurokawa et al., 1993).

The Indian plant, *Pongamia pinnata* Linn. (Papillionaceae) used in the Ayurveda and Siddha traditional medicine systems for the treatment of skin and genital lesions, was evaluated for antiviral properties against HSV-1 and HSV- 2. A crude aqueous extract of *P. pinnata* seeds completely inhibited the growth of HSV-1 and HSV-2 at concentrations of 1 and 20 mg/ml (w/v). This extract was shown to have *in vitro* antiviral activity against HSV-1 and HSV- 2 by the absence of cytopathic effect in Vero cells (Elanchezhiyan et al., 1993).

Two Indian medicinal plants, *Annona muricata* (Annonaceae) and *Petunia nyctaginiflora* (Solanaceae), were screened for their antiviral activity against HSV-1. The clinical isolate of HSV used in this study was obtained from the human keratitis lesion. This study looked at the ability of the extracts to inhibit the cytopathic effect of HSV-1 in Vero cells. The minimum inhibitory concentration of ethanolic extract of *A. muricata* and aqueous extract of *P. nyctaginiflora* was found to be 1 mg/ml (Padma et al., 1998).

Extracts prepared from four medicinal plants native to Bangladesh, *Hemidesmus indicus*, *Paederia foetida*, *Shorea robusta*, and *Embelica officinalis*, were evaluated for antiviral activity against 38 clinical strains of HSV. The median IC_{50} of each extract for the 17 strains of HSV-1 were *H. indicus*, 400 μg/ml; *P. foetida*, 400 μg/ml; *S. robusta*, 20 μg/ml; and *E. officinalis*, 80 μg/ml. The median IC_{50} of the extracts for the 21 strains of HSV-2 were *H. indicus*, 200 μ/ml; *P. foetida*, 200 μg/ml; *S. robusta*, 5.0 μg/ml; and *E. officinalis*, 20 μ/ml, respectively. All four herbal extracts were significantly more active against strains of HSV-2 than against strains of HSV-1, $p<0.01$. Although the active components in the extracts responsible for anti-HSV activity have not been identified, all four extracts inhibited the binding of HSV to cellular receptors (Thompson et al., unpublished data).

Several articles have reported on the screening of medicinal plants used in Nepal to treat viral diseases. A report by Taylor et al. (1996) studied 21 extracts from 20 plant species that were assayed for antiviral activity against three viruses: HSV, Sindbis virus, and poliovirus. Antiviral activities were exhibited by species of *Bauhinia* (Fabaceae), *Carissa* (Apocynaceae), *Milletia* (Fabaceae), *Mallotus* (Fabaceae), *Rumex* (Polygonaceae), *Streblus* (Moraceae), *Terminalia* (Combretaceae), and *Tridax* (Asteraceae). The methanolic extract from *Carissa carandas* was the most active against these three viruses at an inhibitory concentration of 12 μg/ml (Taylor et al., 1996). Another report on the screening of Nepalese plants used in traditional medicine evaluated extracts derived from 23 plant species. The extracts were tested against influenza A virus in Madin-Darby canine kidney (MDCK) cells and HSV in Vero cells. Two species, *Bergenia ligulata* and *Nerium indicum*, showed the highest activity against an influenza A virus with 50% inhibitory dose of 10 μg/ml. Twelve extracts from nine different species of plants were active against HSV. The highest anti-HSV activity was seen with the methanolic extracts of *Holoptelia integrifolia* and *N. indicum* and these extracts were not toxic to Vero cells (Rajbhandari et al., 2001).

Several Thai medicinal plants, used as remedies to treat herpesvirus infection, were investigated for their activity against HSV. *Barleria lupulina* Lindl. and *Clinacanthus nutans* (Burm. f.) Lindau, belonging to the family Acantaceae, are well-known medicinal plants used in Thai folklore medicine. Virucidal effects of organic extracts of these two plants against HSV-2 strain G were investigated. The extracts were assessed for intracellular activities against HSV-2 strain G and five clinical isolates of HSV-2. *B. lupulina* extract exhibited activity against all five clinical isolates but not the HSV strain G while that of *C. nutans* did not show any activity against these viruses when tested in a plaque reduction assay. When the activities were verified by yield reduction assay, anti-HSV-2 activities of *B. lupulina* extract against HSV-2 strain G were seen as well. The results suggest a therapeutic potential of *B. lupulina* but not *C. nutans* against HSV-2 (Yoosook et al., 1999). Extracts of *Centella asiatica* L., *Maclura cochinchinensis* Cornor, and *Mangifera indica* L. were determined by a plaque reduction assay to be active against both HSV-1 and HSV-2. Mixture of *C. asiatica* and *M. indica* gave an additive effect against HSV. The active constituent present in the *C. asiatica* extract was determined to be asiaticoside while the active component in *M. indica* was mangiferin (Yoosook et al., 2000).

Extracts of 30 traditional medicinal plants collected in Indonesia were tested for anti-HSV-1 activity. The extracts of eight plant species showed antiviral activity in a plaque reduction assay at concentrations of 100 μg/ml. The therapeutic efficacy of seven plants was demonstrated by using a mouse HSV-1 infection assay. Both the methanol extracts of the fruit of *Melaleuca leucadendron* (Myrtaceae) and the pericarp of *Nephelium lappaceum* (Sapindaceae) significantly prolonged the development of skin lesions and reduced the mortality in mice (Nawaw et al., 1999).

A total of 267 crude extracts were obtained from 100 Rwandese medicinal plants used by traditional healers to treat infections. These plant extracts were screened for antibacterial, antifungal, and antiviral properties. About 27% of the plant species exhibited antiviral activity against one or more test viruses; 12% against poliovirus, 16% against Coxsackie, 3% against Semliki Forest virus, 2% against measles virus, and 8% against HSV. Although a number of plant extracts had antiviral activity, eight plant extracts exhibited anti-HSV activity; *Clerodendron myricoides* (Verbana), *Crassocephalum multicorymbosum* (Asclepidiaceae), *Dryopteris inaequalis* (Aspidiaceae), *Euphorbia hiria* (Euphorba), *Erythrina abyssinica* (Faba), *Glycine javanica* (Faba), *Markhamia lutea* (Bignoniaceae), and *Rhus vulgaris* (Anacardiaceae) (Vlietinck et al., 1995). In a second study of seven plants used by Rwandan traditional healers, aqueous ethanol extracts were screened for antibacterial, antifungal, and antiviral activities. Only two of the selected plants showed antiviral activity against HSV-1, *C. mimosides* and *I. involucrata*. Also, two saponin mixtures, maesasaponin mixtures A and B were obtained by extraction and chromatography of *Maesa lanceolata*. The maesasaponin mixture A exhibited virucidal activity against HSV-1, HSV-2, and vesicular stomatitis viruses (VSV) (Sindambiwe et al., 1999).

Plants used in both human and veterinary traditional medicine in northern Nigeria were investigated for their antiviral activity. Extracts of 17 plants were tested *in vitro* against poliovirus, astrovirus, parvovirus, and herpesviruses and 10 of the extracts had activity against HSV. Most of the extracts have activity against one or more viruses at concentrations between 1.0 and 4.0 mg/ml. Although some were only minimally active, two extracts, *Anacardium occidentale* and *Sterculia setgera*, were quite active in this *in vitro* assay (Kudi and Myint, 1999).

Antiviral and antimicrobial activities were found in the extracts of 24 medicinal plants used in the treatment of skin infections in Colombia. Thirteen extracts had activity against HSV while none of the extracts displayed active against poliovirus. The antiviral activity was indicated by a total inhibition of viral CPE at a noncytotoxic concentration of the extract. The most potent extract was obtained from *Byrsonima verbascifolia* (L.) HBK., which had anti-HSV activity at a concentration as low as 2.5 μg/ml (Lopez et al., 2001).

Extracts prepared from 54 medicinal plants used in Brazil to treat infections were screened for antiviral properties against five different viruses: HSV-1, HSV-2, poliovirus type 2, adenovirus type 2, and VSV. Of the plant extracts tested, 42.6% showed activity against HSV-1 and HSV-2, 26% against poliovirus, and 24% against VSV. None of the extracts was active against adenovirus. *Trixis praestans* (Vell.)

Cabr. and *Cunila spicata* Benth. extracts had the most antiviral activity against the viruses tested (Simoes et al., 1999).

The screening of selected medicinal plants used in Argentina for the treatment of infectious diseases was conducted using aqueous extracts of five species of plants. The extracts were assayed for antiviral activity against HSV-1, respiratory syncytial virus (RSV), and adenovirus serotype 7 (ADV-7). *Polygonum punctatum*, *Lithraea molleoides*, *Sebastiania brasiliensis*, and *Sebastiania klotzschiana* but not *Myrcianthes cisplatensis* showed anti-HSV activity with 50% effective dose (ED_{50}) ranging from 39 to 169 μg/ml. *P. punctatum*, *L. molleoides*, and *M. cisplatensis* showed antiviral activity against RSV with ED_{50} ranging from 78 to 120 μg/ml. None of the extracts had antiviral activity against ADV-7 (Kott et al., 1999).

Screening of traditional medicinal plants from countries around the world has yielded many extracts that have antiviral activity against HSV and other viruses including sexually transmitted viruses. These extracts are from very diverse plant species in which the antiviral activity and toxicity are quite variable. The molecular structure and modes of action are not known for the active components in most of these extracts, therefore, it is possible that some of the extracts have similar modes of action. Since most of these screening assays used aqueous or organic solvent extracts, the investigators are testing very impure compounds. Purification of these extracts may yield very active compounds for the treatment of HSV.

I.B.1. Plant extracts with activity against HSV

Many investigators have focused their studies on finding specific anti-HSV activity from plants. The following extracts from a single plant species were investigated as anti-HSV compounds or as an antiviral compound against a few viruses.

Meyer et al. (1996) described the inhibition of the cytopathic effect of HSV-1 in human lung fibroblasts. An aqueous extract from *Helichrysum aureonitens* (Asteraceae), a medicinal plant from southern Africa, was evaluated for antiviral activity against HSV-1. The crude aqueous extract from shoots of *H. aureonitens* showed activity against HSV-1 at a concentration of 1.35 mg/ml (Meyer et al., 1996).

The HSV inhibitory effect of five extracts from the Bulgarian medicinal plant *Geranium sanguineum* L. (Geraniaceae) was investigated. A water extract from the aerial roots of the plant inhibited the replication of HSV-1 and HSV-2 with $IC_{50} = 3.6–6.2$ μg/ml and was the least toxic to cell cultures. The inhibition was dose-related, strain-specific, and depended on virus inoculum. The presence of the extract throughout the replicative cycle was necessary for the full antiviral effect. In a preliminary experiment in albino guinea pigs, the extract delayed the development of herpetic vesicles following primary infection with HSV-1 (Serkedjieva and Ivancheva, 1999).

An extract derived from *Echinacea purpurea* was reported to have antiviral activity against HSV. Out of the 40 clinical strains of HSV, 15 strains were resistant to acyclovir (ACV-R) and 25 strains were susceptible to ACV (ACV-S). These data showed that the extract had anti-HSV activity against both HSV-1 and HSV-2. In addition, there was no apparent difference in activity against the ACV-R and ACV-S strains of HSV. The extract was nontoxic in a cell cytotoxicity assay. Although the active components that have anti-HSV activity are not known, this extract has antiviral activity against both ACV-resistant and ACV-susceptible strains of HSV-1 and HSV-2 (Thompson, 1998). In a recent study of *Echinacea*, extracts of eight taxa were found to have antiviral activity against HSV-1 *in vitro* when exposed to visible and UV-A light. *n*-Hexane extracts of roots containing alkenes and amides were more active in general than ethyl acetate extracts containing caffeic acids. The most potent inhibitors of HSV were crude extract of *E. pallida* var. *sanguinea*, MIC = 26 μg/ml; cichoric acid, MIC = 45 μ/ml, and *E. purpurea n*-hexane root extract, MIC = 120 μ/ml (Binns et al., 2002).

The antiviral and cytotoxic activity of an aqueous extract of *Phyllanthus orbicularis* (Euphorbiaceae) was evaluated in cell culture. An extract was prepared from the leaves and stems of this plant and exhibited selective antiviral activity against bovine herpesvirus type 1 (BHV-1) and HSV-2. There was, however, no selective antiviral activity against adenovirus type 5 or mengovirus. Incubation of extract during the infection of cell cultures impaired the productive replication of both herpesviruses in a concentration-dependent manner. The inhibition was also dependent on the multiplicity of infection (MOI) used. These results suggested that antiviral activity of the extract might be partially due to a direct interaction with virus particles or viral entry into the cell, instead of interfering with intracellular virus-specific macromolecular synthesis (del Barrio and Parra, 2000).

The *in vitro* anti-HSV activity of *Maclura cochinchinensis* was investigated using biological activity as a guide to the separated active components. Ethyl acetate and methanol extracts exhibited anti-HSV-2 activity with EC_{50} values of 38.5 and 50.8 μg/ml, respectively. Chromatographic separation of the ethyl acetate extract yielded compound A, identified as morin by a spectroscopic method. Morin exhibited anti-HSV-2 activity at an EC_{50} value of 53.5 μg/ml. These data suggest that free hydroxyl groups are required for anti-HSV activity, as demonstrated previously for the antiviral activity of other flavonoids (Bunyapraphatsara et al., 2000).

Extracts of *Rhus javanica* L. have been shown to exhibit both prophylactic and therapeutic anti-HSV activity *in vitro* and *in vivo* (Kurokawa et al., 1993, 1997). An extract of *R. javanica* was examined for its suppressive efficacy on recurrent genital HSV-2 in guinea pigs infected intravaginally. Prophylactic oral administration of *R. javanica* reduced the incidence, severity, and/or frequency of spontaneous and severe skin lesions as compared with latently infected control guinea pigs that received water. This prophylactic efficacy was confirmed by the crossover administration of *R. javanica* and water to the infected guinea pigs for more than 2 months. When recurrent HSV-2 disease was induced by ultraviolet irradiation 3 months after primary infection, the prophylaxis with *R. javanica* was also effective in reducing the severity of ultraviolet-induced skin lesions (Nakano et al., 1998).

The antiviral activity of a commercial preparation of *Spirulina maxima*, a planktonic blue-green alga, was investigated. The extract inhibited HSV-2, pseudorabies virus (PRV), human cytomegalovirus (HCMV), and HSV-1, with ED_{50} of 69, 103, 142, and 333 μg/ml for each virus, respectively. For adenovirus, the inhibition was less than 20%, and no inhibition was found for measles virus, subacute sclerosing panencephalitis virus (SSPE), VSV, poliovirus 1, and rotavirus SA-11, at concentrations of 2 mg/ml. All four herpesviruses, HSV-1, HSV-2, HCMV, and PRV, were inhibited at the adsorption and penetration steps of the virus life cycle. The isolation and identification of the compound that exhibits the antiviral activity was initiated when several extracts were prepared using solvents with different polarity. Each extract was evaluated in a microplate inhibition assay using HSV-2. The highest anti-HSV activity was detected in methanol–water (3:1), which suggests that the antiviral activity is probably due to highly polar compounds (Hernandez-Corona et al., 2002).

Plantago major L. (Plantagiaceae), a traditional Chinese herbal medicine, has long been used for the treatment of various diseases. An investigation on a series of viruses, HSV-1, HSV- 2, and adenoviruses (ADV-3, ADV-8, and ADV-11) was conducted using an extract of *P. major*. Results showed that the aqueous extract of *P. major* possessed only slight anti-HSV activity. In contrast, certain pure compounds belonging to the five different classes of chemicals found in the extract of this plant exhibited potent antiviral activity. Caffeic acid exhibited the strongest activity against HSV-1, $EC_{50} = 15.3\,\mu g/ml$, SI = 671; HSV-2, $EC_{50} = 87.3\,\mu g/ml$, SI = 118; and ADV-3, $EC_{50} = 14.2\,\mu g/ml$, SI = 727, whereas chlorogenic acid possessed the strongest anti-ADV-11 activity, $EC_{50} = 13.3\,\mu g/ml$, SI = 301. This study concludes that pure compounds of *P. major*, which possess antiviral activities are mainly derived from the phenolic compounds, especially caffeic acid. Its mode of action against HSV-2 and ADV-3 was found to be at the multiplication stages, post infection of both HSV-1 and ADV-3 (Chiang et al., 2002).

Extracts and fractions rich in flavonoids from fruits and leaves of *Vitex polygama* Cham. (Verbenaceae) were tested for antiviral activity using an acyclovir-resistant strain of HSV-1. Both fruit and leaf extracts exhibited a dose-dependent antiviral activity. The extract from the leaves showed intracellular antiviral activity while the extract from the fruits had a virucidal effect. A fraction from the ethyl acetate extract of the leaves inhibited virus replication by blocking HEp-2 cell receptors (Goncalves et al., 2001). These authors present data indicating that a herbal extract contains compound(s) with multiple modes of action, mainly virucidal activity and inhibiting the virus from binding to cellular receptors.

Water extract obtained from coconut husk fiber of *Cocos nucifera* L. and fractions from adsorption chromatography revealed antimicrobial activity against *Staphylococcus aureus*. The crude extract and one of the fractions rich in catechin also showed inhibitory activity against acyclovir-resistant HSV-1-ACVr. Catechin and epicatechin together with condensed tannins were demonstrated to be the components of the water extract (Esquenazi et al., 2002).

An extract of *Ribes nigrum* L., known as blackcurrant in Europe and Kurokarin in Japan, was investigated for its antiherpesvirus activity *in vitro*. The extract inhibited HSV-1 attachment to the cell membrane at a 100-fold dilution. It also inhibited the plaque formation of HSV-1, HSV-2, and VZV by 50% at 400-fold dilution. The inhibition of virus replication was due to the inhibition of protein synthesis in infected cells at an early stage of infection (Suzutani et al., 2003).

Most of these studies used *in vitro* assays for the evaluation of extracts that have antiviral activity against HSV. Although most of these screening studies have not identified the active anti-HSV components present in these extracts, several studies have found extracts with excellent anti-HSV activity. Finding anti-HSV extracts that are nontoxic are important aspects to consider in the development of useful anti-HSV compounds. It is likely that there may be a spectrum of modes of action that could be exploited in the development of new antiviral compounds. Determining the molecular structure and modes of action are the next steps in the development of antiviral compounds. It is likely that there may be a spectrum of modes of action that could be exploited in the development of topical agents. Finding a compound with broad antiviral activity could be quite useful in the development of a topical microbicide. Alternatively, several antiviral compounds with specific antiviral activity could be combined to make a broad-spectrum preparation that could be used in this manner. Although many of these extracts appeared to be nontoxic, the purified compounds need to be evaluated for toxicity *in vitro* as well as for efficacy in animals.

An extract prepared from the seeds of *Licania tomentosa* (Benth.) Fritsch (Chrysobalanaceae) was tested for antiviral activity against a strain of acyclovir-resistant HSV-1 (ACV-R-HSV-1). The extract impaired the productive replication of this virus in a concentration-dependent manner. The extract appeared to be virucidal in addition to interfering with a very early event of infection at a noncytotoxic concentration (Miranda et al., 2002).

I.B.2. Plant compounds with anti-HSV activities

While many published reports demonstrate antiviral or anti-HSV activities in herbal extracts (Kudi and Myint, 1999; Yoosook et al., 1999; del Barrio and Parra, 2000), some investigators have purified and identified the active molecules. Identifying the active components in herbal extracts is an important step in the development of antiviral compounds. Once the compounds are known, it may be possible to improve the stability or activity of the molecule or decrease the toxicity of the molecule thereby improving the desired antiviral effect. This section deals with the discovery and investigation of known plant compounds against HSV and some other viruses.

Thirty-one species of medicinal plants used in the treatment of viral diseases in China were assayed for inhibition of Sindbis and murine cytomegalovirus *in vitro*. Sixteen species displayed antiviral activity and a compound was isolated from the leaves and twigs of *Elsholtzia ciliata* (Lamiaceae) using bioassay-guided fractionation and identified as the polycyclic aromatic hydrocarbon, fluoranthene (Yip et al., 1995). *Rhus javanica*, a medicinal herb, has been shown to have anti-HSV activity in mice. Two major anti-HSV compounds, moronic acid and betulonic acid, were purified from the herbal extract by extraction with ethyl acetate at pH 10 followed by chromatographic separations. Moronic acid was quantitatively a major anti-HSV compound in the ethyl-acetate-soluble fraction. The EC_{50}s for moronic acid and betulonic acid using wild-type HSV-1 were 3.9 and 2.6 μg/ml, respectively. The therapeutic index of moronic acid (10.3–16.3) was larger than that of betulonic acid (6.2). Susceptibility of acyclovir-resistant HSV-1 and wild-type HSV-2 to moronic acid was similar to that of the wild-type HSV-1. Thus, moronic acid was purified as a major anti-HSV compound from the herbal extract of *Rhus javanica* (Kurokawa et al., 1999).

Methanolic extracts from dried *Combretum micranthum* G. Don (Combretaceae) leaves were reported to contain antiviral activity against HSV-1 and HSV-2. The precursors of the active compounds have been identified as condensed catechinic tannins. Under alkaline conditions, the condensed tannins are rapid cleaved and rearranged to catechinic acid by auto-oxidation. The alkaline auto-oxidation products of the methanolic extract of *C. micranthum* and those of the synthetic catechinic acid show similar infrared (IR) and ultraviolet (UV) absorption spectra as well as similar anti-HSV-1 and -HSV-2 activities. The EC_{50}s of the catechinic acid auto-oxidation products against HSV-1 and HSV-2 were 2 and 4 μg/ml, respectively (Ferrea et al., 1993).

A series of dimeric procyanidins and some related polyphenols were investigated as model compounds to compare various biological activities in a study of structure–activity relationships. The antiviral activity of these compounds against HSV and HIV were assessed along with antibacterial, superoxide radical-scavenging, and complement-modulating properties. More antiviral activity was seen with epicatechin-containing dimers against HSV, HIV, and radical scavenging. The presence of ortho-trihydroxyl groups in the B-ring was important in compounds exhibiting anti-HSV, radical-scavenging effects, and complement classical pathway inhibition (De Bruyne et al., 1999).

Galangin (3,5,7-trihydroxyflavone) was isolated from the shoots of *Helichrysum aureonitens* Sch. Bip. (Asteraceae), a medicinal plant from southern Africa. The *in vitro* antiviral activity of galangin, a major antimicrobial compound isolated from *H. aureonitens*, was investigated against HSV-1, Coxsackie B virus type 1 (Cox B1), adenovirus type 31 (Ad31), and reovirus. At concentrations ranging from 12 to 47 μg/ml, galangin showed antiviral activity against HSV-1 and CoxB1, limited activity against reovirus, and no antiviral activity against Ad31 (Meyer et al., 1997).

From the root of *Limonium sinense* (Girard) Ktze, a new (2R,3S)-3,5,7,4'- tetrahydroxy-3',5'-dimethoxyflavanone was isolated and named isodihydrosyringetin, together with nine other known compounds, (−)-epigallocatechin 3-*O*-gallate, Sam B, myricetin, myricetin 3-*O*-α-rhamnopyranoside, quercetin 3-*O*-α-rhamnopyranoside, (−)-epigallocatechin, gallic acid, –*trans*-caffeoyltyramine, and –*trans*-feruloyltyramine. All compounds were examined for their inhibitory effects on HSV-1 replication in Vero cells. Both (−)-epigallocatechin 3-*O*-gallate and Sam B exhibited potent inhibitory activities in HSV-1 replication. Comparison of the IC_{50} values indicated that (−)-epigallocatechin 3-*O*-gallate and Sam B had higher inhibitory activities than the positive control acyclovir (38.6 ± 2.6 vs. 55.4 ± 5.3 μM, $p<0.001$; 11.4 ± 0.9 vs. 55.4 ± 5.3 μM, $p<0.0005$). No cell deaths were observed in Vero cells following 5-day treatments with (−)-epigallocatechin 3-*O*-gallate or Sam B indicating that these compounds are not cytotoxic (Lin et al., 2000).

A prenylated flavonol, sophoflavescenol, together with five known flavonoids, kurarinol, kushenol K, kushenol H, trifolirhizin, and kuraidin, were isolated from the roots of *Sophora flavescens*. The structure of sophoflavescenol was determined by spectroscopic analysis. Among the five known flavonoids, kurarinol, kushenol K, and kushenol H showed weak antiviral activity against HSV-1 and HSV-2 (Woo et al., 1998).

The C4-sulfated isoflavonoid, torvanol A, and the steroidal glycoside, torvoside H, together with the known glycoside, torvoside A, were isolated from a MeOH extract of the fruit of *Solanum torvum*. Upon enzymatic hydrolysis with *β*-glucosidase, torvoside A and torvoside H yielded the corresponding acetal derivatives. Torvanol A, torvoside H, and the acetal derivative of torvoside H exhibited anti-HSV-1 activity with IC_{50} values of 9.6, 23.2, and 17.4 μg/ml, respectively. These compounds showed no cytotoxicity at 50 μg/ml against BC, KB, and Vero cell lines (Arthan et al., 2002).

To investigate basidiomycetes as a source of antiviral compounds, two hot-water-soluble extracts and eight methanol-soluble extracts were prepared from carpophores of *Ganoderma lucidum* (Fr.) Karst. (Ganodermataceae). These extracts were examined for their *in vitro* activities against HSV-1, HSV-2, influenza A virus (Flu A), and VSV. Antiviral activities were evaluated using the inhibition of CPE and a plaque reduction assay. Five extracts, GLhw, GLMe-1, -2, -4, and -7 inhibited the cytopathic effects of HSV and VSV. In the plaque reduction assay, GLhw inhibited plaque formation of HSV-2 with EC_{50}s of 590 and 580 μg/ml in Vero and HEp-2 cells. GLMe-4 did not exhibit cytotoxicity up to 1000 μg/ml, while it exhibited potent antiviral activity on the VSV New Jersey strain (Eo et al., 1999a). The investigation of the anti-HSV activity from *G. lucidum* investigated various protein-bound polysaccharides, GLhw, GLhw-01, GLhw-02, and Glhw-03. These materials were isolated from water-soluble substances of the carpophores. The extracts were examined for their antiviral activities against HSV-1 and HSV-2 by a plaque reduction assay. The acidic protein-bound polysaccharide, GLhw-02, exhibited the most potent anti-HSV activity with EC_{50} of 300–520 μg/ml in Vero and HEp-2 cells. GLhw-02 consisted primarily of polysaccharide (~40.6%) and protein (~7.80%) and had the usual molar ratio (C:H:O = 1:2:1) of carbohydrates by elemental analysis (Eo et al., 1999b). The two protein-bound polysaccharides, a neutral protein-bound polysaccharide (NPBP), and an acidic protein-bound polysaccharide (APBP) were isolated from the water-soluble extract of *G. lucidum* by EtOH precipitation and DEAE–cellulose column chromatography. Their antiviral activities against HSV-1 and HSV-2 were investigated by a plaque reduction assay. APBP exhibited more potent HSV-1 and HSV-2 antiviral activity than NPBP with EC_{50} of 300–520 μg/ml. To determine the possible mode of the antiviral activity of APBP, its virucidal effect, antiviral activity in preincubation, attachment, and penetration assays were tested with HSV-1 and HSV-2. APBP was found to have a direct virucidal effect on HSV-1 and HSV-2. APBP was found to inhibit 50% of the attachment of HSV-1 and HSV-2 at concentrations of 100 and 90 μg/ml, respectively, to Vero cells. APBP was also found to prevent penetration of HSV into Vero cells. These results show that the anti-HSV activity of APBP is the inhibition of binding of HSV-specific glycoproteins responsible for the attachment and penetration of the host cells (Eo et al., 2000).

A compound purified from the fruit of *Melia azedarach* exerted an antiviral effect on HSV-1 in Vero cells. The compound was identified as 28-deacetylsendanin (28-DAS). The IC_{50} of 28-DAS for HSV-1 was 1.46 μg/ml. Electron microscopy showed that low electron-dense cores of newly synthesized nucleocapsids remained in swollen nuclei and no extracellular virus particles were observed at 15 hours post infection. This result was confirmed by a plaque assay that few infectious progeny viruses were released from the 28-DAS-treated virus-infected cells at 24 hours post infection. The synthesis of the viral TK was reduced by 28-DAS at early stage. Therefore, 28-DAS inhibited the replication of HSV-1, reduced the synthesis of HSV-1 TK, and led to the formation of defective nucleocapsids (Kim et al., 1999).

The *in vivo* effect of meliacine, an antiviral compound isolated from the leaves of *Melia azedarach* L., was investigated by corneal inoculation of HSV-1 strain KOS in Balb/c mice. Meliacine was administered topically three times a day for 4 days beginning 1 day before inoculation. Infected animals treated with meliacine or controls were observed for the development of stromal keratitis and the clinical scoring was done 14 days post infection. Histological examination of corneas and viral isolation from eyes of infected mice were also performed. It was found that the treatment of HSV-1-induced ocular disease in Balb/c mice and treatment with meliacine significantly reduced the development of clinical disease and histological damage to the corneas. The viral titers in the eyes of treated mice were two logs lower than those of control animals. Mock-infected and treated mice did not reveal any corneal alteration owing to the administration of meliacine. Meliacine was found to exert a strong antiviral action on HSV-1-induced ocular disease in mice with no evidence of toxic effects (Alche et al., 2000).

The antiviral activity of sulfated galactans extracted from the red seaweed *Bostrychia montagnei* was investigated using HSV-1, strain F, HSV-2 strain G, and the TK-deficient strains Field and B2006. Two crude extracts obtained using cold and hot water as well as some fractions obtained by anion-exchange chromatography, inhibited the replication of these strains HSV as determined by a plaque reduction assay. These compounds inhibited HSV only during the adsorption of HSV to the host cells. These extracts were found to be highly selective antiviral substances and they did not affect Vero cell viability. Additionally, they had no direct inactivating effect on virions in a virucidal assay. The antiviral activity could be correlated with the molecular weight and sulfate content of the polysaccharides (Duarte et al., 2001).

The anti-HSV-1 activity of an aqueous extract of *Achyrocline flaccida* was demonstrated *in vitro* and an activity-guided purification process of the extract indicated that negatively charged polysaccharides were responsible for this anti-HSV activity. This anti-HSV-1 activity was exerted early during the viral replication inhibiting the HSV-1 adsorption to host cells (Garcia et al., 1999).

An antiviral compound was isolated from a Chinese herb, *Prunella vulgaris*, by hot-water extraction, ethanol precipitation, and gel permeation column chromatography. The compound was identified by chemical tests to be an anionic polysaccharide. Using a plaque reduction assay, this polysaccharide was active against HSV-1 and HSV-2 at 100 μg/ml, however, it was inactive against cytomegalovirus, the human influenza virus types A and B, poliovirus type 1, and VSV. The IC_{50} of the polysaccharide for HSV-1 and HSV-2 was 10 μg/ml. Clinical isolates and known acyclovir-resistant (TK-deficient or polymerase-mutant) strains of HSV-1 and HSV-2 were also inhibited by the polysaccharide. Preincubation of HSV-1 with the polysaccharide at 4, 25, or 37 °C completely abrogated the infectivity of HSV-1, but pretreatment of Vero cells with the polysaccharide did not protect cells from infection by the virus. The addition of the polysaccharide at 0, 2, 5.5, and 8 hours post infection with HSV-1 at an MOI of five reduced the 20-hour yield of intracellular infectious virus by 100, 99, 99, and 94%, respectively. These results suggest that the polysaccharide may inhibit HSV by competing for cell receptors as well as by some unknown mechanisms after the virus has entered the host cells. The *Prunella* polysaccharide was not cytotoxic to mammalian cells up to the highest concentration tested, 500 μg/ml, and did not show any anti-coagulant activity (Xu et al., 1999).

Witvrouw and De Clercq (1997) review the antiviral activity of sulfated polysaccharides from seaweed. The inhibitory effects of polyanionic substances on the replication of HSV and other viruses were reported almost four decades ago. Shortly after the identification of HIV as the causative agent of AIDS, heparin and other sulfated polysaccharides were found to be potent and selective inhibitors of HIV-1 replication. The activity spectrum of the sulfated polysaccharides has been shown to extend to various enveloped viruses, including HSV and HCMV. Because sexual transmission is responsible for the large majority of genital herpes and HIV infections, polysulfates may be considered as potentially effective in a vaginal formulation to protect against sexually transmitted viruses (Witvrouw and De Clercq, 1997).

A virucidal saponin mixture (maesasaponin mixture B) was isolated from the MeOH extract of the leaves of *Maesa lanceolata* with anti-HSV-1 activity. The maesasaponin mixture B consisted of six homologous oleanane-type triterpenoid saponins 1-6, identified by 1 H, 13C, and 2D NMR spectroscopy and mass spectrometry. The maesasaponin mixture B showed several biological activities expected for saponins. It exhibited a moderate virucidal activity against enveloped viruses (Sindambiwe et al., 1998).

Eugeniin was purified as an anti-HSV compound from the hot-water extract of *Geum japonicum* and another herb, *Syzygium aromaticum*. The effective concentration for 50% plaque reduction of eugeniin for wild-type HSV-1 on Vero cells was 5.0 μg/ml, 13.9-fold lower than its 50% cytotoxic concentration. Eugeniin also inhibited the growth of acyclovir-phosphonoacetic acid-resistant HSV-1, TK-deficient HSV-1 and wild HSV type 2. Eugeniin, as well as phosphonoacetic acid inhibited viral DNA and late viral protein syntheses in their infected Vero cells, but not cellular protein synthesis at its inhibitory concentrations. Purified HSV-1 DNA polymerase activity was inhibited by eugeniin. The apparent K_i value for eugeniin was 8.2- and 5.8-fold lower than the K_i values of purified human DNA polymerases alpha and beta, respectively. Thus, one of the major target sites of inhibitory action of eugeniin is viral DNA synthesis. The inhibitory action for viral DNA polymerase activity was novel compared with anti-HSV nucleoside analogs (Kurokawa et al., 1998).

The leaves of *Aglaia edulis* were found to contain a new bisamide, aglaiduline, and two new sulfur-containing bisamides, aglaithioduline and aglaidithioduline. The structures of these compounds were established from spectroscopic studies. The sulfur-containing amides exhibited slight antiviral activity against HSV-1 and HSV-2 (Saifah et al., 1999).

Antiviral activities of 13 sesquiterpenes isolated from *Tripterygium wilfordii* Hook fil. var. *regelii* Makino were tested against HSV-1. Of these compounds, only triptofordin C-2 showed a selectivity index (SI) of greater than 10. The compound did not affect either adsorption or penetration of HSV-1 to host cells. The compound did have moderate virucidal activity against several enveloped viruses including HSV-1, HCMV, measles virus, and influenza A virus. Triptofordin C-2 suppressed viral protein synthesis of infected cells when added at the early steps of HSV-1 replication and exerted inhibition of translation of the transcripts of the immediate early genes. When acyclovir and triptofordin C-2 were evaluated in combination for antiviral activity against HSV-1 replication, additive antiviral effects were observed for this virus (Hayashi et al., 1996).

These compounds with anti-HSV activity have a broad range of structures: polyphenols, polysaccharides, tannins, flavinoids, etc. Knowing the molecular structure may be very helpful in the development of an antiviral compound. Once the molecular structure is known, modifications can be made to increase the antiviral activity, increase the solubility, as well as reduce the toxicity. All of these features may be extremely important in the development of a clinically useful antiviral compound. Using some of these herbal extracts and compounds, it may be possible to find a compound or series of compounds with good anti-HSV activity that could be developed as a therapeutic agent or as a topical microbicide.

I.B.3. Modes of action of anti-HSV compounds

In some instances, the mechanism of action of the antiviral compounds has been investigated, for example, an acidic protein-bound polysaccharide from *Ganoderma lucidum* inhibited the attachment and penetration of HSV-1 and HSV-2 to Vero cells (Eo et al., 2000). This section presents information on herbal compounds with known antiviral activity and the mode of action of these compounds has been identified.

Recently, it was shown that meliacine exhibits its antiviral activity against HSV-1 by inhibiting specific infected-cell polypeptides (ICPs) produced late in infection. Some ICPs are involved in DNA synthesis and in the assembly of nucleocapsids. These experiments confirmed that meliacine affects late events in the multiplication cycle of HSV-1. Using HSV-1 that expresses β-galactosidase activity, it was shown that meliacine diminished the synthesis of viral DNA and inhibited the spread of infectious viral particles. Ultrastructural analysis of infected cells showed that meliacine treatment results in a large number of unenveloped nucleocapsids accumulated in the cytoplasm and a minor proportion of mature virus was found in cytoplasmic vesicles. These findings suggest that meliacine exerts an antiviral action on both the synthesis of viral DNA and the maturation and egress of HSV-1 during the infection of Vero cells (Alche et al., 2002; Pifarre et al., 2002).

An extract of the cactus plant *Opuntia streptacantha* inhibited intracellular virus replication and inactivated extracellular virus. Inhibition of virus replication also occurred following preinfection treatment. There was inhibition of both DNA and RNA virus replication including HSV, equine herpesvirus, PRV, influenza virus, RSV, and HIV. The extract appeared to be nontoxic to normal cells at concentrations that were 15-fold in excess of 50% viral inhibitory concentrations. The active inhibitory component(s) of the extract appeared to be protein in nature and resided mainly in the wall of the plant rather than in the cuticle or inner sap. The extract was nontoxic when administered orally to mice, horses, and humans. Also, the extract was nontoxic when 70 mg was administered intravenously to a mouse representing at least fifty tissue culture 50% viral inhibitory dosages (Ahmad et al., 1996).

The inhibitory activities of 10 ethanolic extracts from Chinese herbs were investigated to determine the effects on the replication of HSV-1. Using a bioassay-guided fractionation procedure, Sam B was isolated from *Limonium sinense*. Sam B suppressed HSV-1 multiplication in Vero cells without apparent cytotoxicity. To further localize the point in the HSV-1 replication cycle where inhibition occurred, the regulatory events leading to viral multiplication were examined, including viral immediate-early, early, and late gene expression and DNA replication. The results indicated that levels of gB, gC, gD, gG, ICP5 expression, and gB mRNA expression in Vero cells were impeded by Sam B. PCR results showed that the replication of HSV-1 DNA in Vero cells was arrested by Sam B. Sam B also decreased the DNA polymerase, ICP0, and ICP4 gene expression in Vero cells. Results of an electrophoretic mobility shift assay demonstrated that Sam B interrupted the formation of an alpha-trans-induction factor/C1/Oct-1/GARAT multiprotein complex. The mechanisms of antiviral action of Sam B seem to be mediated, at least in part, by inhibiting HSV-1 alpha gene expression, including expression of the ICP0 and ICP4 genes, by blocking beta transcripts such as DNA polymerase mRNA, and by arresting HSV-1 DNA synthesis and structural protein expression (Kuo et al., 2002).

Resveratrol, a phytoalexin, was found to inhibit HSV-1 and HSV-2 replication in a dose-dependent, reversible manner. The observed reduction in virus yield was not caused by the direct inactivation of HSV or inhibition of virus attachment to the cell. Resveratrol did, however, target an early event in the virus replication cycle since it was most effective when added within 1 hour of infection and less effective if added at 6 hours post infection. Resveratrol was not effective if added 9 hours post infection. Resveratrol was also found to delay the cell cycle at S-G2-M interphase and inhibit reactivation of virus from latently infected neurons. It also reduced the amount of ICP4, a major immediate early viral regulatory protein, which is produced when compared to control cells. These results suggest that a critical early event in the viral replication cycle is being adversely affected (Docherty et al., 1999).

The effects of ethanolic extracts from seven Chinese herbs on HSV-1 replication were investigated. Using a bioassay-guided fractionation procedure, PS-A-6 was isolated from *Psychotria serpens*. The active fractions suppressed HSV-1 replication in Vero cells without apparent cytotoxicity. To further localize the point in the HSV-1 replication cycle where inhibition occurred, a set of key regulatory events leading to viral multiplication was examined, including viral gene expression, DNA replication, and structural protein synthesis. Southern blot analysis showed that HSV-1 DNA replication in Vero cells was arrested by PS-A-6. In addition, PS-A-6 decreased thymidine kinase (TK) and ICP27 mRNA expression in the cells. The mechanisms of antiviral action of PS-A-6 seem to be mediated, at least in part, through inhibition of early transcripts of HSV-1, such as tk and ICP27 mRNAs, arresting DNA synthesis and gB gene expression in Vero cells (Kuo et al., 2001).

The *in vitro* anti-HSV-2 activity of Casuarinin, a hydrolyzable tannin isolated from the bark of *Terminalia arjuna* Linn. (Combretaceae), was investigated. The results showed that the IC_{50} of casuarinin in a XTT and plaque reduction assays were 3.6 ± 0.9 and $1.5 \pm 0.2\,\mu M$, respectively. The 50% cytotoxic concentration for cell growth (CC_{50}) was $89 \pm 1\,\mu M$. Thus, the SI of casuarinin was 25 and 59 for XTT and plaque reduction assays, respectively. In an attachment assay, casuarinin was shown to prevent the attachment of HSV- 2 to Vero cells. Also, casuarinin exhibited an antiviral activity by inhibiting the penetration HSV-2 into the host cells. In addition, casuarinin was virucidal at a concentration of 25 μM, reducing viral titers up to 100,000-fold. Casuarinin continued to exhibit antiviral activity even when added 12 hours after infection. This study concluded that casuarinin possesses anti-HSV activity by inhibiting viral attachment and penetration, and also disturbing the late events of virus replication (Cheng et al., 2002).

In cultured mammalian cells (Vero), different antiviral agents change to differing degrees the ability of HSV-2 to down-regulate gap junctions. Measured by intracellular electrodes, control cell populations showed 49–51% coupling, uninfected populations treated with acyclovir or SDS averaged 43–51% coupling, while populations infected with HSV-2 had coupling reduced to 8%. ACV at a concentration of 1.0 μg/ml, which is adequate to suppress viral replication, failed to prevent this down regulation (final coupling ratio of 11%). A plant extract (250 μg/ml) from *Pilostigma thonningii* offered slightly more protection (final coupling ratio of 22%), while SDS (50 μM) afforded nearly complete protection (final coupling ratio of 40%). While SDS was originally believed to alter the viral coat and prevent entry into the cell, these data are in agreement with recent studies indicating that

SDS-treated viruses can enter into host cells, but in a severely diminished condition. These data suggest that the gap junction antagonist is brought into the cells as part of the entering virus (Musee et al., 2002).

Studies of herbal compounds with anti-HSV activity have indicated that there are several modes of action of the various extracts or compounds. It is likely that there may be a spectrum of modes of action that could be exploited in the development of new antiviral compounds. Determining the molecular structure and modes of action are the next steps in the development of antiviral compounds. Finding an extract or compound with broad antiviral activity could be quite useful in the development of a topical microbicide. Alternatively, several antiviral compounds with specific antiviral activity could be combined to make a broad-spectrum preparation that could be used in this manner.

I.B.4. Topical microbicides

Condoms have been regarded as an effective means for the prevention of many STDs; however, the lack of compliance continues to be a major drawback to their efficacy. Topical microbicides preparations that could be applied to the vagina or rectum may be effective in preventing the transmission of STD pathogens. Topical microbicides may be more effective than condoms because they would be easier to use, therefore promoting greater compliance. An ideal topical microbicide would be safe and nonirritating to the mucosal tissues while having antiviral activity against several common viruses. Alternatively, several microbicides with activity against single infectious agents could be combined to form a very effective topical agent. Recently, the use of broad-spectrum, intravaginal microbicides has emerged as a potential new approach to the control of the worldwide epidemic of STDs (Herold et al., 1999; Howett et al., 1999). The female partner would have control over the use of this type of protection, and the ease of use of a topical preparation might promote a higher degree of compliance. Several topical agents have been investigated including benzalkonium chloride (Wainberg et al., 1990), nonoxynol-9, a nonionic surface-active agent (Jennings and Clegg, 1993; Herold et al., 1999), and sodium dodecyl sulfate (Howett et al., 1999). The increased interest in testing these different broad-spectrum topical microbicides speaks to the logic of this approach in the prevention of STDs, however, the potential toxicity of surface-active agents to epithelial cells suggests that other classes of molecules that can be administered by this route need to be identified. A report of a large study in Africa assessing the efficacy of non-oxynol-9 indicated that the risk of HIV infection was higher in women exposed to nonoxynol-9 than a control preparation (Van Damme et al., 2002). The probable cause of the increased risk of HIV infection may be the cellular toxicity of this surface-active agent and the erosion of the mucosal surface (Fichorova et al., 2001; Van Damme et al., 2002). Additional compounds that have been investigated as topical microbicides for the prevention of STDs are sulfated carbohydrate compounds (Herold, 1997) and polystyrene sulfonate (Herold et al., 2000).

There is considerable interest in developing topical microbicides to be used intravaginally by women for protection against STDs. Many compounds derived from plants have been shown to have antimicrobial properties. Nineteen such compounds were tested *in vitro* by a plaque reduction assay to determine their activity against a common sexually transmitted pathogen, HSV-2. Compounds with an $ED_{50} \leqslant 7.0$ mg/ml were tested for efficacy *in vivo*. Four compounds, carrageenan λ type IV, cineole, curcumin, and eugenol, provided significant protection ($p<0.05$) in a mouse model of intravaginal HSV-2 challenge. Eugenol, which provided the greatest protection in mice was also evaluated using the guinea pig model of genital HSV-2 infection where it also demonstrated significant protection (Bourne et al., 1999).

Sexually transmitted diseases are gaining significant importance at present owing to the rapid spread of the diseases, high cost of treatment, and the increased risk of transmission of other STDs including HIV. Because some sexually transmitted viruses such as HSV, HIV, and HPV result in either a latent state or an integrated state, a virologic cure with these viruses may be extremely difficult. Prevention would therefore seem to be the most logical approach, however, vaccines for HSV and HIV are not currently available to prevent infections. Also, current therapies available for symptomatic treatment of STDs and AIDS are expensive and are associated with the emergence of drug resistance.

Medicinal plants have been used for the treatment of many infectious diseases without any scientific evidence. At present there is more emphasis on determining the scientific evidence and rationalization of the use of these preparations. Research is in progress to identify plants and their active principles possessing activity against sexually transmitted pathogens including HIV with an objective of providing an effective approach for prevention of transmission as well as treatment of these diseases. Plants reported to possess activity or used in traditional systems of medicine for prevention and treatment of STDs including AIDS, herbal formulations for vaginal application, and topical microbicides from herbal origin, have been discussed (Vermani and Garg, 2002). A recent review suggests that the heterosexual transmission of HIV is low with each encounter, however, once the infection is established there is little to be done. Topical microbicides and vaccines that lower the risk of transmission may be the best strategies to prevent infection with HIV (Pope and Haase, 2003).

A number of plant-derived compounds have been evaluated against HSV-2 both *in vitro* and *in vivo* and several compounds were found to be promising as potential topical microbicides (Bourne et al., 2000). Vermani and Garg (2002) reviewed many herbal compounds that have been investigated for their activity against STDs and AIDS. Determining the time course for the inhibition of HSV and other sexually transmitted viruses by the various antiviral compounds presented here will be an important aspect in the development of a topical microbicide. Useful topical microbicides need to be inhibitory shortly after they are applied to the mucosal surface and the antiviral activity needs to be maintained for period of time before the activity diminishes. In addition, compounds that are to be considered as topical microbicides should have a finite life span of effectively inhibiting virus infection before the activity diminishes. Topical antiviral agents need to be turned over at the cell surface otherwise there may be adverse complications from compounds that remain attached to cells for extended periods of time.

II. Discussion

Sexually transmitted diseases are among the most common infectious diseases reported to the CDC. STDs are caused by a variety of infectious agents, most commonly bacteria and viruses, which infect millions of people in the U.S. each year. Genital herpes is one of the common STDs and it affects an estimated 60 million people in the U.S. alone. At current rates, at least one person in four will contract an STD at some point in his or her life. It is now believed that people who have an STD are at increased risk of contracting HIV, and HIV infection alone represents a considerable portion of the STD problems.

Frequently recurring genital herpes can be treated with one of the several systemic antiviral drugs. Daily anti-HSV therapy with these nucleoside analogs helps to control the symptoms but do not eliminate the virus from its latent state. Suppressive antiviral therapy can be used to reduce occurrences and prevent transmission. Women who acquire genital herpes during pregnancy can transmit the virus to their newborn and untreated HSV infection in newborns can result in very serious infection and death.

There are probably no fail-safe measures to prevent STDs except abstinence and this may not be an option for some women who are forced to have nonconsensual sex. Effective female-controlled strategies are needed to enhance the ability of women to avoid infection with sexually transmitted viruses. Although drugs are available to treat some STDs, systemic antiviral therapy is expensive and drug resistance may be a consequence. The requirement for daily antiviral therapy for genital herpes suggests that new drugs and new strategies to treat and prevent infections with HSV should be considered. In some cases, antiviral drugs can control symptoms but cannot eliminate the virus, such as herpesviruses and HIV.

Women may be more susceptible than men to certain STDs and the infections in women may be asymptomatic and often go untreated. Therefore, women may suffer more frequent and more severe STD complications such as pelvic inflammatory disease (PID) and cervical cancer. Human papillomavirus (HPV) is another common sexually transmitted virus and cervical cancer is most often related to infections with certain high-risk strains of HPV.

Vaccines are not available for a number of sexually transmitted viruses including HIV and HSV. Until safe and effective vaccines are available, other means of preventing infection need to be available. To meet the needs of women, topical microbicides have become the focus of many investigations. There is an urgent need for the development of antiviral gels, foams, or creams known as topical microbicides, which women could apply intravaginally before engaging in sexual intercourse. These products would work by one of the several mechanisms including inactivating sexually transmitted viruses, by creating a barrier to the viruses, or by blocking the ability of the virus to bind to or enter the host cells. Effective products would give women greater control over the type of protection they use to prevent infections with many sexually transmitted viruses.

Herbal compounds are being investigated as therapeutic agents for a variety of infectious agents. An important application of these compounds would be their use as topical microbicides in preventing the transmission of STDs. Herbal compounds may be nontoxic and ideally suited for clinically evaluating their ability to prevent the transmission of viral STDs such as HSV. In addition, the use of a topical microbicide could also be evaluated for the treatment of recurrent HSV. Topical antiviral agents would be of benefit to many patients with recurrent HSV infection for several reasons including the ease of use, avoiding systemic exposure to drug, and providing greater levels of antiviral compound at the site of virus replication.

References

Ahmad A, Davies J, Randall S, Skinner GRB. (1996) Antiviral properties of extract of *Opuntia streptacantha*. Antiviral Res 30(2–3):75–85.

Alche LE, Barquero AA, Sanjuan NA, Coto CE. (2002) An antiviral principle present in a purified fraction from *Melia azedarach* L. leaf aqueous extract restrains herpes simplex virus type 1 propagation. Phytother Res 16(4):348–52.

Alche LE, Berra A, Veloso MJ, Coto CE. (2000) Treatment with meliacine, a plant derived antiviral, prevents the development of herpetic stromal keratitis in mice. J Med Virol 61(4):474–80.

Arthan D, Svasti J, Kittakoop P, Pittayakhachonwut D, Tanicharoen M, Thebtaranonth Y. (2002) Antiviral isoflavonoid sulfate and steroidal glycosides from the fruits of *Solanum torvum*. Phytochemistry 59(4):459–63.

Beuscher N, Bodinet C, Neumann-Haefelin D, Marston A, Hostettmann K. (1994) Antiviral activity of African medicinal plants. J Ethnopharmacol 42(2):101–9.

Binns SE, Hudson J, Merali S, Arnason JT. (2002) Antiviral activity of characterized extracts from *Echinacea* spp. (Heliantheae: Asteraceae) against herpes simplex virus (HSV-I). Planta Med 68(9):780–3.

Bourne KZ, Bourne N, Reising SF, Stanberry LR. (1999) Plant products as topical microbicide candidates: assessment of in vitro and in vivo activity against herpes simplex virus type 2. Antiviral Res 42(3):219–26.

Bourne N, Stanberry LR, Kern ER, Holan G, Matthews B, Bernstein DI. (2000) Dendrimers, a new class of candidate topical microbicides with activity against herpes simplex virus infection. Antimicrob Agents Chemother 44(9):2471–4.

Bunyapraphatsara N, Dechsree S, Yoosook C, Herunsalee A, Panpisutchai Y. (2000) Anti-herpes simplex virus component isolated from Maclura cochinchinensis. Phytomedicine 6(6):421–4.

Cheng HY, Lin CC, Lin TC. (2002) Antiherpes simplex virus type 2 activity of casuarinin from the bark of *Terminalia arjuna* Linn. Antiviral Res 55(3):447–55.

Chiang LC, Chiang W, Chang MY, Ng LT, Lin CC. (2002) Antiviral activity of *Plantago major* extracts and related compounds in vitro. Antiviral Res 55(1):53–62.

De Bruyne T, Pieters L, Witvrouw M, De Clercq E, Berghe DV, Vlietinck AJ. (1999) Biological evaluation of proanthocyanidin dimers and related polyphenols. J Nat Prod 62(7):954–8.

del Barrio G, Parra F. (2000) Evaluation of the antiviral activity of an aqueous extract from *Phyllanthus orbicularis*. J Ethnopharmacol 72(1–2):317–22.

Docherty JJ, Fu MM, Stiffler BS, Limperos RJ, Pokabla CM, DeLucia AL. (1999) Resveratrol inhibition of herpes simplex virus replication. Antiviral Res 43(3):145–55.

Duarte ME, Noseda DG, Noseda MD, Tulio S, Pujol CA, Damonte EB. (2001) Inhibitory effect of sulfated galactans from the marine alga *Bostrychia montagnei* on herpes simplex virus replication in vitro. Phytomedicine 8(1):53–8.

Elanchezhiyan M, Rajarajan S, Rajendran P, Subramanian S, Thyagarajan SP. (1993) Antiviral properties of the seed extract of an Indian medicinal plant, *Pongamia pinnata*, Linn., against herpes simplex viruses: in-vitro studies on Vero cells. J Med Microbiol 38(4):262–4.

Eo SK, Kim YS, Lee CK, Han SS. (1999a) Antiherpetic activities of various protein bound polysaccharides isolated from *Ganoderma lucidum*. J Ethnopharmacol 68(1–3):175–81.

Eo SK, Kim YS, Lee SK, Han SS. (1999b) Antiviral activities of various water and methanol soluble substances isolated from *Ganoderma lucidum*. J Ethnopharmacol 68(1–3):129–36.

Eo SK, Kim YS, Lee SK, Han SS. (2000) Possible mode of antiviral activity of acidic protein bound polysaccharide isolated from *Ganoderma lucidum* on herpes simplex viruses. J Ethnopharmacol 72(3):475–81.

Esquenazi D, Wigg MD, Miranda MMFS, Rodrigues JM, Tostes JBF, Rozental S, daSilva AJR, Alviano CS. (2002) Antimicrobial and antiviral activities of polyphenolics from *Cocos nucifera* Linn. (Palmae) husk fiber extract. Res Microbiol 153(10):647–52.

Ferrea G, Canessa A, Sampietro F, Cruciani M, Romussi G, Bassetti D. (1993) In vitro activity of a *Combretum micranthum* extract against herpes simplex virus types 1 and 2. Antiviral Res 21(4):317–25.

Fichorova RN, Tucker LD, Anderson DJ. (2001) The molecular basis of nonoxynol-9-induced vaginal inflammation and its possible relevance to human immunodeficiency virus type 1 transmission. J Infect Dis 184(4):418–28.

Fleming DT, McQuillan GM, Johnson RE, Nahmias A, Aral SO, Lee FK, St. Louis ME. (1997) Herpes simplex virus type 2 in the United States, 1976 to 1994. N Engl J Med 337(16):1105–11.

Garcia G, Cavallaro L, Broussails A, Ferraro G, Martino V, Campos R. (1999) Biological and chemical characterization of the fraction with antiherpetic activity from *Achyrocline flaccida*. Planta Med 65(4):343–6.

Goncalves JL, Leitao SG, Delle Monache F, Miranda MMFS, Santos MGM, Romanos MTV, Wigg MD. (2001) In vitro antiviral effect of flavonoid-rich extracts of *Vitex polygama* (Verbenaceae) against acyclovir-resistant herpes simplex virus type 1. Phytomedicine 8(6):477–80.

Gottlieb SL, Douglas Jr. JM, Schmid DS, Bolan G, Latesta M, Malotte CK, Zenilman J, Foster M, Baron AE, Steiner JF, Peterman TA, Kamb JL. (2002) Seroprevalence and correlates of herpes simplex virus type 2 infection in five sexually transmitted-disease clinics. J Infect Dis 186(10):1381–9.

Hayashi K, Hayashi T, Ujita K, Takaishi Y. (1996) Characterization of antiviral activity of a sesquiterpene, triptofordin C-2. J Antimicrob Chemother 37(4):759–68.

Hernandez-Corona A, Nieves I, Meckes M, Chamorro G, Barron BL. (2002) Antiviral activity of *Spirulina maxima* against herpes simplex virus type 2. Antiviral Res 56(3):279–85.

Herold BC. (1997) Therapeutic blockade of sexually transmitted pathogens. Pediatr Infect Dis J 16(10):925–31.

Herold BC, Bourne N, Marcellino D, Kirkpatrick R, Strauss DM, Zaneveld LJ, Waller DP, Anderson RA, Chany CJ, Barham BJ, Stanberry LR, Cooper MD. (2000) Poly(sodium 4-styrene sulfonate): an effective candidate topical antimicrobial for the prevention of sexually transmitted diseases. J Infect Dis 181(2):770–3.

Herold BC, Kirkpatrick R, Marcellino D, Travelstead A, Pilipenko V, Krasa H, Bremer J, Dong LJ, Cooper MD. (1999) Bile salts: natural detergents for the prevention of sexually transmitted diseases. Antimicrob Agents Chemother 43(4):745–51.

Holmberg SD, Stewart JA, Gerber AR, Byers RH, Lee FK, O'Malley PM, Nahmias AJ. (1988) Prior herpes simplex virus type 2 infection as a risk factor for HIV infection. JAMA 259(7):1048–50.

Howett MK, Neely EB, Chirstensen ND, Wigdahl B, Krebs FC, Malamud D, Patrick SD, Pickel MD, Welsh PA, Reed CA, Ward MG, Budgeon LR, Kreider JW. (1999) A broad-spectrum microbicide with virucidal activity against sexually transmitted viruses. Antimicrob Agents Chemother 43(2):314–21.

Jennings R, Clegg A. (1993) The inhibitory effect of spermicidal agents on replication of HSV-2 and HIV-1 in-vitro. J Antimicrob Chemother 32(1):71–82.

Johnson RE, Nahmias AJ, Magder LS, Lee FK, Brooks CA, Snowden CB. (1989) A seroepidemiologic survey of the prevalence of herpes simplex virus type 2 infection in the United States. N Engl J Med 321(1):7–12.

Kim M, Kim SK, Park BN, Lee KH, Min GH, Seoh JY, Park CG, Hwang ES, Cha CY, Kook YH. (1999) Antiviral effects of 28-deacetylsendanin on herpes simplex virus-1 replication. Antiviral Res 43(2):103–12.

Kott V, Barbini L, Cruanes M, de D. Munoz J, Vivot E, Cruanes J, Martino V, Ferraro G, Cavallaro L, Campos R. (1999) Antiviral activity in Argentine medicinal plants. J Ethnopharmacol 64(1):79–84.

Kudi AC, Myint SH. (1999) Antiviral activity of some Nigerian medicinal plant extracts. J Ethnopharmacol 68(1–3):289–94.

Kuo YC, Chen CC, Tsai W-J, Ho Y-H. (2001) Regulation of herpes simplex virus type 1 replication in Vero cells by *Psychotria serpens*: relationship to gene expression, DNA replication, and protein synthesis. Antiviral Res 51(2):95–109.

Kuo YC, Lin LC, Tsai W-J, Chou C-J, Kung S-H, Ho Y-H. (2002) Samarangenin B from *Limonium sinense* suppresses herpes simplex virus type 1 replication in Vero cells by regulation of viral macromolecular synthesis. Antimicrob Agents Chemother 46(9):2854–64.

Kurokawa M, Basnet P, Ohsugi M, Hozumi T, Kadota S, Namba T, Kawana T, Shiraki K. (1999) Anti-herpes simplex virus activity of moronic acid purified from *Rhus javanica* in vitro and in vivo. J Pharmacol Exp Ther 289(1):72–8.

Kurokawa M, Hozumi T, Basnet P, Nakano M, Kadota S, Namba T, Kawana T, Shiraki K. (1998) Purification and characterization of eugeniin as an anti-herpesvirus compound from *Geum japonicum* and *Syzygium aromaticum*. J Pharmacol Exp Ther 284(2):728–35.

Kurokawa M, Nakano M, Ohyama H, Hozumi T, Kageyama S, Namba T, Shiraki K. (1997) Prophylactic efficacy of traditional herbal medicines against recurrent herpes simplex virus type 1 infection from latently infected ganglia in mice. J Dermatol Sci 14(1):76–84.

Kurokawa M, Ochiai H, Nagasaka K, Neki M, Xu H, Kadota S, Sutardjo S, Matsumoto T, Namba T, Shiraki K. (1993) Antiviral traditional medicines against herpes simplex virus (HSV-1), poliovirus, and measles virus in vitro and their therapeutic efficacies for HSV-1 infection in mice. Antiviral Res 22(2–3):175–88.

Lin LC, Kuo YC, Chou C-J. (2000) Anti-herpes simplex virus type-1 flavonoids and a new flavanone from the root of *Limonium sinense*. Planta Med 66(4):333–6.

Lopez A, Hudson JB, Towers GH. (2001) Antiviral and antimicrobial activities of Colombian medicinal plants. J Ethnopharmacol 77(2–3):189–96.

Meyer JJ, Afolayan AJ, Taylor MB, Engelbrecht L. (1996) Inhibition of herpes simplex virus type 1 by aqueous extracts from shoots of *Helichrysum aureonitens* (Asteraceae). J Ethnopharmacol 52(1):41–3.

Meyer JJ, Afolayan AJ, Taylor MB, Erasmus D. (1997) Antiviral activity of galangin isolated from the aerial parts of *Helichrysum aureonitens*. J Ethnopharmacol 56(2):165–9.

Miranda MM, Goncalves JL, Romanos MTV, Silva FP, Pinto L, Silva MH, Ejzemberg R, Granja LFZ, Wigg MD. (2002) Anti-herpes simplex virus effect of a seed extract from the tropical plant *Licania tomentosa* (Benth.) Fritsch (Chrysobalanaceae). Phytomedicine 9(7):641–5.

Musee J, Mbuy GN, Woodruff RI. (2002) Antiviral agents alter ability of HSV-2 to disrupt gap junctional intercellular communication between mammalian cells in vitro. Antiviral Res 56(2):143–51.

Nakano M, Kurokawa M, Hozumi T, Saito A, Ida M, Morohashi M, Namba T, Kawana T. (1998) Suppression of recurrent genital herpes simplex virus type 2 infection by *Rhus javanica* in guinea pigs. Antiviral Res 39(1):25–33.

Nawawi A, Nakamura N, Hattori M, Kurokawa M, Shiraki K. (1999) Inhibitory effects of Indonesian medicinal plants on the infection of herpes simplex virus type 1. Phytother Res 13(1):37–41.

Padma P, Pramod NP, Thyagarajan SP, Khosa RL. (1998) Effect of the extract of *Annona muricata* and *Petunia nyctaginiflora* on Herpes simplex virus. J Ethnopharmacol 61(1):81–3.

Pifarre MP, Berra A, Coto CE, Alche LE. (2002) Therapeutic action of meliacine, a plant-derived antiviral, on HSV-induced ocular disease in mice. Exp Eye Res 75(3):327–34.

Pope M, Haase AT. (2003) Transmission, acute HIV-1 infection and the quest for strategies to prevent infection. Nat Med 9(7):847–52.

Rajbhandari M, Wegner U, Julich M, Schopke T, Mentel R. (2001) Screening of Nepalese medicinal plants for antiviral activity. J Ethnopharmacol 74(3):251–5.

Saifah E, Suttisri R, Shamsub S, Pengsuparp T, Lipipun V. (1999) Bisamides from *Aglaia edulis*. Phytochemistry 52(6):1085–8.

Serkedjieva J, Ivancheva S. (1999) Antiherpes virus activity of extracts from the medicinal plant *Geranium sanguineum* L. J Ethnopharmacol 64(1):59–68.

Severson JL, Tyring SK. (1999) Relation between herpes simplex viruses and human immunodeficiency virus infections. Arch Dermatol 135(11):1393–7.

Simoes CM, Falkenberg M, Auler Mentz L, Schenkel EP, Amoros M, Girre L. (1999) Antiviral activity of south Brazilian medicinal plant extracts. Phytomedicine 6(3):205–14.

Sindambiwe JB, Calomme M, Cos P, Totte J, Pieters L, Vlietinck AJ, Vande Berghe DA. (1999) Screening of seven selected Rwandan medicinal plants for antimicrobial and antiviral activities. J Ethnopharmacol 65(1):71–7.

Sindambiwe JB, Calomme M, Geerts S, Pieters L, Vlietinck AJ, Vanden Berghe DA. (1998) Evaluation of biological activities of triterpenoid saponins from *Maesa lanceolata*. J Nat Prod 61(5):585–90.

Stanberry LR, Spruance SL, Cunningham AL, Bernstein DI, Mindel A, Sacks S, Tyring S, Aoki FY, Slaout M, Denis M, Vandepapeliere P, Dublin G. (2002) Glycoprotein-D-adjuvant vaccine to prevent genital herpes. N Engl J Med 347(21):1652–61.

Suzutani T, Ogasawara M, Yoshida I, Azuma M, Knox YM. (2003) Anti-herpesvirus activity of an extract of *Ribes nigrum* L. Phytother Res 17(6):609–13.

Taylor RS, Hudson JB, Manandhar NP, Towers GH. (1996) Antiviral activities of medicinal plants of southern Nepal. J Ethnopharmacol 53(2):97–104.

Taylor RS, Manandhar NP, Hudson JB, Towers GH. (1996) Antiviral activities of Nepalese medicinal plants. J Ethnopharmacol 52(3):157–63.

Thompson K, Jabbar S, Datta BK, Khan MTH. Antiviral activity of four Bangladeshi medicinal plant extracts against herpes simplex viruses. Unpublished data.

Thompson KD. (1998) Antiviral activity of *Viracea* against acyclovir susceptible and acyclovir resistant strains of herpes simplex virus. Antiviral Res 39(1):55–61.

Van Damme L, Ramjee G, Alary M, Vuylsteke B, Chandeying V, Rees H, Sirivongrangson P, Mukenge-Tshibaka L, Ettiegne-Traore V, Uaheowitchai C, Abdool Karim SS, Masse B, Perriens J, Laga M. (2002) Effectiveness of COL-1492, a nonoxynol-9 vaginal gel, on HIV-1 transmission in female sex workers: a randomised controlled trial. Lancet 360(9338):971–7.

Vermani K, Garg S. (2002) Herbal medicines for sexually transmitted diseases and AIDS. J Ethnopharmacol 80(1):49–66.

Vlietinck AJ, Van Hoof L, Totte J, Lasure A, Vanden Berghe D, Rwangabo PC, Mvukiyumwami J. (1995) Screening of hundred Rwandese medicinal plants for antimicrobial and antiviral properties. J Ethnopharmacol 46(1):31–47.

Wainberg MA, Spira B, Bleau G, Thomas R. (1990) Inactivation of human immunodeficiency virus type 1 in tissue culture fluid and in genital secretions by the spermicide benzalkonium chloride. J Clin Microbiol 28(1):156–8.

Wald A, Corey L, Cone R, Hobson A, Davis G, Zeh J. (1997) Frequent genital herpes simplex virus 2 shedding in immunocompetent women. Effect of acyclovir treatment. J Clin Invest 99(5):1092–7.

Witvrouw M, De Clercq E. (1997) Sulfated polysaccharides extracted from sea algae as potential antiviral drugs. Gen Pharmacol 29(4):497–511.

Woo ER, Kwak JH, Kim HJ, Park H. (1998) A new prenylated flavonol from the roots of *Sophora flavescens*. J Nat Prod 61(12):1552–4.

Xu F, Schillinger JA, Schillinger, Sternberg MR, Johnson RE, Lee FK, Nahmias AJ, Markowitz LE. (2002) Seroprevalence and coinfection with herpes simplex virus type 1 and type 2 in the United States, 1988–1994. J Infect Dis 185(8):1019–24.

Xu HX, Lee SH, Lee SF, White RL, Blay J. (1999) Isolation and characterization of an anti-HSV polysaccharide from *Prunella vulgaris*. Antiviral Res 44(1):43–54.

Yip L, Hudson JB, Towers GH. (1995) Isolation of the anthropogenic compound fluoranthene in a screening of Chinese medicinal plants for antiviral compounds. Planta Med 61(2):187–8.

Yoosook C, Bunyapraphatsara N, Boonyakiat Y, Kantasuk C. (2000) Anti-herpes simplex virus activities of crude water extracts of Thai medicinal plants. Phytomedicine 6(6):411–9.

Yoosook C, Panpisutchai Y, Chaichana S, Santisuk T, Reutrakul V. (1999) Evaluation of anti-HSV-2 activities of *Barleria lupulina* and *Clinacanthus nutans*. J Ethnopharmacol 67(2):179–87.

M.T.H. Khan and A. Ather (eds.)
Lead Molecules from Natural Products

Pharmacological modulation of cough reflex

GABRIELA NOSALOVA, JURAJ MOKRY, SONA FRANOVA

Abstract

The cough reflex is an attack of powerful expiratory efforts produced by active contractions of the expiratory muscles, preceded by deep inspirations.

Cough is a normal physiological defensive reflex responsible for keeping the airways free of obstruction and harmful substances. However, cough is also the most frequent symptom of acute respiratory system diseases and is the most common reason why sick patients visit physicians. The largest diagnostic group associated with chronic cough is asthma or asthma-like syndromes and chronic obstructive pulmonary disease. Furthermore, gastroesophageal reflux, rhinitis/postnasal drip syndrome and unpleasant side effects accompanying the therapy with angiotensin-converting enzyme inhibitors (ACEI) represent other causes of chronic cough. The cough reflex may be elicited by the activation of receptors of non-myelinated nociceptive Aδ-fibers and C-fibers and receptors of myelinated Aδ-fibers distributed throughout the respiratory tract. In recent times as proper cough receptors are thought to be rapidly adapting receptors (RARs) of myelinated fibers, C-fibers and transient receptor potential vanilloid 1 (TRPV1) localize on the non-myelinated fibers.

As the pathological cough (especially its chronic form) has significant impact on patient's quality of life, observed either in physical activity or psychosocial domain, various treatment attitudes are used for different forms of cough (acute, subacute, chronic, productive, nonproductive, psychogenic, asthma-associated, or painful).

Interest of research in this field is accompanied with the fact, that many antitussive drugs (mainly from the opioid group), which have been the antitussives of choice for decades, are limited by their unpleasant and very often adverse reactions.

In this chapter the authors divided the drugs used in the pharmacological treatment of cough into several groups. These include the drugs acting at the level of receptors, the drugs affecting the propagation of cough impulses in afferent nerves, the drugs modulating the central coordination of cough reflex, the drugs acting at the level of efferent nerves, and the drugs affecting the effectors. Efficacy of many antitussive agents is connected with their influence on more than one level of cough reflex.

Another group used in the therapy of cough are mucoactive substances that change the dry, irritant and nonproductive cough into the so-called productive cough with nonirritating expectoration. The last-mentioned group in this chapter is demulcerative and hydrating drugs, which can create a defense layer protecting the mucous membranes from other irritant stimuli.

Keywords: *cough reflex, antitussive drugs, mucoactive substances, demulcerative drugs*

Abbreviations: ACE, angiotensin-converting enzyme; ACEI, angiotensine-converting enzyme inhibitors; GABA, gamaaminobutyric acid; LPh, laryngopharyngeal area; MAOI, monoaminooxidase inhibitors;

NEP, neutral endopeptidase; NKA, neurokinine A; NKB, neurokinine B; NMDA, N-methyl-D-aspartate; OTC, over-the-counter; RARs, rapidly adapting receptors; REM, rapid eye movement; SP, substance P; TB, tracheobronchial area; TRPV1, transient receptor potential vanilloid 1.

I. Introduction

Exchange of gases while breathing forms the major role of the lungs. As the inspired air is markedly polluted, the airways are fit for numerous defense mechanisms. The defense of the respiratory system against inhaled particles and gases involves the integration of many complex physiological, biochemical and immunological processes that interact directly with the properties of inhaled materials (Korpas and Tomori, 1979). The various defense mechanisms are integrated to provide local degradation and detoxication, as well as mechanical elimination of both exogenous substances and products of pathological processes from the airways. The most important defensive reflex of the airways is cough, which, together with the mucociliary transport system, forms the main mechanism for cleaning of the respiratory tract (Korpas and Nosalova, 1991; Chung and Chang, 2002; Belvisi and Geppetti, 2004).

II. Definition of cough

Korpas and Tomori (1979) characterized the cough reactions as an attack of powerful expiratory efforts produced by active contractions of the expiratory muscles, preceded by deep inspirations. The violent expiration, which provides the high flow, helps to shear away mucus and remove foreign particles from the larynx, trachea and large bronchi.

Cough appears when the membrane lining of the respiratory tract produces excessive mucus or phlegm. These secretions help to protect airways from infections and irritants. Coughing prevents the breathing passages from closing and also prevents infected mucus from falling into lungs and bronchial tubes, which could be very dangerous. In the absence of abnormal sputum production there is likely to be another reason for cough. The probable explanation could be an increased sensitivity of the cough reflex, which leads to an abnormal response to "naturally" inhaled stimuli (Fuller and Jackson, 1990; Reynolds et al., 2004).

III. Epidemiological data

Cough is one of the most frequent symptoms of respiratory system diseases and is the most common reason why sick patients visit physicians in the United States (Bolser, 1996; Ziment and O'Connell, 2002). Cough prevalence in populations depends on smoking prevalence and on the level of air pollution. Statistical data are in the range of 5–40% (Fuller and Jackson, 1990; Irwin and Madison, 2002; Wright et al., 2004). Cough is a common complaint of patients presenting to primary care physicians. Most patients with acute cough have self-limited illnesses like common cold and influenza. Gwaltney (1997) showed that the viral infections represent the most frequent causes of self-limited cough. These viruses include rhinovirus, coronavirus, parainfluenza virus, respiratory syncytial virus, adenovirus and influenza virus. These viruses can cause a spectrum of diseases including common cold, acute infections bronchitis, bronchiolitis and bacterial pneumonia. According to Gwaltney (1997) the highest incidence of cough occurred in patients with influenza (reaching 70% on day 3 of the illness) and rhinovirus colds (approximately 35% over the 7-day period of illness). Antitussive drugs such as codeine and dextromethorphan are among the most commonly used prescription and over-the-counter (OTC) drugs in the world (Choundry and Fuller, 1992). The widespread use of antitussive medications for the treatment of cough associated with upper respiratory tract infection generates many millions of dollars for the pharmaceutical industry (Eccles, 1996). In the UK, the cough/cold market accounted for £350 million, whereas in the USA, well over $2 billion was spent on OTC medicines (Morice, 2001). These facts demonstrate the vast socio-economic consequences of acute cough.

Chronic cough is one of the most common symptoms presenting to respiratory physicians and affects 10–38% of patients (Hsu et al., 1994). The largest diagnostic group is asthma or asthma-like syndromes. A second large group connected with cough is gastroesophageal reflux (Plutinsky et al., 1998). Rhinitis/postnasal drip syndrome represents another cause of chronic cough. Predominantly this kind of cough seems to occur (3–40%) accompanied by unpleasant side effects due to therapy using ACE inhibitors (Kubota et al., 1996).

IV. Physiology of cough reflex

Cough reflex may be elicited by the activation of non-myelinated C-fibers and myelinated Aδ-fibers distributed throughout the respiratory tract (Fox, 1996a,b; Reynolds et al., 2004). These nerve endings have strictly vagal origin. They are located underneath the airways epithelium and demonstrate high sensitivity to inhaled irritants. Coughing is a reflex activity initiated by stimulation of sensory nerves in the lining of the respiratory passages, the tubes we use to breathe. Coughing can be induced from the larynx and tracheobronchial tree, but not directly from structures above or below these sites (Widdicombe, 1998). In a healthy man, cough is present only when, for example, an insufficiency of mucociliar apparatus and alveolar macrophages prevents them from fulfilling their cleaning role in airways. An example of such a situation could be when solid or liquid food particles accidentally get into the airways, or through inhalation of irritant gases (Widdicombe, 1995).

Knowledge of structures participating in the genesis of this defense mechanism is potentially important for antitussive therapy, because the mechanism of both peripherally and centrally acting antitussive drugs may depend on the identity of the afferent pathways involved.

V. Larynx and pharynx – laryngopharyngeal cough

The larynx, being the sentinel of the lungs, possesses abundant sensory innervations, which can produce violent coughing after their activation. Afferent activity may be elicited from the larynx by mechanical and chemical irritants. Most of the sensory traffic from the larynx is conveyed in the superior laryngeal nerves (Fuller and Jackson, 1990; Sant'Ambrogio and Sant'Ambrogio, 1996). Three types of receptors – pressure, drive and cold – take part in a clear respiratory modulation. Pressure receptors respond to changes in trans-laryngeal pressure, drive receptors are stimulated by passive or active motion of the larynx and cold/flow receptors respond to a decrease in laryngeal temperature. Although respiratory-modulated receptors play an important role in the function of the upper airways, they are not generally viewed as a primary factor in the elicitation of cough reflex. There are two types of receptors localized in laryngeal mucus, both of which are connected with the cough reflex. These are:

Rapidly adapting receptors (RARs) or irritant receptors. The cough reflex is probably caused by the stimulation of the irregular firing irritant receptors. They are usually silent in quiet breathing and activated only by mechanical and chemical irritant stimuli (e.g. cigarette smoke, distilled water, CO_2, etc.), the prompt blocking effect of topical anesthetics are thought to have a superficial location (Widdicombe et al., 1988). Laryngeal irritant receptors are stimulated by halothane (which inhibits those receptors in tracheobronchial tree) and by water solutions lacking chloride anions (Anderson et al., 1990; Sant'Ambrogio, 1996), but not by hyposmolal solutions in the trachea (Ventresca et al., 1990).

C-fiber receptors. They are sensitive to mechanical stimulation, cooling, chemical stimulation by capsaicin and phenylbiguanide. However, we must respect the existence of considerable differences in species.

The characteristic signs of the laryngopharyngeal cough are active expirations and several fast and powerful inspiratory movements. It differs from tracheobronchial cough in the presence of the inspiratory component, which is in this case as strong as, or even stronger, than the active expiratory component (Figure 1). From the pharmacological frame of reference, this type of cough is more resistant to cough suppressant agents. From our long-standing results we can conclude that the influence of antitussive agents varies very strongly. We found that some of the substances can suppress the cough from the laryngopharyngeal area more potently than the others, which were more effective in tracheobronchial cough suppression (Table 1).

VI. The tracheobronchial tree – tracheobronchial cough

There are three types of sensory nerve endings in the tracheobronchial region of the respiratory tract:

(a) *Rapidly adapting stretch receptors (RARs) or "irritant"receptors.* Regarding the cough reflex this type of receptor seems to play a very important role. RARs respond to mechanical and chemical stimuli (prostaglandin E_2, histamine, cigarette smoke etc.). All of the stimuli that can induce coughing are able to stimulate RARs, although most of them activate bronchial C-fiber receptor, too (Karlsson, 1996; Reynolds et al., 2004). "Irritant" receptors are situated within the mucosal surface of the major airways in high concentrations, especially at bifurcations. These places are responsible for mucus clearing and expelling foreign materials from the airways. Myelinated fibers in the vagal nerves conduct the sensory information. It is interesting that some viral infections can enhance their sensitivity to stimuli, causing the cough (Empey et al., 1976; Chung et al., 2003).

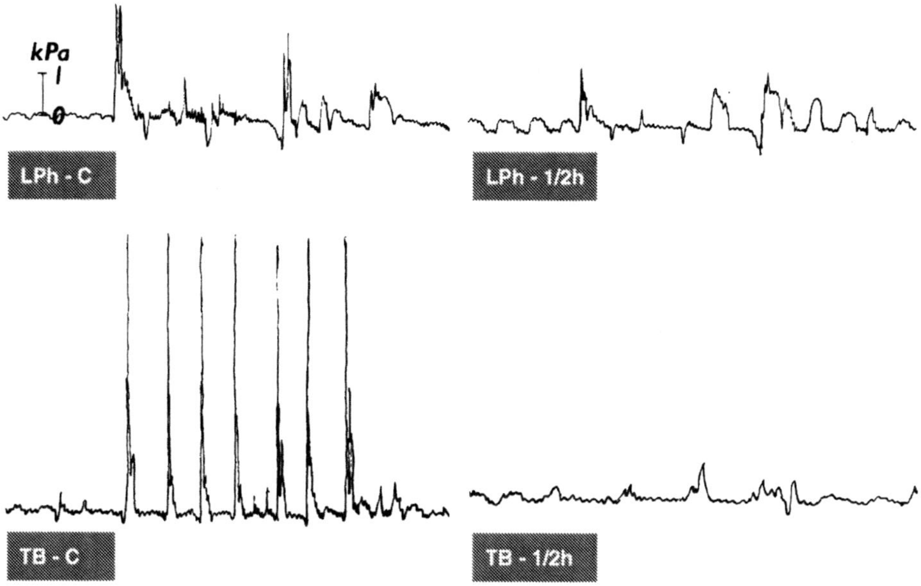

Fig. 1. Cough records from laryngopharyngeal (LPh) and tracheobronchial (TB) areas of the airways before (C, control) and 30 min after (1/2 h) application of codeine in dose of 10 mg/kg i.p.

Table 1
Substances divided according to the area with stronger effect. LPh – laryngopharyngeal, TB – tracheobronchial area of the airways

Higher efficacy in LPh area	Higher efficacy in TB area
Procaine	Tramadol
Azipranon	Tilidin
Glaucine	Pentazocine
Butamirate citrate	Codeine
Dropropizine	L-propoxyphene
Isoprenaline	Dextromethorphan
Guaifenesin	Silomat
Diazepam	Gabalid

(b) *Slowly adapting stretch receptors.* These are located in the membranous posterior wall of conducting airways within the smooth muscle. They are supposed to be important for the act of coughing.

(c) *Tracheobronchial C-fiber receptors.* The stimulation of bronchial C-fiber receptors can cause cough, as well as apnea and bronchoconstriction, presumably in response to different varieties or concentrations of stimulants. All the stimuli used for irritation of C-fiber receptors can also activate RARs. Many cough-inducing chemical stimuli (like bradykinin, sulfur dioxide and capsaicin) may act directly on the RARs, but in addition they can activate C-fiber receptors, which release tachykinins. These will in turn act on postcapillary venules, causing plasma extravasations and stimulation of adjacent RARs. The increase in interstitial liquid volume might also affect the structure of epithelium and stimulate the branches of RARs there (Widdicombe, 1995). In recent times TRPV1 receptors have been identified on Aδ nociceptive fibers, which under normal physiological conditions do not synthesize neuropeptides, but can be activated by capsaicin. Reynolds et al. (2004) suggested that C-fibers do not incite cough per se but might be involved in the sensitization of the cough reflex.

(d) *Pulmonary C-fiber receptors.* Raj et al. (1995) showed that pulmonary C-fiber receptors cause cough in unanesthetized humans. However, further experiments are needed for wider acceptance of this fact.

VII. Central nervous mechanisms in cough

There is clear evidence that airway sensory afferents, irrespective of whether they are myelinated or non-myelinated, terminate within the brainstem, in the nucleus tractus solitarius (Jordan, 1996, 1997; Takahama, 2003). Vagal afferents have been shown to contain mediators such as glutamate and 5-hydroxytryptamine. It is also a well-known fact that NMDA receptors, opiate receptors and 5-HT receptors are able to modulate transmission through the nucleus tractus solitarius. This region is anatomically adjacent to the area postrema, where the blood–brain barrier is weak or absent, so antitussive drugs given peripherally could be acting at the central site, on the sensory side of the system. Other areas, such as nucleus ambiguus and dorsal vagal nuclei, also contain these types of receptors and could be considered as additional sites of action. Apart from this, activation of NMDA receptors and both the opioid and serotonergic systems can modify the central respiratory network (Chung et al., 2003). This is also supported by our results with substances known for their ability to influence these receptors (Figure 2; Nosalova, 1998). The antitussive actions of these agents may occur via this respiratory action (Kamei, 1996).

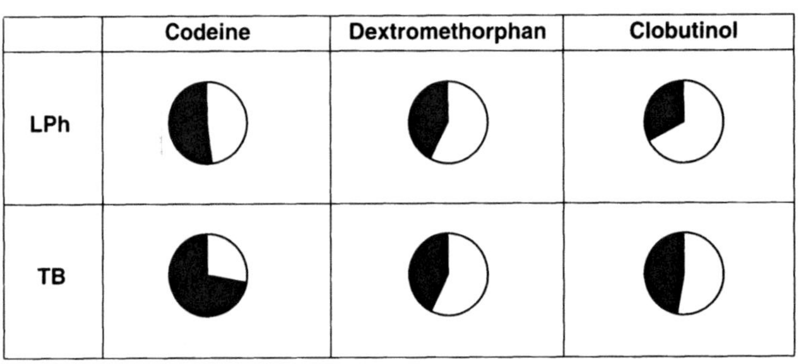

Fig. 2. Centrally acting antitussive drugs – comparison of antitussive activity. Codeine – opiate receptor antagonist 10 mg/kg i.p., Dextromethorphan – NMDA receptor antagonist 2 mg/kg i.p., Clobutinol 20 mg/kg i.p. LPh – laryngopharyngeal and TB – tracheobronchial area of the airways.

Despite huge endeavors of experimental researchers, all of the neurochemical processes participating in the mechanism of action of antitussive agents are still not known. Hedner et al. (1980) highlight an important role of gamaaminobutyric acid (GABA) in tonic modulation of respiratory activity. As the breathing and cough centers are tightly associated, we decided to verify the ability of gabaergic substances to influence the cough reflex. We obtained original priority results (Nosalova et al., 1986a,b, 1987), which support the statement that gabaergic substances are able to suppress cough (Figure 3) and that gabaergic mechanisms can take part in mechanism of action of other cough-suppressing agents.

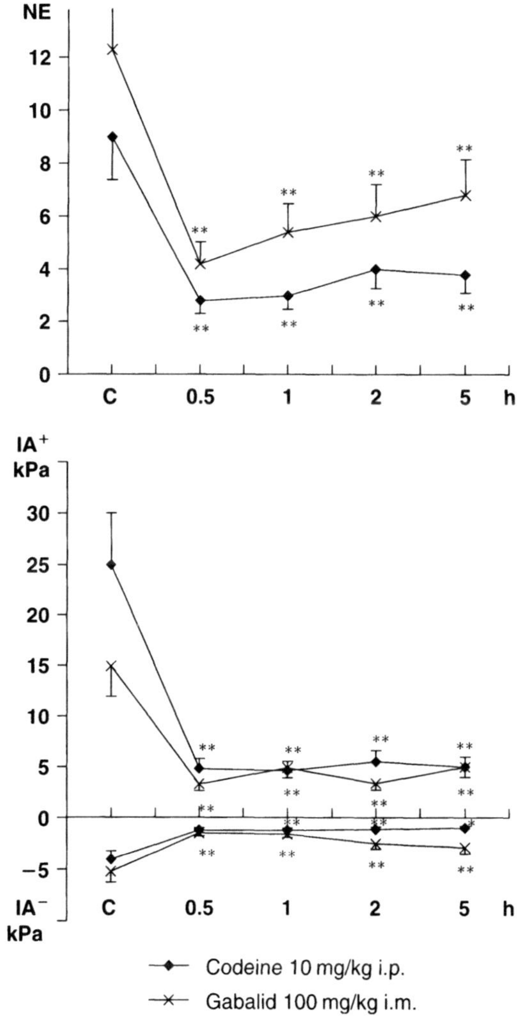

Fig. 3. Antitussive effect of Gabalid (gabaergic receptor agonist) compared to codeine. NE – number of cough efforts after mechanical stimulation of tracheobronchial area of the airways, IA^+ – intensity of cough attack in expirium, IA^- – intensity of cough attack in inspirium, C – control, 0.5, 1, 2, 5 h – time after administration of tested substance.

VIII. Pathophysiology of cough

Cough is considered to be a pathological reflex from the moment it stops fulfilling its cleaning role and starts to burden the patient. Pathological cough (especially its chronic form) has a significant impact on the patient's quality of life, observed either in physical activity or in the psychosocial domain. Coughing patients suffer often from insomnia, exhaustion, nausea and emesis, worsening of performance at work or incontinence. In many of them it can lead to social problems connected to the need for changing or leaving a job or important social activities (Irwin, 2001). In addition to this, the intrathoracal pressure changes during the cough (inspirium −13 kPa and exspirium +30 kPa) can cause costal fractures, cough syncope, pneumothorax, pneumomediastinum, herniation, etc. Patients with coronary lesions may experience anginal pain. Cough also represents one of the ways for transmission of infections (Korpas and Nosalova, 1991).

The most frequent cause of abnormal cough reaction is the presence of pathological processes in the respiratory system, leading to increased irritability of afferent nerve endings or increased sensitivity to different stimuli (Schuligoi et al. 1998; Shinagawa et al., 2000). These conditions can be found during inflammatory diseases of upper and lower airways of either viral or bacterial origin.

IX. Cough and inflammation

Inflammatory process in the airways increases mucus production, which, owing to impaired mucociliar transportation, could be one of the reasons for mechanical irritation. However, more important is the fact that accumulated phlegm rich in inflammatory cells and mediators functions as a cough-provoking chemical stimulus.

Many inflammatory mediators may modulate cholinergic and sensory nerves in the airways through the activation of receptors on nerve terminals (Barnes, 1992; Reynolds et al., 2004). Sensory nerves in the airways contain several neuropeptides – the tachykinins. The main members of this family are substance P (SP), neurokinin A (NKA) and neurokinin B (NKB), among which substance P is the best known. Stimulation of these nerves causes a constellation of responses known as neurogenic inflammation. Distribution of substance-P-containing nerves is close to that of the RARs mediating cough in the airways (Hay, 2001). The most important enzyme for the breakdown of SP in these nerve fibers has been found (Sekizawa, 1996). Neutral endopeptidase (NEP) is a key enzyme in the degradation of tachykinins in the airways. This enzyme is strongly expressed in epithelial cells (Barnes, 2001). In humans, aerosol of SP did not cause cough at the concentration of up to 10^{-5} M in normal subjects, but they did in patients with upper-respiratory tract infection (Katsumata et al., 1989).

Another factor, the attenuation of the barrier function of the airways epithelium, is involved in the inflammation. This way, the irritants can get to the nerve endings more easily. This process is aggravated by total destruction of epithelial cells, leading to loss of another important role – production of NEP.

During inflammation of the airways, damaged epithelium is not able to form NEP, which may facilitate the penetration of substance P. Fox (1996b) showed, that SP causes cough by direct stimulation of rapidly adapting pulmonary stretch receptors (RARs). Apart from this, tachykinins and especially SP also showed indirect activity resulting in cough reflex facilitation (Advenier, 1996). SP stimulates mucus secretion from submucosal glands and mediates the increased plasma exudation and the vasodilator response to tachykinins. Tachykinins also enhance cholinergic neurotransmission by facilitating acetylcholine release at cholinergic nerve terminals and by enhancing ganglion transmission (Watson et al., 1993). A variety of species, including human tachykinins, stimulate C-fiber receptors and rapidly adapting pulmonary stretch receptors and can cause cough (Widdicombe, 1996). Tachykinins may also interact with inflammatory and immune cells.

Besides the mechanism mentioned already, increased sensitivity of the afferent nerve fibers can also cause the genesis of pathological cough. This occurs when in an individual a dry cough is response to such concentrations of tussigen stimuli, which in normal healthy individuals do not provoke cough response. O'Connell (2001) showed that the cough reflex is "upregulated" in persistent dry cough. This upregulation occurs at the level of the afferent cough receptors in the airways. These are the action site of the most important drugs, which lead to downregulation of the abnormally sensitive cough receptors.

Increased as well as decreased reactivity of afferent nerve endings, leading to the so-called ineffective type of cough, is considered to be a pathological state (Laurence et al., 1997; Korpas and Tomori, 1979). The other reasons for this type of cough may be central structures disorders, which take part in cough reflex regulation, or disorder of effectors mechanisms. The consequence of ineffective cough can manifest in development and intensification of pathological processes in the airways (Nishino et al., 1996).

Understanding the mechanisms participating in cough-reflex genesis, (especially the cough with pathological character) enables its appropriate treatment. The cough, fulfilling a physiological role and cleaning the airways of accumulated secretions, phlegm, tissue detritus, dirt and foreign bodies in the airways and the ciliary mechanisms cannot be pharmacologically influenced (Nosalova et al., 1989; Sada, 1997). The best we can do is to help with expectoration (Nosalova et al., 1986a,b, 1998). On the contrary, the pathological type of cough has to be influenced pharmacologically. We can suppress it, hold it at an acceptable level, or modify its effectiveness.

X. Curiosities of cough reflex

The cough reflex with its characteristics is one of the strangest reflexes of all. It is one of the airway-defense reflexes and also can be evoked voluntarily (Lavigne et al., 1991; Lee et al., 2002). Hutching et al. (1993) showed that cough could be not only initiated but also inhibited by conscious control. There is also evidence, that cough can only occur during consciousness and does not occur during REM phase of sleep (Anderson et al., 1996). Cough is the first respiratory response inhibited by general anesthesia without any prior slowing down of respiratory rate.

From the clinical aspect it is particularly important that cough is present as a symptom by most respiratory system diseases (Nosal and Banovcin, 2001; Chung et al., 2003). Cough is often registered as the only symptom of respiratory diseases (Koh et al., 1999).

The particularity of cough reflex is determined also by the fact that if it is disproportionately intensive or persisting, it changes to a pathological reflex and starts to burden the patient. It disturbs daily activities like sleep, food intake, normal breathing and circulation and disables the individual in society by noise or by incontinence. This leads to a significant worsening of the quality of the patient's life. Cough can induce complications, that will jeopardize human life. Therefore, this pathological kind of cough has to be suppressed or held at an acceptable level (Korpas and Nosalova, 1991; Nosalova, 1998; Nosalova et al., 2004).

Another particularity of cough reflex could be its adverse action on normal physiology. The increased activity of muscles involved in breathing leads to increase in intake of oxygen. When oxygen supply is impaired owing to principal respiratory system disease, the tissues are lacking it.

Cough is a reflex that may be initiated voluntarily without any input from afferent vagal fibers. Conversely, cough induced by incoming vagal impulses may be voluntarily suppressed (O'Connell, 2001).

XI. Cough forms

Depending on the cause and duration, cough can be divided into three categories: acute cough, lasting less than three weeks; subacute cough, lasting three to eight weeks; and chronic cough, lasting more than eight weeks. Since all types of cough are acute at the outset, it is the duration of the cough at the time of presentation that determines the spectrum of likely causes. Depending on the phlegm (mucus) production, the cough can be called as dry or wet (productive) (Parvez et al., 1996). Painful cough, the name for the pain-eliciting cough, sometimes accompanied by symptoms of various pathological processes in the respiratory tract such as bronchial asthma, chronic bronchitis, bronchogenic carcinoma, gastroesophageal reflux, etc. It could possibly have a psychogenic basis (Lavigne et al., 1991). Cough can occur as an adverse effect of drug treatments, e.g. after the administration of ACEI (Franova and Nosalova, 1998; Gajdos and Huttova, 2002).

XI.A. Acute cough

Acute cough is mostly mild and lasts for a very short time (up to 21 days). It occurs during acute diseases of respiratory tract caused especially by viral infections, common cold, acute bacterial sinusitis, exacerbations of chronic obstructive pulmonary disease, allergic rhinitis, rhinitis due to environmental irritants and pertussis. In the initial phase it is non-productive (dry), but it can later change into a productive one. Its duration is usually several days and it weakens gradually. This kind of cough often does not need to be modulated therapeutically, as the drugs have no impact on disease outcome (Irwin et al., 1993; Chung et al., 2003). The drugs are recommended if the cough attacks are too frequent and so significant that they extensively lessen the patient's comfort, or if the cough is persistent and painful. In these situations, drugs that suppress and lower the irritation of inflamed mucous membranes are used facilitating daily comfort and tranquil sleep (Braga and Allegra, 1989; Kurz, 1989; Korpas and Nosalova, 1991; Parvez et al., 1996).

Korppi et al. (1991) recommended that the use of antitussives in children should be restricted to such situations where their efficacy has been proven, i.e. in the treatment of chronic non-productive cough.

For treatment of this form of cough we recommend a combination of older-generation H_1 antihistamines and nasal decongestants. For patients who cannot take and tolerate this combination it is possible to use intranasal ipratropium bromide (0.06% spray). The newer generation of antihistamines is ineffective for treatment of cough caused by common cold. These agents are effective when cough is a symptom during histamine-mediated conditions such as allergic rhinitis. In this case we can also use nasal cromolyn and corticosteroids. Irwin and Madison (2000) recommend the usage of antibiotic therapy for patients with exacerbations of chronic obstructive pulmonary disease (if acute cough is accompanied by worsening shortness of breath, wheezing, or both), acute bacterial sinusitis, pneumonia and Bordetella pertussis infection. We deem it very important to say that antibiotic treatment should not be determined generally, such as for 3 weeks, but should be individualized. The duration of the treatment depends on symptoms, remission and normalization of patient's health constitution (well-being).

XI.B. Subacute cough

If the cough lasts for 3–8 weeks, it is considered to be the subacute form of cough. The treatment of this kind of cough is based on the elimination of the cough-provoking cause. Practically, it does not differ from the treatment of chronic form of cough. For elimination of subacute form of cough we use older-generation antihistamines combined with nasal decongestants, ipratropium bromide, bronchodilators from the group of β-agonists, corticosteroids, antitussives and the target antibiotic treatment, if necessary.

XI.C. Chronic cough

Chronic cough is defined as a cough that lasts for more than three weeks (Irwin et al., 1990; Cap, 1999). Many physicians accept a longer limit: such as six to eight weeks. As for diagnosis, this form of cough signals the possibility of serious disease, such as chronic bronchitis, bronchial tumor, lung abscess, blood stasis in small circulation, etc. According to Philp (1997), more than 90% of cases of chronic cough are a result of five common causes: smoking, postnasal drip, asthma, gastroesophageal reflux and chronic bronchitis. We prefer specific therapy in the treatment of chronic cough (Table 2). The success of the specific therapy depends on the correct diagnosis of the cause of the cough mechanisms, that arouse it (Korpas and Nosalova, 1991; Bolser, 1996; Nosalova, 1998; Nosalova, 2001).

Symptomatic treatment of chronic cough (also called nonspecific therapy) is indicated only in cases where the cough does not fulfill its role and the complications could represent real or possible danger for the patient (Korpas and Nosalova, 1991; Irwin et al., 1998). We prefer the drugs from the group of non-narcotic antitussives with peripheral site of action (Table 3). In our conditions these drugs showed lower

Table 2
Management of chronic cough

Cause	Therapy	Remark
Cigarette smoking	Cessation of smoking	Leads to a dramatic decrease in cough within 1 month
Occupational exposure (e.g. dust, fumes, other irritants)	Reduction of exposure	Wearing a face mask, improving air circulation, a change of job may be necessary
Postnasal-drip syndromes Nonallergic rhinitis	Combination of older-generation of antihistamines and decongestants	Newer generation of histamine antagonists are inferior in treating cough but avoid sedation
	Ipratropium (0.06%) nasal spray for 3 weeks	Mainly for patients who cannot take the older-generation of antihistamines
	Nasal steroids	Nasal steroids should be added, if cough is not controlled by antihistamine-decongestant medication or persist 1–2 weeks
	Vasoconstrictor e.g., oxymetazolone	Vasoconstrictor should not be used for more than 5 days
Chronic bacterial sinusitis	Antibiotics Older-generation antihistamine-decongestant combination	If sinus infection is suspected, appropriate antibiotics may also be ordered. The selection and duration of the antibiotic treatment is individual.
Allergic rhinitis	Avoidance of offending allergens, Newer-generation antihistamines	
Hypersensitivity that follows an upper respiratory infection (as in the so-called cough variant asthma)	Antihistamines, inhaled steroids – if unresponsive to treatment with an antihistamine, dextromethorphan or codeine	Present only on a chronic, usually non-productive cough, a positive result on metacholine challenge, physical examination out of periods with acute symptoms is essentially normal
Chronic bronchitis	Discontinuation of smoking Avoidance of enviromental irritants and toxins	In this case we prefer ipratropium in the therapy of cough, because it decreases mucus production, dilates the bronchi and is more effective in the therapy of cough than beta-agonists

Table 2 (*continued*)

Cause	Therapy	Remark
	Preventive health measures (e.g. immunizations with pneumococcal vaccine, annual influenza vaccinations) Treatment of community-acquired respiratory infections Optimal bronchodilatory therapy, Postural drainage and hydration, Correction of malnutrition, Oral steroid therapy, if it is necessary Inhaled ipratropium	
Asthma bronchiaie	Avoidance of allergens, Prophylactic inhalation: Cromolyn, Beta-agonist and/or Steroid inhalers, or Oral corticosteroids, if required	Typical syndromes: —dyspnea —coughing —wheezing Some times it is necessary to have long-term maintenance therapy with anti-inflammatory drugs.
Drugs induction: -Angiotensin-converting enzyme inhibitors (ACEI)	-Withdrawal of drug, -Sulindac, -Indometacin, -Calcium channel blockers (e.g. nifedipine, dilthiazem), -Alternative class of drugs,	Cough occurs in 5–20% of patients treated with ACEI,
Beta blockers	-Withdrawal of drug, -Substitute a drug from a different class	Beta-blockers can cause increased airway resistance resulting from unopposed parasympathetic activity
Gastroesophageal reflux (GERD)	High doses of proton-pump inhibitors, e.g. omeprazole	Omeprazole, in a dose of 80 mg per day
	-Anticholinergic drugs	As is necessary
	-Calcium channel blockers	As is necessary
	-Theophylline	As is necessary
	-Other muscle relaxants	As is necessary
	-Protective agent, e.g. sucralfate	Sucralfate may be helpful in a dose of 1 g taken one hour before meals
	-Prokinetic agent, e.g. metoclopramide or cisapride before meals and at bedtime	Metoclopramide or cisapride may be added before meals and at bedtime, necessary to avoid eating or drinking for 2 h before sleeping

Table 3
Classification of peripherally acting antitussive drugs

Peripherally acting antitussives	
Butamirate citrate	Intussin Sinecod Tussin
Butamirate citrate + guaifenesin	Stoptussin
Dropropizine	Ditustat
Pentoxyverine	Sedotussin
Prenoxdiazine`	Libexin

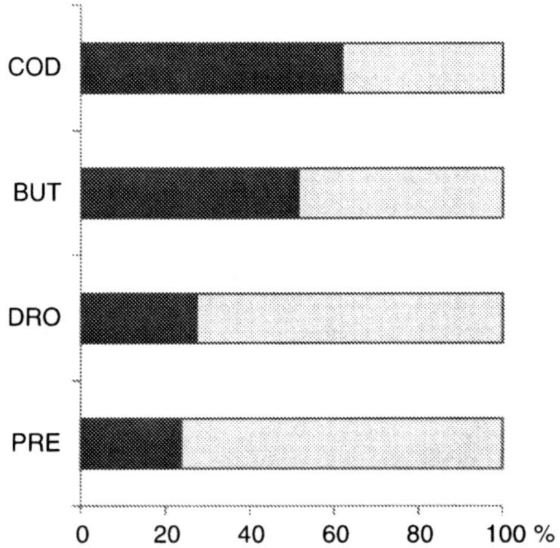

Fig. 4. Antitussive activity of peripherally acting antitussive agents compared to codeine. COD – codeine 10 mg/kg i.p., BUT – butamirate citrate 5 mg/kg i.p., DRO – dropropizine 100 mg/kg i.p., PRE – prenoxdiazine 30 mg/kg i.p.

antitussive activity than the so-called codeine-type agents (Figure 4), but their administration was not associated with unwanted effects. From the centrally acting drugs we prefer agents from the group of synthetic morphine derivatives, or clobutinol (Table 4, groups B and C).

XI.D. Nonproductive cough

Dry, irritant, nonproductive cough significantly burdens the afflicted patient. The cough usually repeats, as rapid air expulsion irritates tracheal as well as tracheobronchial mucous membranes. In this situation the cough must be suppressed by antitussives. Codeine showed especially high efficacy (Figure 5). Adding of expectorants is also very useful. They diminish the cough by increasing the fluid content in the respiratory tract and hence calm the mucous membranes of the airway, helping the patient to breathe (Korpas and Nosalova, 1991).

Table 4
Classification of centrally acting antitussive drugs

Centrally acting antitussives		
A. Opium derivatives and alkaloids	Codeine	Codein
	Combined preparations	Ipecarin
		Benadryl with codein
		Benadryl N with codein
		Codipront
		Kodynal
	Etylmorphine	Diolan
	Pholcodine	Neocodin
		Evaphol
		Galenphol
		Pholcomed
B. Synthetic morphine derivatives	Dextromethorphan	Cosylan
		Pertussin robitussin
		Rhinotussal
C. Other substances	Clobutinol	Silomat

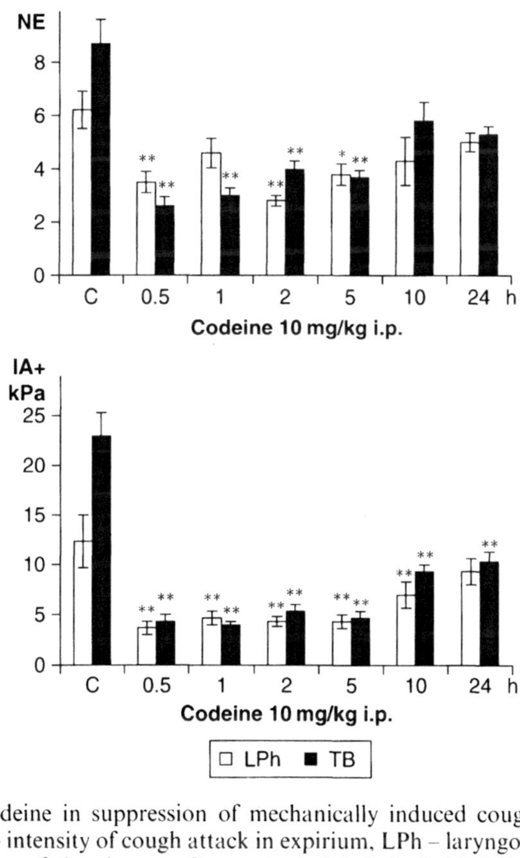

Fig. 5. Effect of codeine in suppression of mechanically induced cough. NE – number of cough efforts, IA^+ – intensity of cough attack in expirium, LPh – laryngopharyngeal and TB – tracheobronchial area of the airways, C – control, 0.5, 1, 2, 5, 10, 24 h – time after administration of codeine in dose of 10 mg/kg i.p.

XI.E. Psychogenic cough

Nonproductive, irritating cough characterizes psychogenic cough. Its incidence is very often in young people during pubescence, often after an underlying disease subsides. It can occur without any clear cause (Hutching et al., 1993). This form of cough does not respond to antitussive or other therapy and is typically not present during eating or talking. If the patient feels that he is being observed, the cough accentuates. Agents from the group of anxiolytics such as guaifenesin (Guajacuran, Figure 6), or combinations of anxiolytics, with antitussives like guaifenesin + butamirate citrate in Stoptussin are used in the treatment of psychogenic cough (Nosalova et al., 1986a; Korpas and Nosalova, 1991; Parvez et al., 1996). We found very strong cough-suppressive effect in benzodiazepine and phenotiazine derivatives (Figure 6). These agents could also be used in the treatment of cough induced by psychogenic factors. In addition, antitussive action could be beneficially used for its tranquilizing effect. At the same time we have to be very careful while administering these drugs in patients who need to enforce expectoration. Special attention must be granted to diazepam ordination in pediatric practice, because in acute respiratory diseases associated with fever it very often used to be a part of polyvalent therapy. Its application could lead to suppression of breathing, consecutive phlegm stasis and global worsening of health of the patient. Psychotherapy, is very useful too. But the essential condition is to take the patient away from the surroundings in which he coughs.

XI.F. Cough in asthma bronchiale

Bronchial asthma is very often characterized by the only one symptom – irritant, nonproductive cough, which very often precedes the classical symptoms picture of asthma. In these cases the use of drugs from the group of beta$_2$-sympathomimetics or corticosteroids (Bush, 2002; Nosal, 2003) that also suppress the cough in allergic patient with bronchial hyperresponsiveness is recommended (Barnes, 1996; Hupka et al., 1996). Beta$_2$ sympathomimetics, similar to other bronchodilators (Zibolen et al., 1999), can reduce bronchial hyperresponsiveness, stimulate mucociliary clearance and increase water and ion flow into bronchial lumen. This leads to enforcing of

%	NE		IA^+		IA^-	
	LPh	TB	LPh	TB	LPh	TB
GUI	51,8	31,8	58,7	47,5	43,5	47,2
DIA	33,4	26,9	38,6	45,4	38,3	11,3
CHL	29,0	39,5	52,1	55,8	44,3	46,5

Fig. 6. Antitussive effects of guaifenesin 100 mg/kg i.p. (GUI), diazepam 0.3 mg/kg i.p. (DIA), and chlorpromazine 5 mg/kg i.p. (CHL) estimated by mechanically induced cough in conscious cats (% of decrease). NE – number of cough efforts, IA^+ – intensity of cough attack in expirium, IA^- – intensity of cough attack in inspirium, LPh – laryngopharyngeal and TB – tracheobronchial area of the airways.

expectoration of viscous secretions. Many authors advise a combination therapy of beta$_2$-sympatomimetics with dextromethorphan in the cough therapy of asthmatics.

XI.G. Productive cough

Productive cough must not be completely suppressed when the retention of secretions could occur, leading to a worsening of the global health constitution of the patient or to the development of pneumonia or atelectasis (Braga et al., 1989; Fox, 1996b; Parvez et al., 1996). In some cases, productive cough can be hyperactive and often repeating. It burdens the patient and disturbs the sleep. The reduction of cough frequency and intensity by expectorants or antitussive–expectorant agents, respectively, is indicated in this situation (butamirate citrate – Sinecod, guaifenesin + butamirate citrate – Stoptussin, ephedrine – Solutan, Ipecarin, etc.). Mucoactive agents are also very useful (bromhexine – Bromhexin, Bisolvon, ambroxol – Mucosolvan, carbocysteine – Mucopront, etc.), as they reduce the phlegm viscosity (Table 5). Furthermore, it is advisable to add agents from the group of secretomotorics to the therapy, which act by inducing of more effective phlegm expulsion from lower parts of the respiratory tract (Braga et al., 1989; Korpas and Nosalova, 1991; Rubin, 2003). Also many plant extracts and derivatives exert their antitussive effect by expectorant and mucolytic activity (see next chapter – Franova et al.: Phytotherapy of the cough).

XI.H. Painful cough

There are certain kinds of cough owing to which the patients feel pain in the throat or in the chest, especially behind the sternum. Painful cough is observed in pleuritis and lung cancer (Homsi et al., 2001). Out of our results (Figure 7) arises the unambiguous fact that in these patients, the group of antitussives with analgesic activity (Nosalova et al., 1985), represents the best therapy for suppression of cough. Alternatively, analgesics with a cough-suppressive effect (tilidine chloride – Valoron, tramadol – Tramal, pentazocine chloride – Fortral, and butorphanol – Beforal) may be used. As in these diseases analgesic agents are often prescribed for pain relief, they can simultaneously and adequately suppress cough as well as pain (Strapkova et al., 1987, 1988).

XII. Choice of drugs

The drugs that selectively inhibit or lessen cough reflex are called antitussive agents. The moderation of cough is a result of the reduction of number or frequency of cough efforts, a decrease in their intensity, or both (Empey, 1996). Pharmacologically, it is very important to know that cough as a reflex can be modulated by agents acting at various levels of the reflex arc: at the level of receptors, afferent nerves, cough center, efferent nerves and effectors, as well as by alteration of bronchial secretion (Reynolds et al., 2004). Some agents are able to influence more levels, with predominance in one of them.

XII.A. Drugs acting at the level of receptors

According to the thinking nowadays, cough reflex is triggered by the irritation of nerve endings that are present under the airway mucous membrane. They are called rapidly adapting irritant receptors and mediate the cough reflex after mechanical stimulation. In smaller bronchi there is another group of epithelial receptors with medially rapid adaptive ability. These receptors mediate the cough evoked by chemical stimuli (Widdicombe, 1995).

The anatomical localization of cough receptors in the airways enables the use of local anesthetics (Nosalova and Strapkova, 1989; Reynolds et al., 2004), which could

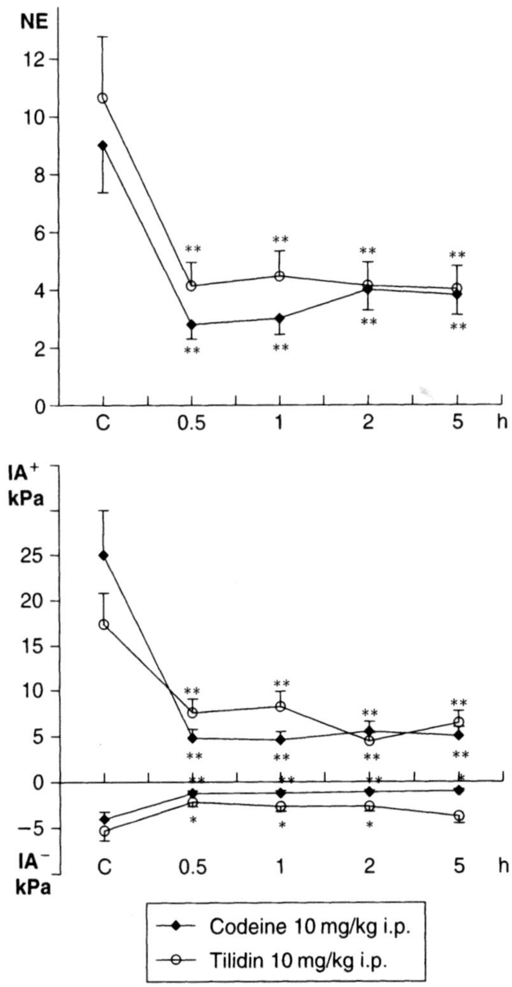

Fig. 7. Antitussive effect of tilidin compared to codeine. NE – number of cough efforts after mechanical stimulation of tracheobronchial area of the airways, IA^+ – intensity of cough attack in expirium, IA^- – intensity of cough attack in inspirium, C – control, 0.5, 1, 2, 5 h – time after administration of tested substance.

be ideal from the pharmacological point of view, but they are not selective enough. The inhalation of anesthetizing aerosols leads to desensitization of cough receptors as well as other segments of mucous membranes in the mouth, and the upper and lower parts of the respiratory system, and increases the risk of aspiration (Sevecova and Calkovska, 2002; Sevecova et al., 2002). Inhalant procaine, lidocaine and bupivacaine inhibit the cough reflex elicited by both mechanical and chemical irritation. In clinical settings, the local anesthetics are used for the elimination of cough before diagnostic procedures such as laryngoscopy, bronchoscopy, intubations, etc. In the mechanism of action of some cough-suppressing agents (benzonatate – Tessalon, dropropizine – Ditustat) a local anesthetic activity takes place apart from other properties (Nosalova and Strapkova, 1989; Korpas and Nosalova, 1991).

XII.B. Drugs affecting the propagation of cough impulses in afferent nerves

Information from cough receptors localized in the airways are spread through various branches of vagal and glossopharyngeal nerves. In cough evoked by unilateral irritation due to bronchial tumor it is possible in certain conditions to use lead anesthesia of vagal nerve stem.

XII.C. Drugs modulating the central coordination of cough reflex

The drugs that suppress the cough reflex by central mechanism are listed in Table 4. In the past, alkaloids and opium derivatives were the only group of drugs the physicians used for cough suppression. The antitussive activity of these agents is associated with their ability to lower the sensitivity of the cough center to nerve impulses coming from airways receptors (Fox, 1996b). Although this group can suppress the cough very potently (Figure 2), the contemporary trend worldwide is to limit their use as they also suppress the breathing center. This unwanted effect occurs especially in children. The other negative property is the diminished activity of bronchial glands, leading to an increase in phlegm viscosity and worsening of expectoration. A very important adverse effect is their ability to induce drug dependence.

Dextromethorphan represents the most important member of the group of synthetic morphine derivatives (Rhinotussal, Romilar), which show, according to our results, excellent antitussive activity (Strapkova et al., 1987). They do not induce drug dependence and their significantly suppressive effect on cough does not suppress the breathing center. Adverse effects, which occur in less than 1% of people, include drowsiness, dizziness, nausea, constipation and abdominal discomfort. Dextromethorphan is contraindicated in a person taking monoamine oxidase inhibitors (MAOI).

Another drug with central cough-suppressing effect is clobutinol (Silomat). Clobutinol does not have the addictive properties of codeine and it does not suppress the breathing center: rather, stimulates it. It is suitable for cough therapy to treat hemoptysis and severe pertussis and to conduct endoscopical procedures in respiratory system.

XII.D. Drugs acting at the level of efferent nerves

Efferent nerves conduct the impulses activating the expiratory skeletal musculature, laryngeal muscles, tracheobronchial smooth muscles and secreting apparatus. Cough can be suppressed by pharmacodynamic elimination of any part. The most significant suppression can be achieved by myorelaxants, which block the expiratory muscles. However, this method is not recommended because of its serious adverse effects. Some authors include ganglioplegics, mostly hexamethonium, in this group.

XII.E. Drugs affecting the effectors

Lately, bronchodilatants have started to be used in the therapy of cough. Bronchodilating action is one effect of dropropizine (Ditustat), L-dropropizine (Levopront) and butamirate citrate (Sinecod), which suppress cough by relaxing the bronchial muscles and facilitating the expectoration. In addition to their bronchospasmolytic and secretomotoric activity, these drugs also have an anti-inflammatory effect. The stimulatory effect on the breathing center and the absence of drug dependence are of great importance (Nosalova et al., 1984; Korpas and Nosalova, 1991).

Another agent with significant bronchodilating activity is pentoxyverine (Sedotussin), which acts at the level of the peripheral parasympathetic nerve endings, similar to atropine.

XII.F. Mucoactive substances

Mucoactive substances are representatives of a specific group of drugs. They do not affect the cough receptors directly, but modify the physical and chemical properties of bronchial secretions in order to facilitate their motion in an oral direction (Rubin, 2003). Their therapeutic goal is to change dry, irritant and nonproductive cough to so-called productive cough with nonirritating expectoration (Strapkova, 2000). The effects of Sirupus althaeae (Nosalova et al., 1992, 1993) and other herbal substances, mentioned in Chapter 7 are especially beneficial. This group of drugs (Table 5) is used in clinical conditions for the treatment of cough with hypersecretion phenomenon (Braga et al., 1989).

XII.G. Demulcerative and hydrating drugs

Demulcerative and hydrating drugs are water-soluble substances with high molecular weight, but without any pharmacological activity. Despite this they are used for the treatment of upper airway inflammations, sore throat and other diseases associated with dryness in mouth, irritation in throat and cough. The representatives of this group are glycerin, liquorices, synthetic cellulose derivatives, sugar syrup, honey, etc. The goal of this kind of treatment is to create a defense layer protecting the mucous membranes from other irritant stimuli (Braga et al., 1989; Franova et al., 1998). Various plants are rich in this kind of agents and are discussed in Chapter 7.

Table 5
Classification of mucoactive drugs according to direct or indirect effects on bronchial secretions (by Braga and Allegra, 1989)

Mucoactive substances		
Directly acting	Drugs destroying mucous polymers	Thiols – *N*-acetylcystein (ACC, Broncholysin, L-Cimexyl, Solmucol), Mesna (Mistabron)
		Enzymes—Trypsin, chymotrypsin, streptokinase other – Urea, citric acid, Hypertonic salt solutions, Anorganic iodides
Indirectly acting	Drugs modifying mucin biochemistry and mucus secretion	Carbocysteine (Fenorin, Mucopront, Mucodyn)
		Bromhexine (Bisolvon, Bromhexin, Flegamina, Paxirasol)
	Drugs modifying the adhesiveness of the gel layer	Ambroxol (Ambrobene, Ambrosan, Bronchopront, Mucosolvan)
		Tyloxapol (Tacholiquin)
	Drugs modifying the sol layer and hydration	Water
		Sodium salts
		Potassium salts
	Volatile inhalants and balsams	Terpenes
		Phenol derivatives
	Drugs stimulating gastropulmonary reflex	Ammonium chloride
		Guaifenesin (Guajacuran)
		Ipecacuanha
		Sodium citrate
	Drugs modifying bronchial glands activity	Sympathomimetics
		Parasymphathomimects
		Corticosteroids
		Antihistamines

References

Advenier C. (1996) Tachykinin antagonists and cough. Abstract of Symposium on Cough: Methods and Mechanism, London.

Anderson CA, Dick TE, Orem J. (1996) Respiratory responses to tracheobronchial stimulation during sleep and wakefulness in the adult cat. Sleep 19:472–8.

Anderson JW, Sant'Ambrogio FB, Mathew OP, Sant'Ambrogio G. (1990) Water-responsive laryngeal receptors in the dog are not specialized endings. Respir Physiol 79:33–44.

Barnes JP. (1992) Modulation of neurotransmission in airways. Physiol Rev 72:699–729.

Barnes PJ. (2001) Neurogenic inflammation in the airways. Respir Physiol 125:145–54.

Belvisi MG, Geppetti P. (2004) Cough 7: current and future drugs for the treatment of chronic cough. Thorax 59 438–440.

Bolser DC. (1996) Mechanisms of action of central and peripheral antitussive drugs. Pulm Pharmacol Therapeut 9:357–64.

Braga PC, Allegra L. (1989) Cough. New York: Raven Press.

Bush A. (2002) Paediatric problems of cough. Pulm Pharmacol Therapeut 15:309–15.

Cap P. (1999) Chronic persistent cough. Medicina 4:7–8.

Choundry NB, Fuller RW. (1992) Sensitivity of the cough reflex in patients with chronic cough. Eur Respir J 5:296–300.

Chung KF, Chang AB. (2002) Therapy of cough: active agents. Pulm Pharmacol Therapeut 15:335–8.

Chung KF, Widdicombe JG, Boushey HA. (2003) Cough: causes, mechanisms, and therapy. London: Blackwell Publishing Ltd.
Eccles, R. (1996). Codeine, cough and upper respiratory tract infection. Abstract of Symposium on Cough: Methods and Mechanism. London.
Empey, D.W. (1996) Antitussive drugs. In: Munson PL, Mueller RA, Breese GR: Principles of Pharmacology. ITP, pp. 615–9.
Empey DW, Laitinen LA, Jacobs L, Gold WM, Nadel JA. (1976) Mechanism of bronchial hyperreactivity in normal subjects after upper respiratory tract infection. Am Rev Respir Dis 113:131–9.
Fox AJ. (1996a) Modulation of airway receptors and cough. Abstract of Symposium on Cough: Methods and Mechanism, London.
Fox AJ. (1996b) Modulation of cough and airway sensory fibres. Pulm Pharmacol 9:335–42.
Franova S, Nosalova G. (1998) ACE inhibitors and defense reflexes of the airways. Acta Physiol Hung 8:359–66.
Franova S, Kardosova A, Kostalova D. (1998) Herbal polysaccharides in the therapy of cough. Bratisl Lek Listy 99:108–10.
Fuller RW, Jackson DM. (1990) Physiology and treatment of cough. Thorax 45:425–30.
Gajdos M, Huttova D. (2002) Angiotensin II AT_1 receptor blockers and diabetes mellitus. Cardiology 11:S35–40.
Gwaltney J. (1997) Clinical and mechanistic perspectives on acute self-limited cough. Int Pharm (FIP) 11(Suppl):5–7.
Hay DWP. (2001) Tachykinins and cough. Abstract of 2nd International Symposium on Cough: Pharmacology and Therapy. London October, 25–27.
Hedner J, Hedner T, Bergman B, Lundberg D. (1980) Respiratory depression by GABA-ergic drugs in the preterm rabbit. J Dev Physiol 2:401–7.
Homsi J, Walsh D, Nelson KA. (2001) Important drugs for cough in advanced cancer. Supportive Care Cancer 9:565–74.
Hsu JY, Stone RA, Logan-Sinclair RB, Worsdel M, Busst CM, Chung KF. (1994) Coughing frequency in patients with persistent cough: assessment using a 24 hour ambulatory recorder. Eur Respir J 7:1246–53.
Hupka V, Kapellerova A, Raskova J. (1996) Bronchial hyperresponsiveness in children after mycoplasmic pneumonia. Stud Pneumol Phtiseol 56:19–21.
Hutching HA, Eccles R, Smith AP, Jawad M. (1993) Voluntary cough suppression as an indication of symptom severity in upper respiratory tract infections. Eur Respir J 6:1449–54.
Irwin RS. (2001) Quality of life in coughers. Abstract of 2nd International Symposium on Cough: Pharmacology and Therapy, London, October.
Irwin RS, Madison JM. (2000) The diagnosis and treatment of cough. New England Journal 343:1715–21.
Irwin RS, Madison JM. (2002) The persistently troublesome cough. Am J Resp Crit Care Med 165:1469–74.
Irwin RS, Curley FJ, Benneti FM. (1993) Appropriate use of antitussives and protussives. Drugs 46:80–91.
Irwin RS, Curley FJ, French CL. (1990) Chronic cough: the spectrum and frequency of causes, key components of the diagnostic evaluation, and outcome of specific therapy. Am Rev Respir Dis 141:640–7.
Irwin RS, Boulet LP, Cloutier MM, Fuller RW, Gold PM, Hoffstein V, Ing AJ, McCool FD, OByrne P, Poe RH, Prakash UBS, Pratter MR, Rubin BK. (1998) Managing cough as a defense mechanism and a symptom. A consensus panel report of the American College of Chest Physicians. Chest 114:1538–918.
Jordan D. (1996) Central nervous mechanisms in cough. Pulm Pharmacol 9:389–92.
Jordan D. (1997) Central nervous control of the airways. In: Jordan D editor. Central nervous control of autonomic function. London: Harwood Academic Press, pp. 63–107.
Kamei J. (1996) Role of opioidergic and serotonergic mechanisms in cough and antitussives. Pulm Pharmacol 9:349–56.
Karlsson JA. (1996) The role of capsaicin-sensitive C-fibre afferent nerves in the cough reflex. Pulm Pharmacol 9:315–22.

Katsumata U, Sekizawa K, Inoue H, Sasaki H, Takishima T. (1989) Inhibitory actions of procaterol, a beta-2 stimulant, on substance P-induced cough in normal subjects during upper respiratory tract infection. Tohoku J Exp Med 158:105–6.
Koh YY, Jeong JH, Park Y, Im CK. (1999) Development of wheezing in patients with cough variant asthma during an increase in airway responsiveness. Eur Respir J 14:302–8.
Korpas J, Nosalova G. (1991) Pharmacotherapy of the cough. Osveta: Martin.
Korpas J, Tomori Z. (1979) Cough and other respiratory reflexes. Progress in Respiration Research, vol.12, Basel: S. Karger.
Korppi M, Laurikainen K, Pietikainen M, Silvasti M. (1991) Antitussives in the treatment of acute transient cough in children. Acta Pediatr Scand 80:869–971.
Kubota K, Kubota N, Pearce GL, Inman WHW. (1996) ACE-inhibitor-induced cough, an adverse drug reaction unrecognized for several years: studies in prescription-event monitoring. Eur J Clin Pharmacol 49:431–7.
Kurz H. (1989) Antitussiva und Expektoranzien. Wissenschaftliche Verlagsgesellschaft mbH Stuttgart.
Laurence DR, Bennett PN, Brown MJ. (1997) Clinical pharmacology, 8th edition, Edinburgh: Churchill Livingstone.
Lavigne JV, Davis AT, Fauber R. (1991) Behavioral management of psychogenic cough. Alternative to the "Bedsheet" and other aversive techniques. Pediatrics, 87:532–7.
Lee PCL, Cotterill-Jones C, Eccles R. (2002) Voluntary control of cough. Pulm Pharmacol Therapeutics 15:317–20.
Morice AH. (2001) Epidemiology of cough. Abstract of 2nd International Symposium on Cough: Pharmacology and Therapy, London, October.
Nishino T, Tagaito Y, Isono S. (1996) Cough and other reflexes in irritation of the airway mucosa in man. Pulm Pharmacol 9:285–93.
Nosal S. (2003) Comparison of the bronchodilatory effects of salbutamol in children with asthma bronchiale and obstructive bronchitis. Acta Pneumol Allergol Pediatr 6:18–23.
Nosal S, Banovcin P. (2001) Selected chapters of pediatrics, 8th edition, JLF UK Martin, pp. 29–31.
Nosalova G. (1998) Mechanism of action of the drugs influencing the cough reflex. Bratisl Lek Listy 99:531–5.
Nosalova G. (2001) Tussis. Vademecum of pediatrics. Osveta: Martin, pp. 1057–62.
Nosalova G, Strapkova A. (1989) Cough and local anesthetic effect of various substances. Ces Fyziol 38:165.
Nosalova G, Strapkova A, Korpas J. (1984) Butamirate citrate action on cough components. Res Pharmac Bratislava. Incheba 12:45–53.
Nosalova G, Strapkova A, Korpas J. (1985) Study of antitussive effect of analgesic tilidin. Bratisl Lek Listy 84:653–8.
Nosalova G, Strapkova A, Kardosova A, Capek P. (1993) Antitussive activity of a rhamnogalacturonan isolated from the roots of Althaea officinalis L., var. Robusta. J Carbohydrate Chem 12:589–96.
Nosalova G, Strapkova A, Korpas J, Crisciulo D. (1989) Objective assessment of cough suppressants under normal and pathological experimental conditions. Drugs Exptl Clin Res 15:77–81.
Nosalova G, Strapkova A, Korpas J, Kubec F. (1986a) Guaifenesin as a component of new Czechoslovak antitussive–expectorant Stoptussin. Res Pharmac Praha Chemapol 14:69–76.
Nosalova G, Strapkova A, Kardosova A, Capek P, Zathurecky L, Bukovska E. (1992) Antitussive action of extracts and polysaccharides of marshmallow (*Althaea officinalis* L., var. *Robusta*). Pharmazie 47:224–6.
Nosalova G, Sutovska M, Strapkova A, Franova S. (2004) The mechanisms of action of drugs affected the cough reflex. Abstract of 3rd International Symposium on Cough: Acute and Chronic, London.
Nosalova G, Varonos D, Papdopoulous-Daifotis Z. (1986b) Cough and central gabaergic mechanism. Bratisl Lek Listy 85:526–32.
Nosalova G, Varonos D, Papdopoulous-Daifotis Z, Visnovsky P, Strapkova A. (1987) GABA-ergic mechanism in the central control of cough. Acta Physiol Hung 70:189–94.

O'Connell F. Central pathways – unanswered questions. (2001) Abstract of 2nd International Symposium on Cough: Pharmacology and Therapy, London, October, 25–27.
Parvez L, Vaidya M, Sakhardande A, Subburaj S, Rajagopalan TG. (1996) Evaluation of antitussive agents in man. Pulm Pharmacol 9:299–308.
Philp EB. (1997) Chronic cough. Am Fam Phys 56:1353–9.
Plutinsky J, Petras D, Galisova Z, Henzel Z, Bitter K. (1998) Differential diagnosis of cough and gastroesophageal reflux. Abstract book, Martin Days of Respiration. Martin.
Raj H, Singh VK, Annand A, Paintal AS. (1995) Sensory origin of lobeline-induced sensations: a correlative study in man and cat. J Physiol (Lond) 482:235–46.
Reynolds S, Mackenzie AJ, Spina D, Page CP. (2004) The pharmacology of cough. TIPS 25:569–76.
Rubin BK. (2003) Mucoactive agents for the treatment of cough. In: Chung KF, Widdicombe JG, Boushey HA editors. Cough: causes, mechanisms, and therapy. London: Blackwell Publishing, pp. 269–81.
Sada E. (1997) The physiology of cough. International Pharmacy (FIP) 11(Suppl):8–10.
Sant'Ambrogio, G. (1996). Role of laryngeal afferents in cough. Abstract of Symposium on Cough: Methods and Mechanism. London.
Sant'Ambrogio G, Sant'Ambrogio FB. (1996) Role of laryngeal afferents in cough. Sensory mechanism in cough. Pulmonary Pharmacol, 9:309–14.
Schuligoi R, Peskar BA, Donnerer J, Amann R. (1998) Bradykinin-evoked sensitization of neuropeptide release from afferent neurons in the guinea-pig lung. Br J Pharmacol 125:388–92.
Sekizawa K. (1996) Role of substance P in cough. Abstract of Symposium on Cough: Methods and Mechanism, London.
Sevecova D, Calkovska A. (2002) Pathophysiological mechanisms of meconium aspiration syndrome. Ces Slov Pediat 57:183–6.
Sevecova D, Calkovska A, Drgova A, Javorka K. (2002) Surfactant lung lavage in rabbits with meconium aspiration – a pilot study. Acta Med Mart 2:9–14.
Shinagawa K, Kojima M, Ichikawa K, Hiratochi M, Aoyagi S, Akahane M. (2000) Participation of tromboxane A_2 in the cough response in guinea-pigs: antitussive effect of ozagrel. Br J Pharmacol 131:266–70.
Strapkova A. (2000) Antioxidant activity of mucolytics. Farm Obzor 49:231–5.
Strapkova A, Nosalova G, Korpas J. (1987) Relationship of antitussic and analgesic activity of various substances. Bratisl Lek Listy 88:538–45.
Strapkova A, Nosalova G, Korpas J. (1988) Antitussive effect of tramadol. Stud Pneumol Phtiseol Cechoslov 48:377–83.
Takahama K. (2003) Mechanisms of actions of centrally acting antitussives – electrophysiological and neurochemical analysis. In: Chung KF, Widdicombe JG, Boushey HA editors. Cough: causes, mechanisms, and therapy. London: Blackwell Publishing, pp. 225–36.
Ventresca PG, Nichol GM, Barnes PJ, Chung KF. (1990) Inhaled furosemide inhibits cough induced by low chloride content solutions but not by capsaicin. Am Rev Respir Dis 142:143–6.
Watson N, Maclagan J, Barnes JP. (1993) Endogenous tachykinins facilitate transmission through parasympathetic ganglia in guinea-pig trachea. Br J Pharmacol 109:751–9.
Widdicombe JG. (1995) Neurophysiology of the cough reflex. Eur Respir J 8:1193–202.
Widdicombe JG. (1996) Sensory mechanisms. Pulm Pharmacol 9:383–7.
Widdicombe JG. (1998) Afferent receptors in the airways and cough. Respir Physiol 114:5–15.
Widdicombe JG, Sant'Ambrogio G, Mathew OP. (1988) Nerve receptors of the upper airway. In: Mathew OP, Sant'Ambrogio G editors. Respiratory function of the upper airway. New York: Marcel Dekker, pp. 193–232.
Wright CE, Thompson RH, Hull D, Morice AH. (2004) Assessment of antitussive efficacy of dextromethorphan in smoking related cough: objective versus subjective. Thorax 59:4.
Zibolen M, Banovcin P, Nosal S. (1999) Selected chapters of pediatric. 5th edition, JLF UK Martin, pp. 45–9.
Ziment I, O'Connell F. (2002) Clinical needs for cough therapy. Pulm Pharmacol Therapeut 15:293–4.

M.T.H. Khan and A. Ather (eds.)
Lead Molecules from Natural Products

Phytotherapy of cough

Sona Franova, Gabriela Nosalova, Juraj Mokry

Abstract

The problems emerging from the treatment of cough during many types of respiratory diseases by conventional opioid antitussive agents, such as codeine and codeine-like compounds, are well known. In recent years, much effort has been made to create drugs that exhibit minimum side effect on the organism. One of them is the medicinal plants, which are potential source of substances with high-antitussive efficiency with minimal unwanted effects. Recent trends of modern phytotherapy include specification of active substances responsible for therapeutic effect as well as their quantification in the healing drugs, which enables the treatment rationalization, especially the dosing and pursuing of adverse effects. The purpose of this chapter is to give the overview of some medicinal plants and their active compounds with cough-suppressing activity. The common information about antitussive efficiency of selected herbal products are replenished with results of our ongoing research program related to search for potentially antitussive active herbal polysaccharides.

Keywords: *cough, antitussive activity, herbal antitussives, herbal polysaccharides*

I. Introduction

Phytotherapy has a very long tradition in the treatment of respiratory-tract diseases. In time, when medicine had no reliable diagnostic instruments, the diagnostics as well as therapy were based on symptoms, of which cough was one of the most important. A large number of herbal preparations is empirically used in the therapy of cough. Recent trends of modern phytotherapy include specification of active substances responsible for therapeutic effect as well as their quantification in the healing drugs, which enables the treatment rationalization, especially the dosing and pursuing of adverse effects. A permanent hunt for new antitussives from the plant kingdom has its substantiation related to adverse effects of opioid antitussives. The administration of these centrally acting cough-suppressing agents has its importance in some indications, associated especially with painful cough. But during long-lasting applications, the risk of dependence is rising as well as increasing phlegm viscosity, depression of the cough center, and other adverse effects.

The basic mechanism of action of herbal antitussives may be the same as those of orthodox antitussives, most of which originate from herbal predecessors. The mechanisms may include:

- central cough suppression – narcotic, non-narcotic;
- peripheral cough suppression – anesthetic, depletion of substance P;
- receptor/sensory nerve suppressive effect;
- bronchodilator/sensory nerve suppressive effect;
- influence of mucociliary clearance;
- antibacterial, antiviral activity.

In common inflammations of the airways associated with the cough, the administration of herbal products is very beneficial, as they can stimulate expectoration via higher production and expulsion of phlegm with protective effect on the airway mucous membrane and they can also correct the phlegm properties.

The plant expectorants can act in the following ways:

- *Secretolytic* – by increasing phlegm secretion, influencing serous cells of submucosal glands. This leads to production of the phlegm with lower viscosity and its easier to remove by mucociliar clearance. Thin phlegm can easily scavenge the bacterial and corpuscular particules, which are then expectorated faster.
- *Mucolytic* – by modulation of the physical and chemical properties of the phlegm. This leads to the fission of the disulfide bonds in phlegm glycoprotein chains, affecting some of the transport systems, which regulate phlegm composition, thereby decreasing its viscosity.
- *Secretomotoric* – by increasing the cilia motion in the airway ciliar epithelium. This mechanism is responsible for better effectivity of mucociliar clearance, which together with other defense reflexes controls a clean airway (Rang et al., 1999).

The following active substances are responsible for antitussive and expectorant effect in herbal medicinal products.

Saponins. The mechanism of action of saponins belongs to the best-clarified mechanisms of substances from herbal drugs, which can modulate the cough parameters and phlegm quality. Saponins are heterosides and are made of glycid and non-glycid parts. The non-glycid part, the so-called aglycone, is responsible for its pharmacological effects. After peroral administration of therapeutic doses, the saponins irritate vagal nerves reflexively (Baltina, 2003). This leads to increased phlegm secretion in the airways (Korpas and Nosalova, 1991). Additionally, the breathing and the cough center are irritated, resulting in more frequent expectoration. However, higher doses of saponins can irritate the mucous membrane of stomach and intestine leading to emesis, diarrhea, and bleeding.

The best-known drugs containing saponins are represented by *Radix primulae, Herba thymi, R. saponariae*, and *Folium Hederae helicis*.

The administration of *H. helix* leaves extract confirmed under clinical conditions, using double-blind trial, the ability to suppress the cough reflex in patients with chronic bronchitis. The antitussive activity of *F. H. helicis* extract was comparable to cough-suppressive activity of ambroxol (Meyer-Wegener et al., 1993). It is based both on high content of saponins and on flavonoids or tannins presence, respectively.

The mechanism of action of some alkaloids, increasing the phlegm secretion by reflexive mechanism, is similar to saponins. Both emetin and cefaelin are present in the root of South African plant *Uragoga ipecacuanhae*. After long-lasting administration and overdose they can evoke arrhythmias, heart muscle impairment, and convulsions.

Flavonoids. The flavonol glycosides and their aglycones are generally termed flavonoids. There are several types of flavonoids divided according to their basic structure: flavanes (katechin), flavones (luteolin, apigenin, diosmin), isoflavones (genistein, daidazin), flavanols (rutin, quercetin, quercitrin, kaempferol, myricetin, hyperoside), flavanones (hesperidin, naringenin) (Mojzis and Mojzisova, 2001). Flavonoids can inhibit oxidative and reductive processes and decrease the activity of cholinesterase and xanthinoxidase. Both rutin and quercetin inhibit the metabolism of catechol-*O*-methyltransferase (COMT), and so prolong the pharmacodynamic effect of norepinephrine. Rutin slows down the ascorbic acid oxidation and protract the effect of vitamin C in organisms (Mika, 1991). Therapeutically, the effect of flavonoids is used in the treatment of cardiovascular diseases, thromboembolic complications, and renal diseases (Ruf, 1999).

In respiratory systems, the flavonoids show spasmolytic activity. Antiflogistic and antiallergic effect of flavonoids is enhanced by concomitant administration of vitamin C (Chang et al., 1994). Quercetin, pinocembrin, possesses significant bacteriostatic effect to Gram-positive as well as Gram-negative bacteria (Takaisi-Kikuni and Shilcher, 1994). Ramnezin, fizetin, and related antocyans inhibit the growth and replication of tuberculous bacilli. Most of flavonoids mark out by significant antioxidant action (Kelly et al., 1995). All of mentioned flavonoid properties together with antitussive-expectorant activity participate probably in positive and beneficial effect of drugs such as *H. helix*, *Plantago lanceolata*, *Malva sylvestris*, *Polygonium aviculare*, *Primula veris*, *Verbascum densiflorum*, and others in the therapy of respiratory tract diseases.

Essences (aetheric oils). Essences are compounds containing fragrant terpenes. They are volatile agents, causing irritation in many tissues of the organism, such as airway epithelium, by direct stimulation of secreting cells. Simultaneously they can accelerate the movement of ciliar epithelium and have antibacterial and antiphlogistic effects (Burrow et al., 1983). The essence drugs are obtained from *Fructus anisi*, *F. foeniculi*, *F. melissae*, *H. seu*, *F. thymi*. Therapeutically, more effective are the spring outgrowths of the coniferous trees *Pinus mugo*, *P. silvestris*, *Abies alba*. The adverse effects, which can occur after administration of aetheric oils, are nausea, allergic reactions, and renal parenchyma impairment.

Mucilage. In inflammations of upper airways associated with dry irritating cough, the so-called slime drugs are currently very often used. The best known are *Radix*, *Folium et Flos althaeae*, *Folium et Flos malvae*, *Folium plantaginis*. The slime consists of lyophilic colloids of water-soluble macromolecular polysaccharides, and according to mutual ratio creates thin sol up to jelly gel. The slime drugs contacting the airway mucous membrane produce a protective layer on its surface, which diminishes the irritation of cough receptors (rapidly adapting cough receptors, RARs) on myelinated fibers of vagal nerves as well as irritation of nerve endings of non-myelinated C-fibers (Nosalova, 1998). This leads to decreased irritation to cough induced by inflammatory mediators or foreign bodies on impaired mucous membrane.

The single active polysaccharides are probably responsible for antitussive activity of slime. The polysaccharides showed significant cough-suppressive effects under experimental conditions, overlapping the effects of peripherally acting antitussives (Table 1).

Except for significant antitussive activity, glucuronoxylans from herb of *Rudbeckia fulgida* and stem of *Mahonia aquifolium* showed in experimental conditions are immunomodulating activity on innate cellular and humoral immunity (Bukovsky et al., 1998; Kostalova et al., 2001; Ebringerova et al., 2002).

So far no significant adverse effects were observed after administration of slime drugs.

Gums. Gums are natural plant hydrocolloids, translucent, amorphous substances that are frequently produced in higher plants as a protective after injury. An effort has been made to distinguish between mucilage and gums on the basis that gums readily dissolve in water, whereas mucilages form slimy masses. Gums are typically heterogeneous in composition. Upon hydrolysis, arabinose, galactose, glucose, mannose, xylose, and various uronic acids are the most frequently observed components (Tyler et al., 1988). The herbal gums have significant antitussive effect after peroral administration. Krajkovicova et al. (2002) followed the antitussive effect of gum isolated from peach. The results showed that this substance had in dose of 50 mg/kg body wt. higher antitussive effect (38.2%) when was compared to efficacy of peripherally acting antitusives dropropizine (27.4%) and prenoxdiazine (23.7%). The mechanism of cough-suppressing action is probably the same as for mucilage.

Pectins. Pectin is a purified carbohydrate product obtained from the dilute acid extract of the inner portion of the rind of citrus fruit or from the apple pomace. It is a natural hydrophilic colloid, consisting chiefly of partially methoxylated polygalacturonic acids; the main carbohydrate component is linear, 1-4-linked

Table 1
Comparison of the antitussive activity of the mixtures of polysaccharides and rhamnogalacturonans from medicinal plants with clinically used antitussives codeine, prenoxdiazine, and dropropizine

Substance	Antitussive activity (%)[a]	References
Polysaccharide complex (50 mg/kg b.w.) Root of *R. fulgida*	23.5	Nosalova et al. (2000)
Polysaccharide complex (50 mg/kg b.w.) Herb of *R. fulgida*	46.5	Kardosova et al. (1997)
Glucuronoxylan (50 mg/kg b.w.) Herb of *R. fulgida*	48.2	Franova et al. (1998)
Glucuronoxylan (50 mg/kg b.w.) Stems of *M. aquifolium*	40.1	Kardosova et al. (2002)
Codeine (10 mg/kg b.w.)	61.8	
Prenoxdiazine (30 mg/kg b.w.)	23.7	Korpas and Nosalova (1991)
Dropropizine (100 mg/kg b.w.)	27.4	

[a]The antitussive activity is expressed as percentage decrease of all evaluated cough parameters (number of cough efforts, cough frequency, intensity of cough attack during exspirium and inspirium) after drug administration.

D-galacturonan (Tyler et al., 1988). Pectin is classified as a protector of the gastric mucous membrane. In the upper-intestinal tract, pectin possesses a surface area composed of ultramicroscopic particles (micelles) that have the property of colloidal absorption of toxins. The pectins are able to accelerate the blood clotting and have significant antiseptic properties. The mechanism of antitussive action of pectins is not precisely clarified, but under experimental conditions the pectins isolated from citrus fruits (30.2%) showed (dose of 50 mg/kg b.w.) antitussive effect comparable with the effect of peripherally acting antitussives – prenoxdiazine (23.7%) and dropropizine (27.4%) (Franova et al., 1995).

II. *Althaea officinalis* L. (Fam. Malvaceae)

The roots of marshmallow are used internally for soothing sore throats, laryngitis, tonsillitis, and cough. Root tea and syrup are recommended for treating mild inflammation of the gastrointestinal mucous membrane. Marshmallow mucilage also helps to eliminate anaerobic pathogens from the gastrointestinal tract. It is used as a poultice for healing wounds and skin inflammation through external administration (Duke, 1997a).

Forms. Marshmallow whole root, powdered root, root tea, syrup, complex extract of polysaccharides.

Active components. The marshmallow root contains up to 35% of slime (mucilage), 37% of starch, approximately 10% of glucose, 2% of asparagin and betaine, pectin, flavonoids, and minerals.

Therapeutic use. The slime content in *A. officinalis* is responsible for classification to the group of the so-called mucilagineous drugs. The slime is made up of acid polysaccharides containing arabinose, glucose, rhamnose, galactose, and galacturonic acid. The phytotherapy uses slime drugs as a shield agent protecting the nerve endings and relieving the pain. They are useful especially for irritated mucous membranes. The calming and protective effect of slime agents is apparent also in inflamed mucous membranes of upper airways, leading to reflexive irritation that elicits cough. In this case, they are prescribed as antitussives. The antitussive activity is caused probably by the presence of active substances in the marshmallow slime. One of these active substances is acid heteropolysaccharide, rhamnogalacturonan, which forms approximately 30% of the slime from root. Nosalova et al. (1992) followed the antitussive activity of rhamnogalacturonan under experimental conditions and compared it with cough-suppressing effect of *Sirupus althaeae*, complex water extract, and slime (mucilage) from the root of this plant (Figure 1). The cough was induced by mechanical stimulation of laryngopharyngeal and tracheobronchial mucous areas of the airways in conscious cats, using thin nylon fiber. The cough parameters were recorded 0.5, 1, 2 and 5 h after administration of the drugs.

In these experiments, rhamnogalacturonan caused a statistically significant decrease in the number of cough efforts, and intensity of cough attacks, in expirium and inspirium, from both laryngopharyngeal and tracheobronchial regions. Administration of rhamnogalacturonan noticeably decreased all cough parameters, except for intensity of maximum cough efforts. These findings are considered significant

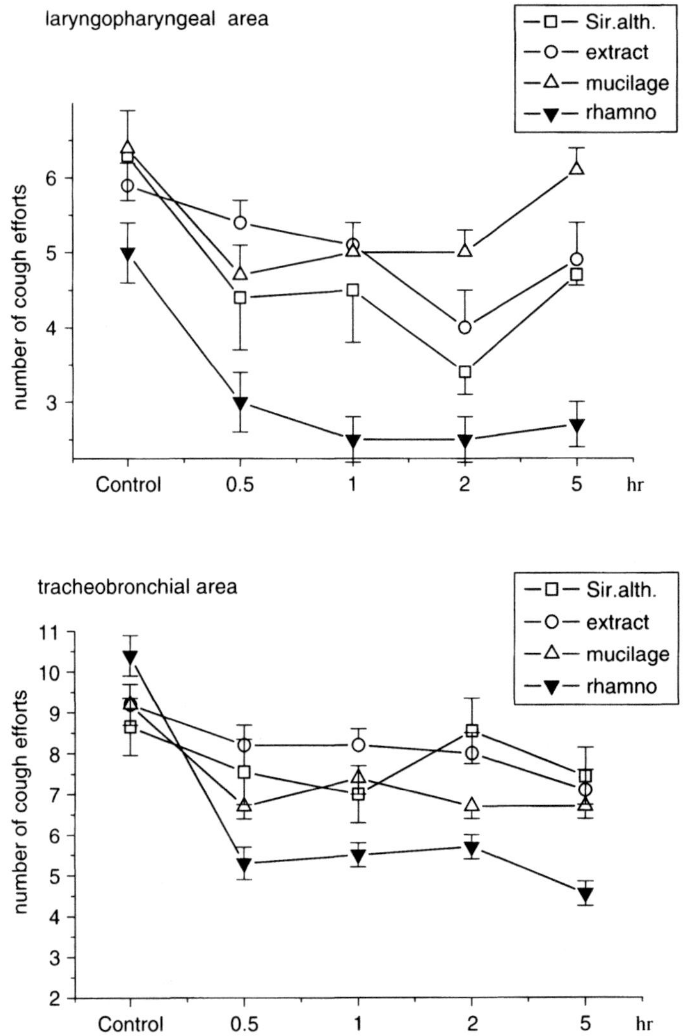

Fig. 1. Number of cough efforts after administration of *S. althaeae* (Sir. alth. – dose 1 g/kg b.w.), water root extract of *A. officinalis* (extract – dose 100 mg/kg b.w.), mucilage (dose 100 mg/kg b.w.), and rhamnogalacturonan (rhamno – dose 50 mg/kg b.w.).Symbols represent the average values; the range represents the standard error of means (±SEM).

from clinical point of view, because they indicate that the tested compound suppressed the cough reflex, but promote the expectoration.

The antitussive activity of single components of *A. officinalis* was compared to antitussive activity of antitussives usually prescribed in clinical practice (Figure 2). According to these results, the cough-suppressing effect of mucilage from *S. althaeae*, was comparable to the activity of peripherally active antitussive dropropizine. The administration of rhamnogalacturonan was associated with significantly higher antitussive activity compared to slime, *S. althaeae* and peripheral antitussives, although it did not reach the effect of opioid antitussive codeine (Nosalova et al., 1993).

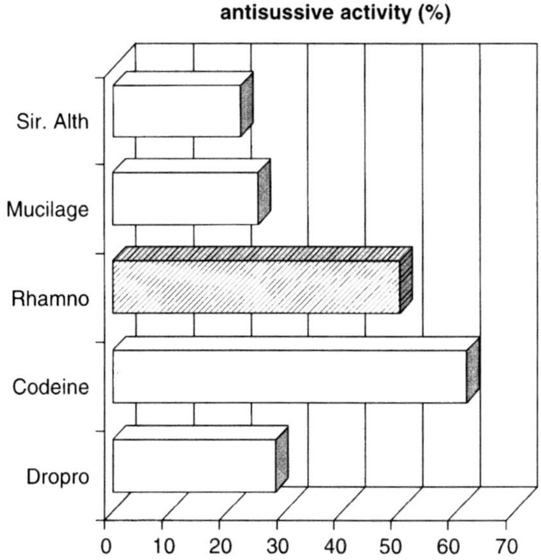

Fig. 2. Comparison of antitussive activity of *S. althaeae* (Sir. alth. – dose 1 g/kg b.w.), mucilage (dose 100 mg/kg), rhamnogalacturonan (rhamno – dose 50 mg/kg b.w.) with codeine (codeine – dose 10 mg/kg b.w.), and dropropizine (dropro – dose 100 /mg kg b.w.) on the mechanically induced cough reflex under experimental conditions.

Side effects and interactions. So far there were no significantly adverse effects observed after the administration of products from marshmallow root. Therefore, their use is recommended in pediatric practice.

Marshmallow mucilage can interfere with the absorption of other medicines within the digestive tract if they are taken at the same time. One has to take prescriped medications to consuming marshmallow tea at an alternating time.

III. *Emblica officinalis* (Fam. Euphorbiaceae)

E. officinalis (syn. *Phyllanthus emblica*) is a tree growing in subtropical and tropical parts of China, India, Indonesia, and Malaya Peninsula. The fruits of *E. officinalis* are highly nutritious and is an important source of vitamin C (Nandi et al., 1997). Its ascorbic acid content ranges from 1100 to 1700 mg/100 g. Amla fruits are acrid, cooling, diuretic, and laxative. The dried fruit is useful in hemorrhage, diarrhea and dysentery, and has anabolic and antibacterial properties. In combination with iron, it is used as a remedy for anemia. *E. officinalis* possesses expectorant, cardiotonic, antipyretic, antioxidative, and antiviral properties (Asmawi et al., 1993).

Forms. Fresh fruit, aqueous extract of fruit.

Active components. The fruit contains ascorbic acid, amino acids (glutamic acid, proline, aspartic acid, alanine, and lysine), gallic acid, 1% tannin, 35% sugar, 14% gum, 13% albumin, 17% crude cellulose, 4% minerals (chromium, zinc, copper) and 4% moisture. Fruits also contain phyllemblin and curcuminoides.

Therapeutic use. Extracts of leaves and fruits of *E. officinalis* are used in Indian traditional system of medicine, Ayurveda, for antiinflammatory, antipyretic properties or in the treatment of pancreatic disorders for their spasmolytic activities. Other effects of this plant, such as action against free radicals, damage induced during stress, protective effects against chemical carcinogenesis, protection against genotoxicity induced by aluminum, lead, nickel chloride better than ascorbic acid, have all been reported (Dhir et al., 1993). Among other effects are the inhibition of lipid peroxidation, antibacterial effects, antiatherosclerotic, and hypolipidemic activity. The water fraction of methanol extract inhibited migration of human polymorphonuclear cells and platelets at relatively low concentration (Ihantola-Vormisto et al., 1997).

Ethanol extracts of fruits of *E. officinalis* exhibit significant antitussive activity. The antitussive activity of this fruit extract was tested in conscious cats by mechanical stimulation of the laryngopharyngeal and tracheobronchial mucous areas of the airways (Nosalova et al., 2003). The results showed that the peroral dose (200 mg/kg b.w.) of this substance was effective in decreasing the cough parameters. The cough-suppressing activity of *E. officinalis* was lower than that of centrally acting codeine, but higher than the antitussive activity of the commonly used non-opioid antitussives dropropizine and prenoxdiazine (Figure 3). It is supposed that this activity of extract of *E. officinalis* is related to inhibition of prostaglandin and leukotriene synthesis. Some experimental data suggest that the plant contains as yet unidentified polar compounds, which inhibit both prostanoid and leukotriene synthesis. Among pharmacological properties, which could play a role in the antitussive efficacy of this plant extract, antioxidant, spasmolytic, and antibacterial properties are the chief

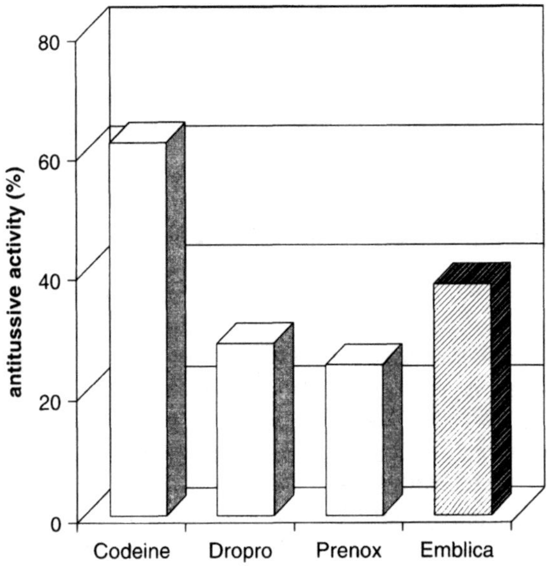

...rison of antitussive activity of *E. officinalis* (Emblica – dose 200 mg/kg b.w.), ... mg/kg b.w.), dropropizine (dropro – dose 100 mg/kg b.w.), and prenoxdiazine ... mg/kg b.w.) on the mechanically induced cough reflex under experimental

Ebringerova A, Kardosova A, Hromadkova Z, Malovikova A, Hribalova V. (2002) Immunomodulating activity of acidic xylans in relation to their structural and molecular properties. Int J Biol Macromol 30:1–6.

Foster S, Duke JA. (1990) Celandine in medicinal plants. New York: Houghton Mifflin Co., p. 92.

Franova S, Nosalova G, Strapkova A, Capek P, Kardosova A. (1995) Pectins and antitussive activity. Cs fysiol 44:30–1.

Franova S, Nosalova G, Kardosova A, Kostalova D. (1998) Herbal polysaccharides in the therapy of cough. Bratisl Lek Listy 99(2):108–10.

Gajdos M, Cagan S, Spustova V, Dzurik R. (1996) Atherosclerosis prevention and therapy. Bratisl Lek Listy 97(7):382–7.

Gonzales BM, Yslas N, Reyes E, Quiroz V, Santana J, Jimenez G. (2001) Clinical effect of Mexican sanquinaria extract (*Polygonum aviculare* L.) on gingivitis. J Ethnopharmacol 74(1):45–51.

Hiller KO, Ghorbani M, Schilcher H. (1998) Antispasmodic and relaxant activity of chelidonine, protopine, coptisine, and *Chelidonium majus* extracts on isolated guinea-pig ileum. Planta Med 64:758–60.

Hoffmann D. (1990) The new holistic herbal, 3rd edition. Shaftesbury, Dorset, UK: Element.

Ihantola-Vomisto A, Summanen J, Kankaanranta H, Vuorela H, Asmawi ZM, Moilanen E. (1997) Anti-inflammatory activity of extracts from leaves of *Phyllanthus emblica*. Planta Med 63(6):518–24.

Inouye S, Takizawa T, Yamaguchi H. (2001) Antibacterial activity of essential oils and their major constituents against respiratory tract pathogens by gaseous contact. J Antimicrob Chemother 47(5):565–73.

Kardosova A, Kostalova D, Capek P, Patoprsty V, Franova S. (1997) Water-extractable polysaccharide complex of *Rudbeckia fulgida*, var. Sullivantii (Boyton et Beadle). Possesses antitussive activity. Chem Papers 51:52–9.

Kardosova A, Malovikova V, Patoprsty V, Nosalova G, Matakova T. (2002) Structural characterization and antitussive activity of a glucuronoxylan from *Mahonia aquifolium* (Pursh Nutt.). Carbohydrate Polymers 47:27–33.

Kelly GE, Joannou GE, Reeder AY, Nelson C, Waring MA. (1995) The variable metabolic response to dietary isoflavones in human. In: Proceeding of the Society of Experimental Biology and Medicine, Vol. 208, pp. 40–43.

Korpas J, Nosalova G. (1991) Farmakoterapia kašl'a. Osveta: Martin.

Kostalova D, Kardosova A, Hajnicka V. (2001) Effect of *Mahonia aquifolium* stem bark crude extract and one of its polysaccharide componet on production of IL-8. Fytoterapia 72:802–6.

Krajkovicova D, Nosalova G, Franova S, Kardosova A. (2002) Plant polysaccharides in the treatment of the cough. Bratisl Lek Listy 103:319.

Leung AY, Foster S. (1996) Encyclopedia of common natural ingredients used in food, drugs, and cosmetics. New York: Wiley.

Malini T, Vanithakumari G, Megala N, Anusya S, Devi K, Elango V. (1985) Effect of *Foeniculum vulgare* Mill. Seed extract on the genital organs of male and female rats. Indian J Physiol Pharmacol 29(1):21–6.

Meyer-Wegener J, Liebsher K, Hettich M, Kastner HG. (1993) Efeu versus ambroxol bei chronisher bronchitis. Zeitschrift fur Allgemeinmedizin 69(3):6–61.

Mika K. (1991) Fytoterapia pre lekarov. Osveta: Martin, p. 379.

Mojzis J, Mojzisova G. (2001) Flavonoidy a ich biologicke ucinky. Vienala: Kosice, pp. 11–13.

Murai M, Tamayama Y, Hishibe S. (1995) Phenylethanoids in the herb of *Plantago lenceolata* and inhibitory effect on arachidonic acid – induced mouse ear edema. Plant Med 61(5):479–80.

Nandi P, Talukder G, Sharma A. (1997) Dietary chemoprevention of clastogenic effects 3,4-benzo(a)pyrene by *Emblica officinalis* Gaertn. Fruit extract. Br J Cancer, 76(10):1279–83.

Nosalova G, Strapkova A, Kardosova A, Capek P, Zathurecky L, Bukovska E. (1992) Antitussive action of extracts and polysaccharides of marshmallow (*Althaea officinalis* L., var. robusta). Pharmazie 47:224–6.

Nosalova G, Strapkova A, Kardosova A, Capek P. (1993) Antitussive activity of a rhamnogalacturonan isolated from the roots of *Althaea officinalis* L., var. robusta. J Carbohydrate Chemistry 12(4–5):589–96.

Nosalova G, Capek P, Kardosova A, Strapkova A. (1994) Changes in cough reflex induced by administration of polysaccharides from flowers of *Malva mauritiana* L. Pharm Pharmacol Lett 3:245–8.

Nosalova G, Strapkova A, Kardosova A, Capek P. (1992) Antitussive activity of an D-glucan isolated from the roots of *Althaea officinalis* L., var. robusta. Pharm Pharmacol Lett 2:195–7.

Nosalova G. (1998) Mechanism of action of the drugs influencing the cough reflex. Bratisl Lek Listy 99(10):531–5.

Nosalova G, Kardosova A, Franova S. (2000) Antitussive activity of a glucuronoxylan from *Rudbeckia fulgida* compared to the potency of two polysaccharide complexes from the same herb. Pharmazie 55:65–8.

Nosalova G, Mokry J, Khan MTH. (2003) Antitussive activity of the fruit extract of Emblica officinalis Gaertn. (Euphorbiaceae). Phytomedicine 10:583–9.

Paulsen E. (2002) Contact sensitization from compositae-containing herbal remedies and cosmetics. Contact Dermatitis 47:189–98.

Pfeifer S, Hanus V. (1965) On the alkaloids from *Papaver rhoeas*. Pharmazie 20(6):394.

Prahoveanu E, Esanu V. (1987) Immunomodulation with natural products. Effect of an aqueous extract of *Raphanus sativus* niger on experimental influenza infection in mice. Virologie 38(2):115–20.

Prahoveanu E, Esanu V. (1990) The effects of aqueous extracts of *Raphanus niger* on an experimental influenza infection in mice and on the enzyme polymorphism in lung tissue extracts. Rev Roum Virol 41(2):113–7.

Rang HP, Dale MM, Ritter JM. (1999) Pharmacology, 4th edition. Edinburgh: Churchill Livingstone, p. 830.

Rhee JK, Baek BK, Ahn BZ. (1985) Structural investigation on the effects of herbs in rabbits. Am J Chin Med 13(1–4):119–25.

Romaguera C, Grimalt F, Vilaplana J. (1985) Occupational dermatitis from Gordolobo (Mullein) contact. Dermatitis 12:176.

Ruf JC. (1999) Wine and polyphenols related to platelet aggregation and atherosthrombosis. Drugs Exp Clin Res 25:125–31.

Rumball W, Keogh RG, Lane GE, Miller JE, Claydon RB. (1997) Grassland Lencelot' plantain (Plantago lanceolata L). New Zealand Journal of Agricultural Research 40:373–7.

Schwartz HJ, Jones RT, Rojas AR, Squillace DL, Yunginger JW. (1997) Occupational allergic rhinoconjunctivitis and asthma due fennel seed. Ann Allergy Asthma Immunol 78(1): 37–40.

Skidmore-Roth L. (2001) Mosby's handbook of herbs and natural supplements. St. Luis: Harcourt Health Sciences Company, p. 689.

Srivastava S, Gupta PP, Prasad R, Dixit KS, Palit G, Ali B, Misra G, Saxena RC. (1999) Evaluation of antiallergic activity (type I hypersensitivity)of racemosa in rats. Indian J Physiol Pharmacol 43:235–41.

Stickel F, Poschl G, Seitz HK, Waldherr R, Hahn EG, Schuppan D. (2003) Acute hepatitis induced by greater celandine (*Chelidonium majus*). Scand J Gastroenterol 38:565–8.

Sufka KJ, Roach JT, Chambliss Jr. WG, Broom SL, Feltenstein MW, Wyandt CM, Zeng L. (2001) Anxiolytic properties of botanical extracts in the chick social separation-stress procedures. Psychopharmacology (Berl) 153(2):219–24.

Taborska E, Bochorakova H, Dostal J, Pavlova H. (1995) The greater celandine (*Chelidonium majus* L.) – review of present knowledge. Cesko-Slov Farm 44(2):71–5.

Takaisi-Kikuni NB, Schilcher H. (1994) Electron microscopy and microcalorimetric investigations of the possible mechanism of the antibacterial action of a defined propolis provenance. Planta Med 60:222–7.

Tunon H, Olavsdotter C, Bohlin L. (1995) Evaluation of anti-inflammatory activity some Swedish medicinal plants. Inhibition of prostaglandin biosynthesis and PAF-induced exocytosis. J Ethnopharmacol 48(2):61–76.

Tyler VE, Brady LR, Robbers JE. (1988) Pharmacognosy, 9th edition. Philadelphia, PA: Lea and Febiger, pp. 26–45.
Weiss RF. (1985) Herbal medicine. (Trans. Meuss AR) Gothenburg. Sweden: Ab Arcanum and Beaconsfield, Bieconsfield Publishers Ltd.
Wichtl M, Bisset NG. (1994) Primula radix – primula root. Herbal drugs and phytopharmaceuticals. Stuttgart: CRC Press, pp. 390–395.
Wichtl M, Bisset NG. (1994) Primula flos – primula flower. Herbal drugs and phytopharmaceuticals. Stuttgart: CRC Press, pp. 388–389.
Wichtl M, Bisset NG. (1994) Foeniculi fructus – fennel. Herbal drugs and phyto-pharmaceuticals. Stuttgart: CRC Press, pp. 200–202.
Zgorniak-Nowosielska I, Grzybek J, Manolova N, Serkedjieva J, Zawlinska B. (1991) Antiviral activity of flos verbasci infusion against influenza and Herpes simplex viruses. Arch Immunol Ther Exp 39(1–2):103–8.

M.T.H. Khan and A. Ather (eds.)
Lead Molecules from Natural Products

The medicinal potential of black seed (*Nigella sativa*) and its components

HALA GALI-MUHTASIB, NAHED EL-NAJJAR, REGINE SCHNEIDER-STOCK

Abstract

The seeds of *Nigella sativa* L., commonly known as black seed, have been used in traditional medicine by many Asian, Middle Eastern and Far Eastern Countries to treat headache, coughs, abdominal pain, diarrhea, asthma, rheumatism and other diseases. The seeds of this plant are the most extensively studied, both phytochemically and pharmacologically. The aqueous and oil extracts of the seeds have been shown to possess antioxidant, antiinflammatory, anticancer, analgesic and antimicrobial activities. Thymoquinone, the most abundant constituent of black seed essential oil, has been shown to be the active principle responsible for many of the seed's beneficial effects. This review paper describes the seed, its chemical components and popular uses in traditional medicine. The paper also discusses the medicinal potential and therapeutic values of some of the individual components present in the extracts of the seeds.

Keywords: *medicinal plants, nigella sativa, black seed, thymoquinone, N. sativa oil*

Abbreviations: BP, benzo(a)pyrene; TQ_2, dithymoquinone; DOX, doxorubicin; GC, gas chromatography; HPLC, High-performance liquid chromatography; IL, interleukin; NSO, Nigella sativa oil; PGE2, prostaglandin E2; TLC, thin layer chromatography; THQ, thymohydroquinone; TOH, thymol; TQ, thymoquinone; TNF, tumor necrosis factor.

I. Introduction

Plants are natural factories for the production of chemical compounds, many of which are used to promote health and fight diseases and some of them are marketed as food or herbal medicines (Dubick, 1986). Herbal medicines have long been viewed as a source of curative remedy based on religious and cultural traditions (Huxtable, 1992; Ghazanfer, 1994). The use of indigenous plant medicines in developing countries became a World Health Organization policy since 1970. Of the 520 new drugs approved in the period 1983–1994 by either the US Food and Drug Administration or comparable entities in other countries, 30 drugs came directly from natural product sources, 173 were either semi-synthetics or synthetics originally modeled on a natural parent product (De Smet, 1997).

Nigella sativa is an annual herb of the Ranunculaceae family, which grows in countries bordering the Mediterranean Sea, Pakistan and India. This widely distributed plant is native to Arab countries and other parts of the Mediterranean region (Jansen, 1981). For thousands of years, this plant has been used in many Asian, Middle Eastern and Far Eastern Countries as a spice and food preservative as well as a protective and health remedy in traditional folk medicine for the treatment of numerous disorders (Chopra et al., 1956; Nadkarni, 1976). The seed of this plant is commonly known as black seed and is referred to by the prophet Mohammed as having healing powers. The seeds are commonly eaten alone or in combination with honey and in many food preparations. The oil prepared by compressing the seeds of *N. sativa* is used for cooking. Black seed is also identified as the curative black cumin in the Holy Bible, and is described as the Melanthion of Hippocrates and Discroides and as the Gith of Pliny (Chopra et al., 1956; Nadkarni, 1976). Other names for the seed include black caraway seed, Habbatu Sawda and Habatul Baraka "the Blessed Seed".

N. sativa plant is one of the most extensively studied, both phytochemically and pharmacologically. The extracts of *N. sativa* seeds have been used by patients to suppress coughs (Mahfouz et al., 1960), disintegrate renal calculi (Hashem and El-Kiey, 1982), retard the carcinogenic process (Hassan and El-Dakhakhny, 1992; Worthen et al., 1998), treat abdominal pain, diarrhea, flatulence and polio (Enomoto et al., 2001), exert choleretic and uricosuric activities (El-Dakhakhny, 1965), anti-inflammatory (Chakravarty, 1993; Houghton et al., 1995) and antioxidant effects (Mansour et al., 2002). Besides, the essential oil was shown to have antihelminthic (Agarwal et al., 1979), antinematodal (Akhtar and Riffat, 1991), antischistosomal (Mahmoud et al., 2002), antimicrobial (Hanafy and Hatem, 1991; Aboul-Ela et al., 1996) and antiviral (Salem and Hossain, 2000) effects. In addition, the crude oil prepared from the seeds produce a variety of pharmacological actions such as anti-histaminic (El-Dakhakhny, 1965; Mahfouz et al., 1965; Chakravarty, 1993), diuretic and antihypertensive (El-Tahir et al., 1993b; Zaoui et al., 2000), hypoglycemic (Al-Hader et al., 1993), antioxytocic (Aqel and Shaheen, 1996), antinociceptive (Abdel-Fattah et al., 2000), respiratory stimulation (El-Tahir et al., 1993a), hematological (Enomoto et al., 2001) hepatoprotective (Daba and Abdel-Rahman, 1998) and immunopotentiating (Swamy and Tan, 2000) effects. The latter pharmacological properties appear to be involved in the beneficial effects of *N. sativa* oil on headache, flatulence, blood homeostasis abnormalities, rheumatism and related inflammatory diseases (Boulos, 1983). Moreover, the seeds are believed to have carminative, stimulatory and diaphoretic properties and are used in the treatment of bronchial asthma and eczema (Boulos, 1983). This chapter will review the medicinal potential of *N. sativa* seed extracts and emphasize the reasons for its long history of use in folklore medicine in Mediterranean countries.

II. Chemical constituents and active principles in *N. sativa* seeds

Millions of people in the Mediterranean region and on the Indian subcontinent use the oil from the seed of *N. sativa* daily as a natural protective and curative remedy.

The seeds are very rich and diverse in chemical composition. They contain amino acids, proteins, carbohydrates, fixed and volatile oils (Khan, 1999). Many of the pharmacological activities mentioned above have been attributed to quinone constituents in the seed. As early as 1956, Chopra et al. found that thymoquinone (TQ) (Figure 1) is the main active constituent of the volatile oil of the black seed. Mahfouz and El-Dakhakhny (1960) were the first to report on the isolation of 'nigellone' from the oil of *N. sativa* seed, using Girard's reagent. Nigellone was later found to possess antihistaminic properties in relatively low concentrations (Mahfouz et al., 1965). El-Dakhakhny (1963) was able to isolate the constitutive components of *N. sativa* seeds from its essential oil, among which TQ was later shown to be the main constituent of the volatile oil (Houghton et al., 1995). In addition, El-Dakhakhny determined that the 'nigellone' isolated earlier was a dimer of TQ, which was later named dithymoquinone (TQ_2) (Figure 1). The latter compound was shown to be formed via photodimerization of TQ as a consequence of exposure to sunlight during separation and extraction of the quinones from the seed. El-Fatatry (1975) reported the isolation of thymohydroquinone (THQ) from *N. sativa* seed volatile oil. In another study (Aboutabl et al., 1986), the chemical composition of the black seed of *N. sativa* was found to contain a fixed oil (30%) and a volatile oil (average 0.5%, maximum 1.5%). The volatile oil was found to contain 54% TQ and many monoterpenes such as p-cymene and α-pinene, TQ_2 and THQ.

In recent years, the seeds of *N. sativa* have been subjected to a range of phytochemical investigations. They have been shown to contain more than 30% (w/w) of a fixed oil with 85% of total unsaturated fatty acid (Houghton et al., 1995). The seeds also contain alkaloids of unknown pharmacological actions, such as nigellidine,

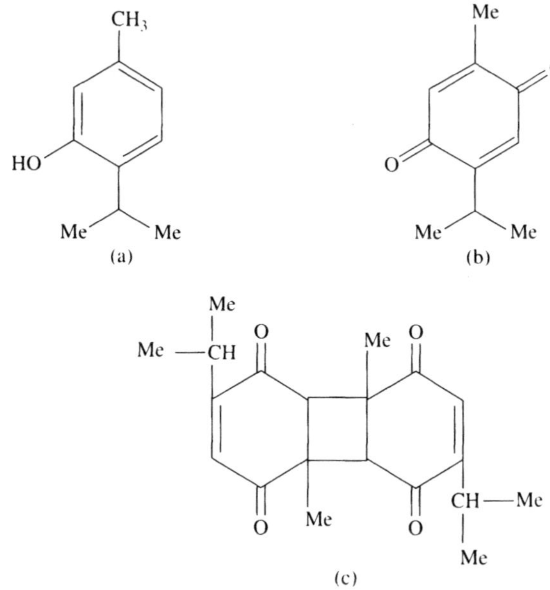

Fig. 1. Chemical structures of thymol (a), thymoquinone (b) and dithymoquinone (c) (Me: methyl group).

nigellimine and nigellicine (ur-Rahman et al., 1985), saponins and crude fiber as well as minerals such as calcium, iron, sodium and potassium. Other constituents of the volatile oil include thymol (TOH) (Figure 1) (Aboutabl et al., 1986). Recently, the presence of TQ, TQ_2 and TOH in *N. sativa* seed was confirmed using thin layer chromatography (TLC) and normal phase high-performance liquid chromatography (HPLC) methods (Abou-Basha et al., 1995; Aboul-Enein and Abou Basha, 1995). The content of TQ in *N. sativa* seed oil samples, obtained from different origins, was measured by gas chromatography (GC) analysis and found to be in the range of 0.13–0.17% w/v of the oil (Houghton et al., 1995). The seeds are also rich in proteins; when whole *N. sativa* seeds were fractionated using SDS-PAGE, they were found to contain a number of protein bands ranging from 10 to 94 kDa molecular mass (Haq et al., 1999). An HPLC method for quantifying the putative pharmacologically active constituents (TQ, TQ_2, THQ and TOH) in the oil of *N. sativa* seed was recently described by Ghosheh et al. (1999). In this procedure, the four compounds mentioned were separated and quantified in commercial *N. sativa* seed oil with good resolution, reproducibility and sensitivity. Both heat and light are known to affect the levels of the constituents in the oil. Since various storage and manufacturing conditions are expected to make a difference in the amounts of the quinone constituents of the oil, the analytical HPLC method described by Ghosheh et al. (1999) can be used to quantify the levels of the above constituents in the oil and seed extracts of *N. sativa* under different manufacturing conditions. The protocol is also useful as a quality control method for the determination of pharmacologically active quinones in *N. sativa* seed oil. Using TLC, the oil of black seed was found to contain TQ and the terpenoid components carvacrol, t-anethole and 4-terpineol (Burits and Bucar, 2000). GC-MS analysis of the essential oil obtained from six different samples of *N. sativa* seeds and from a commercial fixed oil showed that the qualitative composition of the volatile compounds was almost identical. Differences were mainly restricted to the quantitative composition (Burits and Bucar, 2000).

In conclusion, *N. sativa* seeds contain fixed oils and volatile oils, which are rich sources of quinones, unsaturated fatty acids, amino acids and proteins and contain traces of alkaloids and terpenoids. Most of the studies on the biological effects of *N. sativa* have dealt with its crude extracts in different solvents; however, some studies used its active principles. Among the components isolated from the volatile oil of *N. sativa*, TQ has been shown to be the principal active ingredient (Mahfouz and El-Dakhakhny, 1960) and thus is the most studied of all. In what follows, the physiological, antioxidant, antimicrobial, analgesic, antiinflammatory and chemopreventive effects of black seed with a special emphasis on TQ will be discussed.

III. Physiological effects of *N. sativa* and its component TQ

The oil extract of black seed has been shown to exert effects on various systems including the respiratory, cardiovascular, gastric and uterine and smooth muscle. The effects of intravenous administration of volatile oil and of TQ were investigated on the respiratory system of the guinea pig (El-Tahir et al., 1993a). The latter compounds were found to increase the intratracheal pressure in the dose range of 4–32 μl/kg and 1.6–6.4 mg/kg, respectively. Although *N. sativa* oil (NSO) significantly increased the respiratory rate of guinea pigs, TQ was without any effect. The effects of NSO were significantly antagonized by treatment of the animals with antihistamines such as atropine and reserpine, suggesting that the oil-induced respiratory effects were mediated via the release of histamine and indirect activation of muscarinic and cholinergic mechanisms (El-Tahir et al., 1993a). This also suggested that the removal of TQ from black seed oil might provide a potential centrally acting respiratory stimulant (El-Tahir et al., 1993a). This same group demonstrated that the intravenous administration of NSO (4–32 μl/kg) or TQ (0.2–1.6 mg/kg) to rats decreased the arterial blood pressure and the heart rate in a dose-dependent manner (El Tahir et al., 1993b), suggesting that the oil may possess antihypertensive effects. The cardiovascular depressant effects of the oil were significantly antagonized by atropine and cyproheptadine, suggesting that these effects were mediated mainly centrally via indirect and direct mechanisms that involved both 5-hydroxy tryptaminergic and muscarinic mechanisms (El-Tahir et al., 1993b). NSO has also been shown to increase bile secretion in dogs and uric acid in rats as well as protect guinea pigs against histamine-induced bronchospasm (El-Dakhakhany, 1982). The fatty and petroleum extracts shortened bleeding time and inhibited fibrinolytic activity in rabbits (Ghoneim et al., 1982). In a recent study, the crude extract of *N. sativa* seeds was found to exhibit spasmolytic and bronchodilator activities mediated possibly through calcium channel blockade and this activity was concentrated in the organic fraction of the extract (Gilani et al., 2001).

Traditionally *N. sativa* plant has been in use in many Middle Eastern countries as a natural remedy for diabetes. Significant reduction in blood glucose and cholesterol levels in humans following the use of the plant was reported by Bamosa et al. (1997). The oil of this plant has a great potential in the treatment of diabetic animals because of its combined hypoglycemic (Al-Hader et al., 1993; Zaoui et al., 2002a) and immunopotentiating properties (Haq et al., 1999). A plant extract mixture comprising *N. sativa*, myrrh, gum Olibanum, gum asafetida and aloe was found to lower blood glucose in streptozotocin diabetic rats (Al-Awadi et al., 1991). In an attempt to elucidate the mechanism of this antidiabetic action, the rate of gluconeogenesis in isolated hepatocytes as well as the activity of pyruvate carboxylase and phosphoenol pyruvate carboxykinase in rat liver homogenates was examined. It was found that the plant extracts significantly decreased hepatic gluconeogenesis, suggesting that it may prove to be a useful therapeutic agent in the treatment of non-insulin-dependent diabetes mellitus. Similar insulinotropic effects of NSO were recently observed in streptozotocin plus nicotinamide-induced diabetes mellitus in hamsters (a model of type 2 diabetes) orally fed with the oil (Fararh et al., 2002). In this study, positive immunoreactivity for the presence of insulin was observed in the pancreases from oil-treated vs. non-treated hamsters using immunohistochemical staining, suggesting that the hypoglycemic effect of NSO resulted, partly, from a stimulatory effect on beta cell function with consequent increase in serum insulin level. The ability of NSO to lower blood glucose concentrations was later confirmed in streptozotocin diabetic rats following 2, 4 or 6 weeks of treatment (El-Dakhakhny et al., 2002b). In addition, the effects of NSO, nigellone and TQ were studied on insulin secretion of isolated rat pancreatic islets. The blood glucose-lowering effect of NSO was not paralleled by a stimulation of insulin release. The data indicated that the hypoglycemic effect of NSO might be mediated by extrapancreatic actions, to be elucidated, rather than by stimulated insulin release (El-Dakhakhny et al., 2002b).

In many Arab countries *N. sativa* and its derived products are consumed abusively for traditional treatment of blood homeostasis abnormalities and as a treatment for dyslipidemia (Zaoui et al., 2002a). Several studies support the use of NSO extract for the treatment of thrombosis and dyslipidemia (Labhal et al., 1997; Enomoto et al., 2001; Zaoui et al., 2002a). The purified components (2-(2-methoxypropyl)-5-methyl-1,4-benzenediol, thymol and carvacrol) obtained from the methanol-soluble portion of NSO showed inhibitory effects on arachidonic acid-induced platelet aggregation and blood coagulation. Interestingly, some aromatic compounds present in the extract were found to be more potent than aspirin, which is well known as a remedy for thrombosis (Enomoto et al., 2001). In addition, an aqueous suspension of *N. sativa* seeds was found to decrease the serum total lipids and body weight in Psammomys obesus sand rat (Labhal et al., 1997). Analogous results, accompanied by decreases in serum lipid levels have also been observed in rats chronically treated with *N. sativa* fixed oil (Zaoui et al., 2002b). Animals were treated daily with an oral dose of 1 ml/kg body weight of the *N. sativa* seed fixed oil for 12 weeks. The serum cholesterol, triglycerides and the count of leukocytes and platelets decreased significantly by 15.5%, 22%, 35% and 32%, respectively, compared to the control values. Hematocrit and hemoglobin levels increased significantly by 6.4% and 17.4%, respectively (Zaoui et al., 2002a), suggesting that the oil influences blood homeostasis.

N. sativa is also used in Arab folk medicine as a diuretic and hypotensive plant. In an attempt to experimentally support the above traditional uses of the plant, a study was conducted on the diuretic and hypotensive effects of the dichloromethane extract of *N. sativa* seeds in the spontaneously hypertensive rat (Zaoui et al., 2000). An oral dose of either *N. sativa* extract (0.6 mL/kg/day) or furosemide (5 mg/kg/day) significantly increased diuresis by 16% and 30%, respectively, after 15 days of treatment. The urinary excretions of Cl^-, Na^+, K^+ and urea were also increased after 15 days of treatment. In the same rat study, a comparison between *N. sativa* and nifedipine found mean arterial pressure to be decreased by 22% and 18% in the *N. sativa*- and nifedipine-treated rats, respectively, suggesting that *N. sativa* extract may play a role in decreasing blood pressure.

Evidence indicates that NSO has a protective role against gastric ulcers (El-Dakhakhny et al., 2000b). Oral administration of NSO for 2 weeks in rats produced a significant increase in gastric mucin content and glutathione level and a significant decrease in gastric mucosal histamine content without significant changes in free acidity and peptic activity of the gastric juice (El-Dakhakhny et al., 2000b). Ethanol administration, however, produced 100% ulcer induction accompanied by a reduction in free acidity, mucin content and glutathione level without any significant changes in peptic activity. When animals were pretreated with NSO before ulcer induction by ethanol, a protection ratio of 53.56% was noted as compared to the ethanol group (El-Dakhakhny et al., 2000b). The protective action of NSO was believed to be through the increase of the cytoprotective mucin content and/or decrease of histamine.

A final physiological effect of NSO includes its potential as an antioxytocic agent. Aqel and Shaheen (1996) tested the effects of NSO on the uterine smooth muscle of rats and guinea pigs *in vitro* using isolated uterine horns. The volatile oil was found to inhibit the spontaneous movements of rat and guinea pig uterine smooth muscle and also the contractions induced by oxytocin stimulation (Aqel and Shaheen, 1996). These effects were concentration dependent and reversible by tissue washing.

IV. Antimicrobial and antiparasitic effects of *N. sativa* oil

Extracts of *N. sativa* have shown promising effects against bacteria, fungi, viruses, parasites and worms. In 1975, the purified compound THQ from NSO was found to have high antimicrobial effect against Gram positive microorganisms (El-Fatatry, 1975). In later studies, seed extracts of *N. sativa* were found to inhibit the growth of *Escherichia coli*, *Bacillus subtilis* and *Streptococcus feacalis* (Saxena and Vyas, 1986). The antimicrobial activity of *N. sativa* was further established against several species of pathogenic bacteria and yeast (Topozada et al., 1965; Hanafy and Hatem, 1991). In the latter study, filter paper discs impregnated with the diethyl ether extract of *N. sativa* seeds caused concentration-dependent inhibition of Gram-positive *Staphylococcus aureus* and Gram-negative *Pseudomonas aeruginosa* and *E. coli* and a pathogenic yeast *Candida albicans*. The extract showed antibacterial synergism with streptomycin and gentamicin and showed additive antibacterial action with spectinomycin, erythromycin, tobramycin, doxycycline, chloramphenicol, nalidixic acid, ampicillin, lincomycin and sulfamethoxyzole–trimethoprim combination. Interestingly, the extract successfully eradicated a non-fatal subcutaneous staphylococcal infection in mice when injected at the site of infection (Hanafy and Hatem, 1991). Recently, crude extracts of *N. sativa* showed promising antimicrobial effects against bacterial isolates with multiple resistances against antibiotics (Morsi, 2000). The most effective extracts were the crude alkaloid and water extracts.

The antiparasitic actions of NSO have been well documented by several researchers (Agarwal et al., 1979; Akhtar and Riffat, 1991; Abdel-Salam et al., 1993; Mahmoud et al., 2002). The antihelminthic activities of NSO were studied by Agarwal et al. (1979) who reported that the essential oil from the seeds of *N. sativa* showed pronounced activity even in 1:100 dilutions against tapeworms and earthworms. Anticestodal effects of *N. sativa* seeds were studied in children infected naturally with the respective worms. A single oral administration of 40 mg/kg of *N. sativa* ethanolic extract reduced the percentage of the fecal eggs without producing any adverse side effects in the doses tested (Akhtar and Riffat, 1991). When given orally to *Schistosoma mansoni*-infected mice, a 2-week treatment with NSO reduced the number of *S. mansoni* worms in the liver and decreased the total number of ova deposited in both the liver and the intestine (Mahmoud et al., 2002). Furthermore, it increased the number of dead ova in the intestinal wall and reduced the granuloma diameters markedly (Mahmoud et al., 2002). When NSO was administered in combination with praziquantel, the drug of choice for the treatment of schistosomiasis, the most prominent effect was a further lowering of the dead ova number over that produced by praziquantel alone. These changes were correlated mainly with the ability of NSO to improve liver function and the immunological system of infected mice and partly to its antioxidant effects (Mahmoud et al., 2002). The protection is also due to the ability of NSO and TQ to reduce the cytogenetic damage induced by schistosomiasis infection (Aboul-Ela, 2002). Karyotyping of bone marrow and spleen cells of infected mice showed that the main chromosomal abnormalities were gaps, fragments and deletions especially in chromosomes 2, 6 and some in chromosomes 13 and 14. Treatment with NSO or TQ for 7 days was found to reduce the percentage of chromosomal aberrations and the incidence of deletions and tetraploidy compared to the control level. Thus, NSO may be improving the therapeutic efficacy of *S. mansoni* infection by decreasing the induced chromosomal abnormalities.

The antiviral effect of NSO was only recently investigated using murine cytomegalovirus as a model (Salem and Hossain, 2000). The cytomegalovirus is a herpesvirus that causes disseminated and fatal disease in immunodeficient animals (Reynolds et al., 1993) similar to that caused by human cytomegalovirus in immunodeficient humans (Smith and Brennessel, 1994). Intraperitoneal administration of NSO to mice strikingly inhibited the virus titers in spleen and liver on day 3 of infection. The difference in the viral load in spleen and liver of the control and NSO-treated mice was very high, 45×10^4 vs. 7×10^4 and 23×10^3 vs. 3×10^3 for liver and spleen, respectively. This antiviral effect coincided with an increase in serum level of interferon-gamma and increased numbers of $CD4^+$ helper T cells and suppressor function and numbers of macrophages. On day 10 of infection, the virus titer was undetectable in spleen and liver of NSO-treated mice, while it was detectable in control mice (Salem and Hossain, 2000). The antiviral effects of NSO were more potent than the action of Chinese traditional herbal medicine hochuekki – against murine cytomegalovirus (Hossain et al., 1999).

V. Anticancer effects of *N. sativa* and its components

The active principles in NSO have been found to exert antineoplastic effects both *in vitro* and *in vivo* using various models of carcinogenesis. In what follows, the anticancer effects of NSO and its components will be discussed.

V.A. In vitro effects

Black seed preparations (TQ and TQ_2) have been demonstrated to have significant antineoplastic activity against various tumor cells *in vitro* (Salomi et al., 1991, 1992; Swamy and Tan, 2000). The active principles of *N. sativa* showed 50% cytotoxicity against Ehrlich ascites carcinoma, Dalton's lymphoma ascites and Sarcoma-180 cells at a concentration of 1.5, 3 and 1.5 μg, respectively, with little activity against lymphocytes (Salomi et al., 1991). *In vitro* cytotoxicity was also demonstrated against human pancreatic adenocarcinoma, uterine sarcoma and leukemic cell lines (Salomi et al., 1992). The growth inhibitory activity was found to be related to the extract's ability to inhibit DNA synthesis as measured by the incorporation of tritiated thymidine into cells. These findings were later confirmed by Worthen et al. (1998) who assayed the *in vitro* cytotoxicity of a crude gum, a fixed oil and two purified components of *N. sativa* seed, TQ and TQ_2, on several parental and multidrug resistant human tumor cell lines. Although as much as 1% w/v of the gum or oil was devoid of cytotoxicity, both TQ and TQ_2 were cytotoxic for all of the tested cell lines (IC_{50}: 78 to 393 μM). Interestingly, the multidrug resistant cell variants that are over 10-fold more resistant to the standard antineoplastic agents doxorubicin and etoposide were sensitive to TQ and TQ_2 (Worthen et al., 1998). The ethyl acetate fraction of *N. sativa* seeds (identified as CC-5) was later found to exhibit significant growth inhibition on a variety of cancer cell lines without inhibiting the growth of normal human endothelial cells (Swamy and Tan, 2000). The ED_{50} values of the extract showed increased sensitivity towards Hep G2, LL/2 and Molt4 cell lines compared with SW620 and J82 cell lines. Badary and Gamal El-Din (2001) also showed that

TQ inhibited the survival of fibrosarcoma cells with IC_{50} of 15 μM by inhibiting the incorporation of ^{3}H thymidine into cells. The cellular mechanism of antineoplastic activity of TQ was only recently investigated (Shoieb et al., 2003). In this study, the cellular mechanisms of TQ-induced cytotoxicity in parental and cisplatin-resistant osteosarcoma human breast adenocarcinoma, human ovarian adenocarcinoma and Madin–Darby canine cell lines have been examined. The cisplatin-resistant cells were the most sensitive to TQ treatment, while the canine cell lines were the least sensitive. A dose of 25 μM of TQ induced apoptosis of osteosarcoma cells 6 h after treatment. This dose also decreased the number of cells in S-phase and increased cells in G_1-phase, indicating cell cycle arrest at G_1. These results suggest that TQ induces cell cycle arrest and apoptosis in cancer cells. Interestingly, non-cancerous cells are relatively resistant to the apoptotic effects of TQ (Shoieb et al., 2003). In our laboratories, we have recently investigated the effects of TQ on the proliferation and cytotoxicity of a panel of primary, benign and malignant mouse and human epidermal keratinocytes and colon cells. Although lower doses of TQ were found to exert no effects on the morphology or proliferation of normal cells, they inhibited cellular proliferation of benign and malignant cells, confirming the selectivity of this compound to cancer cells (Gali-Muhtasib et al., 2004). The growth-inhibitory effects of TQ against colon cancer cells were found to be mainly due to the ability of this compound to induce G1 cell cycle arrest and apoptosis. The apoptotic effects of TQ are modulated by Bcl-2 protein and are linked to and dependent on p53 (Gali-Muhtasib et al., 2004). Our data support the strong potential for using the agent TQ in the prevention or therapy against colon cancer. We are presently testing the potency of TQ in the 1,2-dimethyl hydrazine mouse model of colon carcinogenesis by administering it in drinking water, intraperitoneally or by gavage.

V.B. In vivo effects

Several studies have shown that NSO and TQ retard the carcinogenic process in animals. The active principles of *N. sativa* seeds containing fatty acids were found to completely inhibit the Ehrlich ascites carcinoma in mice (Salomi et al., 1991, 1992). A dose of 100 mg/kg body weight (b.w.) of *N. sativa* extract delayed the onset of papilloma formation and reduced the mean number of papillomas per mouse (Salomi et al., 1991). Intraperitoneal administration of *N. sativa* (10 mg/kg b.w.) 30 days after subcutaneous administration of 20-methylcholanthrene-induced soft tissue sarcoma restricted tumor incidence to 33.3% compared to 100% in methylcholanthrene-treated controls (Salomi et al., 1991). *In vivo* Ehrlich ascites carcinoma tumor development was completely inhibited by the active principle at the dose of 2 mg per mouse per day for 10 days (Salomi et al., 1992). Furthermore, NSO was reported to possess a protective effect on chemical-induced carcinogenesis in hamster cheek pouch (Hassan and El-Dakhakhny, 1992). In another study, the administration of a dose of 1 mg of TQ twice weekly for 4 weeks demonstrated powerful chemopreventive effects against benzo(a)pyrene (BP)-induced forestomach tumors (Badary et al., 1999). TQ inhibited both BP-induced forestomach tumor incidence and multiplicity by 70% and 67%, respectively. More recently, this same group (Badary and Gamal El-Din, 2001) demonstrated that the administration of 0.01% of TQ in drinking water 1 week before and after 20-methylcholanthrene treatment significantly inhibited fibrosarcoma tumor incidence and tumor burden by 43% and 34%, respectively. Moreover, TQ delayed the onset of methylcholanthrene-induced fibrosarcoma tumors that appeared at 12 weeks and produced less methylcholanthrene-induced mortality. The possible modes of anticarcinogenic actions of TQ in the above two studies were suggested to be through its antioxidant and antiinflammatory activities, coupled with enhancement of detoxification processes.

In a recent study, the effect of CC-5 (ethyl acetate fraction of NSO) was evaluated for its *in vivo* antitumor activity against intraperitoneally implanted murine P388 leukemia and subcutaneously implanted Lewis lung carcinoma cells in BDF1 mice (Kumara and Huat, 2001). At doses of 200 and 400 mg/kg b.w., the fraction prolonged the life span of these mice by 153% compared to DMSO-treated control mice. The antitumor activity of a 21-day treatment of CC-5 against subcutaneously implanted LL/2 was tested and found to produce a 60–70% tumor inhibition rate. A triterpene saponin was isolated from the CC-5 fraction and identified to be α-hederin. This compound was found to exert more potent anticancer effects compared to the commonly used anticancer drug, cyclophosphamide. When α-hederin was given i.p. at doses of 5 and 10 mg/kg b.w. to mice with formed tumors, it produced significant dose-dependent tumor inhibition rate values of 50% and 71%, respectively, on day 15, compared to 42% on day 15 in the cyclophosphamide (CP)-treated group. The underlying mechanism(s) of antitumor activity of α-hederin is not defined yet (Kumara and Huat, 2001). The protective effect of Nigella grains on carcinogenesis induced by methylnitrosourea in Sprague Dawley rats was recently investigated (Mabrouk et al., 2002). When given orally (0.2 g ground Nigella grains) alone or with honey, a 6-month treatment reduced MNU-induced inflammatory reaction in lung and skin and MNU-induced colon adenocarcinomas by 80% (Mabrouk et al., 2002). There was an associated elevation of malondialdehyde and nitric oxide in sera obtained from methylnitrosourea-treated animals, which was lowered by ingestion of *N. sativa* grains. Interestingly, combined oral treatment of honey and *N. sativa* grains protected 100% against methylnitrosourea-induced oxidative stress, carcinogenesis and abolished the nitric oxide and malondialdehyde elevations shown in sera of animals that did not receive these nutrients (Mabrouk et al., 2002).

TQ has also been shown to improve the therapeutic index of several anticancer agents and to protect non-tumor tissues from chemotherapy-induced damage. TQ protected against ifosfamide-induced Fanconi syndrome in rats and enhanced its antitumor activity in Ehrlich ascites carcinoma-bearing mice (Badary, 1999). The disease Fanconi syndrome is characterized by wasting off glucose, electrolytes and organic acids along with elevated serum creatinine and urea as well as decreased creatinine clearance rate (Brade et al., 1986). The changes in renal function observed in the rat model of Fanconi syndrome correlate well with the nephrotoxic effects of ifosfamide observed in man. Oral supplementation of TQ (5 mg/kg/day) with drinking water rendered rats significantly less susceptible to ifosfamide-induced renal abnormalities. It also corrected for the damage induced by ifosfamide on phosphorus, glucose, serum creatinine and urea levels and significantly normalized creatinine clearance rate. This effective dose of TQ was found to be very safe (Badary et al., 1998). TQ protected the kidney against ifosfamide-induced damage through an antioxidant mechanism, since it significantly prevented ifosfamide-induced renal glutathione depletion and lipid peroxide accumulation. In mice bearing Ehrlich ascites carcinoma xenograft, TQ (10 mg/kg/day) administered in drinking water significantly enhanced the antitumor effect of ifosfamide. Furthermore, mice treated with ifosfamide in combination with TQ showed less body weight loss and mortality rate compared to ifosfamide single therapy. This finding is in full agreement with previous findings that TQ potentiates cisplatin antitumor activity and protects against cisplatin-induced nephrotoxicity in mice and rats (Badary et al., 1997; El-Daly, 1998), carbon tetrachloride-induced hepatotoxicity and lipid peroxidation (Al-Gharably et al., 1997; Nagi et al., 1999) in mice and doxorubicin (DOX)-induced cardiotoxicity (Al-Shabanah et al., 1998) in mice. In this context, *N. sativa* seed extract was shown to protect against cisplatin-induced myelosuppression in mice (Nair et al., 1991). Moreover, recent investigations by Nagi and Mansour (2000) showed that oral administration of TQ (10 mg/kg/day) with drinking water starting 5 days before a single i.p. injection of DOX (15 mg/kg) and continuing during the experimental period ameliorated the DOX-induced cardiotoxicity in rats. TQ also protected against the nephropathy and oxidative stress induced by DOX in rats (Badary et al., 2000). Although DOX is a potent cancer chemotherapeutic agent against several malignancies, its clinical efficacy is limited because of severe cytotoxic side effects, the most serious being cardiotoxicity (Cortes et al., 1975). Experimentally DOX induces hyperlipidemic nephropathy in rats associated with hypoalbuminemia, hypoproteinemia, elevated serum urea, hyperlipidemia and a high urinary excretion of protein and albumin. The nephropathy observed in this model resembles histologically and clinically the focal and segmental glomerulosclerosis that occurs in humans (Zima et al., 1997). There is increasing evidence that free radical generation by DOX is involved in the primary pathogenic mechanism of DOX-induced nephropathy in rats (Bertani et al., 1986). Treatment of rats with TQ (10 mg/kg per day) supplemented with the drinking water for 5 days before DOX, and daily thereafter significantly lowered serum urea and serum and kidney levels of triglycerides and total cholesterol. It also suppressed DOX-induced proteinuria and albuminuria (Badary et al., 2000). In both studies, TQ's protective effects against DOX damage to the heart and kidney was found to be mainly due to its superoxide-scavenging and antilipid peroxidation effects.

In conclusion, the ability of TQ to enhance the therapeutic index of anticancer drugs and provide protection from cytotoxicity induced by these agents strengthens the potential use of this readily available drug as a cytoprotective agent. This protection documented by several investigators enforces its preclinical evaluation in combination with anticancer agents.

VI. Antiinflammatory and immunomodulatory effects of *N. sativa*

N. sativa and its derived products have been traditionally used as a treatment for rheumatism, liver diseases and related inflammatory disorders. The effect of black seed on the immune system has been investigated by several researchers (Houghton et al., 1995; El-Dakhakhny et al., 2000a; Haq et al., 1995; Al-Ghamdi, 2001). All studies have shown that the oil and its most abundant component, TQ, inhibit many inflammatory mediators, and, thus, may be useful in ameliorating inflammatory and autoimmune conditions. Chakravarty (1993) reported that the *N. sativa*-derived nigellone, the carbonyl polymer of TQ, was very effective at low concentrations in inhibiting histamine release from rat peritoneal mast cells *in vitro*. He suggested that the mechanism of action is mainly due to the ability of TQ to decrease intracellular calcium by inhibiting protein kinase C and partly due to its ability to inhibit oxidative energy metabolism.

Several studies point to the effect of *N. sativa* on the human immune system (El-Kadi and Kandil, 1986; El-Kadi et al., 1987). The seeds were found to produce an increase in the ratio of helper to suppressor T cells and to enhance natural killer cell activity in normal volunteers (El-Kadi and Kandil, 1986). *In vitro* studies showed that the crude fixed oil and pure TQ were potent inhibitors of eicosanoid generation, namely thromboxane B_2 and leucotriene B_4, by inhibiting both cyclooxygenase and lipoxygenase, respectively (Houghton et al., 1995). Thromboxane B_2 has been implicated in the mechanism of hepatocyte plasma membrane bleb formation, which is an early event in hepatocyte injury when exposed to oxidative stress (Horton and Wood, 1990). In another study, *N. sativa* enhanced the production of IL-3 by human lymphocytes and had a stimulatory effect on macrophages (Haq et al., 1995). Besides, the immunomodulatory effect of *N. sativa* purified proteins was found in mixed lymphocyte cultures and caused increased secretion in the levels of the cytokines IL-1β and IL-8 (Haq et al., 1999). Moreover, the fixed oil increased the release of PGE_2, inhibited the release of leukotrienes and histamine from normal and sensitized guinea pig lungs. Other pieces of evidence include the inhibition of TNF-α production in murine septic peritonitis by TQ (El-Dakhakhny et al., 2000a) and the unique immunomodulatory properties of the ethyl acetate (CC-5) fraction of *N. sativa* at non-cytotoxic doses (Swamy and Tan, 2000). The ability of TQ to modulate cytokines and enhance the immune system has been implicated as the main reason for its protective effect against schistosome egg infection in the liver (Mahmoud et al., 2002).

In an attempt to determine the immunomodulatory role of TQ, the effect of this compound on the production of nitric oxide (NO) by rat peritoneal macrophages was investigated (El-Mahmoudy et al., 2002). It was found that it reduced production of NO in supernatants of lipopolysaccharide-stimulated macrophages without affecting the cell viability. The protein and mRNA levels of inducible nitric oxidesynthase in peritoneal macrophages were also decreased by TQ. Immunofluorescence staining of inducible nitric oxide synthase in macrophages showed decreased immunoreactivity for inducible nitric oxide synthase after TQ treatment.

The antiinflammatory effect of *N. sativa* has been found to be comparable to that of 100 mg/kg aspirin (Al-Ghamdi, 2001). In conclusion, the pharmacological activities of *N. sativa* documented by several researchers support its use in folk medicine to reduce inflammation.

VII. Antioxidant and hepatoprotective effects of *N. sativa*

Health food stores sell *N. sativa* seeds as a natural remedy for a variety of complaints including liver diseases (Boulos, 1983). The hepatoprotective effects of TQ have been well documented and have been found to be related to its strong antioxidant potentials. In fact, the antioxidant and free radical scavenging properties of many plants have been found to play an important role in their hepatoprotective activity (Kiso et al., 1984; Valenzuela et al., 1986; Navaro et al., 1993; Thabrew et al., 1995). Oxidant stress can increase the susceptibility to irreversible injury by oxidative intoxication and by free radicals that can result in lipid peroxidation, protein oxidation, protein inactivation, disturbance in calcium homeostasis and consequent loss of cell viability (Masaki et al., 1989; Shertzer et al., 1994). Most of the hepatoprotective drugs belong to the group of free radical scavengers and their mechanism of action involves membrane stabilization, neutralization of free radicals and immunomodulation. The flavanolignan mixture, silymarin and its most active constituent, silybin, obtained from the plant *Silybum marianum* have been studied intensively for their antihepatotoxic effects (Vogel, 1977; Miguez et al., 1994). They are now used clinically in the treatment of many liver diseases (Fernandez et al., 1995).

The oil of *N. sativa* and TQ are known to possess strong antioxidant activities (Aboul-Enein et al., 1999; Nagi and Mansour, 2000; Meral et al., 2001; El-Dakhakhny et al., 2002a; Mahmoud et al., 2002); TQ has been shown to inhibit non-enzymatic peroxidation in ox brain phospholipid liposomes (Houghton et al., 1995) with a potency that is 10 times higher than NSO. Using TLC screening methods, Burits and Bucar (2000) showed that TQ and NSO components, namely, carvacrol, t-anethole and 4-terpineol possess strong radical-scavenging properties. Moreover, TQ showed extremely high superoxide anion radical-scavenging abilities (as effective as superoxide dismutase against superoxide) in pure chemical systems (Nagi and Mansour, 2000). This high scavenging power of TQ was responsible for its protective effects against DOX-induced cardiotoxicity in rats (Nagi and Mansour, 2000). In a recent study, TQ was observed to be metabolized by liver DT diaphorase to dihydrothymoquinone, a phenolic metabolite that acts as a radical scavenger and inhibits lipid peroxidation *in vitro* (Mansour et al., 2002). TQ and dihydrothymoquinone acted not only as superoxide anion scavengers, but also as general free radical scavengers with IC_{50} values in the nanomolar and micromolar ranges, respectively (Mansour et al., 2002). Treatment of mice with TQ orally for 5 successive days produced significant reductions in hepatic superoxide dismutase, catalase, and glutathione peroxidase activities (Mansour et al., 2002). Moreover, TQ significantly reduced hepatic and cardiac lipid peroxidation as compared with the respective control group. The most comprehensive evidence on the antioxidant effects of NSO and its components came from the studies conducted by Kruk et al. (2000) and El-Dakhakhny et al. (2002a). They showed that TOH, TQ and TQ_2 exhibit antioxidant properties and acted as scavengers of various reactive oxygen species. TOH, for example, acted as $^1O_2$ quencher, while TQ and TQ_2 showed superoxide dismutase-like activity removing $O_2^{\cdot-}$. The same group (El-Dakhakhny et al., 2002a) also showed that NSO as well as nigellone and TQ exert inhibitory actions on the production of leukotriene-type mediators of inflammation *in vitro*. Whereas TQ exerts a strong inhibitory activity (IC_{50}: 0.3 mg/ml), nigellone is far less active (IC_{50}: 12 mg/ml), possibly due to the loss of antioxidative activity through polymerization. The high antioxidative action of NSO and its components suggests their importance for the treatment of various diseases occurring with participation of reactive oxygen species.

In an attempt to evaluate the hepatoprotective effects of TQ, Daba and Abdel-Rahman (1998) studied its ability to protect against oxidative stress caused by tert-butyl hydroperoxide in isolated rat hepatocytes and compared it to the effects of the known hepatoprotective agent silybin. The toxicity of tert-butyl hydroperoxide was manifested by the loss of cell viability and the progressive depletion of intracellular glutathione and leakage of cytosolic enzymes, alanine transaminase and aspartic transaminase in isolated rat hepatocytes treated with this compound. Preincubation of cells with 1 mM of either TQ or silybin resulted in protection against tert-butyl hydroperoxide-induced toxicity as evidenced by decreased leakage of alanine transaminase and aspartic transaminase and increased cell viability. Silybin was slightly more potent in preventing loss of cell viability and enzyme leakage, but both compounds prevented tert-butyl hydroperoxide-induced depletion of glutathione to the same extent (Daba and Abdel-Rahman, 1998). Hepatoprotective effects of TQ were also documented against carbon tetrachloride-induced toxicity (Mansour, 2000). In this study, oral administration of TQ in drinking water, starting 5 days before carbon tetrachloride injection and continuing during the experimental period, ameliorated the hepatotoxicity induced by carbon tetrachloride, as evidenced by a significant reduction in the elevated levels of serum enzymes as well as a significant decrease in the hepatic malonaldehyde content and a significant increase in the total sulfhydryl content. While oral administration of TQ in a single dose (100 mg/kg) resulted in significant hepatoprotection against carbon tetrachloride-induced toxicity in male Swiss albino mice, dihydrothymoquinone (IC_{50}: 0.34), the reduction metabolite of TQ, was found to be more potent than TQ (IC_{50}: 0.87) in protecting against carbon tetrachloride-induced hepatotoxicity (Nagi et al., 1999). Similar hepatoprotective effects in the same system (carbon tetrachloride-induced hepatotoxicity) were obtained following a 4-weeks' oral intake of NSO in male albino rats (El-Dakhakhny et al., 2000c). Recently, it was shown that *N. sativa* seeds given orally every day for 2 months decreased the lipid peroxidation, increased the antioxidant defense system and prevented the lipid peroxidation-induced liver damage in experimentally induced diabetic rabbits (Meral et al., 2001), suggesting that the seed may be used in diabetic patients to prevent lipid peroxidation.

VIII. Analgesic and antinociceptive effects of *N. sativa*

The analgesic and antinociceptive effects of *N. sativa* were only recently reported (Abdel-Fattah et al., 2000; Al-Ghamdi, 2001) and the mechanisms by which they occur are not fully understood. Evidence, however, points to the potential of using the aqueous extract of *N. sativa* as an analgesic agent. *N. sativa* crude aqueous suspension was found to produce significant increase in the hot plate reaction time in mice (indicating analgesic effects); however, it had no effect on yeast-induced pyrexia. The absence of antipyretic effect suggests that the constituents of these seeds may not inhibit the synthesis of prostaglandins (Al-Ghamdi, 2001). This was in agreement with the findings of Abdel-Fattah et al. (2000) who showed that the oral administration of NSO (50–400 mg/kg) dose-dependently suppressed the nociceptive responses caused by thermal, mechanical and chemical nociceptive stimuli in mice. In this study, the systemic administration and the i.c.v. injection of NSO attenuated the response in not only the early phase, but also the late phase of the chemical test. In another study, upon using several opioid receptor antagonists, it was demonstrated that NSO and TQ produce antinociceptive effects through indirect activation of the supraspinal opioid systems (Abdel-Fattah et al., 2000). It remains unclear if TQ antinociception in the chemical test is due to its direct interaction with opioid receptors, since no information is available regarding the *in vitro* opioid receptor binding of TQ. However, the difference in receptor antagonist sensitivity of TQ antinociception between the early and late phases raises the possibility that the antinociceptive action of TQ in the chemical test is mediated by mechanisms other than direct stimulation of opioid receptors located in the central nervous system. Other mechanisms that could explain the antinociceptive effects of NSO include its inhibitory effect on the inflammatory mediators. Further experiments are needed to clarify the mechanisms underlying the antinociceptive action of NSO and TQ.

IX. Are *N. sativa* seeds or its components safe to consume?

The toxicity properties of TQ and THQ were investigated in male rats whereby the drugs were dissolved in propylene glycol, injected i.p. into 30 male rats and LD_{50} determined (El-Dakhakhny, 1965). Using this protocol, TQ (LD_{50}: 10 mg/kg b.w.) was found to be more toxic than THQ (LD_{50}: 25 mg/kg b.w.).

In more recent studies, the oral administration of aqueous extracts of the seeds of *N. sativa* for 14 days has been shown to cause no toxicity symptoms in male Sprague-Dawley rats (Tennekoon et al., 1991). The safety of consuming *N. sativa* seeds was also recently reported by Al-Homidan et al. (2002) whereby the seeds did not affect the growth of 7-day-old Hibro broiler chicks when fed to them at 20 and 100 g/kg of the diet for 7 weeks.

Although several studies have reported the safety of consuming *N. sativa* seeds, a recent comprehensive investigation has shown that the plant is relatively unsafe if consumed for prolonged periods of time (Zaoui et al., 2002b). LD_{50} values obtained by single doses, orally and intraperitoneally administered in mice were 28.8 and 2.06 ml/kg b.w., respectively. Treatment of animals with a daily oral dose of 1 ml/kg b.w. of NSO for 12 weeks resulted in significant slowdown of the body weight in *N. sativa*-treated animals compared to untreated control animals. Changes in key hepatic enzymes levels and histopathological modifications (heart, liver, kidneys and pancreas) were not observed in rats treated with *N. sativa* after 12 weeks. However, the serum cholesterol, triglyceride and glucose levels and the count of leukocytes and platelets decreased significantly, compared to control values, while hematocrit and hemoglobin levels increased significantly. The decrease in body weight in *N. sativa*-treated rats was thought to be related to the decrease in serum lipids and glucose levels as a consequence of a possible reduction in food intake by NSO administration (Zaoui et al., 2002b). Interestingly, no evidence of toxicity was noted in 10 times this dose in mice, suggesting only a seeming margin of safety for the used therapeutic doses of *N. sativa*.

In this regard, it is worth mentioning that TQ is both an irritant and a potent elicitor of allergic contact dermatitis (Steinmann et al., 1997; Zedlitz et al., 2002).

X. Conclusions

The use of ethnobotanical drugs among Asians as complementary medicine is prevalent and is also gaining increasing popularity in the West. More than 25% of currently used drugs are derived directly from plants; while the other 25% are chemically altered natural products. Evidence indicates that *N. sativa* seeds have a potential medicinal value and are relatively safe to consume. Future research should focus on the mechanisms by which *N. sativa* seeds exert their medicinal effects. With the increased understanding of its mechanism of bioactivity, the incorporation of this medicinal herb as complementary medicine into mainstream medical science can be achieved in the future.

Acknowledgments

We thank the Deutsche Forschungsgemeinschaft for supporting the *in vitro* and *in vivo* testing of the role of thymoquinone in colon cancer prevention and therapy (Germany: DFG 477/6-1 and 477/7-1).

References

Abdel-Fattah AM, Matsumoto K, Watanabe H. (2000) Antinociceptive effects of *Nigella sativa* oil and its major component, thymoquinone in mice. Eur J Pharmacol 400:89–97.

Abdel-Salam IM, Hassanein AG, El-Sayed MM, Mokhtar NM. (1993) The protective effect of *Nigella sativa* on bilharzial-induced liver injury in mice. Arab J Lab 19:127–38.

Abou Basha LI, Rashed MS, Aboul-Enein HY. (1995) TLC assay of thymoquinone in black seed oil (*Nigella sativa* Linn) and identification of dithymoquinone and thymol. J Liquid Chrom 18:105–15.

Aboul Ela MA, El-Shaer NS, Ghanem NB. (1996) Antimicrobial evaluation and chromatographic analysis of some essential and fixed oils. Pharmazie 51:993–4.

Aboul-Ela EI. (2002) Cytogenetic studies on *Nigella sativa* seeds extract and thymoquinone on mouse cells infected with schistosomiasis using karyotyping. Mutat Res 516:11–7.

Aboul-Enein YH, Abou-Basha IL. (1995) Simple HPLC method for the determination of thymoquinone in black seed oil (*Nigella sativa* Linn). J Liquid Chrom 18:895–902.

Aboul-Enein HY, Kruk I, Lichszteld K. (1999) Antioxidant action of several β-adrenergic blockers and related compounds. Toxicol Environ Chem 68:349–61.

Aboutabl EA, El-Azzouny AA, Hammerschmidt FJ. (1986) Aroma volatiles of *Nigella sativa* L. seeds. In: Brunke EJ editor. Progress in essential oil research. Proceedings of the 16th International Symposium on essential oils. Berlin: DeGruyer, pp. 49–55.

Agarwal R, Kharya MD, Shrivastava R. (1979) Antimicrobial and anthelminthic activities of the essential oil of *Nigella sativa* Linn. Indian J Exp Biol 17:1264–5.

Akhtar MS, Riffat S. (1991) Field trial of *Saussurea lappa* roots against nematodes and *Nigella sativa* seeds against cestodes in children. J Pak Med Assoc 41:185 7.

Al-Awadi F, Fatania H, Shamte U. (1991) The effect of a plant's mixture extract on liver gluconeogenesis in streptozotocin-induced diabetic rats. Diabetes Res 18:163–8.

Al-Ghamdi MS. (2001) The anti-inflammatory, analgesic and antipyretic activity of *Nigella sativa*. J Ethnopharmacol 76:45–8.

Al-Gharably NM, Badary OA, Nagi MN, Al-Shabanah OA, Al-Sawaf HA, Al-Rikabi AC, Al-Bekairi AM. (1997) Protective effect of thymoquinone against carbon tetrachloride-induced hepatotoxicity in mice. Res Commun Pharmacol Toxicol 2:41–50.

Al-Hader AA, Aqel MB, Hasan ZA. (1993) Hypoglycemic effects of the volatile oil of *Nigella sativa* seeds. Int J Pharmacogn 31:96–100.

Al-Homidan A, Al-Qarawi AA, Al-Wally SA, Adam SE. (2002) Response of broiler chicks to dietary *Rhazya stricta* and *Nigella sativa*. Br Poult Sci 43:291–6.

Al-Shabanah OA, Badary OA, Nagi MN, Al-Gharably NM, Al-Rikabi AC, Al-Bekairi AM. (1998) Thymoquinone protects against doxorubicin-induced cardiotoxicity without compromising its antitumor activity. J Exp Clin Cancer Res 17:193–8.
Aqel M, Shaheen R. (1996) Effect of the volatile oil of *Nigella sativa* seeds on the uterine smooth muscle of rat and guinea pig. J Ethnopharmacol 52:23–6.
Badary OA. (1999) Thymoquinone attenuates ifosfamide-induced Fanconi syndrome in rats and enhances its antitumor activity in mice. J Ethnopharmacol 1:135–42.
Badary OA, Abdel-Naim AB, Abdel-Wahab MH, Hamada FM. (2000) The influence of thymoquinone on doxorubicin-induced hyperlipidemic nephropathy in rats. Toxicology 143:219–26.
Badary OA, Al-Shabanah OA, Nagi MN, Al-Bekairi AM, Elmazar MA. (1998) Acute and subchronic toxicity of thymoquinone in mice. Drug Dev Res 44:56–61.
Badary OA, Al-Shabanah OA, Nagi MN, Al-Rikabi AC, Elmazar MM. (1999) Inhibition of benzo(a)pyrene-induced fore-stomach carcinogenesis in mice by thymoquinone. Eur J Cancer Prev 8:435–40.
Badary OA, Gamal El-Din AM. (2001) Inhibitory effects of thymoquinone against 20-methylcholanthrene-induced fibrosarcoma tumorigenesis. Cancer Detect Prev 25:362–8.
Badary OA, Nagi MN, Al-Shabanah OA, Al-Sawaf HA, Al-Sohaibani MO, Al-Bekairi AM. (1997) Thymoquinone ameliorates the nephrotoxicity induced by cisplatin in rodents and potentiates its antitumor activity. Can J Physiol Pharmacol 75:1356–61.
Bamosa AO, Ali BA, Sowayan SA. (1997) Effect of oral ingestion of *Nigella sativa* seeds on some blood parameters. Saudi Pharm J 5:126–9.
Bertani T, Cutillo F, Zoja C, Broggini M, Remuzzi G. (1986) Tubulo-interstitial lesions mediate renal damage in adriamycin glomerulopathy. Kidney Int 30:488–96.
Boulos L. (1983) Medicinal plants of North Africa. Algonac, MI: Reference Publications, p. 103.
Brade WP, Seeber S, Herdrich K. (1986) Comparative activity of ifosfamide and cyclophosphamide. Cancer Chemother Pharmacol 18:S1–9.
Burits M, Bucar F. (2000) Antioxidant activity of *Nigella sativa* essential oil. Phytother Res 14:323–8.
Chakravarty N. (1993) Inhibition of histamine release from mast cells by nigellone. Ann Allergy 70:237–42.
Chopra RN, Nayar SL, Chopra IC. (1956) Glossary of Indian medicinal plants. New Delhi: CSIR, p. 175.
Cortes EP, Lutman G, Wanta J, Pickren J, Wallace J, Holland J. (1975) Adriamycin (NSC 123127) cardiotoxicity: a clinicopathologic correlation. Cancer Chemother Rep 6:215–25.
Daba MH, Abdel-Rahman MS. (1998) Hepatoprotective activity of thymoquinone in isolated rat hepatocytes. Toxicol Lett 16:23–9.
De Smet PAGM. (1997) The role of plant-derived drugs and herbal medicines in health care. Drugs 54:801–40.
Dubick MA. (1986) Historical perspectives on the use of herbal preparations to promote health. J Nutr 116:1348–54.
El-Dakhakhny M. (1963) Studies on the chemical constitution of the Egyptian *Nigella sativa* L. seeds. II: The essential oil. Planta Medica, 12:465–70.
El-Dakhakhny M. (1965) Studies on the Egyptian *Nigella sativa* L. IV Some pharmacological properties of seeds active principle in comparison to its dihydro compound and its polymer. Arzneim Forsch (Drug Res Germ) 15:1227–9.
El-Dakhakhny M, El-Halim M, Aly SM, Barakat MK. (2000a) Protective effect of *Nigella sativa* oil and clindamycin against production of TNF-α and liver injury in murine septic peritoritis. Naunyn Schmiedeberg's Arch Pharmacol 362:R17 (abstract no. 12).
El-Dakhakhny M, Barakat M, Abdel El-Halim M, Aly SM. (2000b) Effects of *Nigella sativa* oil on gastric secretion and ethanol-induced ulcer in rats. J Ethnopharmacol 72:299–304.
El-Dakhakhny M, Mady NI, Halim MA. (2000c) *Nigella sativa* L. oil protects against induced hepatotoxicity and improves serum lipid profile in rats. Arzneim Forsch 50:832–6.

El-Dakhakhny M, Mady NJ, Lembert N, Ammon HP. (2002a) *Nigella sativa* oil, nigellone and derived thymoquinone inhibit synthesis of 5-lipoxygenase products in polymorphonuclear leukocytes from rats. J Ethnopharmacol 81:161–4.

El-Dakhakhny M, Mady N, Lembert N, Ammon HP. (2002b) The hypoglycemic effect of *Nigella sativa* oil is mediated by extrapancreatic actions. Planta Med 68:465–6.

El-Dakhakhany M. (1982) Some pharmacological properties of some constituents of *Nigella sativa* L. seeds. The carbonyl fraction of the essential oil. In: Proceedings of the Second International Conference on Islamic Medicine, 12th April. Kuwait: Studies in Islamic Medicine and Advantages of Herbal treatment, pp. 246–431.

El-Daly ES. (1998) Protective effect of cysteine and vitamin E, *Crocus sativus* and *Nigella sativa* extracts on cisplatin-induced toxicity in rats. J Pharmacol (Belgium) 53:87–93.

El-Fatatry HM. (1975) Isolation and structure assignment of an antimicrobial principle from the volatile oil of *Nigella sativa* L seeds. Pharmazie 30:109–11.

El-Kadi A, Kandil O. (1986) Effect of *Nigella sativa* (the black seeds) on immunity. Proceedings of the Fourth International Conference on Islamic Medicine. Bull. Islamic Med, 4:344–8.

El-Kadi A, Kandil O, Tabuni AM. (1987) Nigella cell-mediated immunity. Arch AIDS Res 1:232–3.

El-Mahmoudy A, Matsuyama H, Borgan MA, Shimizu Y, El-Sayed MG, Minamoto N, Takewaki T. (2002) Thymoquinone suppresses expression of inducible nitric oxide synthase in rat macrophages. Int Immunopharmacol 2:1603–11.

El-Tahir KE, Ashour MM, Al-Harbi MM. (1993a) The respiratory effects of the volatile oil of the black seed (*Nigella sativa*) in guinea pigs: elucidation of the mechanism(s) of action. Gen Pharmacol 24:1115–22.

El-Tahir KE, Ashour MM, Al-Harbi MM. (1993b) The cardiovascular actions of the volatile oil of the black seed (*Nigella sativa*) in rats: elucidation of the mechanism of action. Gen Pharmacol 24:1123–31.

Enomoto S, Asano R, Iwahori Y, Narui T, Okada Y, Singab AN, Okuyama T. (2001) Hematological studies on black cumin oil from the seeds of *Nigella sativa* L. Biol Pharm Bull 24:307–10.

Fararh KM, Atoji Y, Shimizu Y, Takewaki T. (2002) Insulinotropic properties of *Nigella sativa* oil in streptozotocin plus nicotinamide diabetic hamster. Res Vet Sci 73:279–82.

Fernandez ER, Carvalho FD, Remiao FG, Bastos ML, Pinto MM, Gottlieb OR. (1995) Hepatoprotective activity of xanthones and xanthonolignoids against *tert*-butylhydroperoxide-induced toxicity in isolated hepatocytes – comparison with silybin. Pharmacol Res 12:1756–60.

Gali-Muhtasib H, Diab-Assaf M, Boltze C, Al-Hmaira J, Hartig R, Roessner A, Schneider-Stock R. (2004) Thymoquinone extracted from black seed triggers apoptotic cell death in human colorectal cancer cells via a p53-dependent mechanism. Int J Oncol 25(4):857–66.

Ghazanfer SA. (1994) Handbook of Arabian medicinal plants. Boca Raton, FL: CRC Press, p. 180.

Ghoneim MT, El-Gindy AR, El-Alami E, Shoukry R, Yaseen S. (1982) Possible effect of some extracts of *Nigella sativa* L. seeds on blood coagulation system and fibrinolytic activity. Proceedings of the Second International Conference on Islamic Medicine, Kuwait, 12th April.

Ghosheh OA, Houdi AA, Crooks PA. (1999) High-performance liquid chromatographic analysis of the pharmacologically active quinones and related compounds in the oil of the black seed. J Pharm Biomed Anal 19:757–62.

Gilani AH, Aziz N, Khurram IM, Chaudhary KS, Iqbal A. (2001) Bronchodilator, spasmolytic and calcium antagonist activities of *Nigella sativa* seeds (Kalonji): a traditional herbal product with multiple medicinal uses. J Pak Med Assoc 51:115–20.

Hanafy MS, Hatem ME. (1991) Studies on the antimicrobial activity of *Nigella sativa* seed (black cumin). J Ethnopharmacol 34:275–8.

Haq A, Adbullatif M, Labo P, Khabar K, Sheth K, AlSedairy S. (1995) *Nigella sativa*: effect on human lymphocytes and polymorphonuclear leucocyte phagocytic activity. Immunopharmacology 30:147–55.

Haq A, Lobo PI, Al-Tufail M, Rama NR, Al-Sedairy ST. (1999) Immunomodulatory effect of *Nigella sativa* proteins fractionated by ion exchange chromatography. Int J Immuno pharmacol 21:283–95.

Hashem FM, El-Kiey MA. (1982) *Nigella sativa* seeds of Egypt. J Pharm Sci United Arab Republic 3:121–33.

Hassan M, El-Dakhakhny M. (1992) Effect of some *Nigella sativa* constituents on chemical carcinogenesis in hamster cheek pouch. J Egyptian Soc PharmacolExp Therap 11:675–7.

Horton AA, Wood JM. (1990) Prevention of thromoboxane B_2-induced hepatocyte plasma membrane bleb formation by certain prostaglandins and a proteinase inhibitor. Biochim Biophys Acta 1022:319–24.

Hossain S, Takimoto H, Hamano S, Yoshida H, Ninomiya T, Minamishina Y, Kimura G, Nomoto K. (1999) Protective effects of hochu-ekki-to, a Chinese traditional herbal medicine against murine cytomegalovirus infection. Immunopharmacology 41:169–81.

Houghton PJ, Zarka R, De las Heras B, Hoult RS. (1995) Fixed oil of *Nigella sativa* and derived thymoquinone inhibit eicosanoid generation in leukocytes and membrane lipid peroxidation. Planta Medica 61:33–6.

Huxtable RJ. (1992) The pharmacology of extinction. J Ethnopharmacol 37:1–11.

Jansen PCM. (1981) Spices, condiments and medicinal plants in Ethiopia, their taxonomy and agricultural significance. Addis Ababa: Center for Agricultural Publishing and Documentation, pp. 76–85.

Khan MA. (1999) Chemical composition and medicinal properties of *Nigella sativa* Linn. Inflammopharmacology 7:15–35.

Kiso Y, Tohkin M, Hikine H, Hattori M, Sakamoto T, Namba T. (1984) Mechanism of antihepatotoxic activity of glycerrhizin. Effect of free radical generation and lipid peroxidation. Planta Med 50:298–302.

Kruk I, Michalska T, Lichszteld K, Kladna A, Aboul-Enein HY. (2000) The effect of thymol and its derivatives on reactions generating reactive oxygen species. Chemosphere 41:1059–64.

Kumara SS, Huat BT. (2001) Extraction, isolation and characterisation of antitumor principle, α-hederin, from the seeds of *Nigella sativa*. Planta Med 67:29–32.

Labhal A, Settaf A, Bennani-kabchi N, Cherrah Y, Slaoui A, Hassar M. (1997) Actio antiobesite, hypocholesterolemiante et hypotriglyceridemiante de *Nigella sativa* chez le Psammomys obesus. Caducee 27:26–8.

Mabrouk GM, Moselhy SS, Zohny SF, Ali EM, Helal TE, Amin AA, Khalifa AA. (2002) Inhibition of methylnitrosourea (MNU)-induced oxidative stress and carcinogenesis by orally administered bee honey and Nigella grains in Sprague Dawley rats. J Exp Clin Cancer Res 21:341–6.

Mahfouz M, Abdel Maguid R, El-Dakhakhny M. (1960) Effectiveness of nigellone in asthma in adults. Alexandria Med J 6:543–7.

Mahfouz M, Abdel-Maguid R, El-Dakhakhny M. (1965) The effect of "nigellone therapy" on the histaminopexic power of the blood sera in asthmatic patients. Arzneim Forsch (Drug Res.), 15:1230–4.

Mahfouz M, El-Dakhakhny M. (1960) The isolation of a crystalline active principle from *Nigella sativa* L. seeds. J Pharm Sci United Arab Republic 1:9–19.

Mahmoud MR, El-Abhar HS, Saleh S. (2002) The effect of *Nigella sativa* oil against the liver damage induced by Schistosoma mansoni infection in mice. J Ethnopharmacol 79:1–11.

Mansour MA. (2000) Protective effects of thymoquinone and desferrioxamine against hepatotoxicity of carbon tetrachloride in mice. Life Sci. 66:2583–91.

Mansour MA, Nagi MN, El-Khatib AS, Al-Bekairi AM. (2002) Effects of thymoquinone on antioxidant enzyme activities, lipid peroxidation and DT-diaphorase in different tissues of mice: a possible mechanism of action. Cell. Biochem. Funct. 20:143–51.

Masaki N, Kyle ME, Farber JL. (1989) Tert-butyl hydroperoxide kills cultured hepatocytes by peroxidizing membrane lipids. Arch Biochem Biophys 269:390–9.

Meral I, Yener Z, Kahraman T, Mert N. (2001) Effect of *Nigella sativa* on glucose concentration, lipid peroxidation, anti-oxidant defence system and liver damage in experimentally induced diabetic rabbits. J Vet Med A Physiol Pathol Clin Med 48:593–9.

Miguez M, Anundi I, Sainz-Pardo L, Lindros KO. (1994) Hepatoprotective mechanism of silymarin: no evidence for involvement of cytochrome P450 2EI. Chem Biol Interact. 91:51–63.

Morsi NM. (2000) Antimicrobial effect of crude extracts of *Nigella sativa* on multiple antibiotics-resistant bacteria. Acta Microbiol Pol 49:63–74.

Nadkarni K. (1976) Crocus sativus, *Nigella sativa*. In: Nadkarni KM editor. Indian materia medica. Bombay, India: Popular Prakashan, pp. 386–411.

Nagi MN, Alam K, Badary OA, Al-Shabanah OA, Al-Sawaf HA, Al-Bekairi AM. (1999) Thymoquinone protects against carbon tetrachloride hepatotoxicity in mice via an antioxidant mechanism. Biochem Mol Biol Int 47:153–9.

Nagi MN, Mansour MA. (2000) Protective effect of thymoquinone against doxorubicin-induced cardiotoxicity in rats: a possible mechanism of protection. Pharmacol Res 41:283–9.

Nair SC, Salomi MJ, Panikkar B, Panikkar KR. (1991) Modulatory effects of Crocus sativus and *Nigella sativa* extracts on cisplatin-induced toxicity in mice. J Ethnopharmacol 31:75–83.

Navaro MC, Montilla MP, Martin A, Jimenez J, Utrilla MP. (1993) Free radical scavenger and antihepatotoxic activity of Rosemarinus tomentosus. Planta Med 53:312–4.

Reynolds P, Rahija J, Schenkman I, Richter B. (1993) Experimental murine cytomegalovirus infection in severe combined immunodeficient mice. Lab Anim Sci 43:291–5.

Salem ML, Hossain MS. (2000) Protective effect of black seed oil from *Nigella sativa* against murine cytomegalovirus infection. Int J Immunopharmacol 22(9):729–40.

Salomi NJ, Nair SC, Panikkar KR. (1991) Inhibitory effects of *Nigella sativa* and saffron (Crocus sativus) on chemical carcinogenesis in mice. Nut Cancer 16:67–72.

Salomi NJ, Nair SC, Jayawardhanan KK, Varghese CD, Panikkar KR. (1992) Antitumour principles from *Nigella sativa* seeds. Cancer Lett 31:41–6.

Saxena A, Vyas K. (1986) Antimicrobial activity of seeds of some ethnopharmacinal plants. J Econ Taxon Botany 8:291–9.

Shertzer HG, Bannenberg H, Zhu RM, Moldeus P. (1994) The role of thiols in mitochondrial susceptibility to iron and tert-butyl hydroperoxide-mediated toxicity in cultured hepatocytes. Chem Res Toxicol 7:358–66.

Shoieb AM, Elgayyar M, Dudrick PS, Bell JL, Tithof PK. (2003) *In vitro* inhibition of growth and induction of apoptosis in cancer cell lines by thymoquinone. Int J Oncol 22:107–13.

Smith A, Brennessel J. (1994) Cytomegalovirus. Infect Dis Clin North Am 8:427–38.

Steinmann A, Schatzle M, Agathos M, Breit R. (1997) Allergic contact dermatitis from black cumin (*Nigella sativa*) oil after topical use. Contact Dermatitis 36:268–9.

Swamy SM, Tan BK. (2000) Cytotoxic and immunopotentiating effects of ethanolic extract of *Nigella sativa* L seeds. J Ethnopharmacol 70:1–7.

Tennekoon K, Jeevathayaparan S, Kurukulasooriya A, Karunanayake E. (1991) Possible hepatotoxicity of *Nigella sativa* seeds and Dregea volubilis leaves. J Ethnopharmacol 31:283–9.

Thabrew MI, Gove CD, Hughes RD, McFarlane IG, Williams R. (1995) Protective effects of Osbeckia octandra against galactosamine and tert-butyl hydroperoxide-induced hepatocyte damage. J Ethnopharm 49:69–76.

Topozada HH, Mazolum HA, El-Dakhakhny M. (1965) The antibacterial properties of *Nigella sativa* seeds. Active principle with some clinical applications. J Egypt Med Assoc 48:187–202.

ur-Rahman A, Malik S, Cunheng H, Clardy J. (1985) Isolation and structure of determination of nigellicine, a novel alkaloid from the seeds of *Nigella sativa*. Tetrahedron Lett 26:2759–62.

Valenzuela A, Guera R, Videla LA. (1986) Antioxidant properties of the flavonoids silybin and +/− cyanidanol-3: comparison with butylated hydroxyanisole and butylated hydroxytoluene. Planta Med 52:438–40.

Vogel G. (1977) Natural substances with effects on the liver. In: Wagner H, Wolff P editors. New natural products and plant drugs with pharmacological, biological or therapeutic activity. New York: Springer, pp. 249–63.

Worthen D, Ghosheh O, Crooks P. (1998) The in vitro anti-tumor activity of some crude and purified components of black seed, *Nigella sativa* L. Anticancer Res 18:1527–32.

Zaoui A, Cherrah Y, Alaoui K, Mahassine N, Amarouch H, Hassar M. (2002a) Effects of *Nigella sativa* fixed oil on blood homeostasis in rat. J Ethnopharmacol 79:23–6.

Zaoui A, Cherrah Y, Lacaille-Dubois MA, Settaf A, Amarouch H, Hassar M. (2000) Diuretic and hypotensive effects of *Nigella sativa* in the spontaneously hypertensive rat. Therapie 5:379–82.

Zaoui A, Cherrah Y, Mahassini N, Alaoui K, Amarouch H, Hassar M. (2002b) Acute and chronic toxicity of *Nigella sativa* fixed oil. Phytomedicine 9:69–74.

Zedlitz S, Kaufmann R, Boehncke WH. (2002) Allergic contact dermatitis from black cumin (*Nigella sativa*) oil-containing ointment. Contact Dermatitis 46:188.

Zima T, Tesar V, Stipek S, Crkovska J, Poledne R, Teminova J, Platenik J, Rychlik I, Merta M, Nemecek K. (1997) The influence of cyclosporin on lipid peroxidation and superoxide dismutase in adriamycin nephropathy in rats. Nephron 75:464–8.

M.T.H. Khan and A. Ather (eds.)
Lead Molecules from Natural Products

Cyclin-dependent kinase inhibitors from natural sources: recent advances and future prospects for cancer treatment

HALA GALI-MUHTASIB

Abstract

Natural products have long been a source of cure for cancer, a disease of major cause of deaths in many countries. Despite the advances in cancer treatment, there is a continuing need for the development of new anticancer drugs by exploring the pool of natural products. Of the more than 2000 protein kinases that regulate cellular functions, *cyclin-dependent kinases (CDKs)*, have been extensively studied because of their essential role in the regulation of cell proliferation, transcription and apoptosis. Intensive screening has led, in the last few years, to the identification of a series of naturally occurring chemical inhibitors of CDKs, some of which display high selectivity and efficiency. These inhibitors are antiproliferative agents that arrest cells in G_1 and/or G_2/M phases of the cell cycle. In addition, many CDK inhibitors facilitate or trigger programmed cell death (apoptosis) in proliferating cells. The potential use of these inhibitors is being extensively evaluated for cancer chemotherapy. This paper reviews naturally derived CDK inhibitors with promising anticancer potential, mainly flavopiridol, staurosporines and indirubins, and the recent advances in understanding their mechanism of action and structure–function relationships at the molecular, cellular and physiological levels.

Keywords: *cancer therapy, cell cycle, cyclin-dependent kinases, cyclin-dependent kinase inhibitors, natural drugs*

Abbreviations: CDKs, cyclin-dependent kinases; CAK, CDK-activating kinases.

I. Introduction

Naturally based compounds play an essential role in the primary health care of the majority of the world's population. Of the 250,000 species of plants, more than one thousand have been found to contain agents with significant anticancer properties. While many molecules obtained from nature have shown potent cancer preventive and therapeutic effects, there is a huge number of molecules that remains to be trapped or studied in detail by the medicinal scientists. There is at present much optimism about the possibility of discovering anticancer drug treatment strategies that modulate cell cycle regulatory molecules. Candidate targets for such strategies include the cyclin-dependent kinases (CDKs). CDKs are central players that drive the progression through the individual phases of the cell cycle. They determine the timing of key events in the cell cycle, and may also regulate other important cellular functions (Murray, 1994; Morgan, 1995; Li and Blow, 2001). There has been a great deal of advances in cancer therapy using natural products (Mukherjee et al., 2001). Several attempts to screen extracts from the National Cancer Institute National Institutes of Health Natural Products Repository has led to the discovery of anticancer agents. In this chapter, I have not attempted to be exhaustive but have given a current overview on naturally derived CDK modulators used in the clinic for the treatment of cancer with an emphasis on flavopiridol, staurosporines and indirubins.

II. Traditional vs. advanced therapies of cancer

The goal of cancer therapy is to eliminate malignant cells while sparing normal cells. Most of the anticancer drugs discovered in the past act by inhibiting DNA synthesis in one way or the other. Drugs that bind or damage DNA were discovered as antiproliferative agents without specific consideration to their site of action. Thus, these drugs tend to be nonselective – while treating the cancer they also attack a number of normal cells and this, in turn, limits their usefulness in the clinic. As a consequence of the lack of drug selectivity, cancer chemotherapy has been often accompanied by a variety of sometimes devastating short- or long-term side effects. Some adverse effects include mutagenic, teratogenic and bone marrow toxicity effects. Drugs with low bone marrow toxicity such as bleomycin, steroid hormones and vincristine are often used in combination chemotherapy with drugs that have high bone marrow toxicity. The nonselective toxicity of anticancer drugs has stressed the need to identify key genes, components of signaling pathways, or cellular processes, which are altered in human cancer, as potential intervention points or targets that could be used in the design of new cancer drugs.

An exciting new approach to drug development includes the discovery of small molecules that are able to specifically attack the aberrant genetic alterations and deregulated biochemical pathways that are responsible for cancer while sparing healthy tissues. By targeting cancer cells, this new generation of anticancer agents promises to be more selective and less toxic than current drugs used for cancer prevention and treatment. In selecting drug targets for novel therapies, interest has been focused on cell cycle molecular targets (Buolamwini, 2000), particularly in those pathways that are most frequently deregulated in cancers. The CDKs are frequently deregulated in cancers; mutations and overexpression of these kinases, mainly CDK4, have been reported and proposed to be oncogenic events.

The opportunity of molecular targeting by drugs has been generated by knowledge gained through recent advances in multiple research disciplines. First, the combination of structural and functional genomics and proteomics research with specific studies in human molecular oncology has led to a detailed understanding at the cellular and molecular levels of genes and proteins that are responsible for cancer causation and progression. This coupled with the emerging field of bioinformatics that manages the generated information has led to a revolution in the identification of new therapeutic targets. Second, the use of combinatorial chemistry with high-throughput screening for identifying and optimizing drugs complemented with advanced structural biology has paved the way for a whole new approach to cancer drug design and discovery. Finally, the improved ties between laboratory and clinical research have allowed the integration of drug discovery, development and clinical testing. In such a cooperative setting, researchers can now effectively (a) identify molecular targets in the cell, (b) find drugs that will "hit" the targets, (c) test these drugs for safety and efficacy in the laboratory and in animal studies and (d) test the usage of successful candidate drugs in clinical trials.

III. Cell cycle control and CDKs

During the last two decades, it has become possible to identify the molecular mechanisms that regulate the cell cycle, and, thereby, cell division. Before a cell divides, it grows in size (gap 1 phase or G_1), duplicates its chromosomes (S phase), checks that DNA replication is completed (gap 2 phase or G_2) and finally, in mitosis (M phase), separates the chromosomes for exact distribution between the two daughter cells. After division, the cells are back in G_1 and the cell cycle is completed. Cells in the first cell cycle phase (G_1) do not always continue through the cycle. Instead, they can exit from the cell cycle and enter a resting stage (G_0), a stage at which basic cellular metabolism, including transcription and translation, are depressed. Other cases of deviation from the cell cycle include programmed cell death (apoptosis) and differentiation. For all living eukaryotic organisms it is essential that the different phases of the cell cycle are precisely coordinated and the processes of cell growth, differentiation and apoptosis are balanced. The signaling pathways that control these processes are central to the functioning of all multicellular life and any defects in cell cycle control may lead to chromosome alterations, loss of cellular growth control and the induction of cancer.

The identification of key molecules that regulate the cell cycle has opened new possibilities for cancer treatment. It is believed that the next 5 to 10 years will reveal the extraordinary potential for advances in cell cycle control-based therapies for the treatment of human cancers. Among the main player, proteins in eukaryotic cells that control the passage of a cell through the cell cycle are the CDKs. The CDKs are a family of heterodimeric serine/threonine protein kinases, each consisting of a catalytic CDK subunit and an activating cyclin subunit. CDKs are like engines driving progression through each of the individual phases of the cell cycle. To be active, these proteins form complexes with cyclins, and together they act as major control switches for the cell cycle, causing the cell to move from the G_1 to S phase or the G_2 to M phase. So far, 25 different cyclins and 13 different CDKs have been reported (Knockaert et al., 2002). Their levels are invariant throughout the cell cycle, but as mentioned earlier, CDK activities are modulated by their interaction with the cyclins whose levels fluctuate. Their full activation requires phosphorylation of a conserved threonine residue near the kinase active site (Morgan, 1997). Some CDKs, the CDK-activating kinases (CAK) exemplified by CDK7, control the activity of other cell cycle CDKs (Fisher and Morgan, 1994; Kaldis, 1999); CDK7 complexes with cyclin H to activate cell cycle CDKs by phosphorylating critical residues for the kinase activity. In contrast to CAKs, some other kinases are responsible for phosphorylations that inactivate CDKs. These include the Wee1/Myt1 family kinases that inactivate CDK1 (Parker and Piwnica-Worms, 1992; Mueller et al., 1995). These inactivated phosphorylated residues can be removed by the Cdc25 family of protein phosphatases (Coleman and Dunphy, 1994; Draetta and Eckstein, 1997). CDK activity is also regulated through their interaction with members of the Cip/Ink4 family of cell cycle inhibitory proteins.

CDKs phosphorylate many substrates critical to cell cycle progression although some members of the CDK family are involved in other processes as well. Among these, CDK5 is involved in controlling the differentiation of postmitotic neural and muscle cells (Gervasi and Szaro, 1995; Philpott et al., 1997), and CDK7, 8, 9 are involved in controlling basal transcription by RNA polymerase II (Bregman et al., 2000). Other CDKs that are directly involved in cell cycle control include CDK2 which operates primarily in the S phase through its interaction with cyclin As and Es, and Cdc2/CDK1, which is involved in progression through M phase (by complexing with cyclin B). CDK4 and CDK6 are important regulators of entry into and exit from the cycle in G_1, interacting with cyclin Ds, and phosphorylating Rb, thereby releasing its growth-suppressive functions (Ewen et al., 1993; Kato Matsushime, Hiebert Ewen Sherr, 1993).

IV. Naturally derived CDKs used in cancer therapy

The search for CDK inhibitors was initially started because of their antitumor potential and was mostly based on the use of CDK1/cyclin B as a molecular target. From this search 10 specific inhibitors have been identified (Table 1): olomoucine (Vesely et al., 1994), roscovitine (de Azevedo et al., 1997; Meijer et al., 1997), purvalanol (Gray et al., 1998; Chang et al., 1999), CVT-313 (Brooks et al., 1997), toyocamycin (Lee CH et al., 1996), CGP60474 (Zimmermann, 1995), indirubin-3′-monoxime (Hoessel et al., 1999), the paullones (Schultz et al., 1999; Zaharevitz et al., 1999), γ-butyrolactone (Kitagawa et al., 1993) and hymenialdisine (Meijer et al., 2000; Tasdemir et al., 2002). They are all derived from structure/activity studies and from molecular modeling based on the crystal structure of the inhibitor in complex with CDK2. Despite their chemical variety, these inhibitors all act by competing with ATP at the ATP-binding site of the catalytic subunit of the kinase. A few less selective inhibitors, namely, flavopiridol (Sedlacek et al., 1996), 6-dimethylaminopurine, isopentenyladenine, suramin, staurosporine, UCN-01 and 9-hydroxyellipticine have been also described (Meijer, 1996). Of the known direct CDK inhibitors, six are derived from natural sources (Table 1). These are toyocamycin, indirubin-3′-monoxime, γ-butyrolactone, staurosporine, flavopiridol and hymenialdisine. In this paper, emphasis will be on the three of the above-mentioned inhibitors, namely, flavopiridol, staurosporines and indirubins.

Table 1
A list of some direct CDK modulators and their mode of action

Drug	Source	Mode of action	Type of cancer	Selectivity
Flavopiridol	Plant alkaloid rohitukine found in *Amoora rohituka* and *Dysoxylum binacteriferum*	General CKI; induces apoptosis; transcriptional control	Leukemia, prostate breast, lung head and neck, neuroblastoma	+ +
Staurosporine	Microbial alkaloid from *Streptomyces*	PKC inhibitor; inhibits CDK1, 2; induces apoptosis	Fibroblast, lung prostate, cervical, others	Poor
Indirubins	From the plant *Indigofera tinctoria* and *Isatis tinctoria*	Inhibits CDK1, 2, 5; inhibits GSK-3β	Breast, others	+ + +
Hymenialdisine	Marine sponge *Axinella carteri*	Inhibits CDKs; inhibits GSK-3β; inhibits MAPK	Unknown	+ +
Toyocamycin	Mycelium of *Streptomyces toyocaensis*	Inhibits rRNA processing	Limited use due to unknown toxicity	
γ-Butyrolactone	Found in microbes, wines and mung beans	Inhibits CDKs; inhibits p21	Breast, prostate, others	+ + +
Roscovitine	Synthetic	Inhibits CDK1, 2, 5; inhibits MAPK; induces apoptosis	Head, neck, cervical, others	+ + + +
Olomoucine	Synthetic	Inhibits CDK1, 2, 5: inhibits MAPK; induces apoptosis	Head, neck, cervical, others	+ + + +
Kenpaullone	Synthetic	Inhibits CDKs; inhibits GSK-3β; induces apoptosis	Carcinomas	+ + + +
Alsterpaullone	Synthetic	Inhibits CDKs and GSK-3β; induces apoptosis	Carcinomas	+ + + +
Purvalanol A and B	Synthetic	Inhibits CDK1, 2; inhibits MAPK	Several cancers	+ + + +
CVT-313	Synthetic	Inhibits CDK1, 2; inhibits Rb phosphorylation	Unknown	Unknown
9-Hydroxy-ellipticine	Metabolite of the plant alkaloid ellipticine	Inhibits telomerase and p53 phosphorylation; induces apoptosis	Unknown	Poor

IV.A. Flavopiridol

Flavopiridol is a very promising small molecule that is currently one of the most advanced CDK inhibitors in clinical development. Flavopiridol is a flavonoid related to rohitukine, an alkaloid isolated from a plant from India. It is the subject of extensive investigations because of its promising anticancer effects in many cancer models (Senderowicz, 2000, 2002; Tan and Swain, 2002; Zhai et al., 2002). This flavonoid can cause cell cycle arrest at G_1 or G_2 and can inhibit the activation and activity of several CDKs, specifically CDK1, CDK2, CDK4 and CDK6 (Losiewicz et al., 1994; Carlson et al., 1996). This inhibition is thought to occur through its ability to dock in the ATP-binding site of all CDKs. Moreover, flavopiridol inhibits CDK7, a CAK, leading to the loss of the activation phosphorylation of most CDKs (Sedlacek, 2001). It has also been shown to inhibit CDK5, expressed mostly in neurons (Rapella et al., 2002), and to associate almost irreversibly with CDK9, a member of the positive transcription elongation factor b (pTEFb), required for elongation control of RNA polymerase II (Chao et al., 2000, 2001).

Although flavopiridol is generally defined as a CDK inhibitor, studies have shown that it may have various other effects which may be mediated by different mechanisms of action. One such mechanism is the depletion of cyclin D1; in breast carcinoma cells exposed to flavopiridol, cyclin D1 promoter activity decreases leading to the loss of cyclin D1 mRNA, and subsequently decreased protein levels (Carlson et al., 1996). Flavopiridol is also thought to mediate its anticancer effects through the inhibition of angiogenesis, since it prevents the hypoxia-induced vascular endothelial growth factor upregulation in human monocytes and neuroblastoma cells (Melillo et al., 1999; Rapella et al., 2002).

Several reports have shown that flavopiridol induces apoptosis and tumor regression in xenografts (Konig et al., 1997; Arguello et al., 1998; Patel et al., 1998). In human leukemia cells, flavopiridol was found to cause cytochrome c release from mitochondria independently of caspase-8 activation (Decker et al., 2001), while in lung carcinomas, the flavopiridol effect was mediated by caspase-8, and was found to be independent of changes in Bcl-2 (Achenbach et al., 2000). It was also reported that flavopiridol inhibits gene expression broadly and globally (Lam et al., 2001), and that it interferes with glycogen degradation in various tumor cells (Kaiser et al., 2001). Flavopiridol is currently undergoing advanced clinical trials as both mono and combination therapy, and with different regimens of administration. The first clinical trial that dates back to 1998 (Senderowicz et al., 1998) was conducted at the National Cancer Institute. Flavopiridol has been administered in 72-hour continuous infusions every 2 weeks in phase I/II clinical trials against gastric, colorectal, renal and lung carcinomas (Senderowicz et al., 1998; Stadler et al., 2000; Schwartz et al., 2001; Shapiro et al., 2001). It has also been administered in a 1-hour infusion mode for 3–5 days every 2 weeks against several neoplasms (Senderowicz et al., 1998). It is currently being investigated in combination therapies with several other anticancer drugs, both in laboratories and clinical trials. Breast and gastric carcinoma cells exposed to a combination of flavopiridol and paclitaxel showed enhanced activation of caspase 3 and poly (ADP-ribose) polymerase (PARP) cleavage (Motwani et al., 1999). When administered with the histone deacetylase suberoylanilide hydroxamic acid (SAHA), a 63% increase in cell death was observed, together with increases in caspase-3 and -8 activation and cytochrome c release (Almenara et al., 2002).

IV.B. Staurosporines

Staurosporines include staurosporine and its derivative 7-hydroxystaurosporine (UCN-01), both of which are natural product kinase inhibitors that were originally identified as potent protein kinase C (PKC) inhibitors. Although staurosporines inhibit certain PKC isozymes, their strong antiproliferative effects involving CDK inhibition and induction of apoptosis are thought to be unrelated to PKC inhibition (reviewed in de Azevedo et al., 1996; Brusselbach et al., 1998; Gescher, 2000).

Staurosporine is a microbial alkaloid isolated from a Streptomyces species that was first characterized in 1986 (Tamaoki et al., 1986). Subsequently, staurosporine was found to be a potent and nonspecific inhibitor of several protein kinases (Meggio et al., 1995; Konstantinidis et al., 1998), with the exception of casein kinases 1 and 2 that are resistant to inhibition by staurosporine. This PKC inhibitor has been found to cause G_1 arrest in a variety of human cell lines (Gadbois et al., 1995; Kwon et al., 1997), and this arrest is dependent on a functional pRb protein (Schnier et al., 1996). Although staurosporine is nonselective and too toxic for use as a therapeutic agent, it has proved to be useful in cellular studies as a cytostatic agent that protects normal cells from the toxic effects of chemotherapeutic agents (Chen et al., 2000). Treatment of human prostatic cancer cells with staurosporine induced remarkable inhibition of cell proliferation, G_1 arrest and suppression of CDK2 activity with an increase in CDK2-bound p21 and CDK2-bound p27 (Shimizu et al., 2001). An analysis of staurosporine-induced G_1 cell cycle arrest in various tumor cell lines has shown that the CDK inhibitor protein p27/Kip1 accumulated after staurosporine treatment (Nishi et al., 1998), suggesting that p27 is involved in staurosporine-mediated G_1 arrest. In several human tumor cell lines, staurosporine treatment in combination with other agents induced cells to enter apoptosis (Jacobson et al., 1994). A recent study has shown that the selective expression of p57/Kip2, potentiated staurosporine-induced apoptosis in HeLa cells, suggesting a role for p57/Kip2 in the response of tumor cells to staurosporine (Samuelsson et al., 2002).

UCN-01, isolated from Streptomyces, was reported in 1987 as a more selective protein kinase inhibitor than staurosporine (Takahashi et al., 1987) and was later found to possess potent antitumor activities in several in vitro and in vivo preclinical models (Akinaga et al., 1991; Seynaeve et al., 1993; Wang et al., 1995). UCN-01 has cytostatic properties (reviewed in Gescher, 1998; Senderowicz, 2000) and has been shown to inhibit checkpoint kinase 1 (Chk1), abrogate the G_2 checkpoint (Wang et al., 1996), enhance radiation toxicity in human cancer cell lines (Playle et al., 2002) and sensitize tumor cells to various DNA damaging agents (Bunch and Eastman, 1996; Hsueh et al., 1998; Sugiyama et al., 2000). Studies on the mechanism(s) of action of UCN-01 suggested that induction of apoptosis and G_1 phase accumulation were important for its anticancer activity (Sugiyama et al., 1999). UCN-01-induced G_1 phase accumulation was found to be mediated by direct and indirect inhibition of Rb kinase(s), such as CDK2, accompanied by its dephosphorylation (Akiyama et al., 1997). Also, the decrease in expression level of cyclin A was found to play an important role in the G_1 phase accumulation induced by UCN-01 (Sugiyama et al., 1999). Although the mechanism of UCN-01-induced apoptosis is still unknown (Wang et al., 1995), several reports demonstrate that, in some in vitro models, it can downregulate some antiapoptotic proteins, similar to flavopiridol (Kitada et al., 2000). UCN-01 has also been shown to possess favorable pharmacokinetic and toxicological properties (Senderowicz, 2000; Sausville et al., 2001). In view of these properties, UCN-01 is now being developed as an anticancer agent in the USA and Japan. The first clinical trial with this drug was recently completed at the National Cancer Institute (Sausville et al., 2001). Phase I clinical trials using a combination of cytotoxic agents (cisplatin, 5-fluorouracil and fludarabine) with UCN-01 are ongoing. Potential targets of UCN-01 are currently under investigation.

IV.C. Indirubins

The bis-indole indirubin is the active ingredient of the traditional Chinese medicine recipe Danggui Longhui Wan that has been described more than 35 years ago as being clinically active against chronic myelocytic leukemia (Han, 1994; Hoessel et al., 1999). Indirubins have already been used in clinical evaluation for cancer treatment; Phases I and II clinical trials in cancer patients are underway (Damiens and Meijer, 2000). Indirubin shows poor solubility, low absorption and presents gastrointestinal toxicity. Studies to reduce the toxic side effects, improve the pharmacokinetic properties and increase their antitumor activity of indirubin have led to the synthesis of several indirubin analogs with better pharmacological properties and reduced toxicity, such as N-methyl isoindigo, 5-chloro-indirubin and indirubin-3′-monoxime (Liu et al., 1996). The antitumoral properties of indirubins appear to correlate with their antimitotic and CDK inhibitory effects (Marko et al., 2001). Indirubins are potent inhibitors of CDK2, CDK5/p25 and CDK1/cyclin B (Hoessel et al., 1999). The crystal structure of CDK2 in complex with indirubin derivatives revealed that indirubin interacts with the kinase's ATP-binding site through van der Waals interactions and three hydrogen bonds (Hoessel et al., 1999). Treatment of human mammary carcinoma MCF-7 cells with indirubins inhibited the proliferation of these cells and induced G_2/M arrest, an effect that was mediated by the inhibition of CDK1 and CDK1/cyclin B activity (Hoessel et al., 1999; Marko et al., 2001). Recently, a cell-permeable indirubin-3′-monoxime was found to induce G_2 arrest in M phase synchronized human HBL-100 breast cells by inducing endoreplication in these cells leading to polyploidy, followed by aneuploidy and cell death by necrosis (Damiens et al., 2001), a mechanism that may also contribute to the antitumoral properties of these drugs. In addition, indirubins constitute the first family of low nanomolar inhibitors of GSK-3β to be described (Leclerc et al., 2001). Upon testing a series of indoles and bis-indoles against GSK-3β, CDK1/cyclin B and CDK5/p25, only indirubins were found to inhibit these kinases (Leclerc et al., 2001). Indirubins did bind to the ATP-binding pocket of GSK-3β in a way similar to their binding to CDKs, the details of which were recently revealed by crystallographic analysis (Damiens et al., 2001). Similar to paullones, indirubin-3′-monoxime inhibited the hyperphosphorylation of the microtubule-binding protein tau both in vitro and in vivo at Alzheimer's disease-specific sites, suggesting its potential use for the treatment of neurodegenerative disorders (Leclerc et al., 2001).

V. Conclusions

Nature has provided many of the effective anticancer agents in current use, such as the microbiologically derived drugs dactinomycin, bleomycin and doxorubicin, and the plant-derived drugs vinblastine, irinotecan, topotecan, etoposide and paclitaxel. Among the wealth of anticancer drugs of natural origin, CDK inhibitors represent an important group. The new approaches to drug development will hopefully lead to the discovery of novel selective and powerful inhibitors in the near future with efficient applications in various human pathologies.

References

Achenbach TV, Muller R, Slater EP. (2000) Bcl-2 independence of flavopiridol-induced apoptosis. Mitochondrial depolarization in the absence of cytochrome c release. J Biol Chem 275:32089–97.

Akinaga S, Gomi K, Morimoto M, Tamaoki T, Okabe M. (1991) Antitumor activity of UCN-01, a selective inhibitor of protein kinase C, in murine and human tumor models. Cancer Res 51:4888–92.

Akiyama T, Yoshida T, Tsujita T, Shimizu M, Mizukami T, Okabe M, Akinaga S. (1997) G_1 phase accumulation induced by UCN-01 is associated with dephosphorylation of Rb and CDK2 proteins as well as induction of CDK inhibitor p21/Cip1/WAF1/ Sdi1 in p53-mutated human epidermoid carcinoma A431cells. Cancer Res 57:1495–501.

Almenara J, Rosato R, Grant S. (2002) Synergistic induction of mitochondrial damage and apoptosis in human leukemia cells by flavopiridol and the histone deacetylase inhibitor suberoylanilide hydroxamic acid (SAHA). Leukemia 16:1331–43.

Arguello F, Alexander M, Sterry JA, Tudor G, Smith EM, Kallavar NT, Greene JF, Koss W, Morgan CD, Stinson SF, Siford TJ, Alvord WG, Klabansky RL, Sausville EA. (1998) Flavopiridol induces apoptosis of normal lymphoid cells, causes immunosuppression, and has potent antitumor activity in vivo against human leukemia and lymphoma xenografts. Blood 91:2482–90.

Bregman DB, Pestell RG, Kidd V. (2000) Cell cycle regulation and RNA polymerase II. Front Biosci 5:D244–57.

Brooks EE, Gray NS, Joly A, Kerwar SS, Lum R, Mackman RL, Norman TC, Rosete J, Rowe M, Schow SR, Schultz PG, Wang X, Wick MM, Shiffman D. (1997) CVT-313, a specific and potent inhibitor of CDK2 that prevents neointimal proliferation. J Biol Chem 272:29207–11.

Brusselbach S, Nettelbeck DM, Sedlacek HH, Muller R. (1998) Cell cycle-independent induction of apoptosis by the anti-tumor drug flavopiridol in endothelial cells. Int J Cancer 77:146–52.

Bunch RT, Eastman A. (1996) Enhancement of cisplatin-induced cytotoxicity by 7-hydroxystaurosporine (UCN-01), a new G_2-checkpoint inhibitor. Clin Cancer Res 2:791–7.

Buolamwini JK. (2000) Cell cycle molecular targets in novel anticancer drug discovery. Curr Pharm Des 6:379–92.

Carlson BA, Dubay MM, Sausville EA, Brizuela L, Worland PJ. (1996) Flavopiridol induces G_1 arrest with inhibition of cyclin-dependent kinase (CDK) 2 and CDK4 in human breast carcinoma cells. Cancer Res 56:2973–8.

Chang YT, Gray NG, Rosania GR, Sutherlin DP, Kwon S, Norman TC, Sarohia R, Leost M, Meijer L, Schultz PG. (1999) Synthesis and application of functionally diverse 2,6,9-tri-substituted purine libraries as CDK inhibitors. Chem Biol 6:361–75.

Chao S, Fujinaga K, Marion JE, Taube R, Sausville EA, Senderowicz AM, Peterlin BM, Price DH. (2000) Flavopiridol inhibits P-TEFb and blocks HIV-1 replication. J Biol Chem 275:28345–8.

Chen X, Lowe M, Herliczek T, Hall MJ, Danes C, Lawrence DA, Keyomarsi K. (2000) Protection of normal proliferating cells against chemotherapy by Staurosporine-mediated, selective, and reversible $G_{(1)}$ arrest. J Natl Cancer Inst 2:1999–2008.

Coleman TR, Dunphy WG. (1994) Cdc2 regulatory factors. Curr Opin Cell. Biol, 6:877–82.

Damiens E, Baratte B, Marie D, Eisenbrand G, Meijer L. (2001) Anti-mitotic properties of indirubin-3′-monoxime, a CDK/GSK-3 inhibitor: induction of indoreplication following prophase arrest. Oncogene 20:3786–97.

Damiens E, Meijer L. (2000) Chemical inhibitors of cyclic-dependent kinases: preclinical and clinical study. Pathol Biol (Paris) 3:340–51.

de Azevedo Jr. WF, Mueller-Dieckmann HJ, Schulze-Gahmen U, Worland PJ, Sausville E, Kim SH. (1996) Structural basis for specificity and potency of a flavonoid inhibitor of human CDK2, a cell cycle kinase. Proc Natl Acad Sci USA 93:2735–40.

de Azevedo WF, Leclerc S, Meijer L, Havlicek L, Strnad M, Kim S-H. (1997) Inhibition of cyclin-dependent kinases by purine analogues: crystal structure of human cdk2 complexed with roscovitine. Eur J Biochem 243:518–26.

Decker RH, Dai Y, Grant S. (2001) The cyclin-dependent kinase inhibitor flavopiridol induces apoptosis in human leukemia cells (U937) through the mitochondrial rather than the receptor-mediated pathway. Cell Death Differ 8:715–24.

Draetta G, Eckstein J. (1997) Cdc25 protein phosphatases in cell proliferation. Biochim Biophys Acta 1332:M53–63.

Ewen ME, Sluss HK, Sherr CJ, Matsushime H, Kato J, Livingston DM. (1993) Functional interactions of the retinoblastoma protein with mammalian D-type cyclins. Cell 73:487–97.

Fisher RP, Morgan DO. (1994) A novel cyclin associates with MO15/CDK7 to form the CDK-activating kinase. Cell 78:713–24.

Gadbois DM, Peterson S, Bradbury EM, Lehnert BE. (1995) CDK4/cyclin D1/PCNA gene expression and enhances 5-fluorouracil-induced apoptosis in a sequence-dependent manner. Clin Cancer Res 4:2201–6.

Gervasi C, Szaro BG. (1995) The Xenopus laevis homologue to the neuronal cyclin-dependent kinase (cdk5) is expressed in embryos by gastrulation. Brain Res Mol Brain Res 33:192–200.

Gescher A. (1998) Analogs of staurosporine: potential anticancer drugs? Gen Pharmacol 31:721–8.

Gescher A. (2000) Staurosporine analogues—pharmacological toys or useful antitumour agents? Crit Rev Oncol Hematol 34:127–35.

Gray N, Wodicka L, Thunnissen AM, Norman T, Kwon S, Espinoza FH, Morgan DO, Barnes G, Leclerc S, Meijer L, Kim SH, Lockhart DJ, Schultze P. (1998) Exploiting chemical libraries, structure, and genomics in the search for new kinase inhibitors. Science 281:533–8.

Han R. (1994) Highlight on the studies of anticancer drugs derived from plants in China. Stem Cells 1:53–63.

Hoessel R, Leclerc S, Endicott J, Noble M, Lawrie A, Tunnah P, Leost M, Damiens E, Marie D, Marko D, Niederberger E, Tang W, Eisenbrand G, Meijer L. (1999) Indirubin, the active constituent of a Chinese antileukaemia medicine, inhibits cyclin-dependent kinases. Nature Cell Biol 1:60–7.

Hsueh CT, Kelsen D, Schwartz GK. (1998) UCN-01 suppresses thymidylate synthase hydroxystaurosporine (UCN-01), a new G_2-checkpoint inhibitor. Clin Cancer Res 2:791–7.

Jacobson MD, Burn JF, Raff MC. (1994) Programmed cell death and Bcl-2 protection in the absence of a nucleus. EMBO J 13:1899–910.

Kaiser A, Nishi K, Gorin FA, Walsh DA, Bradbury EM, Schnier JB. (2001) The cyclin-dependent kinase (CDK) inhibitor flavopiridol inhibits glycogen phosphorylase. Arch Biochem Biophys 386:179–87.

Kaldis P. (1999) The cdk-activating kinase (CAK): from yeast to mammals. Cell Mol Life Sci 55(2):284–96.

Kato J, Matsushime H, Hiebert SW, Ewen ME, Sherr CJ. (1993) Direct binding of cyclin D to the retinoblastoma gene product (pRb) and pRb phosphorylation by the cycline D-dependent kinase CDK4. Genes Dev. 7:331–42.

Kitada S, Zapata JM, Andreeff M, Reed JC. (2000) Protein kinase inhibitors flavopiridol and 7-hydroxy-staurosporine down-regulate antiapoptosis proteins in B-cell chronic lymphocytic leukemia. Blood 96:393–7.
Kitagawa M, Okabe T, Ogino H, Matsumoto H, Suzuki-Takahashi I, Kokubo T, Higashi H, Saitoh S, Taya Y, Yasuda H, Ohba Y, Nishimura S, Tanaka N, Okuyama A. (1993) Butyrolactone I, a selective inhibitor of cdk2 and cdc2 kinase. Oncogene 8:2425–32.
Knockaert M, Greengard P, Meijer L. (2002) Pharmacological inhibitors of cyclin-dependent kinases. Trends Pharmacol Sci 23:417–25.
Konig A, Schwartz RM, Mohammad RM, Al-Katib A, Gabrilove JL. (1997) The novel cyclin-dependent kinase inhibitor flavopiridol downregulates bcl2 and induces growth arrest and apoptosis in chronic B cell leukemia lines. Blood 90:4307–12.
Konstantinidis AK, Radhakrishnan R, Gu F, Rao RN, Yeh W-K. (1998) Purification, characterisation and kinetic mechanism of cyclin D1-CDK4, a major target for cell regulation. J Biol Chem 273:26506–15.
Kwon TK, Buchholz MA, Ponsalle P, Chrest FJ, Nordin AA. (1997) The regulation of p27Kip1 expression following the polyclonal activation of murine G0 T cells. J Immunol 158:5642–8.
Lam LT, Pickeral OK, Peng AC, Rosenwald A, Hurt EM, Giltnane JM, Averett LM, Zhao H, Davis RE, Sathyamoorthy M, Wahl LM, Harris ED, Mikovits JA, Monks AP, Hollingshead MG, Sausville EA, Staudt LM. (2001) Genomic-scale measurement of mRNA turnover and the mechanisms of action of the anti-cancer drug flavopiridol. *Genome Biol* 2(10): RESEARCH0041.
Leclerc S, Garnier M, Hoessel R, Marko D, Bibb JA, Snyder GL, Greengard P, Biernat J, Wu YZ, Mandelkow EM, Eisenbrand G, Meijer L. (2001) Indirubins inhibit glycogen synthase kinase-3 beta and CDK5/p25, two protein kinases involved in abnormal tau phosphorylation in Alzheimer's disease. A property common to most cyclin-dependent kinase inhibitors? J Biol Chem 276:251–60.
Lee CH, Cheon JY, Park JS, Youl Lee KY, Park SG, Lee SK, Lee YH. (1996) A specific inhibitor of cyclin-dependent protein kinases, CDC2 and CDK2. Mol Cells 6:679–83.
Li A, Blow JJ. (2001) The origin of CDK regulation. Nat Cell Biol 3:E182–4.
Liu XM, Wang LG, Li HY, Ji XJ. (1996) Induction of differentiation and down-regulation of c-myb gene expression in ML-1 human myeloblastic leukemia cells by the clinically effective anti-leukemia agent meisoindigo. Biochem Pharmacol 51:1545–51.
Losiewicz MD, Carlson BA, Kaur G, Sausville EA, Worland PJ. (1994) Potent inhibition of CDC2 kinase activity by the flavonoid L86-8275. Biochem Biophys Res Commun 201:589–95.
Marko D, Schatzle S, Friedel A, Genzlinger A, Zankl H, Meijer L, Eisenbrand G. (2001) Inhibition of cyclin-dependent kinase 1 (CDK1) by indirubin derivatives in human tumour cells. Br J Cancer 84:283–9.
Meggio F, Donella-Deana A, Ruzzene M, Brunati AM, Cesaro L, Guerra B, Meyer T, Mett H, Fabbro D, Furet P, Dobrowolska G, Pinna LA. (1995) Different susceptibility of protein kinases to staurosporine inhibition. Eur J Biochem 234:317–32.
Meijer L, Borgne A, Mulner O, Chong JPJ, Blow JJ, Inagaki N, Inagaki M, Delcros JG, Moulinoux JP. (1997) Biochemical and cellular effects of roscovitine, a potent and selective inhibitor of the cyclin-dependent kinases cdc2, cdk2 and cdk5. Eur J Biochem 243:527–36.
Meijer L, Thunissen AMWH, White A, Garnier M, Nikolic M, Tsai LH, Walter J, Cleverley KE, Salinas PC, Wu YZ, Biernat J, Mandelkow EM, Kim S-H, Pettit GR. (2000) Inhibition of cyclin-dependent kinases, GSK-3b and casein kinase 1 by hymenialdisine, a marine sponge constituent. Chem Biol 7:51–63.
Meijer L. (1996) Chemical inhibitors of cyclin-dependent kinases. Trends Cell Biol 6:393–7.
Melillo G, Sausville E, Cloud K, Lahusen T, Varesio L, Senderowicz AM. (1999) Flavopiridol, a protein kinase inhibitor, down-regulates hypoxic induction of vascular endothelial growth factor expression in human monocytes. Cancer Res 59:5433–7.
Morgan DO. (1995) Principles of CDK regulation. Nature 374:131–4.
Morgan DO. (1997) Cyclin-dependent kinases: engines, clocks, and microprocessors. Annu Rev Cell Dev Biol 13:261–91.

Motwani M, Delohery TM, Schwartz GK. (1999) Sequential dependent enhancement of caspase activation and apoptosis by flavopiridol on paclitaxel-treated human gastric and breast cancer cells. Clin Cancer Res 5:1876–83.

Mueller PR, Coleman TR, Dunphy G. (1995) Cell cycle regulation of a Xenopus Wee1-like kinase. Mol Biol Cell 6:119–34.

Mukherjee AK, Basu S, Sarkar N, Ghosh AC. (2001) Advances in cancer therapy with plant based natural products. Curr Med Chem 8:1467–86.

Murray AW. (1994) Cyclin-dependent kinases: regulators of the cell cycle and more. Chem Biol 1:191–5.

Nishi K, Schnier JB, Bradbury EM. (1998) The accumulation of cyclin-dependent kinase inhibitor p27kip1 is a primary response to staurosporine and independent of G_1 cell cycle arrest. Exp Cell Res 243:222–31.

Parker LL, Piwnica-Worms H. (1992) Inactivation of the p34cdc2-cyclin B complex by the human WEE1 tyrosine kinase. Science 257:1955–7.

Patel V, Senderowicz AM, Pinto Jr. D, Igishi T, Raffeld M, Quintanilla-Martinez L, Ensley JF, Sausville EA, Gutkind JS. (1998) Flavopiridol, a novel cyclin-dependent kinase inhibitor, suppresses the growth of head and neck squamous cell carcinomas by inducing apoptosis. J Clin Invest 102:1674–81.

Philpott A, Porro EB, Kirschner MW, Tsai LH. (1997) The role of cyclin-dependent kinase 5 and a novel regulatory subunit in regulating muscle differentiation and patterning. Genes Dev 11:1409–21.

Playle LC, Hicks DJ, Qualtrough D, Paraskeva C. (2002) Abrogation of the radiation-induced G_2 checkpoint by the staurosporine derivative UCN-01 is associated with radiosensitisation in a subset of colorectal tumour cell lines. Br J Cancer 87:352–8.

Rapella A, Negrioli A, Melillo G, Pastorino S, Varesio L, Bosco MC. (2002) Flavopiridol inhibits vascular endothelial growth factor production induced by hypoxia or picolinic acid in human neuroblastoma. Int J Cancer 99:658–64.

Samuelsson M, Pazirandeh A, Okret S. (2002) A pro-apoptotic effect of the CDK inhibitor p57(Kip2) on staurosporine-induced apoptosis in HeLa cells. Biochem Biophys Res Commun 296:702–9.

Sausville EA, Arbuck SG, Messmann R, Headlee D, Bauer KS, Lush RM, Murgo A, Figg WD, Lahusen T, Jaken S, Jing X, Roberge M, Fuse E, Kuwabara T, Senderowicz AM. (2001) Phase I trial of 72-hour continuous infusion UCN-01 in patients with refractory neoplasms. J Clin Oncol 19:2319–33.

Schnier JB, Nishi K, Goodrich DW, Bradbury EM. (1996) G_1 arrest and down-regulation of cyclin E/cyclin-dependent kinase 2 by the protein kinase inhibitor staurosporine are dependent on the retinoblastoma protein in the bladder carcinoma cell line 5637. Proc Natl Acad Sci USA 93:5941–6.

Schultz C, Link A, Leost M, Zaharevitz DW, Gussio R, Sausville EA, Meijer L, Kunick C. (1999) The paullones, a series of cyclin-dependent kinase inhibitors: synthesis, evaluation of CDK1/cyclin B inhibition, and in vitro antitumor activity. J Med Chem 42:2909–19.

Schwartz GK, Ilson D, Saltz L, O'Reilly E, Tong W, Maslak P, Werner J, Perkins P, Stoltz M, Kelsen D. (2001) Phase II study of the cyclin-dependent kinase inhibitor flavopiridol administered to patients with advanced gastric carcinoma. J Clin Oncol 19:1985–92.

Sedlacek HH, Czech J, Naik R, Kaur G, Worland P, Losiewicz M, Parker B, Carlson B, Smith A, Senderowicz A, Sausville E. (1996) Flavopiridol (L86 8275; NSC 649890), a new kinase inhibitor for tumor therapy. Int J Oncol 9:1143–68.

Sedlacek HH. (2001) Mechanisms of action of flavopiridol. Crit Rev Oncol Hematol 38920:139–70.

Senderowicz AM, Headlee SF, Stinson SF, Lush RM, Kalil N, Villalba L, Hill K, Steinberg SM, Figg WD, Tompkins A, Arbuck SG, Sausville EA. (1998) Phase I trial of continuous infusion flavopiridol, a novel cyclin-dependent kinase inhibitor in patients with refractory neoplasms. J Clin Oncol 16:2986–99.

Senderowicz AM. (2000) Small molecule modulators of cyclin-dependent kinases for cancer therapy. Oncogene 19:6600–6.

Senderowicz AM. (2002) The cell cycle as a target for cancer therapy: basic and clinical findings with the small molecule inhibitors flavopiridol and UCN-01. Oncologist 7:12–9.

Seynaeve CM, Stetler-Stevenson M, Sebers S, Kaur G, Sausville EA, Worland PJ. (1993) Cell cycle arrest and growth inhibition by the protein kinase antagonist UCN-01 in human breast carcinoma cells. Cancer Res 53:2081–6.

Shapiro GI, Supko JG, Patterson A, Linch C, Lucca J, Zacarola PF, Muzikansky A, Wright JJ, Linch Jr TJ, Rollins BJ. (2001) A phase II trial of the cyclin-dependent kinase inhibitor flavopiridol in patients with previously untreated stage IV non-small cell lung cancer. Clin Cancer Res 7:1590–9.

Shimizu T, Takahashi N, Tachibana K, Takeda K. (2001) Complex regulation of CDK2 and G_1 arrest during neuronal differentiation of human prostatic cancer TSU-Prl cells by Staurosporine. Anticancer Res 21:893–8.

Stadler WM, Vogelzang NJ, Amato R, Sosman J, Taber D, Liebowitz D, Vokes EE. (2000) Flavopiridol, a novel cyclin-dependent kinase inhibitor, in metastatic renal cancer: a University of Chicago phase II consortium Study. J Clin Oncol 18:371–5.

Sugiyama K, Akiyama T, Shimizu M, Tamaoki T, Courage C, Gescher A, Akinaga S. (1999) Decrease in susceptibility toward induction of apoptosis and alteration in G_1 checkpoint function as determinants of resistance of human lung cancer cells against the antisignaling drug UCN-01 (7-hydroxystaurosporine). Cancer Res 59:4406–12.

Sugiyama K, Shimizu M, Akiyama T, Tamaoki T, Yamaguchi K, Takahashi I, Eastman A, Akinaga S. (2000) UCN-01 selectively enhances mitomycin C cytotoxicity in p53 defective cells which is mediated through S and/or G_2 checkpoint abrogation. Int J Cancer 85:703–9.

Takahashi I, Kobayashi E, Asano K, Yoshida M, Nakano H. (1987) UCN-01 is a selective inhibitor of protein kinase C from Streptomyces. J Antibiot 40:1782–4.

Tamaoki T, Nomoto H, Takahashi I, Kato Y, Morimoto M, Tomita F. (1986) Staurosporine: a potent inhibitor of phospholipid Ca^{2+} dependent protein kinase. Biochem Biophys Res Commun 135:397–402.

Tan AR, Swain SM. (2002) Review of flavopiridol, a cyclin-dependent kinase inhibitor, as breast cancer therapy. Semin Oncol 29:77–85.

Tasdemir D, Mallon R, Greenstein M, Feldberg LR, Kim SC, Collins K, Wojciechowicz D, Mangalindan GC, Concepcion GP, Harper MK, Ireland CM. (2002) Aldisine alkaloids from the Philippine sponge Stylissa massa are potent inhibitors of mitogen-activated protein kinase kinase-1 (MEK-1). J Med Chem 45:529–32.

Vesely J, Havlicek L, Strnad M, Blow JJ, Donella-Deana A, Pinna L, Letham DS, Kato JY, Detivaud L, Leclerc S, Meijer L. (1994) Inhibition of cyclin-dependent kinases by purine derivatives. Eur J Biochem 224:771–86.

Wang Q, Fan S, Eastman A, Worland PJ, Sausville EA, O'Connor PM. (1996) UCN-01: a potent abrogator of G_2 checkpoint function in cancer cells with disrupted p53. J Natl Cancer Inst 88:956–65.

Wang Q, Worland PJ, Clark JL, Carlson BA, Sausville EA. (1995) Apoptosis in 7-hydroxystaurosporine-treated T lymphoblasts correlates with activation of cyclin-dependent kinases 1 and 2. Cell Growth Differ 6:927–36.

Zaharevitz D, Gussio R, Leost M, Senderowicz AM, Lahusen T, Kunick C, Meijer L, Sausville EA. (1999) Discovery and initial characterization of the paullones, a novel class of small-molecule inhibitors of cyclin-dependent kinases. Cancer Res 59:2566–9.

Zhai S, Senderowicz AM, Sausville EA, Figg WD. (2002) Flavopiridol, a novel cyclin-dependent kinase inhibitor, in clinical development. Ann Pharmacother 36:905–11.

Zimmermann, J. (1995). Pharmacologically active pyrimidine derivatives and processes for the preparation thereof. PCT Ciba-Geigy. WO 95/09853.

M.T.H. Khan and A. Ather (eds.)
Lead Molecules from Natural Products

Anticancer and medicinal properties of essential oil and extracts of East Mediterranean sage (salvia triloba)

HALA GALI-MUHTASIB

Abstract

The East Mediterranean Sage (Salvia triloba) is a native plant of the Mediterranean, which has been used in traditional medicine by many Asian and Middle Eastern countries to treat several ailments. The leaves of the plant are boiled as an herbal tea for the relief of headaches, stomachaches, abdominal pain and many other disorders. The aqueous and oil extracts of sage have been shown to possess antioxidant, anti-inflammatory, anticancer and antimicrobial activities. It has recently been shown that the oil extract of S. triloba has potent chemopreventive abilities in the DMBA/TPA mouse model of skin carcinogenesis. This review describes the popular uses of the East Mediterranean Sage in traditional medicine with an emphasis on the anticancer properties of the essential oil extract of the plant.

Keywords: *anticancer, essential oil, medicinal plants, Salvia triloba, sage*

I. Introduction

Since primitive ages, people have learned to use a variety of plants as medicines for different purposes (Houghton, 1995). Recent years have witnessed a renewed interest in plants as pharmaceuticals in the Western world. This interest is channeled into the discovery of new biologically active molecules by the pharmaceutical industry and into the adoption of crude extracts of plants for self-medication by the general public. Natural products and plant-derived products continue to be excellent sources of new drug candidates. The process of drug development from botanicals involves the identification and extraction of active components of whole plants or crude extracts and, in some cases, synthesis of equivalent active compounds.

Among plants that are largely used in traditional medicine include several species of *Salvia.* The word *Salvia* is from the Latin salvere, meaning, "to heal," "save" or "to be safe and unharmed" indicating the medicinal value of the plant. Although 500 species of *Salvia* and many varieties and chemotypes exist, only a few types of sage are commercially important, among which is *S. triloba* (Gali-Muhtasib et al., 2000b). The widespread use of *S. triloba* as a popular household remedy with healing properties is of scientific interest. *S. triloba* has been used not only in therapy but also as a spice to flavor many foods (Parry, 1969; Rosengarten, 1973; Stahl, 1973; Morton, 1976). The dried leaves and essential oil of *S. triloba* and other species of sage are employed as seasonings for sausages, ground meats, fish, honey, salads, soups, and stews. Sage is used, along with rosemary and thyme, to preserve a number of foods, including meats and cheeses. It is also used as a flavoring and antioxidant in cheeses, pickles, vegetables, processed foods, and beverages. The oil of *S. triloba* is used to extend the keeping quality of fats and meats. Sage is used in perfumes and cosmetics and as a natural insect repellent. *S. triloba* can be purchased as whole leaf, ground, sliced, or cut (Leung and Foster, 1995).

In the Lebanese traditional medicine, the oil and water extract of *S. triloba* are used in several towns and villages. The water extract of the plant is either internally used as infusions or is inhaled in steam baths or applied externally to heal fractured bones. The herbalists of Lebanon, Syria, and Jordan consider this species as a "panacea" that is a universal drug, so they sell this plant in the market. It is noteworthy to mention that *S. triloba* is the number one plant sold in Lebanon (personal communication). The plant is used in Jordan for the treatment of ulcer pains and indigestion (Karim and Quraan, 1986). In Turkey, the same species is used for kidney and gall bladder stones and for the relief from colds, coughs, and influenza (Baser et al., 1986).

The first and only investigation of the effectiveness of *S. triloba* at inhibiting the promotion and progression phases of cancer was conducted in our laboratories on mouse skin. We are presently testing the anti-proliferative effects of all the components in sage oil against cultured cancer cells. Our ultimate goal is to identify the active component(s) in sage oil responsible for the observed in vivo anti-tumor effects. If *S. triloba* exerts potent in vitro anticancer effects, the feasibility of propagation and the potential for mass cultivation of high-yielding phenotypes would need to be assessed in the future. This chapter summarizes the therapeutic value of the sage plant, *S. triloba*, with special emphasis on its anticancer effects.

II. The sage plant

The East Mediterranean Sage plant is an aromatic shrub, which belongs to the mint family, Lamiaceae (Labiatae). The plant is also named *S. fruticosa* (Millet et al., 1981), and was formerly known as *S. triloba* L., *S. triloba* L.*f.*- *ssp. cypria*, *S. triloba* L.-*ssp. libanotica* (Mouterde, 1970; Meikle, 1985). It is clearly different from *S. officinalis* because of its trifoliate leaves and is thus commonly known as three-lobed sage. The height of the plant varies from 20 to 100 cm high. The lower part of this plant is woody and the upper part of square stems is covered with hairs. The leaves are 3–10 cm long and 1.5–5 cm broad, opposite, ovate and elongated, greenish-gray with hairs. The flower color is light blue to violet blue, 2–3 cm long with short upper lip; arranged in axillary whorls of 4–8 flowers. Sage is a native plant of the Mediterranean, but it is widely cultivated now (Gali-Muhtasib et al., 2000b). It is found in dry rocky limestone soils or the edges of pine forests, riverbeds, and roadsides. It grows in an altitude from 100 to 800 m. Geographically the plant is distributed in Lebanon, Syria, Palestine, Crete, Cyprus, Turkey, Greece, and in the South of Italy and Sicily (Mouterde, 1970). The maximum density of this plant is in Lebanon.

III. Essential oils of East Mediterranean sage

"Essential oil" is a term used to distinguish between fatty oils and more volatile oils collected by the steam distillation of the plant. In some countries, the essential oils are also called olea aetherea, or essences, a name that nominates the alcoholic solution of volatile oil (Gennaro, 1990). The volatile oil of the sage plant is extracted from the leaves. It is produced in the secretory cells at the base of glandular trichomes of the epidermal cells, and then stored in the epidermal extracellular spaces of the plant. Such spaces are formed by the separation of the cell walls from the covering cuticle. Although oil does not accumulate in the cytoplasm, it will not be lost to the surrounding environment unless the thick, wax-coated cuticle is physically damaged (Guenther, 1949; Hay and Svoboda, 1993).

Sage (*S. triloba*) has a camphor-like scent and its taste is very bitter. The odor and aromatic taste of sage are due to its volatile oil. Oil of sage is obtained by distilling the leaves with water; it is a yellowish or greenish-yellow liquid, having a penetrating characteristic. The essential oil extracted by steam distillation constitutes 1.2% to 2.5% of dry leaves. The oil extract consists of hydrocarbons, alcohols, acids, esters, aldehydes, ketones, phenols, phenol esters, lactones, and various nitrogen and sulfur organic compounds (Gennaro, 1990; Waterman, 1993). The hydrocarbons of principal significance are mostly the terpenes (Guenther, 1949; Gennaro, 1990). The important alcohols in sage oil include borneol (cyclic), linalool (acyclic), and terpineol. Thujone and camphor are two important ketones present in the oil extracts of the plant. Oxides such as 1,8-cineole and esters such as linalylacetate also occur in sage oil (Gennaro, 1990).

The quality and quantity of the *S. triloba* oil differs by geographic region and by the part of the plant used (Bellomaria et al., 1992; Arnold and Bellomaria, 1993). The leaves give the maximum product because of the glandular hairs present in leaves. The oil yield is supreme in the post-flowering period, when the weather is dry in the Mediterranean region (Bellomaria et al., 1992). Generally, oil yield is high in a long dry season (Pitarevic et al., 1985). Furthermore, the content of β-thujone and borneol is higher in the full-blooming period (Verzar-Petri et al., 1985). The range of percentage composition of various sage oil components is summarized below (Farhat et al., 2001).

α-Pinene:	3.1–6.6	β-Thujone:	1.1–1.8
Camphene:	3.0–4.8	Camphor:	7.7–12.3
β-Pinene:	5.1–9.8	Linalool:	0.7–1.7
Limonene:	2.2–6.1	Linalylacetate:	0.8–1.2
Cineole:	47.7–57.4	Borneol:	2.6–3.9
α-Thujone:	1.0–1.9		

Similar variation in the yield and quality of essential oil has been reported in other species of *Salvia*. Of the 50 compounds that were identified in essential oils from stems and leaves of *S. officinalis* growing in northern Portugal, α-thujone was the major compound, representing about 55%, 30%, and 18% of the essential oils from stems, leaves, and flowers, respectively (Santos-Gomes and Fernandes-Ferreira, 2001). The scientific and medicinal interest in *S. triloba* stems from the fact that its thujone levels (1–2%), which account for most cases of human intoxication (see below), are much lower than those of *S. officinalis* (30–50%). Significant variations in the composition of essential oils are also found in *S. officinalis* sampled over the year. A decrease of oxygenated monoterpenes (MO) from 70% to 40% from December to April in stems of *S. officinalis* grown in Portugal was recently reported (Santos-Gomes and Fernandes-Ferreira, 2001). During the same time interval, the percentage of monoterpene hydrocarbons (MH) rose from 10% to 20%. The compounds that mostly accounted for the essential oil composition variation were α-pinene, β-pinene and camphene, α-thujone and camphor, α-humulene and β-caryophyllene, and viridiflorol. In another study, several factors were found to affect the oil yield and quality of essential oils from Dalmatian sage (*S. officinalis* L.) (Perry et al., 1999). These include plant ecotype, plant part, and season. Gas chromatography analyses of oil from *S. officinalis* revealed the presence of three chemotypes with different proportions of α- and β-thujone (α/β 10:1, 1.5:1, and 1:10). Flowering parts of *S. officinalis* had higher oil contents and β-pinene levels than leaves and lower thujone levels. Major seasonal changes were found in the composition of oil; total thujone levels were lowest around flowering in spring and summer and highest during autumn or winter (Perry et al., 1999).

Commercial preparations of essences of sage have caused human intoxication with convulsions of central nervous system origin being the major symptom (Millet et al., 1981). Such toxicity has been mainly attributed to the thujone and camphor components of the oil (Millet et al., 1981). The dose limit from which the intoxication is subclinical was found to be 0.3 g/kg for *S. officinalis* oil. Above 1.25 g/kg of oil, the convulsions appeared and became lethal. The toxicity of sage commercial oil appeared to be related to the presence of camphor while that of sage Dalmatian oil was due to camphor and thujone. The convulsant properties of camphor are well known and the neurotoxicity of thujone is demonstrated in rats (Millet et al., 1981). Other toxic effects of sage oil include epileptic reactions (Arnold, 1988; Kbayssi, 1993), loss of equilibrium, tachycardia, and other problems related to the nervous system (Elisabetsky et al., 1995). Because the sage plant is widely used in traditional medicine, any wrong use could cause various complications, due to the established toxicity of the essential oil of this plant (Millet et al., 1979, 1981; Geller et al., 1984; Hooser, 1990; Leushner, 1997). The essential oil should always be used with great care, as even small doses can be poisonous. The herb should be avoided during pregnancy because it is a uterine stimulant. Patients suffering from epilepsy are not supposed to use sage. Women who are breast-feeding should only use sage in medicinal amounts if they want to dry up the flow of milk. *Salvia* should be avoided when fever is present. Cheilitis and stomatitis follow some cases of sage tea ingestion (British Pharmaceutical Codex, 1923). Thus, the essential oils of *Salvia* must be considered as a drug and handled with precaution. There has been no reported toxicity of the water extract of this plant.

IV. Therapeutic uses of East Mediterranean sage

Sage has one of the longest histories of use as a medicinal herb. The plant has been used as a folk remedy against colds, diarrhea, enteritis, venereal disease, excessive perspiration, snakebites, sore throats, and toothaches (Willershausen et al., 1991). The plant is reported to have a wide range of biological activities, such as antibacterial (Shelef et al., 1990; Hefnawy et al., 1993; Pattnaik et al., 1997), fungistatic, virustatic (Dobrynin et al., 1976; Chervatyi et al., 1980; Farag et al., 1989). The oil extract of sage has been found to possess antimicrobial activities against a wide range of bacteria (Ivanic and Savin, 1976), which are mainly due to the cineole component (Santos and Rao, 1997). In vitro, sage oil has been shown to be effective against both Gram-positive and Gram-negative bacteria including *Escherichia coli* and *Salmonella* species, and against filamentous fungi (Valnet, 1986) and yeast such as *Candida albicans*. Such antimicrobial properties have been shown to reduce plaque growth, inhibit gingival inflammation and exert positive effects on caries prophylaxis (Willershausen et al., 1991). It has been reported to act as a bactericide and is used in mouthwashes and gargles. The plant is also used as an agitator and antisecretory agent, and as salvin (Murakami et al., 1990), a preparation of leaves used as an antimicrobial, anti-inflammatory agent in treating oral cavity disease (Willershausen et al., 1991). One of the significant activities of sage is its antimutagenic potential on *E. coli* repair proficient strains (Baricevic et al., 1996; Filipic and Baricevic, 1997, 1998).

Sage is an antidiarrheal, antiseptic, carminative, digestive, diuretic, expectorant, hemostatic, laxative, sedative, spasmolytic, hypoglycemic, nervine, and tonic and uterine stimulant (Perfumi et al., 1991). It relaxes peripheral blood vessels, reduces perspiration, and blood sugar levels (reviewed in Gali-Muhtasib et al., 2000b). It has been employed to treat excessive perspiration and to dry up milk when a woman is no longer breast-feeding. Based on this anti-perspiration and drying effect, sage is also used for women who are sweating due to menopausal hot flushes (Leung, 1980). Containing sclereol, which stimulates the body to produce its own estrogen, sage may nutritionally support the body during the childbearing years and menopause. Its estrogenic effects may also be used to treat some cases of dysmenorrheal and menstrual irregularity or amenorrhea (reviewed in Gali-Muhtasib et al., 2000b).

Salvia has antispasmodic actions (Newall et al., 1996), which reduces tension in smooth muscle, and it can be used in a steam inhalation for asthma attacks (Pitchford, 1993). It is an excellent remedy for helping to remove mucous congestion in the airways and for checking or preventing secondary infection (Gali-Muhtasib et al., 2000b). Moreover, due to the anti-viral activity of its water and alcohol extracts, sage is used as an ingredient in combined plant preparations for the treatment of acute and chronic bronchitis, and is officially approved for clinical use in Bulgaria (Manolova et al., 1995). Its bitter component stimulates upper digestive secretions, intestinal mobility, bile flow, and pancreatic function (Todorov et al., 1984), the volatile oil also seems to have a general relaxant effect, so that the plant is suitable in the treatment of nervousness, excitability, and dizziness. It helps to fortify a generally debilitated nervous system (Elisabetsky et al., 1995).

The leaves of the sage plant are well known for their anti-oxidative properties (Farag et al., 1989; Schwarz and Ternes, 1992a, b; Baricevic and Bartol, 2000; Yildirim et al., 2000) and anti-inflammatory activities. In fact the potent anti-inflammatory effects of sage extract have been attributed to its antioxidant effects. Triterpenes oleanolic and ursolic acids or diterpene carnosol are some of the plant constituents, which were shown to have anti-inflammatory properties (Huang et al., 1994; Liu, 1995). The *n*-hexane and the chloroform extracts of *S. officinalis* were also found to exert anti-inflammatory activities. The extract of *S. officinalis* was found to dose-dependently inhibit croton oil-induced ear edema in mice. In the same study, the anti-inflammatory effects of the chloroform extract of a sage plant population grown in Slovania was found to contain ursolic acid with activities more potent than indomethacin, a non-steroidal anti-inflammatory drug (Baricevic et al., 2001). A new abietane diterpenoid, 12-*O*-methyl carnosol (2), was isolated from the leaves of *S. officinalis* L. that was found to exhibit remarkably strong activity, comparable to that of α-tocopherol (Miura et al., 2002).

The sage plant has been documented to improve the memory (Zhu et al., 1996, 1997). Old European books on medical herbs document the use of *S. officinalis* (sage) for memory improving properties (Perry et al., 1999). Sage has been recently shown to suppress the activity of the enzyme acetylcholinesterase, an enzyme linked to Alzheimer's disease (Perry et al., 2001). By suppressing the enzyme, the oil extract inhibits the breakdown of acetylcholine, a chemical messenger in the brain, which is suggested to be a potentially useful drug for the treatment of this disease. It is worth mentioning that the use of complementary medicines, particularly plant extracts for the improvement of memory varies according to the different cultural traditions. In the West, in contrast with the Far East, pharmacological properties of traditional cognitive or memory enhancing plants have not been widely investigated (Perry et al., 1998). Other species of *Salvia* have been also found to possess activities relevant to the treatment of Alzheimer's disease. The *Salvia lavandulaefolia* Vahl. (Spanish sage) essential oil and individual mono-terpenoid constituents have been shown to inhibit the enzyme acetylcholinesterase in vitro and in vivo, which is relevant to the treatment of Alzheimer's disease (Perry et al., 2001)

V. Mode of action of sage essential oil

It has been suggested that volatile oils, either inhaled or applied to the skin, act by means of their lipophilic fraction reacting with the lipid parts of the cell membranes, and as a result, modify the activity of the calcium ion channels. At certain dosage, the volatile oils saturate the membranes and show effects similar to those of local anesthetics. They can interact with the cell membranes by means of their physio-chemical properties and molecular shapes, and can influence their enzymes, carriers, ion channels and receptors (Buchbauer and Jirovetz, 1994). Various studies concentrated on the physiological effects of sage oil on humans. These include brain stimulation, anxiety-relieving sedation and antidepressant activities, as well as increasing the cerebral blood flow. The studies also describe the effects of odors on cognition, memory, and mood. The fragrance compounds are absorbed by inhalation and are able to cross the blood–brain barrier and interact with receptors in the central nervous system. Bioassays used for the description and explanation of volatile oil action, are usually carried out on mice, rats, and toads, e.g. the influence of peppermint oil on intestinal transport (Beesley et al., 1996); the effect of volatile oils on the skin penetration (Abdullah et al., 1996); the effect on skeletal muscle fibers (Fogaca et al., 1997); the screening for analgesic properties (Aydin et al., 1996).

Increasing numbers of aromatherapists and physiotherapists are using essential oils both in private practice and within hospitals, and their reports in all the main aromatherapy journals stress the positive effects of oils.

VI. Anticancer effects of East Mediterranean sage

Anticancer drugs counter cellular proliferation and tumor growth by targeting the macromolecular components of the cell that function in these processes. The traditional source of anticancer agents has been natural products that were obtained from animals or plants, and identified through biological assays. To date, natural products have proven to be the most effective in terms of their ability to alter the function of proteins relevant to cancer. Most of the anticancer drugs used in clinical practice today target the genome, either directly, through covalent modification or indirectly, through interference with nucleotide metabolism or chromatin dynamics (Hickman, 1992). Because many agents do not target the genomes of transformed cells with a high degree of selectivity, they tend to kill all rapidly dividing cells indiscriminately and exhibit high systemic toxicity at therapeutic dosages (Gali-Muhtasib and Bakkar, 2002). The poor therapeutic index of such cytotoxic agents is thus linked with their mechanism of action. For this reason, nonspecific targeting of the genome appears to hold little promise for the development of safer

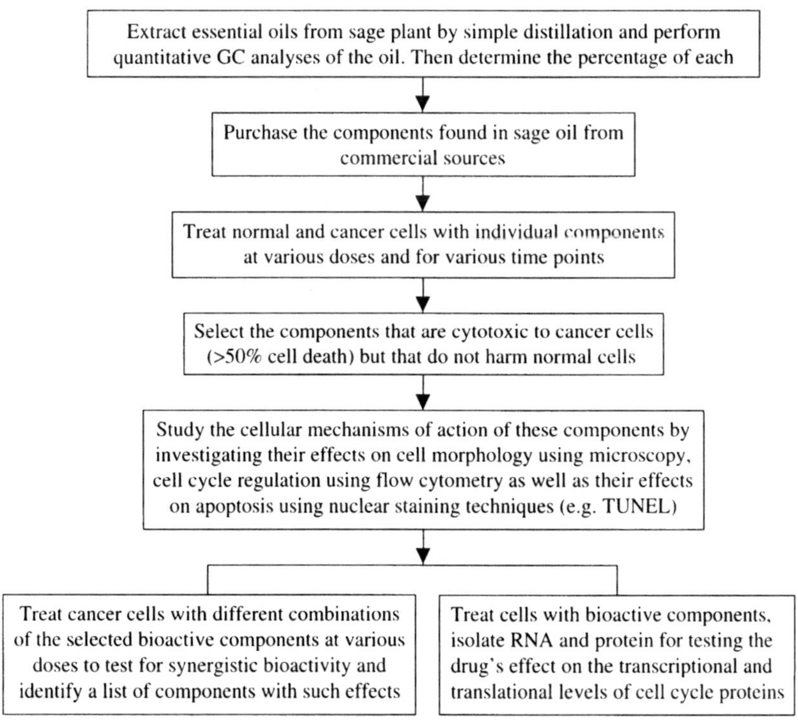

Fig. 1. Schematic representation of the steps and experiments required to identify the bioactive anticancer agent in sage essential oil.

and more effective anticancer agents. Recent scientific progress in understanding signaling pathways and cell cycle regulation has provided a wealth of potential new targets for anticancer drugs (Gali-Muhtasib and Bakkar, 2002). Targeting the signaling pathways involved in cancer cell growth may make it possible to treat cancer with fewer side effects than are caused by conventional cytotoxic therapies.

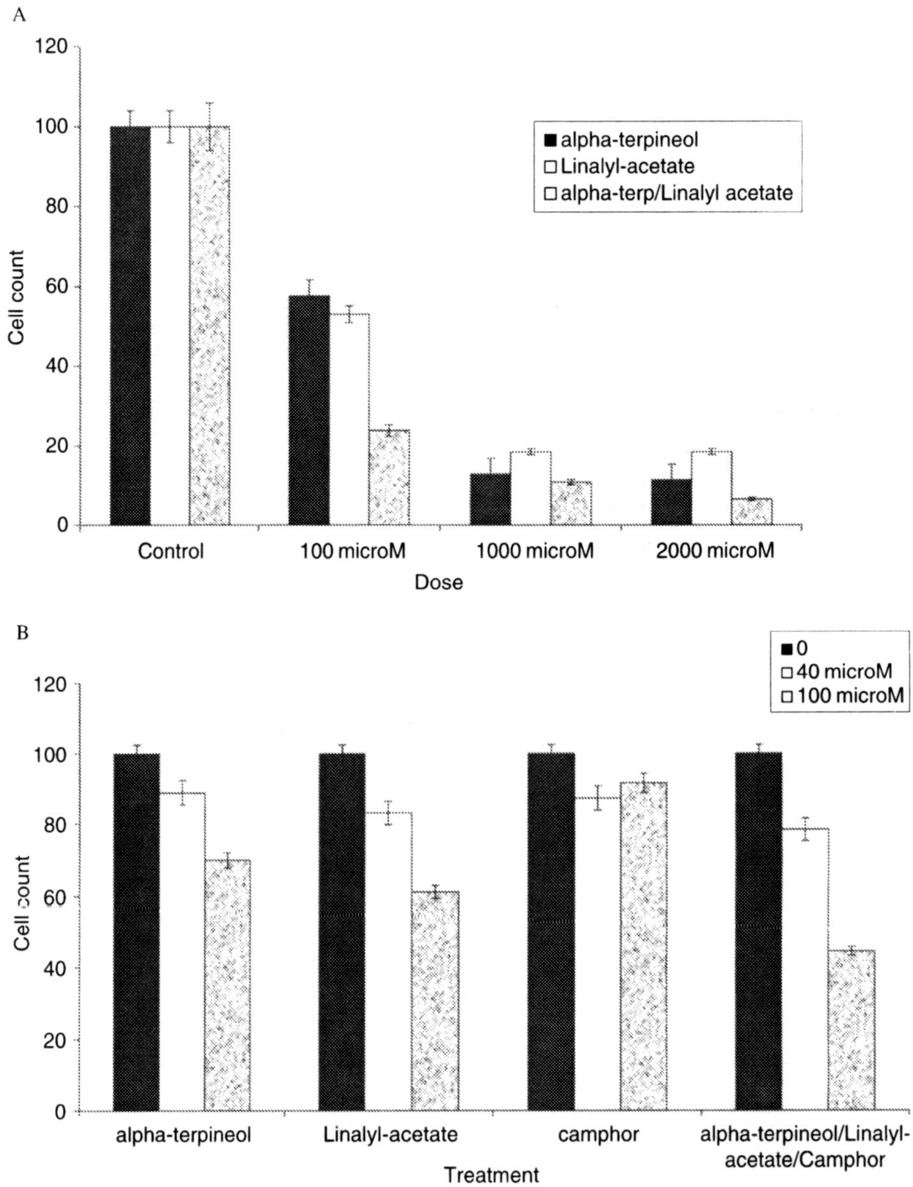

Fig. 2. The effect of graded concentrations of α-terpeniol, linalylacetate, and camphor when applied alone or in combination on the proliferation of human colon cancer cells HCT-116.

But in order to exploit these pathways fully, researchers must be able to find molecules that can modulate protein expression and function.

We have shown that the essential oil extracted from the Lebanese sage plant (*S. triloba)* has potent chemopreventive effects against skin tumor promotion in mice (Gali-Muhtasib and Affara, 2000a). When applied topically to the skin of Balb-c mice, the oil was found to delay tumor appearance by 4 weeks and inhibit tumor multiplicity by almost 78% and reduce the weight of tumors by 80%, with topical application on mouse skin being the preferred route for such inhibition (Gali-Muhtasib and Affara, 2000a). Intraperitoneal injections of the oil in small amounts (8% w/v) significantly decreased tumor weight and volume. This oil was also found to be a potent anti-inflammatory agent in the skin of treated mice. Using GC analyses, we identified 11 different individual components in sage essential oil (Farhat et al., 2001). In an attempt to determine the nature of the agent responsible for the observed anti-tumor promoting effects, we have recently received funding from the International Foundation for Science to test for the antiproliferative activity of the individual components in sage oil against human and rodent cancer cell lines, according to the plan presented in Figure 1. Components in sage oil are purchased commercially and applied alone or in combination, and the activity of the component is assayed by the inhibition of the proliferation rate of tumor cells in vitro, which may result from altered or inhibited DNA synthesis, cell cycle, or induction of apoptosis. The isolation of an active compound is the first stage in the development of a new agent, which might be developed as a drug for advancement to clinical trials and possibly to commercial use. The rationale for this approach is that by reducing or eliminating the variability of chemical composition and concentration that exists in crude plants, precise doses of known compounds can be given to patients. Human colon cancer cells (HCT-116) were cultured and incubated with various concentrations of sage components for 24 hours, and the effects of these components on cell proliferation were determined. Interesting synergisms were noted upon the treatment with combinations of α-terpeniol, linalylacetate and camphor (Figure 2). This study is still ongoing; upon completion of the study, we hope to be able to verify the anticancer potential of sage and determine the bioactive component(s) responsible for such an activity, information that would enhance our understanding of the medicinal and historical usage of Lebanese sage. This information is also instrumental for the design of effective anticancer therapy using the sage plant.

Acknowledgments

We thank the COMSTECH/International Foundation for Science Programme for funding our project to test the anticancer potential of sage essential oil.

References

Abdullah D, Ping QN, Liu GJ. (1996) Enhancing effects of essential oils on the penetration of 5-fluorouracil through rat skin. Acta Pharm Sin 31:214–21.

Arnold WN. (1988) Vincent Van Gogh and the thujone connection. JAMA 260:3042–4.

Arnold N, Bellomaria B. (1993) Observations morpho-anatomique et histochimique de *Salvia willeana* (Holmboe) Hedge et *Salvia fruticosa* Mill. de Chypre. Flora Mediterranea 3:283–97.
Aydin S, Ozturk Y, Beis R, Baser KHC. (1996) Investigation of origanum onites, sideritis congesta, satureja cuneifolia essential oils for analgesic activity. Phytotherapy Research 10:342–4.
Baricevic D, Bartol T. (2000) The biological/pharmacological activity of the *Salvia* genus V., Pharmacology. In: Kintzios SE, editor. Sage: the genus *Salvia*. Abingdon, Marston: Hawood Academic Publishers.
Baricevic D, Filipic M, Tornazin E, Cinc M, Zupancic A. (1996) Evaluation of autochthonous and/or introduced genotype of medicinal and aromatic plants in Slovenia (*Salvia officinalis* L.). In: Franz Ch., Schulz H, Schumann G, Sharma, JR, Tatlioglu T, editors. Proceedings, Federal centre for Breeding research on cultivated plants, Quelinburg.
Baricevic D, Sosa S, Della Loggia R, Tubaro A, Simonovska B, Krasna A, Zupancic A. (2001) Topical anti-inflammatory activity of *Salvia officinalis* L. leaves: the relevance of ursolic acid. J Ethnopharmacol 75:125–32.
Baser K, Honda G, Miki W. (1986) Herb drugs and herbalists in Turkey. Tokyo: Institute for the study of languages and cultures of Asia and Africa.
Beesley A, Hardcastle J, Hardcastle PT, Taylor CJ. (1996) Influence of peppermint oil on absorptive and secretory processes in rat small intestine. Gut 39:214–9.
Bellomaria B, Arnold N, Valentini G. (1992) Contribution to the study of the essential oils from three species of *Salvia* growing wild in the Eastern Mediterranean region. J Essential Oil Res 4:607–14.
British Pharmaceutical Codex. (1923) An Imperial Dispensatory for the use of Medical Practitioners and Pharmacists. London: The Pharmaceutical Press.
Buchbauer G, Jirovetz L. (1994) Aromatherapy – use of fragrances and essential oils as medicamens. Flav Fragr J 9:217–22.
Chervatyi VS, Vashenko TN, Shishkov GZ. (1980) Comparative evaluation of the antibacterial action of different extracts from *Salvia officinalis*. Rastitel'nye Resur 16:137–9.
Dobrynin VN, Kolosov MN, Chemov BK, Derbentseva NA. (1976) Antimicrobial substance of *Salvia officinalis*. Khimiya Prirodynkh Soedineii 5:686–7.
Elisabetsky E, Marschner J, Souza DO. (1995) Effects of linalool on glutamatergic system in the rat cerebral cortex. Neurochem Res 20:461–5.
Farag RS, Badei AZMA, Hewedi FM, El-Baroty GSA. (1989) Antioxidant activity of some spice essential oils on linoleic acid oxidation in aqueous media. J Am Oil Chem Soc 66:792–9.
Farhat GN, Affara NI, Gali-Muhtasib HU. (2001) Seasonal changes in the composition of the essential oil extract of East Mediterranean sage (*Salvia libanotica*) and its toxicity in mice. Toxicon 39:1601–5.
Filipic M, Baricevic D. (1997) Antimutagenic activity of *Salvia officinalis* extracts against UV-induced mutation in *E. coli* strains. Mutat Res 379:182.
Filipic M, Baricevic D. (1998) Inhibitory effect of *Salvia officinalis* extracts on SOS functions induced by UV-irradiation. Abstract book of the 28th Annual Meeting of the European Environmental Mutagen Society (EEMS), Salzburg.
Fogaca RTH, Cavalcante ADA, Serpa AKL, Sousa PJC, Coelho-de-Souza AN, Leal Cardosa JH. (1997) The effects of essential oil of *Mentha x villosa* on skeletal muscle of the toad. Phytother Res 11:552–7.
Gali-Muhtasib HU, Affara NI. (2000a) Chemopreventive effects of sage oil on skin papillomas in mice. Phytomedicine 7:129–36.
Gali-Muhtasib HU, Hilan C, Khater C. (2000b) Traditional uses of *Salvia libanotica* (East Mediterranean Sage) and the effects of its essential oils. J Ethnopharmacol 71:513–20.
Gali-Muhtasib HU, Bakkar N. (2002) Modulating cell cycle: current applications and prospects for future drug development. Curr Cancer Drug Target 2:309–36.
Geller RJ, Spyker DA, Garretson LK, Rogol AD. (1984) Camphor toxicity: development of a triage strategy. VetHuman Toxicol 26:8–10.
Gennaro AR. (1990) Remington's pharmaceutical sciences, 18th edition. Easton, PA: Mack Publishing Company.
Guenther E. (1949) The essential oils, Vol. I. New York: D. Van Nostrand.

Hay RK, Svoboda KP. (1993) Botany. In: Hay RKM, Waterman PG editors. Volatile oil crops. England: Longman Scientific and Technical, pp. 5–22.
Hefnawy Y, Moustafa S, Marth E. (1993) Sensitivity of listeria monocytogenes to selected spices. J Food Protection 56:876–8.
Hickman JA. (1992) Apoptosis induced by anticancer drugs. Cancer Metastasis Rev 11:121–39.
Hooser SB. (1990) Toxicology of selected pesticides, drugs and chemicals. D-Limonene, linalool, and crude citrus oil extracts. Vet Clin N Am. Small, 20:383–5.
Houghton PJ. (1995) The role of plants in traditional medicine and current therapy. J Altern Complement Med 1:131–43.
Huang MT, Ho CT, Wuang ZY, Ferraro T, Luo YR, Stauber K, Ma W, Georgiadis C, Laskin JD, Conney AH. (1994) Inhibition of skin tumorigenesis by rosemary and its constituents carnosol and ursolic acid. Cancer Res 54:701–8.
Ivanic R, Savin K. (1976) A comparative analysis of essential oils from several wild species of *Salvia*. Planta Med 30:25–31.
Karim F, Quraan S. (1986) Medicinal plants of Jordan. Centre for Jordanian studies. Jordan: Yarmouk University Irbid.
Kbayssi H. (1993) Encyclopedia of weeds and medicinal plants. Scientific Books, Beirut, Lebanon, p. 271.
Leung AY. (1980) Encyclopedia of common natural ingredients: used in foods, drugs, cosmetics and medicine. New York: John Wiley & Sons.
Leung AY, Foster S. (1995) Encyclopedia of common natural ingredients: used in food drugs and cosmetics. New York: John Wiley & Sons, pp. 688.
Leushner J. (1997) Reproductive toxicity studies of D-camphor in rats and rabbits. Arzneimittelforschung 47:124–8.
Liu J. (1995) Pharmacology of oleanolic and ursolic acid. J Ethnopharmacol 49:57–68.
Manolova N, Serkedjieva J, Ivanova V. (1995) Anti-influenza activity of the plant preparation "Broncho Pam". Fitoterapia LVI:223–6.
Meikle RD. (1985) Flora of Cyprus, Vol. 2. The Bentham-Moxon Trust, pp. 1287–99.
Millet Y, Jouglard J, Steinmetz MD, Tognetti P, Joanny P, Arditti J. (1981) Toxicity of some essential plant oils. Clinical and experimental study. Clin Toxicol 18:1485–98.
Millet Y, Tognetti P, Lavaire-Perlovisi M, Steinmetz MD, Arditti J, Jouglard J. (1979) Experimental study of the toxic convulsant properties of commercial preparations of essences of sage and hyssop. Rev Electroencephalogr Neurophysiol Clin 9:12–8.
Miura K, Kikuzaki H, Nakatani N. (2002) Antioxidant activity of chemical components from sage (*Salvia officinalis* L.) and thyme (*Thymus vulgaris* L.) measured by the oil stability index method. J Agric Food Chem 50:1845–51.
Morton JF. (1976) Herbs and Spices. New York: Golden Press.
Mouterde P. (1970) Nouvelle Flore du Liban et de la Syrie. Beyrouth: Dar El Machrek.
Murakami S, Kijima H, Isobe Y, Muramatsu M, Aihara H, Otomo S, Li L, Ai C. (1990) Effect of salvianolic acid A, a depside from roots of *Salvia miltiorrhiza*, on gastric H^+, K^+ ATPase. Planta Med 56:360–3.
Newall CA, Anderson LA, Phillipson JD. (1996) Herbal medicine, a guide for health-care professionals. London: The Pharmaceutical Press, pp. 231–232.
Parry JW. (1969) Spices: morphology, histology, chemistry. New York: Chem. Publishing Company.
Pattnaik S, Subramanyan VR, Bapaji M, Kole CR. (1997) Antibacterial and antifungal activity of aromatic constituents of essential oils. Microbios 89:39–46.
Perfumi M, Arnold N, Tacconi R. (1991) Hypoglycemic activity of *Salvia fructicosa* mill. From Cyprus. J Ethnopharmacol 34:135–40.
Perry NB, Anderson RE, Brennan NJ, Douglas MH, Heaney AJ, McGimpsey JA, Smallfield BM. (1999) Essential oils from dalmatian sage (*Salvia officinalis* l.): variations among individuals, plant parts, seasons, and sites. J Agric Food Chem 47:2048–54.
Perry NS, Houghton PJ, Sampson J, Theobald AE, Hart S, Lis-Balchin M, Hoult JR, Evans P, Jenner P, Milligan S, Perry EK. (2001) In-vitro activity of *S. lavandulaefolia* (Spanish sage) relevant to treatment of Alzheimer's disease. J Pharm Pharmacol 53:1347–56.

Perry EK, Pickering AT, Wang WW, Houghton P, Perry NS. (1998) Medicinal plants and Alzheimer's disease: Integrating ethnobotanical and contemporary scientific evidence. J Altern Complement Med 4:419–28.

Pitarevic I, Kustrac D, Kuftine J, Blazevic N. (1985) Influence of ecological factors on the content and composition of the essential oil in *Salvia officinalis*. In: Svendsen B, Scheffer JJC editors. Essential oils and aromatic plants. Economic Botany. Junk Publishers: Dordrecht, The Netherlands, pp. 203–7.

Pitchford P. (1993) Healing with whole foods. Berkley, CA: North Atlantic Books.

Rosengarten F. (1973) The book of spices. New York: Pyramid Books.

Santos-Gomes PC, Fernandes-Ferreira M. (2001) Organ- and season-dependent variation in the essential oil composition of *Salvia officinalis* L. cultivated at two different sites. J Agric Food Chem 49:2908–16.

Santos FA, Rao VS. (1997) Mast cell Involvement in the rat paw oedema response to 1,8-Cineole, the main constituent of eucalyptus and rosemary oils. Eur J Pharmacol 331:253–8.

Schwarz K, Ternes W. (1992a) Antioxidant constituents of *Rosmarinus officinalis* and *Salvia officinalis*. II. Isolation of carnosic acid and formation of other phenolic diterpenes. Z Lebensm Unters Forsch 195:99–103.

Schwarz K, Ternes W. (1992b) Antioxidant constituents of *Rosmarinus officinalis* and *Salvia officinalis*. III. Stability of phenolic diterpenes of rosemary extracts under thermal stress as required for technological processes. Z Lebensm Unters Forsch 195:499–504.

Shelef NA, Naglik OA, Bogen BW. (1990) Sensitivity of some common food borne bacteria to spices, sage, rosemary and all spice. J Food Sci 45:1042–4.

Stahl E. (1973) Drug analysis by chromatography and microscopy: a practical supplement to pharmacopoeias. Ann Arbor, MI: Ann Arbor Science.

Todorov S, Philianos S, Petkov V, Harvala C, Zamfirova R, Olimpiou H. (1984) Experimental pharmacological study of three species from genus *Salvia*. Acta Physiol Pharmacol Bulg 10:13–20.

Valnet J. (1986) The practice of armotherapy. Essex, C.W. Daniel Co. Ltd, pp. 143–144.

Verzar-Petri G, Then M, Meszaros S. (1985) Formation of essential oil in clary sage under different conditions. In: Svendsen B, Scheffer JJC editors. Essential oils and aromatic plants. Dordrecht, The Netherlands: Junk Publishers, pp. 199–202.

Waterman PG. (1993) The chemistry of volatile oils. In: Hay RKM, Waterman PG editors. Volatile Oil Crop. England: Longman Scientific and Technical, pp. 41–61.

Willershausen B, Gruber I, Hamm G. (1991) The influence of herbal ingredients on the plaque index and bleeding tendency of the Gingiva. J Clin Dentistry 2:75–8.

Yildirim A, Mavi A, Oktay M, Kara AA, Algur OF, Bilaloglu V. (2000) Comparison of antioxidant and antimicrobial activities of tilia (Tilia argentea Desf Ex DC), sage (*Salvia triloba* L.) and black tea (*Camellia sinensis*) extracts. J Agric Food Chem 48:5030–4.

Zhu M, Bowery NG, Greengrass PM, Phillipson JD. (1996) Application of radioligand receptor binding assays in the search of CNS active principles from Chinese medical plants. J Ethnopharmacol 54:153–64.

Zhu M, Phillipson JD, Yu P, Greengrass PM, Norman NG. (1997) Application of radioligand receptor binding assays in the search for the active principles of the traditional Chinese medicine "Gouteng". Phytother Res 11:231–6.

M.T.H. Khan and A. Ather (eds.)
Lead Molecules from Natural Products

Marine organisms from Brazil as source of potential anticancer agents

Letícia Veras Costa-Lotufo, Cláudia Pessoa, Maria Elisabete Amaral de Moraes, Adaíla Marta Paixão Almeida, Manoel Odorico de Moraes, Tito Monteiro da Cruz Lotufo

Abstract

The marine environment is considered a very rich source of extremely potent bioactive compounds. Brazil is the largest country in South America, with more than 8500 km of shoreline along the Atlantic Ocean, comprising a vast marine biodiversity. The studies on the chemistry of marine organisms in Brazil started in the beginning of the 1960s, but at that time the purpose for investigating the chemistry of marine organism was only that of a "phytochemistry" nature. It was only in the last decade that the effective collaboration between chemists and pharmacologists took place, and today, the studies on marine natural products in Brazil have focused two major points: the biomedical potential applications and the ecological role of the isolated substances. The aim of this chapter is to present the anticancer potential of Brazilian marine organisms. This potential will be discussed upon the great biodiversity observed in Brazilian coast and the studies performed by Brazilian groups on marine pharmacology. Screening programs with this purpose started just a few years ago, but some very promising results were obtained so far.

Keywords: *marine natural products, Brazil, marine biodiversity*

I. Introduction

The study of marine organisms as a source of biologically active compounds is considered a very promising field for the discovery of pharmacological tools and new medicines (Bhakuni, 1994; Munro et al., 1999; Faulkner, 2000a; Newman and Cragg, 2004). The work of Bergman and Feeney in the beginning of the 1950s initiated the studies on marine natural products and in the last few decades, an increasing number of new compounds have been isolated from marine organisms (Bergman and Feeney, 1951; Bhakuni, 1994; Faulkner, 2000b, 2001). Many authors considered that the improvement on the isolation and chemical identification techniques, the collaboration between chemists and pharmacologists and, most recently, the interest of the pharmaceutical industries were determinant in the development of the marine natural products research (Faulkner, 2000b). In fact, many aspects contributed to increase the interest in this research field, and the great biodiversity found in the marine environment is one of the most important (De Vries and Beart, 1995; Munro et al., 1999). The estimates have pointed out that about half of the total number of living species is restricted to the oceans, but only a few hundreds were studied in the search of natural products (Faulkner, 1995). Another interesting point is the development of a very efficient secondary metabolism in marine organisms, apparently related to evolutive pressures observed in these ecosystems. The secondary metabolism seems to be very important to the adaptive success of living species. Many authors believed in an ecological role to natural products, suggesting that these bioactive substances participate in chemical defence mechanisms (De Vries and Beart, 1995; Faulkner, 2000a, 2000b).

In Brazil, the studies on the chemistry of marine organisms started in the beginning of the 1960s with the identification of cholesterol in the sea urchin *Echinometra lucunter* by Tursch et al. (1963). As observed in other countries, the purpose for investigating the chemistry of marine organism was only that of a "phytochemistry" nature. A brief review of the literature published until the 1980s by Brazilian researchers showed many papers on the isolation of steroids, diterpenes and sesquiterpenes, mainly from echinoderms, cnidarians, sponges, tunicates, molluscs and algae (reviewed by Kelecom, 1997). Some papers also described the chemical composition of algae species used in human diet or in industry because of their economical relevance (Yokoyama and Guimarães, 1977; Caldas et al., 1983). There are also some old studies conducted by De Jorge et al. (1965, 1966, 1967) on the biochemistry of marine invertebrates.

Probably, the first report on the biomedical potential of Brazilian organisms was the work of Pinheiro-Vieira and Caland-Noronha (1971) evaluating the antimicrobial activity of some seaweed collected at the northeastern Brazilian coast. The authors assessed the toxicity of the aqueous extract of 30 algae against *Escherichia coli*, *Proteus vulgaris* and *Staphylococcus aureus* and found that the species of red algae (Rodophyta) presented the most promising results (Pinheiro-Vieira and Caland-Noronha, 1971). After that, in the decade of 1970, some papers evaluating the activity of extracts on isolated muscles were published. Freitas (1977) described the action of extracts of the midgut gland of the sea hare *Aplysia brasiliana* (Gastropoda, Aplysiidae) on some cholinoceptive structures. In the same year, Lunetta and Umiji (1977) performed a pharmacological study of the serum of the moray eel *Gymnothorax* sp. (Teleostei, Muraenidae), and showed the presence of a serotonin-like agent in this specie.

Despite these isolated efforts in evaluating the biomedical potential of marine organisms by some pharmacologists, it was only in the last decade that the effective collaboration between chemists and pharmacologists took place. Nowadays, the studies on natural products in Brazil have focused on two major points: the biomedical potential applications and the ecological role of the isolated substances.

The aim of this chapter is to present the anticancer potential of Brazilian marine organisms. This potential will be discussed on the great biodiversity observed in Brazilian coast and the studies performed by Brazilian groups on marine pharmacology. Screening programs with this purpose started only a few years ago, but some very promising results were obtained so far.

II. Brazilian marine biodiversity

Brazil is the largest country in South America, with more than 8500 km of shore line along the Atlantic Ocean. Along the coast there are many different biomes, each one with its own biological community. Most of the country is in the tropical region, but the extreme south is characterized by a more subtropical or temperate climate regime.

Beginning on the south toward the north, the southern coast is almost exclusively formed by a long and exposed sandy beach, washed by the cold waters of the South Atlantic. From Santa Catarina to Espirito Santo State, the shore is marked by a more irregular contour, with a set of bays and islands, alternating sandy beaches with rocky coasts. Luxurious mangrove forests developed along many estuaries and coral reefs can be found in the Northeast region.

Considering all these features, it becomes evident that the country is privileged in terms of biodiversity. Until the 1960s, the Brazilian coast was frequently included in the West Indies Region, comprising the Caribbean Sea and all the American Tropical Atlantic (Ekman, 1953; Forest, 1966; Briggs, 1974). The Caribbean Sea was much more explored then by scientific expeditions, leading to a large amount of species described. At that time, information about Brazilian marine fauna and flora was scanty. When the animals and plants collected along Brazilian shore were initially identified, many similarities were found with Caribbean species. In the last decades, the revisions of many groups of plant and animals have shown an increasing degree of endemism, leading to the establishment of four different biogeographical marine provinces in Brazil (Briggs, 1974; Coelho et al., 1977/78, 1980; Palacio, 1982). The extreme south is included in the Argentinean Province, with its northern limit at the State of Santa Catarina, where begins the Paulista Province, with transitional characteristics. The Paulista Province extends far to Espirito Santo State, the southern limit of the Brazilian Province, which comprises most of the Brazilian coast. Part of the north Brazilian shelf belongs to the Guyanese Province. Even today there are many animal groups with barely any information, resulting from a recurrent problem in many parts of the globe: the decreasing number of specialists (Migotto, 2000). There is a good knowledge about marine vertebrates (fishes, mammals and reptiles) and some taxa of invertebrates, as crustaceans and molluscs. Other invertebrates such as cnidarians, sponges, annelids, echinoderms and tunicates, just to mention a few, are insufficiently known in many regions along Brazilian coast. A few papers published recently have shown an increasing number of new records and new species, so that in the years to come a better evaluation of Brazilian marine biodiversity will be possible.

Beginning with the Rio'92 world meeting, a rapid increase in effort and funding of projects with the aim on the biodiversity could be noticed. In Brazil, it would not be different and today there are many projects dealing with this subject. One of the most important is the REVIZEE (Programme for Assessing the Sustainable Potential of Living Resources of the Exclusive Economic Zone). In the State of São Paulo, a group of researchers from different universities created the project BIOTA-FAPESP, now transformed into the Virtual Institute of Biodiversity and funded with resources from the government. Another project at the federal level was the

PRONABIO (National Programme on Biological Diversity), created to stimulate the research on biodiversity in every ecosystem.

As a preliminary result of these different programs, a large amount of the information available was organized, providing a sharper picture about the actual knowledge about Brazilian biodiversity. As to the marine environment, the lack of information concerning many animal phyla became evident. Regarding the knowledge on the different taxa, the quality and quantity of the information is a simple result of the presence of a specialist. Records about sponges are very sparse, accounting to a few lists of species of a few localities (Hajdu et al., 1999). The cnidarians are better known, especially scleractinians, but other anthozoans such as the sea anemones are still ignored in most of the coast. Migotto (1996) and Marques (2001) have provided revisions and species' lists of Hydrozoans and both authors are improving considerably the knowledge on the group. Information about bryozoans in Brazil is almost nonexistent. The most important papers on Brazilian bryozoans are those of Marcus, between 1937 and 1962, restricted to the State of Sao Paulo (Rocha and d'Hondt, 1999).

The ascidian fauna is poorly known in most of the tropical coast, but in the last decade, a group of new taxonomists has been trained, establishing three research groups in Brazil, covering tropical, transitional and subtropical regions.

The groups mentioned above are those regarded as with more potential in terms of prospecting new drugs in the sea. The need of new taxonomists trained in groups like bryozoans, sponges, cnidarians and ascidians is clear, in order to give support to further steps in new drug discovery.

III. Current research on marine natural products

The literature of marine natural products shows that cytotoxicity is the most common activity related to these compounds (De Vries and Beart, 1995). In fact, studies performed at the United States National Cancer Institute have shown that marine invertebrates have a higher incidence of cytotoxic compounds than any other group (Munro et al., 1999). While 1.9% of tested extracts obtained from marine macroorganisms were cytotoxic, only 0.3% of tested extracts obtained from plants were active (Munro et al., 1999). In those screening data, Porifera, Bryozoa and Chordata are among the most promising phyla as sources of new active principles for drug development (Faulkner, 2001).

According to Kelecom (1997), studies in marine natural chemistry are focused on invertebrates and algae and, until 1997, around 60 different organisms have yielded over 110 identified metabolites. Most of these metabolites are described as unique compounds but no mention of any related biological activity is found. In this review, studies performed with toxins, proteins such as lectins and sulfated polysaccharides were not considered, thus the total amount of organisms and compounds in the literature is much greater. It is worthwhile to mention that these compounds generally present biological properties, such as hemagglutinating or anticoagulant activities.

The studies of hemagglutinins in Brazilian algae started in the last decade, and since then several papers have been published (Ainouz et al., 1995; Freitas et al., 1997;

Benevides et al., 1998, 1999; Costa et al., 1999; Calvete et al., 2000; Nagano et al., 2002). Ainouz and Sampaio (1991) showed that 10 out of 20 tested algae extracts agglutinated rabbit erythrocytes. Ainouz et al. (1992) screened 27 Brazilian algae for the presence of hemagglutinins and obtained positive results for 15 species: twelve red, two green and one brown. These agglutinins are lectins, glycoproteins that specifically recognize and bind some carbohydrates. Many authors believed that lectins might have an important role on cancer diagnosis. Melo et al. (1997) also described antifungal properties of agglutinins from the red algae *Hypnea musciformis*, suggesting another possible therapeutic role to these compounds.

There are also many studies with sulfated polysaccharides isolated from algae, sponges, echinoderms and tunicates (Pavao, 1996, 2000; Farias et al., 2000; Zierer and Mourao, 2000). The interest in these compounds is mainly related to their anticoagulant and antithrombotic properties.

As mentioned before, the search of antitumoral compounds in Brazilian marine organism is a very recent field. The total number of studies is inexpressive if taken into account the Brazilian marine biodiversity. Table 1 showed the species of marine organisms in which the presence of cytotoxic compounds were reported in the literature. As observed, the total number of species is 24, out of which 80% belong to sponges and tunicates groups, and the remaining 20% correspond to studies performed with cnidarians and fish. In fact, the total number of studied species is greater, reaching almost 50 when included those in which the search for cytotoxic compounds gave negative results. It is interesting to notice that despite the absence of cytotoxic studies on algae group, there are many studies describing the occurrence of metabolites involved in the chemical defenses of seaweed (Pereira and Teixeira, 1999; Pereira et al., 2000, 2002). Unfortunately, these compounds have probably never been tested for cytotoxicity.

In 1986, Freitas and Sawaya demonstrated that the purine caissarone (Figure 1), isolated from the Brazilian endemic anemone *Bunodosoma caissarum*, induced polyspermy in sea urchin eggs. The purine inhibited the detachment of the vitelline layer from the sea urchin egg plasma membrane after fertilization and this effect led to polyspermy. Various abnormalities were detected at various embryonic stages, from multipolar egg division through unequal cleavages and exogastrulation up to teratogenic effects on the sea urchin larvae (pluteus). The studies of alterations in sea urchin egg development is a suitable model for detecting cytotoxic, teratogenic and antineoplastic activities of new compounds, and it has also been extensively used as a model for developmental toxicology evaluation (Jacobs and Wilson, 1986; Costa-Lotufo et al., 2002). Few years later, Malpezzi and Freitas (1990) suggested the presence of different cytotoxic compound in *B. caissarum* that has a significant inhibitory effect on cell division. These studies with *B. caissarum* probably were the first reports of cytotoxicity related to Brazilian marine macroorganisms.

On the other hand, there are some papers mentioning the presence of neurotoxins in the *Bunodosoma* genus. Three neurotoxins were isolated from *B. caissarum* venom, and their characterization indicated high homology with Type 1 long sea anemone neurotoxins (Malpezzi et al., 1993). Santana et al. (1998, 2001) isolated a new neurotoxic peptide, called granulotoxin, from *Bunodosoma granulifera*. The granulotoxin induced seizures after hippocampal injection with a similar profile of pilocarpine.

Table 1
Brazilian marine organisms studied for cytotoxicity as a suggestion of anticancer potential

Specie/group	Substance/extract	Cytotoxicity against	Reference
Sponges			
A. brasiliensis	Arenosclerins A–C and Haliclonacyclamine E	HL-60 (leukemia); L929 (fibrosarcoma); B16 (melanoma); U138 (colon)	Torres et al. (2002a)
Polymastia janeirensis	Methanol/toluene and aqueous extracts	HT29 (colorectal)	Monks et al. (2002)
l Haliclona aff. tubifera	Methanol/toluene and aqueous extracts	HT29 (colorectal)	Monks et al. (2002)
M. arcuiris	Methanol/toluene and aqueous extracts	HT29 (colorectal)	Monks et al. (2002)
Raspailia (Syringella) sp.	Methanol/toluene and aqueous extracts	HT29 (colorectal)	Monks et al. (2002)
Aaptos sp.	Ethanol extract	Sea urchin egg	Rangel et al. (2001)
Axinella aff. corrugata	Ethyl acetate extract	Sea urchin eggs	Rangel et al. (2001)
Geodia corticostylifera	Ch2Cl2/methanol extract	Sea urchin eggs	Rangel et al. (2001)
Mycale laxisima	Aqueous extract	Sea urchin eggs	Rangel et al. (2001)
R. elegans	CH2Cl2 extract	Sea urchin eggs	Rangel et al. (2001)
A. viridis	Halitoxin	Sea urchin eggs	Berlinck et al. (1996)
Cnidarians – Anthozoa			
B. caissarum	Methanol extract and caissarone	Sea urchin eggs	Malpezzi and Freitas, 1990 Freitas and Sawaya, 1986
C. riisei	Riisein A and B	HCT-116 (colon)	Maia et al. (2000)
Tunicates			
C. delleichiajei	Sebastianines A and B	HCT-116 (colon)	Torres et al. (2002b)
Dideminidae	Tamandarins A and B	BX-PC3 (pancreas); DU-145 (protasta); UMSCC10b (head and neck)	Vervoort et al. (2000)
D. granulatum	Granulatimide and isogranulatimide	G2 checkpoint inhibition activity on MCF-7 mp53 (breast)	Roberge et al. (1998) Berlinck et al. (1998)
P. nigra	Methanol extract	T47D (breast); Sea urchin eggs	Costa et al. (1996)
Eudistoma vannamei	*Methanol extract*	HCT-8 (colon); CEM (leukemia); HL-60 (leukemia); B-16 (melanoma)	Jimenez et al. (2003)
Euherdmania sp.	Methanol extract	HCT-8 (colon); CEM (leukemia); HL-60 (leukemia); B-16 (melanoma)	Jimenez et al. (2003)

Table 1 (*continued*)

Specie/group	Substance/extract	Cytotoxicity against	Reference
Didemnum psammatodes	Methanol extract	Sea urchin eggs	Jimenez et al. (2003)
Polysyncraton sp.	Methanol extract	Sea urchin eggs	Jimenez et al. (2003)
Fish			
C. spinosus	Skin secretion	Sea urchin eggs	Malpezzi et al. (1997)
D. hystrix	Skin secretion	Sea urchin eggs	Malpezzi et al. (1997)

In further studies, the same group started to search for antimitotic compounds in marine organisms. Costa et al. (1996) demonstrated the presence of cytotoxic compounds in a methanol extract obtained from the visceral organs of the sea squirt *Phallusia nigra*. The cytotoxicity was measured as the ability of this extract in inhibiting the sea urchin eggs development and the proliferation and DNA synthesis on T47D human breast cells.

In the same year, Berlinck et al. (1996) described the occurrence of halitoxin complex (Figure 1) in *Amphimedon viridis* (Porifera). In fact, Schmitz et al. (1977, 1978) had already detected the presence of halitoxin complex in several *Amphimedon/Haliclona* species including *A. viridis*. This halitoxin complex presented lethality, and also hemolytic and cytotoxic activity (Schmitz et al., 1977, 1978; Berlinck et al., 1996). Berlinck et al. (1996) suggested that the halitoxin complex effect on sea urchin eggs did not seem to be related to the mitotic system itself but with an unspecific activity on cell membrane.

Malpezzi et al. (1997) described the occurrence of cytotoxins in the skin of the puffer-fishes *Ciclichthys spinosus* and *Diodon hystrix*. Both secretions had cytotoxic effects on sea urchin eggs, inhibiting the cleavages and inducing anomalies in a dose-dependent manner. The authors also discussed that these effects were not related to the presence of tetrodotoxin in the skin of these fishes, once the neurotoxic effects of this secretions were the induction of transient depolarizations and blockage of crustacean nerve conduction after prolonged exposure.

Berlinck et al. (1998) described the occurrence of very strong cytotoxic aromatic alkaloids, granulatimide and isogranulatimide (Figure 1), in the Brazilian ascidian *Didemnum granulatum*. These alkaloids were the first examples of a new class of G2-specific cell cycle inhibitors. The use of G2 checkpoint inhibitors in combination with a DNA-damaging agent has been proposed as a strategy for cancer therapy, especially of solid tumors lacking p53 function (Roberge et al., 1998). According to the authors, the inhibition of this inherent system of DNA repair should dramatically increase the killing of p53 lacking cancer cells, once these cells are unable to activate the G_1 checkpoint in response to DNA damage. Another interesting point is that the G_2 checkpoint inhibitors alone seem to have no effect on normal or cancer cells, attenuating the collateral effects of these drugs. Isogranulatimide is probably the most promising drug isolated from a Brazilian marine organism. It is worthwhile to mention that this paper is a result of a rational screening program involving Brazilian (Universidade de São Paulo, Brazil) and Canadian (University of British

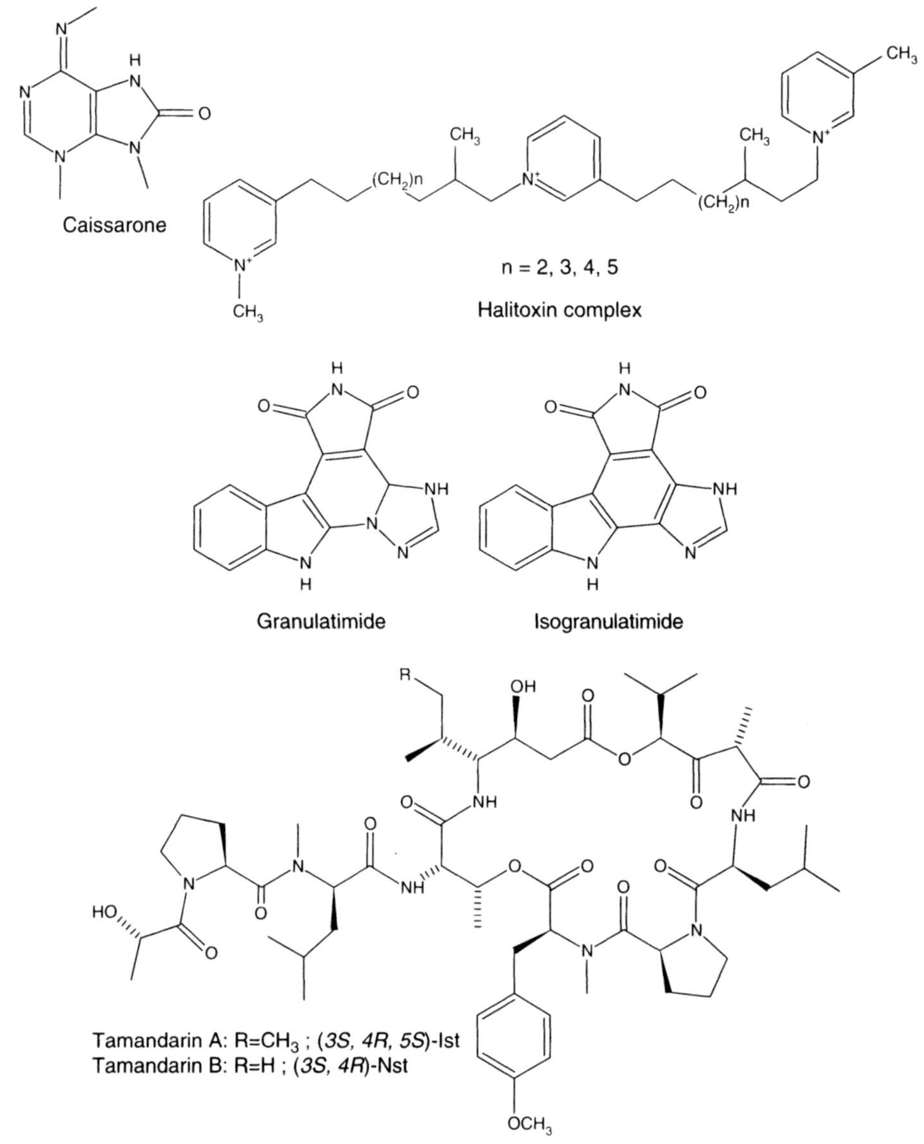

Fig. 1. Chemical structures of secondary metabolites isolated from Brazilian marine organisms.

Columbia) Universities. A few years later, another alkaloid had been isolated from the same ascidian species, 6-bromogranulatimide (Britton et al., 2001).

The tamandarins A and B (Figure 1) are new depsipeptides isolated also from a Brazilian ascidian of the family Didemnidae (Vervoort et al., 2000). The depsipeptides are very common in ascidians, and the most studied ascidan metabolite, didemnin B, belongs to this chemical class. As discussed later, didemnin B underwent clinical trials but it showed neuromuscular toxic effects resulting as not very effective in humans as an antineoplastic drug. Tamandarins A and B showed strong cytotoxicity to human cancer cell lines, and seemed to be a more potent inhibitor of protein synthesis when compared to didemnin B (Vervoort et al., 2000). Despite the strong cytotoxicity of tamandarins, there are no data regarding their toxicity, and it is too soon to have an idea of their therapeutic applications. Those studies are part of a collaborative agreement between Universidade Federal Fluminense (Rio de Janeiro, Brazil) and Scripps Institution of Oceanography (San Diego, United States).

Maia et al. (2000) described the isolation of new cytotoxic sterol glycoside from the octocoral *Carijoa riisei*, collected in Rio de Janeiro, Brazil. Riseiin A and B (Figure 2) showed strong in vitro cytotoxicity (IC_{50} values below 2 μg/mL) against HCT-116 human colon adenocarcinoma cells. The authors hypothesized that these compounds functioned as chemical defenses.

In fact, there are other few studies on the natural product chemistry of octocorallia in Brazil (Fernandes and Kelecom, 1995; Martins & Epifanio, 1998; Maia et al., 1999; Epifanio et al., 2000). However, they did not focus on the biomedical potential of the isolated compounds, instead they generally evaluated the ecological significance of these compounds as chemical defenses. Fernandes and Kelecom (1995) described the occurrence of the naardosinane sesquiterpene, 12-hydroxynardosin-1(10),11(13)-diene in the Brazilian endemic gorgonian *Phyllogorgia dilatata*. A few years later, Martins and Epifanio (1998) described another sesquiterpene, germacra-1(10),4(15),7(11)-trien-5-ol-8-one from *P. dilatata*. Maia et al. (1999) isolated a new sesquiterpene lactone heterogorgiolide from *Heterogorgia uatumani* and proposed a fish feeding deterrent role for this substance based on field bioassays. Epifanio et al. (2000) studied the chemical defenses of the endemic Brazilian gorgonian *Lophogorgia violacea*, and demonstrated that lophotoxin, a neurotoxin originally isolated from *L. rigida*, was the most potent fish deterrent. Until this moment, there are no data regarding compounds that present cytotoxic activity.

Torres et al. (2000) described new tetracyclic alkaloids from the Brazilian endemic Haplosclerida sponge *Arenosclera brasiliensis*, namely Arenosclerins A–C and Haliclonacyclamine E (Figure 2). Further studies have demonstrated that these alkaloids present antibacterial activity against resistant bacteria and also cytotoxicity against human HL-60 leukemia, human L929 fibrosarcoma, murine B16 melanoma and human H138 colon adenocarcinoma (Torres et al., 2002a).

Recently, Torres et al. (2002b) isolated two cytotoxic pyridoacridine alkaloids, Sebastianines A and B (Figure 2) from the Brazilian ascidian *Cystodytes delleichiajei*. These alkaloids displayed cytotoxicity against HCT-116 adenocarcinoma cells, and this activity seemed to be dependent of a p53 mechanism.

In the recent years, Brazilian groups started some screening programs to verify the biomedical potential of marine organisms, optimizing the efforts on the isolation and characterization of bioactive metabolites. Some papers have already been published describing the results of these screening programs. Rangel et al. (2001) studied the cytotoxic and neurotoxic activities in extracts of 24 sponges from the southeastern Brazilian coast, showing that 54% of the tested sponges exhibited from median to high toxicity in at least one of the assays performed. *Axinella aff. corrugata*, *Mycale*

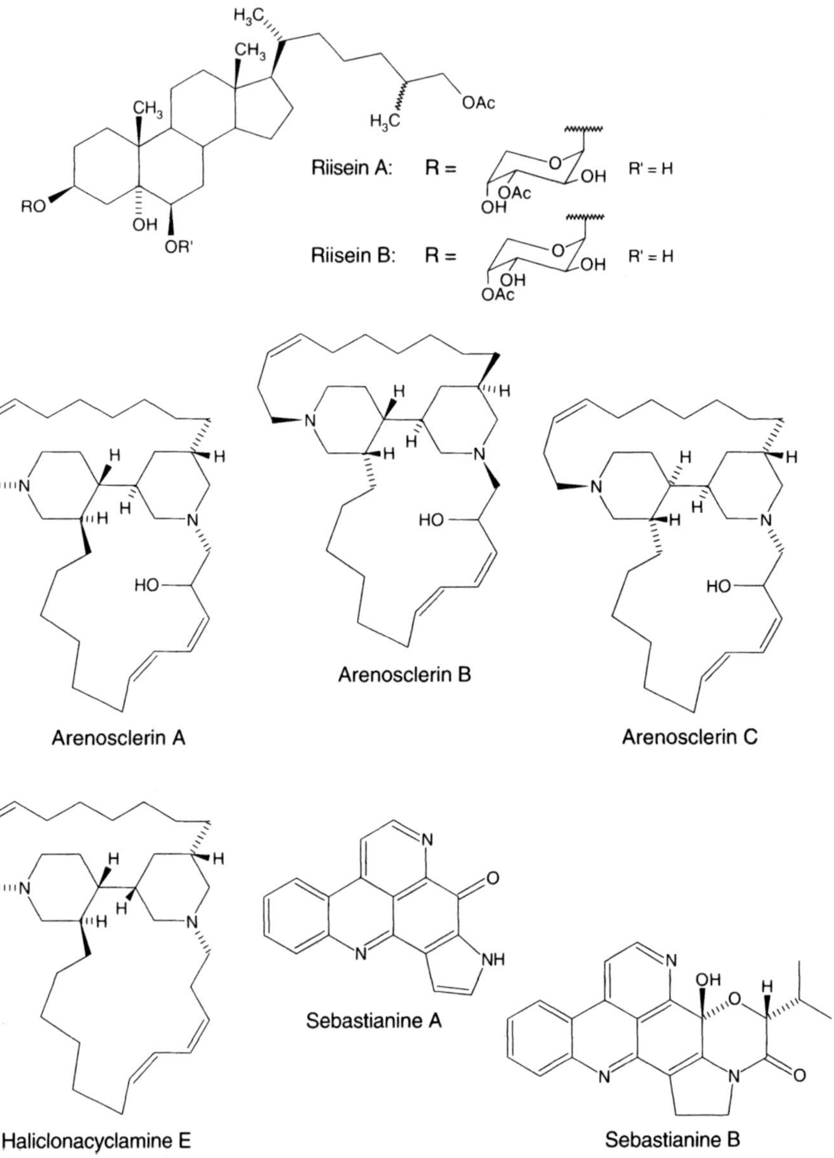

Fig. 2. Chemical structures of secondary metabolites isolated from Brazilian marine organisms (Continuation...).

laxissima, *A. brasiliensis* and *Raspailla elegans* inhibited mitosis and sea urchin egg development.

Monks et al. (2002) reported the results of the in vitro screening of 10 sponges collected off the coast of Santa Catarina, Southern Brazil. The extracts were tested for anticancer, antichemotactic and antimicrobial activities and 80% of the extracts tested were active in at least one bioassay. Organic extracts of *Polymastia janeirenses*, *Haliclona aff. tubifera*, *Mycale arcuiris* and *Raspailia* sp. were cytotoxic toward

HT29 human adenocarcinoma cells, NCI-H460 human large cell lung carcinoma cells and U373 human glioblastoma astrocytoma cells.

In the northeastern Brazilian coast, a screening of ascidians revealed that 5 among 10 species tested were cytotoxic to sea urchin eggs and to cancer cell lines (human HL60 and CEM leukemia cells, murine B16 melanoma and human HCT-8 colon adenocarcinoma cells) (Jimenez et al., 2003). Another very interesting point is that only four species were previously known to science, suggesting a high degree of endemism in this region. Five of them are undescribed species from Didemnidae and Polycitoridae families and one belonged to an undescribed genus of the Holozoidae family.

In a recent review on the status of marine natural products chemistry in Brazil, Berlinck et al. (2004) discussed data on the isolation, structure elucidation and evaluation of biological activities of natural products isolated from sponges, ascidians, octocorals and Opistobranch mollusks. His group had conducted a screening with more than 300 crude extracts for antitumor, antibacterial, antituberculosis and leishmanicidal activities, finding that marine sponges afforded the highest number of active extracts.

IV. Clinical perspectives

Concerning clinical treatments, only the derivates of the nucleosides isolated from the sponge *Cryptotethya crypta*, ARA-A (Vidarabin, Vidarabin Thilo®) and ARA-C (Cytarabin, Alexan®, Udicil®) have been actually used against viruses and tumors until the present moment (Koning and Wright, 1996; Newman and Cragg, 2004). Nevertheless, other compounds are at the preclinical or clinical trial phase for treatments of diseases, and are regarded as promising candidates for therapeutic uses.

As to neoplasies treatment, the studies are in an advanced stage of progress. The briostatin I, isolated from the bryozoan *Bugula neritina*; ecteinascidin 743, isolated from the tunicate *Ecteinascida turbinata*; aplidine, isolated from the tunicate *Aplidium albicans*; dolastatin 10 and derivatives, isolated from the sea hare *Dolabella auricularia*; halichodrin B and derivatives, first isolated from the sponge *Halichondria okadai*; discodermolide, isolated from the sponge *Discodermia dissoluta*; kahalalide F, first isolated from the mollusk *Eylsia rufescens*; spisulosine, isolated from the marine clam *Spisula polynoma* and squalamine, isolated from the dogfish shark *Squalus acanthias*, are marine-derived compounds that have entered clinical trials as antitumor agents (Newman and Cragg, 2004).

Didemnin B was the first marine natural product to undergo human anticancer trials. Unfortunately, the consensus resulted from the phase II clinical trials was that didemnin B showed little efficacy but significant toxicities and it had been withdrawn from antitumor clinical trials in the mid-1990s by National Cancer Institute (NCI) from United States (Vera and Joullié, 2002; Newman and Cragg, 2004). In fact, the neuromuscular toxicity of didemnin was dose limiting. The depsipeptide aplidine belongs to the same chemical class as that of didemnin, and appears to be more active and less toxic than didemnin. The Phase I trials using aplidine suggested an efficient antitumor activity in patients with advanced solid tumors (Schwartsmann et al., 2001; Vera and Joullié, 2002). The alkaloid ecteinascidin 743 had its therapeutic potential confirmed in Phases I and II trials (Schwartsmann et al., 2001). Concerning bryostatins, the results from its clinical trials suggested that this lactone has antitumor activity on patients with malignant melanoma, lymphoma and ovarian carcinoma, but had no significant activity on patients with advanced solid tumors (Schwartsmann et al., 2001). Dolastatin was tested in patients with a range of solid tumors, however, it did not present significant antitumor activity (Schwartsmann et al., 2001).

Many other drugs are in a stage of preclinical studies, including the aplironine A, isolated from the mollusc *Aplysia kurodai*; the criptoficin 1, isolated from *Nostoc* sp. and the crambecidin 816, isolated from the sponge *Crambe crambe* (Flam, 1994; Koning and Wright, 1996; Rinehart, 2000). According to Schwartsmann et al. (2001), the aromatic alkaloids isogranulatimide and granulatimide, isolated from the Brazilian *D. granulatum*, have been synthesized and are prepared to undergo on preclinical development.

Despite the interest in antitumor activity as the most promising in the study of marine natural products, other drugs are showing a great therapeutic potential in the treatment of inflammatory processes. A few examples of drugs in preclinical studies stage can be mentioned, such as the sesteterpenoid manoalide, isolated from the sponge *Luffariella variabilis*, which acts as a strong inhibitor of phospholipase A_2 and the pseudopterosines, isolated from the gorgonian *Pseudopterogorgia elisabethae*, also with antiinflammatory activity (Freitas et al., 1984; Koning and Wright, 1996).

V. Concluding remarks

The interest in marine natural products increased through the years, and the potential of the sponges, ascidians and bryozoans as a prolific source of new secondary metabolites with therapeutic applications had become clear. Nevertheless, the use of sophisticated mechanism-based screens using specific enzymes, receptors and recombinant whole cells represented a significant improvement in the efficacy of the development on natural-based medicines.

The pharmacological potential of Brazilian marine organism is enormous, and the efficient exploration of this resource depends on the organization of screening programs with a multidisciplinary team (biologists, pharmacologists and chemists). As discussed in this chapter, Brazil possesses a great biodiversity and Brazilian scientists have recently joined efforts to use this richness in a rational manner.

References

Ainouz IL, Sampaio AH. (1991) Screening of Brazilian marine algae for hemagglutinins. Bot Mar. 34:211–4.

Ainouz IL, Sampaio AH, Benevides NMB, Freitas ALP, Costa FHF, Carvalho MR, Pinheiro-Joventino F. (1992) Agglutination of enzyme treated erythrocytes by Brazilian marine algal extracts. Bot Mar 35:475–9.

Ainouz IL, Sampaio AH, Freitas ALP, Benevides NMB, Mapurunga S. (1995) Comparative study on hemagglutinins from the red algae *Bryothamnion seaforthii* and *Bryothamnion triquetrum*. Rev Bras Fisiol Veg 7:15–9.

Benevides NMB, Holanda ML, Melo FR, Freitas ALP, Sampaio AH. (1998) Purification and partial characterization of the lectin from the marine red algae *Enantiocladia duperreyi* (C. Agardh) Falkenberg. Bot Mar 41:521–5.
Benevides NMB, Oliveira SRM, Holanda ML, Melo FR, Freitas ALP, Sampaio AH. (1999) Seasonal variations in hemagglutinating activity and chemical composition of two red marine algae *Gracilaria domingensis* and *Gelidium pusillum*. Rev Bras Fis Veg 11:91–5.
Bergman W, Feeney RJ. (1951) Nucleosides of sponges. J Org Chem 16:981–7.
Berlinck RGS, Britton R, Piers E, Lim L, Roberge M, Rocha RM, Andersen RJ. (1998) Granulatimide and isogranulatimide, aromatic alkaloids with G2 checkpoint inhibition activity isolated from the Brazilian ascidian *Didemnum granulatum*: structure elucidation and synthesis. J Org Chem 63:9850–6.
Berlinck RGS, Hajdu E, Rocha RM, Oliveira JHHL, Hernandez ILC, Seleghim MHR, Granato AC, Almeida EVR, Nunez CV, Muricy G, Peixinho S, Pessoa C, Moraes MO, Cavalcanti BC, Nascimento GGF, Thiemann O, Silva M, Souza AO, Silva CL, Minarini PRR. (2004) Challenges and rewards of research in marine natural products chemistry in Brazil. J Nat Prod 67:510–22.
Berlinck RGS, Ogawa CA, Almeida AMP, Sanxhez MAA, Malpezzi ELA, Costa LV, Hajdu E, Freitas JC. (1996) Chemical and pharmacological characterization of halitoxin from *Amphimedon viridis* (Porifera) from the Southeastern Brazilian coast. Comp Biochem Physiol 115C:155–63.
Bhakuni DS. (1994) Bioactive marine alkaloids. J Indian Chem Soc 71:329–40.
Briggs JC. (1974) Marine zoogeography. New York: McGraw-Hill, pp. 475.
Britton R, Oliveira JHHL, Andersen RJ, Berlinck RS. (2001) Granulatimide and 6-Bromogranulatimide, minor alkaloids from the Brazilian ascidian *Didemnum granulatum*. J Nat Prod 64:254–5.
Caldas MJT, Ferreira-Correia MM, Lopes MJS, Correa MM. (1983) Informes preliminaries sobre alguns components químicos de algas marinhas bentônicas do litoral da ilha de São Luís (Estado do Maranhão, Brasil). Bol Lab Hidrobiol São Luís 5/6:9–18.
Calvete JJ, Costa FHF, Saker-Sampaio S, Murciano MPM, Nagano CS, Cavada BS, Grangeiro TB, Ramos MV, Bloch Jr. C, Silveira SB, Freitas BT, Sampaio AH. (2000) The amino acid sequence of the agglutinin isolated from the red marine alga *Bryothamnion triquetrum* defines a novel lectin structure. Cell Mol Life Sci 57:343–50.
Coelho PA, Ramos-Porto M, Koening ML. (1977/78) Crustáceos Marinhos do Brasil, do Uruguai e da Argentina (ao norte de Mar del Plata). Considerações biogeográficas. Anais da Universidade Federal Rural de Pernambuco 2/3:227–56.
Coelho PA, Ramos-Porto M, Koening ML. (1980) Biogeografia e bionomia dos crustáceos do litoral equatorial brasileiro. Trabalhos Oceanográficos da Universidade Federal de Pernambuco 15 7–138.
Costa FHF, Sampaio AH, Neves SA, Rocha MLA, Benevides NMB, Freitas ALP. (1999) Purification and characterization of a lectin from the red marine algae *Amansia multifida*. Physiol Mol Biol Plants 5:53–61.
Costa LV, Malpezzi ELA, Matsui DH, Machado-Santelli GM, Freitas JC. (1996) Cytotoxic activity of a methanol extract of *Phallusia nigra* (Tunicata, Ascidiacea). Braz J Med Biol Res 29:367–73.
Costa-Lotufo LV, Cunha GMA, Farias PAM, Viana GSB, Cunha KMA, Pessoa C, Moraes MO, Silveira ER, Gramosa NV, Rao VSN. (2002) The cytotoxic and embryotoxic effects of kaurenoic acid, a diterpene isolated from *Copaifera langsdorffii* oleo-resin. Toxicon 40:1231–4.
De Jorge FB, Petersen JA, Sawaya P. (1966) Biochemical studies in a Eunicid (Polichaete) of the littoral of São Paulo, Brazil. Comp Biochem Physiol 17:535–51.
De Jorge FB, Petersen JA, Sawaya P. (1967) Biochemical studies on the enteropneust *Balanoglossus gigas* (FR. Muller, 1898). Comp Biochem Physiol 22:467–75.
De Jorge FB, Sawaya P, Petersen JA. (1965) Iodine: accumulation by *Balanoglossus gigas*. Science 150:1182–3.
De Vries DJ, Beart PM. (1995) Fishing for drugs from the sea: status and strategies. TiPS 16:275–9.

Ekman S. (1953) Zoogeography of the Sea. London: Sidgwick and Jackson 417pp.
Epifanio RA, Maia LF, Fenical W. (2000) Chemical defenses of the endemic Brazilian gorgonian *Lophogorgia violacea* pallas (Octocorallia, Gorgonacea). J Braz Chem Soc 11:584–91.
Farias WRL, Valente AP, Pereira MS, Mourao PAS. (2000) Structure and anticoagulant activity of sulfated galactans. J Biol Chem 275:29299–307.
Faulkner DJ. (1995) Marine natural products. Nat Prod Rep 12:223–69.
Faulkner DJ. (2000a) Marine pharmacology. Antonie van Leeuwenhoek 77:135–45.
Faulkner DJ. (2000b) Marine natural products. Nat Prod Rep 17:7–55.
Faulkner DJ. (2001) Marine natural products. Nat Prod Rep 18:1–49.
Fernandes L, Kelecom A. (1995) A further nardosinane sesquiterpene from the gorgonian *Phyllogorgia dilatata* (Octocorallia, Gorgonacea). An Acad Bras Ci 67:171–3.
Flam F. (1994) Chemical prospectors scour the seas for promising drugs. Science 266:1324–5.
Forest J. (1966) Résultats scientifiques des campagnes de la Calypso VII. XXIX. Campagne de la Calypso au large des côtes atlantiques de l'Amerique de Sud (1961–1962) (première partie).I. Compte rendu et liste des stations. Ann Inst océanogr 44:329–50.
Freitas ALP, Teixeira DIA, Costa FHF, Farias WRL, Lobato ASC, Sampaio AH, Benevides NMB. (1997) A new survey of Brazilian marine algae for agglutinins. J Appl Phycol 9:495–501.
Freitas JC. (1977) Action of crude and fractioned homogenates of the midgut gland of the sea hare *Aplysia brasiliana* Rang, 1828 on some cholinoceptive structures. Comp Biochem Physiol 56C:57–61.
Freitas JC, Sawaya MI. (1986) Anomalies in sea-urchin egg development induced by a novel purine isolated from the sea anemone, *Bunodosoma caissarum*. Toxicon 24:751–5.
Freitas JC, Blankemeier LA, Jacobs RS. (1984) In vitro inactivation of the neurotoxic action of β-bungarotoxin by the marine natural product, monoalide. Experientia 40:864–5.
Hajdu E, Berlinck RGS, Freitas JC. (1999) Porifera. In: Migotto AE, Tiago CG, editors. Biodiversidade do Estado de São Paulo, Brasil. 3. Invertebrados Marinhos. São Paulo, Brazil: Fapesp, pp. 21–31.
Jacobs RS, Wilson L. (1986) Fertilized sea urchin egg as a model for detecting cell division inhibitors. In: Aszalor A, editor. Modern analysis of antibiotics. New York: Marcel Dekker, pp. 481–493.
Jimenez PC, Fortier SC, Lotufo TMC, Pessoa C, Moraes MEA, Moraes MO, Costa-Lotufo LV. (2003) Biological activity in extracts of ascidians (Tunicata, Ascidiacea) from the northeastern Brazilian coast. J Exp Mar Biol Ecol 287:93–101.
Kelecom A. (1997) Marine natural products in Brazil. Part 1. Isolation and structure determination. Ciência e Cultura 49:321–30.
Koning GM, Wright AD. (1996) Marine natural products research: current directions and future potencial. Planta Medica 62:193–211.
Lunetta JE, Umiji S. (1977) Pharmacological study of the serum of the moray eel, *Gymnothorax* sp. (Pisces, teleostei, Muraenidae). Bol Fisiol Animal Univ S Paulo 1:21–8.
Maia LF, Epifanio RA, Eve T, Fenical W. (1999) New fish feeding deterrents, including a novel sesquiterpenoid heterogorgiolide, from the Brazilian gorgonian *Heterogorgia uatumani* (Octocorallia, Gogornacea). J Nat Prod 62:1322–4.
Maia LF, Epifanio RA, Fenical W. (2000) New cytotoxic sterol glycisides from the ocotoral *Carijoa (Telesto) riisei*. J Nat Prod 63:1427–30.
Malpezzi ELA, Freitas JC. (1990) Antimitotic effect of an extract of the sea anemone *Bunodosoma caissarum* on the sea urchin egg development. Braz J Med Biol Res 23:811–4.
Malpezzi ELA, Freitas JC, Muramoto K, Kamiya H. (1993) Characterization of novel peptides in sea anemone venom collected by a novel procedure. Toxicon 31:853–64.
Malpezzi ELA, Freitas JC, Rantin FT. (1997) Occurrence of toxins, other than paralyzing type, in the skin of tetraodontiformes fish. Toxicon 35:57–65.
Marques AC. (2001) O gênero Eudendrium (Cnidaria, Hydrozoa, Anthomedusae) no Brasil. Papéis Avulsos de Zoologia, São Paulo 41:329–405.
Martins DL, Epifanio RA. (1998) A new germacrane sesquiterpene from the Brazilian endemic gorgoniam *Phyllogorgia dilatata* Esper. J Braz Chem Soc 9:586–90.

Melo VMM, Medeiros DA, Rios FJB, Castelar LIM, Carvalho FFU. (1997) Antifungal properties of proteins (agglutinins) from the red alga *Hypnea musciformis* (Wulfen) Lamouroux. Bot Mar 40:281–4.

Migotto AE. (1996) Benthic shallow-water hydroids (Cnidaria, Hydrozoa) of the coast of São Sebastião, Brazil, including a checklist of Brazilian hydroids. Zoologische Verhandelingen 306:1–125.

Migotto AE. (2000) Avaliação do estado do conhecimento da biodiversidade de invertebrados marinhos do Brasil. Brasil: Ministério do Meio Ambiente 75pp.

Monks NR, Lerner C, Henriques AT, Farias FM, Schapoval EES, Suyenaga ES, Rocha AB, Schwartsmann G, Mothes B. (2002) Anticancer, antichemotactic and antimicrobial activities of marine sponges collected off the coast of Santa Catarina, southern Brazil. J Exp Mar Biol Ecol 281:1–12.

Munro MHG, Blunt JW, Dumdei EJ, Hickford SJH, Lill RE, Li S, Battershill CN, Duckworth AR. (1999) The discovery and development of marine compounds with pharmaceutical potential. J Biotech 70:15–25.

Nagano CS, Moreno FBMB, Bloch C, Prates MV, Calvete JJ, Saker-Sampaio S, Farias WRL, Tavares TD, Nascimento KS, Grangeiro TB, Cavada BS, Sampaio AH. (2002) Purification and characterization of new lectin from the red marine alga *Hypnea musciformis*. Prot Pep Lett 9:159–65.

Newman DJ, Cragg GM. (2004) Marine natural products and related compounds in clinical and advanced preclinical trials. J Nat Prod 67:1216–38.

Palacio FJ. (1982) Revisión zoogeográfica marina del sur del Brasil. Bol Inst Oceanograf, São Paulo 31(1):69–92.

Pavao MSG. (1996) Unique sulfated polysaccharides from ascidians (Chordata, Tunicata). Braz J Med Biol Res 29:1227–33.

Pavao MSG. (2000) Structure and anticoagulant properties of sulfated glycosaminoglycans from primitive Chordates. An Acad Bras Ciênc 74:105–12.

Pereira RC, Teixeira VL. (1999) Sesquiterpenos das algas marinhas *Laurencia* Lamouroux (Ceramiales, Rhodophyta). 1. Significado ecológico Química Nova 22:369–74.

Pereira RC, Donato R, Teixeira VL, Cavalcanti DN. (2000) Chemotaxis and chemical defenses in seaweed susceptibility to herbivory. Rev Bras Biol 60:405–14.

Pereira RC, Pinheiro MD, Teixeira VL, Gama BAP. (2002) Feeding preferences of the endemic gastropod *Astraea latispina* in relation to chemical defenses of Brazilian tropical seaweeds. Braz J Biol 62:33–40.

Pinheiro-Vieira F, Caland-Noronha MC. (1971) Atividade antibiótica de algumas algas marinhas do estado do Ceará. Arq Ciênc Mar 11:91–3.

Rangel M, Sanctis B, Freitas JC, Polatto JM, Granato AC, Berlinck RGS, Hajdu E. (2001) Cytotoxic and neurotoxic activities in extracts of marine sponges (Porofera) from the southeastern Brazilian Coast. J Exp Mar Biol Ecol 262:31–40.

Rinehart KL. (2000) Antitumor compounds from tunicates. Med Res Rev 20:1–27.

Roberge M, Berlinck RGS, Xu L, Anderson HJ, Lim LY, Curman D, Stringer CM, Friend SH, Davies P, Vincent I, Haggarty SJ, Kelly MT, Britton R, Piers E, Andersen RJ. (1998) High-throughput assay for G2 checkpoint inhibitors and identification of the structurally novel compound isogranulatimide. Cancer Res 58:5701–6.

Rocha RM, d'Hondt JL. (1999) Filo Ectoproca ou Bryozoa. In: Migotto AE, Tiago CG, editors. Biodiversidade do Estado de São Paulo, Brasil. 3. Invertebrados Marinhos. São Paulo, Brazil: Fapesp, pp. 241–249.

Santana ANC, Leite AB, França MSF, França L, Vale OC, Cunha RB, Ricart CAO, Sousa MV, Carvalho KM. (1998) Partial sequence and toxic effects of granulitoxin, a neurotoxic peptide from the sea anemone *Bunodosoma granulifera*. Braz J Med Biol Res 31:1335–8.

Santana ANC, Trindade-Filho EM, Cunha RB, Sousa MV, Cavalheiro EA, Carvalho KM. (2001) Behavioral and electroencephalographic analysis of seizures induced by intrahippocampal injection of granulitoxin, a neurotoxic peptide from the sea anemone *Bunodosoma granulifera*. Braz J Med Biol Res 34:797–801.

Schmitz FJ, Campbell DC, Hollenbeak K, Vanderah DJ, Ciereske LS, Stendler P, Eckstrand JD, van der Helm D, Kaul P, Kulkarni S. (1977) Chemistry related to the search from drug from the sea. In: Faulkner DJ, Fenical WH, editors. Marine natural products chemistry. Naito conference series IV, New York and London: Plenum Press, 293–310.

Schmitz FJ, Hollenbeak KH, Campbell DC. (1978) Marine natural products: halitoxin, toxic complex of several marine sponges of the genus *Haliclona*. J Org Chem 43:3916–22.

Schwartsmann G, Rocha AB, Berlinck RGS, Jimeno J. (2001) Marine organisms as a source of new anticancer agents. Lancet Onco 2:221–5.

Torres YR, Berlinck RGS, Magalhaes A, Schefer AB, Ferreira AG, Hajdu E, Muricy G. (2000) Arenosclerins A–C and Haliclonacyclamine E, new tetracyclic alkaloids from a Brazilian endemic haplosclerid sponge *Arenosclera brasiliensis*. J Nat Prod 63:1098–105.

Torres YR, Berlinck RGS, Nascimento GGF, Fortier SC, Pessoa C, Moraes MO. (2002a) Antibacterial activity against resistant bacteria and cytotoxicity of four alkaloids isolated from the marine sponge *Arenosclera brasiliensis*. Toxicon 40:885–91.

Torres YR, Bugni TS, Berlinck RGS, Ireland CM, Magalhaes A, Ferreira AG, Rocha RM. (2002b) Sebastianines A and B, novel biologically active pyridoacridine alkaloids from the Brazilian ascidian *Cystodytes delleichiajei*. J Org Chem 26:5429–32.

Tursch B, Barreto H, Sharapin N. (1963) Occurrence of cholesterol in Renilla reniformis and Echinometra lucunter. Bull Soc Chim Belges 72:807–8.

Vera MD, Joullié MM. (2002) Natural products probes of cell biology: 20 years of Didemnin research. Med Res Rev 22:102–45.

Vervoort H, Fenical W, Epifanio RA. (2000) Tamandarins A and B: new cytotoxic depsipeptides from a Brazilian ascidian of the family Didemnidae. J Org Chem 65:782–92.

Yokoyama MY, Guimarães O. (1977) Alterations in the chemical composition of some seaweeds from Saí island, Paraná, Brasil. Acta Biol Par 6:67–73.

Zierer MS, Mourao PAS. (2000) A wide diversity of sulfated polysaccharides are synthesized by different species of marine sponges. Carbohydrate Res 328:209–16.

M.T.H. Khan and A. Ather (eds.)
Lead Molecules from Natural Products

Anticancer potential of Northeast Brazilian plants

Cláudia Pessoa, Letícia Veras Costa-Lotufo, Albert Leyva, Maria Elisabete Amaral de Moraes, Manoel Odorico de Moraes

Abstract

Natural products have been isolated as biologically active compounds with great therapeutic potential, providing the molecular basis for most of drugs currently in clinical use, especially for cancer and infectious diseases. The aim of this chapter is to show the worth of Northeast Brazilian plants as a source of potential anticancer drugs. Brazil's vast biodiversity alone justifies the great potential of natural products research in the discovery of new medicines. Furthermore, some studies published by Brazilian research groups have substantiated this potential. Some data from these studies have been presented here. The rational use of the anticancer properties of northeastern Brazilian plants depends on the establishment of a screening program with a multidisciplinary team. A narrow collaboration between chemists, pharmacologists and biologists is essential to elucidate the active principles of medicinal plants.

Keywords: *anticancer natural products, medicinal plants, Brazil*

I. Natural product-derived anticancer drug discovery

Natural products are secondary metabolites, produced by plants, fungi, bacteria, protozoans and animals in response to external stimuli such as nutritional changes, infection and competition (Strohl, 2000). Many of these metabolites have been isolated as biologically active compounds with great therapeutic potential. In fact, these compounds have been used for centuries in the treatment of human diseases. Reports of the use of medicinal plants go back to ancient times. Written records date back at least 5000 years to the Sumerians, and archeological records suggest even earlier use of medicinal plants (Raskin et al., 2002).

The number of new chemical entities based on natural products or natural products-derived compounds approved by the United States agency, Food and Drug Administration (FDA), corresponds to approximately 50% of the total considering all diseases and countries in the period from 1981 to 2002 (Newman et al., 2003). Approximately 20% of all medicines contain plant extracts or active principles derived from higher plants, resulting in 119 drugs currently in use (Cragg et al., 1999). Some examples that stand out are the cardiac glycoside digoxin obtained from

Digitalis purpurea, the anticholinergics alkaloids from *Atropa beladona*, the analgesics codeine and morphine from *Papaver somniferum*, the antihypertensive reserpine from *Rauwolfia serpentina*, the antineoplastics vinblastine and taxol respectively from *Cantharanthus roseus* and *Taxus brevifolia* (Cragg et al., 1997; Cragg and Newmann, 1999; Newman et al., 2000). These drugs represent 25% of prescriptions in the North American market resulting in US$16 billion for the pharmaceutical industries in 1990 (Pezzuto, 1997; Rates, 2001).

Plants have a long history of use in the treatment of cancer. The greatest recent impact of plant-derived drugs was probably felt in this area, where taxol, vinblastine, vincristine and camptothecin have dramatically improved the effectiveness of chemotherapy against some of the deadliest cancers (Table 1). In fact, over 60% of currently used anticancer agents are derived from nature (Cragg and Newmann, 2005).

Some of the extraordinarily exciting potential for the discovery of new naturally occurring anticancer drugs can be easily recognized, since there is an enormous number of living species available for investigation. Fewer than 10% of the higher plants have received even a cursory effort to detect antineoplastic constituents. A multitude of animal, microorganism and plant anticancer drugs still await discovery (Zhang, 2002).

In Brazil, the importance of cancer to health authorities has increased through the years since other diseases have been controlled and diagnostic techniques have improved. According to the estimates, more than 300,000 new cases of cancer will be expected each year, leading to 100,000 deaths (Ministério da Saúde, 1999–2000). In the northeast region, cancer represents the third leading cause of death, corresponding to 6.34% of total deaths, just 0.02 points behind infectious diseases. It is worthwhile to mention that this region has a complicated economical situation, and many people do not have any health assistance (Silva, 1982; Ministério da Saúde, 1999–2000).

Consequently, efforts in seeking new anticancer drugs, in fact, increased through the years. These included compounds isolated from natural sources as well as synthetic compounds usually produced for some other purpose. An important impetus to this work was the large screening program of the National Cancer Institute, which

Table 1
Representative plant-derived drugs in clinical use or development (modified from Schwartsmann et al., 2002)

Drug class	Example	Source plant
Vinca alkaloids	Vinblastine, vincristine, vinorelbine and vindesine	Catharanthus roseus
Lignans	Etoposide, tenoposide and NK611	Podophyllum species
Taxanes	Paclitaxel, docetaxel	Taxus species
Camptothecins	Topotecan, irinotecan HCl and rubitecan	Camptotheca accuminata
Cephalataxanes	Homoharringtonine	Cephalataxus harringtonia
Flavones	Flavopiridol (synthetic based on rohutikine)	Dysaxylum brinectariferum
Stilbenes	Cambrestatin prodrug	Combretum caffrum
Naphthoquinone	Lapachol and β-lapachone	Tabebuia impetiginosa

during this time evaluated approximately 400,000 materials, both pure compounds and crude extracts from synthetic and natural sources (Sausville and Feigal, 1999). Moreover, the evolution of new technologies, such as ultra high-throughput screening, combinatorial chemistry and, most recently, technologies based on genomics, has dramatically increased the efficacy of screening programs in the drug discovery process (Drews, 2000). Certainly the potential for the discovery of new plants for the treatment of human cancer is truly extraordinary and offers the promise of many curative approaches to the cancer problem.

The aim of this chapter is to show the worth of Northeast Brazilian plants as a source of potential anticancer drugs. Brazil has been described as the country with the greatest biodiversity in the entire world. The total number of species found in Brazilian biomes is still unknown. In reference to higher plants, Brazil is host to about 22% of the world's species with a high degree of endemism (Elisabetsky and Costa-Campos, 1996). Brazil's vast biodiversity alone justifies the great potential of natural products research in the discovery of new medicines. Furthermore, some studies published by Brazilian research groups have substantiated this potential.

Most of Northeast Brazil's territory is covered by xerophytic vegetation, with a variable floristic and physiognomic pattern, called "caatinga." Caatinga is a native American term meaning "white forest" which describes the appearance of a leafless forest in the dry season. It is a deciduous, thorny thicket well adapted to the unpredictable rains and arid conditions of the interior of northeastern Brazil. It represents 70% of the northeast area, covering around 1,000,000 km^2. The vegetation is composed mainly of herbaceous species with a high degree of endemism (around 30%). Several works have been carried out describing the caatinga's fauna and flora and the occurrence of 596 species have been reported, including 180 cases of endemism. The most common families were: Caesalpinaceae, Mimosaceae, Euphorbiaceae, Fabaceae and Cactaceae. Among caatinga species, there are many plants currently in use by the indigenous population because of their medicinal properties (Brito and Brito, 1993). Among them, the following stand out: *Myracrodruon urundeuva* Fr. All. (astringent), *Annona muricata* L. (anti-diarrheal), *Allamanda cathartica* L. (cathartic), pau-ferro (anti-asthmatic and antiseptic), *Costus spicatus* Jacq. (anti-diarrheal), *Solanum cernum* Vell (anti-febrile), *Anadenanthera odubrina* (astringent), *Salvia officinallis* L. (chest pains) and *Zizyphus joazeiro* Mart. (stomach pains). In the following section the species, which have been studied for the evaluation of their antitumor potential is discussed.

II. Current research on the antitumoral potential of Northeast Brazilian plants

As previously mentioned the use of medicinal plants is widely distributed in Brazil. Most people depend on remedies locally available, and they are generally based on biologically active plants. The studies on Brazilian medicinal plants have been recently reviewed in some books, showing the therapeutic and pharmacological utilization of approximately 1500 species and varieties of plants (Mors et al., 2000; Yunes and Calixto, 2001; Lorenzi and Matos, 2002).

In Brazil, there are many examples of plants used in cancer therapy, such as *Symphytum officinalis*, *Tabebuia avellandae*, *Aloe* spp. and *Euphorbia tirucalli* (Rates, 2001). Unfortunately, in most cases, there are no scientific data confirming these anticancer properties.

Probably, the first report on the antitumor potential of northeastern plants was the study of the extract obtained from the stem bark of *Tabebuia avellandae* (Santana et al., 1968). This extract and its component, the benzoquinone lapachol, presented strong antineoplastic properties. Lapachol inhibited 82% of tumor growth of Yoshida Sarcoma and 50% of Walker carcinosarcoma in rats (Santana et al., 1968). Further, the same group evaluated the efficacy of lapachol on human patients with cancer and verified a reduction of lesions and pain after four administrations of 20–30 mg/kg/day (Santana et al., 1980–1981). Nowadays, *Tabebuia avellanedae* extract is commercialized as a "miracle" drug in the treatment of cancer and other health problems in Japan (Nakamura, 2000).

The Department of Antibiotics at the Federal University of Pernambuco started a screening program back in 1952 to detect biological activity in plant extracts . They conducted some in vitro studies with plants from the state of Pernambuco to determine the percent of inhibition of tumor cell growth (Nascimento et al., 1985, 1990). Nascimento et al. (1984–1985) evaluated the cytotoxicity of 37 species against KB cells, showing positive results (IC_{50} values less than 10 μg/mL) for the following species: *Amanoa* spp., *Dalbergia hecstophyllum*, *Desmodium canun*, *Combretium fructicosum*, *Ternstroemia brasiliensis*, *Capraria biflora*, *Esmbeckia leiocarpa*, *Allamanda blanchetti*, *Indigofera microcarpa*, *Cassia pudibunda* and *Cassia semicordata*. Continuing these studies, in 1990, another 30 species were tested for the presence of cytotoxic agents in their ethanol extracts. The results showed that 37% of the plants tested possessed cytotoxic activity, inhibiting cell growth by more than 70%. Some species, such as *Cleone aculeate*, *Combretum duarteanum* and *Guazuma ulmifolia* showed a cytotoxic activity greater than 90% (Nascimento et al., 1990).

Moraes et al. (1980) performed a blind screening of 57 samples of northeastern medicinal plants on Walker Carcinoma 256, grafted by intramuscular route. In this study, the hydroalcoholic extract of *Croton mucronifolius* was the most active, presenting the highest inhibition of tumor growth of around 73%, and also reducing the frequency of metastasis (Moraes et al., 1980, 1981–1982). Furthermore, Moraes et al. (1983) also described the antitumor activity of some species belonging to the genus *Lippia*, and showed that the hydroalcoholic extract of *L. aristata* leaves was effective in inhibiting Walker carcinoma cell growth by 51%.

In recent years, the Department of Organic and Inorganic Chemistry and the Department of Physiology and Pharmacology at the Federal University of Ceará have joined forces to promote a natural products drug discovery and development program. The phytochemical analysis of plants commonly used in Brazilian traditional medicine has uncovered compounds with pharmacological activity (Pessoa et al., 1993; Rao et al., 1994, 1997; Silveira et al., 1995; Lima et al., 1996). In addition, the studies of crude extracts of Brazilian medicinal plants have demonstrated antitumor activity in animals (Moraes et al., 1997). Table 2 lists the plants already studied for the presence of antitumor activity in vivo arranged according to family, in alphabetical order.

Table 2
Anticancer activity of hydroalcoholic extracts of plants from Northeast Brazil in experimental tumor (modified from Moraes et al., 1997)

Family	Botanical name	Popular name	Plant part	Daily dose (mg/kg; i.p.)	Inhibition (%)	Stimulation (%)	Tumor
Amarantaceae	*Gomphrena cearensis* HUB	Ervanço	Limbs	122		21	Ehrlich
Anacardiaceae	*Astronium urudeuva* Engl.	Aroeira	Bark	206		7	Walker
	Schinopsis brasiliensis Eng	Brauna	Leaves	149		2	Walker
	Spondias tuberosa Arr. Caml	Cajá-Umbu	Bark	125	18		Walker
	Anacardium occidentale	Cajueiro	Leaves	48	32		Walker
	Mangifera indica L.	Mangueira	Leaves	52		24	Walker
Annonaceae	*Annona squamosa* L.	Ateira	Leaves	129	56[a]		Ehrlich
Apocinaceae	*Plumeria bracteata*	Janaguba	Bark	98	52[a]		Walker
Asteraceae	*Matricaria chamomila*	Camomila	Leaves	192		11	Walker
	Pectis apodocephala	Chá de moça	Limbs	187		4	Walker
	Tagetes minuta	Cravo-de Defunto-miudo	Leaves	56	51		Walker
	Eupatorium spp.	Erva de Sesão	Leaves	184		3	Ehrlich
	Ageratum conyzoides L.	Mentrasto	Leaves	113	43[a]		Walker
	Vanillosmoposis arborrea L.	Csndeeiro	Leaves	151	23	2	Walker
Boraginaceae	*Auxemma oncocalyx*	Pau Branco	Bark	73	48[a]		Ehrlich
Cesalpinaceae	*Platymenia reticulata*	Candeia	Stem wood	52		42[b]	Yoshida
	Hymenae coubaril L.	Jatobá	Bark	106		8	Walker
	Caesalpinia ferrea Mart et Tul	Jucá	Leaves	206		14	Walker
	Copaifera langsdorffi Desf.	Pau-D'-Oleo	Leaves	44	35		Walker
Chenopodiaceae	*Chenopodium ambrosides* L.	Mastruço	Leaves	43	11		Yoshida

Table 2 (*continued*)

Family	Botanical name	Popular name	Plant part	Daily dose (mg/kg; i.p.)	Inhibition (%)	Stimulation (%)	Tumor
Cleomaceae	*Cleone spinosa* Jacq.	Mussambê	Leaves	77	34		Walker
Cochlospermaceae	*Cochlospermum regium* st.	Pacotê	Leaves	204	2		Walker
Crassulaceae	*Bryophyllum pinnatum* Kurtz	Courama	Leaves	122	1		Walker
Cucurbitaceae	*Wilbrandia verticilla ta*	Cabeça de Negro	Leaves	115	84[a]		Walker
Euphorbiaceae	*Croton zehntneri* Pax et Hoff	Canelinha	Leaves	198		30	Walker
	Croton jacobinensis Bill	Marmeleiro–Branco	Leaves	132	13		Walker
	Croton sonderianus Muell. Arg	Marmeleiro Preto	Leaves	128	10		Yoshida
	Croton nepetifolius Muell. Arg	Marmeleiro Sabiá	Leaves	130	6		Walker
	Croton rhamnifolius H.B.K	Quebra-Faca	Leaves	196		28	Walker
	Phyllantus niuri L.	Quebra Pedra	Leaves	142		23	Walker
	Croton argyrophylloides Muell. Arg	Marmeleiro	Leaves	159	33		Walker
	Croton spp.	Velame	Leaves	47	21		Walker
	Croton regelianus Muell. Arg	Velame de Cheiro	Leaves	41	75[a]		Walker
	Euphorbia tirucalli	Aveloz	Bark	02	60[a]		Ehrlich
	Ricinus communis	Mamoeira	Seed	10		9	Yoshida
Fabaceae	*Hymenolobium petraeum* Ducke	Angelim Pedra	Wood	51	71		Walker
	Luetzelburgia auriculata Ducke	Pau Mocó	Stem wood	126		25	Yoshida
	Pterodon polygaliflorus Benth	Sucupira Branca	Leaves	158	8		Walker
	Votaireamacrocarpa Ducke	Amargoso	Wood	172		21	Walker

	Torresea cearensis Fr. All	Cumaru	Bark	83	29		Walker
Iridaceae	*Cypella caerulea* Steud	Alho Bravo	Leaves	103	10		Walker
Lamiaceae	*Rosmarinus officinalis* L.	Alecrim	Limbs	72	23		Walker
	Hyptis martiana Benth	Cidreira Comum	Leaves	131		34	Walker
	Ocimum basilicum L.	Manjericão	Leaves	89		25	Yoshida
Lauraceae	*Nectandraleucantha* Nees.	Canela	Leaves	213		61[b]	Walker
	Endlicheria hirsuta Nees.	Canela Branca	Leaves	95		30	Walker
Malpighiaceae	*Byrsonima crassifolia* H.B.K	Murici	Leaves	75	26		Yoshida
Miliaceae	*Cederla odorata* L.	Cedro	Bark	125	31		Walker
Mimosaceae	*Mimosa caesalpinifolia* Benth	Sabiá	Leaves	179	28	21	Walker
	Anadenanthera macrocarpa Benth	Angico	Leaves	150			Walker
	Stryphnodendron coriaceum Benth	Barbatimão	Leaves	197		88[b]	Walker
Myrtaceae	*Eugenia prasina* Berg.	Batinga	Leaves	150	14		Walker
	Eucaliptus critidora Hook.	Eucalipto	Leaves	75	24		Walker
	Eucaliptus globulus Labil	Eucalipto	Leaves	70	18		Walker
	Myrcia polyantha DC	Pau Santo	Leaves	121	37		Walker
Palmae	*Orbygnia phlerata* Mart.	Babaçu	Fruit	207	59[a]		Ehrlich
Passifloraceae	*Passiflora foetida* L.	Maracujá Fedorento	Leaves	92	16		Walker
Piperacea	*Piper tuberculatum* Jacq.	Pimenta de Macaco	Leaves	143	34[a]		Walker
Plumbaginaceae		Louro	Leaves	112		17	Walker

Table 2 (*continued*)

Family	Botanical name	Popular name	Plant part	Daily dose (mg/kg; i.p.)	Inhibition (%)	Stimulation (%)	Tumor
	Plumbago scandens L.						
Rubiaceae	*Coutarea hexandra* Schum	Quina-Quina	Leaves	77	12		Ehrlich
Rhizophoraceae	*Rhizophora mangue* L.	Mangue	Leaves	69	19		Walker
Rhamnaceaeae	*Zizyphus joazeiro* Witt	Juazeiro	Leaves	188	12		Walke
Rutaceae	*Pilocarpus jaborandi* Holmes	Jaborandi	Leaves	39	37		Walker
	Pilocarpus microphyllus Stapf.	Jaborandi da Viçosa	Leaves	78	17		Yoshida
	Ruta graveolens L.	Arruda	Leaves	123		15	Walker
	Citrus limonia Osbeck	Limão	Leaves	41		29	Yoshida
	Zanthoxylum moifolium Engl.	Limãozinho	Leaves	132	29		Walker
Verbenaceae	*Lippia gracilis* H.B.K	Alecrim da Chapada	Limbs	88	33		Ehrlich
	Lippia thymoides Mart et Schau.	Alecrim Miúdo de Cheiro	Leaves	193	4		Walker
	Lippia microphylla Cham.	Alecrim de Tabuleiro	Leaves	27		10	Walker
	Lippia sidoides Cham.	Alecrim pimenta	Limbs	33		11	Walker
	Lantana câmara L.	Camará	Leaves	30		10	Walker
Zingiberaceae	*Hedychium coronarium* Koeing	Borboleta	Leaves	189	21	15	Walker
	Alpinia speciosa Schum	Colônia	Leaves	62			Walker

The inhibition for the cyclophosphamide group, used as positive control, was always above 90%.
[a]Significant inhibition.
[b]Significant stimulation ($p<0.01$ vs. control, ANOVA-Tukey test).

In the screening of 75 Brazilian plants, Moraes et al. (1997) showed that crude extracts of several species produced tumor growth inhibition of at least 50%. The extracts were considered to be active when the inhibition was above 40% in vivo. Ten extracts showed initial results, a good indication of biological activity. These extracts, with the respective percent of tumor inhibition, were from *Plumeria bracteata* (52%), *Tagetes minuta* (51%), *Hymenolobium petraeum* (71%), *Ageratum conyzoides* (43%), *Annona squamosa* (56%), *Obrygnia phalerata* (59%), *Euphorbia tirucalli* (60%), *Croton regelianus* (75%), *Wilbrandia verticillata* (84%) and *Auxemma oncocalyx* (48%). Despite *Wilbrandia verticillata* showing the highest activity, the result obtained with *Croton regelianus* was considered to be more important by the authors because this species is endemic to Northeast Brazil and are largely used by the indigenous population to treat diseases of different kinds including malignant tumors and inflammation.

The studies with *Wilbrandia verticulata* have been continued leading to the isolation of two nor-cucurbitacin glucosides. A fraction containing these compounds demonstrated potent anti-inflammatory, antitumor and antifertility effects (Rao et al., 1991; Almeida et al., 1992). The authors showed that this fraction exhibited an antitumor effect on rats bearing Walker 256 carcinosarcoma and displayed cytotoxicity against KB cells in vitro. Some undesirable effects, such as tremors, diarrhea and weight loss, were observed in treated animals after 10 days (Rao et al., 1991). At that time, chemical structural modifications were suggested to increase antitumor activity.

A few years later, Pessoa et al. (1999) investigated the antitumor and cytotoxic activity of jatrophone, a macrocyclic diterpenoid, isolated from *Jatropha elliptica*. Jatrophone showed strong in vitro cytotoxicity towards CEM and HL60 leukemic cells, and CCD 922 fibroblasts. In vivo data showed that jatrophone produced a dose-dependent inhibition of Sarcoma 180 and Ehrlich ascites tumor growth; however, the maximum therapeutic effect was achieved at the dose causing 10% lethality. In fact, other species belonging to the genus *Jatropha*, such as *J. gossypiifolia* and *J. curca*, have been described as promising antitumor species, since their extracts were found to inhibit the proliferation of KB cells. The authors suggested that these species also possessed jatrophone as the active compound (Muanza et al., 1995; Sinha and Dhasan, 2002).

Recently, Pessoa et al. (2000) evaluated the antiproliferative effects of compounds obtained from plants of Northeast Brazil. Among 10 phytochemicals tested, four demonstrated cytotoxicity (IC_{50} values less than 10 μg/mL) against CEM leukemia, SW1573 lung tumor and CCD922 normal skin fibroblasts, oncocalyxones A and C (anthracenediones) from *Auxemma oncocalyx*, 12-acetoxy-hawtriwaic acid lactone and ternatine from *Egletes viscosa*. All but ternatine induced DNA damage and inhibition of DNA synthesis. In a further study, the authors compared the cytotoxicity of oncocalyxone A and C with conventional anticancer agents doxorrubicin and mitoxantrone, concluding that 1,4-anthracenediones may be a promising novel class of chemotherapeutic agents effective against multidrug-resistant tumors (Leyva et al., 2000). Costa-Lotufo et al. (2002a) demonstrated that the quinone fraction obtained from *A. oncocalyx* which contained oncocalyxone A is strongly toxic showing antimitotic activity, as observed by its effect on sea urchin egg development.

Other studies have described the occurrence of cytotoxic compounds in Northeast Brazilian plants. Table 3 showed many species and their isolated compounds studied

Table 3
Growth inhibitory effects of hydroalcoholic extracts and phytochemicals from Northeast Brazilian plants

Family	Botanical name	Popular name	Plant part	Cell line (IC_{50} μg/mL)	References
Apocynaceae	*Aspidosperma pyrifolium* Mart.	Pereiro	Leaves	HL60 (62)	Not published
			Stem	HL60 (> 100)	
Anacardiaceae	*Myracrodruon urundeuva* Fr. All.	Aroeira do Sertão	Leaves	HL60 (7.4)	Not published
				SW1573 (8.5)	
			Stem	HL60 (25.5)	
Boraginaceae	*Auxemma oncocalyx* Taub	Pau Branco	Leaves	HL60 (33)	Leyva et al. (2000) and Pessoa et al. (2000)
			Stem	HL60 (> 100)	
	Oncocalyxone A			CEM (0.76)	
	Oncocalyxone C			SW1573 (7.0)	
				CEM (1.5)	
				SW1573 (7.5)	
	Cordia trichotoma	Frei Jorge	Root	HL60 (> 100)	
			Stem	HL60 (> 100)	
	Cordia curassavica Jacc.	Erva Baleira	Root	HL60 (> 100)	Not published
			Stem	HL60 (> 100	
Combretaceae	*Combretum leprosum* Mart.	Mufumbo	Leaves	HL60 (18)	Not published
				SW1573 (18.5)	Not published
			Stem	HL60 (32)	
Compositae	*Egletes viscose* Less	Macela			
	2 acetoxy-hawtriwaic acid lactone		Flower buds	CEM (6.25)	Leyva et al. (2000)
	Ternatin			SW1573 (5.8)	
				CEM (1.9)	
				SW1573 (1.4)	
Convolvulaceae	*Ipomoea bahiensis* Willd		Root	KB (30)	Nascimento et al. (1985)
Cucurbitacine	Wilbrandia *verticillata*	Cabeça de Negro			Rao et al. (1991)
Euphorbiaceae	Cucurbitaceae		Root	KB (6)	
	Croton sonderianus Muell.Arg.	Marmeleiro Preto	Leaves	HL60 (27.0)	Not published
			Stem	HL60 (> 100)	
			Root	CEM (18.0)	
	Cróton micans Muell.Arg.			SW1573 (18.0)	
		Buique	Leaves	KB (> 100)	Not published
	Jatropha curcas L.		*Bark*	KB (> 100)	
		Pinhão Manso			Pessoa et al. (1999)
	Jatropha elliptica Muell.Arg.			CEM (0.098)	

	Jatrofona			HL60 (0.14)	
				CCD922 (0.22)	
		Mamoneira	Seeds	KB (> 100)	
	Ricinus communis L.				Not published
Fabaceae	*Torresea cearensis* Fr. All.	Cumaru	Bark	**KB (> 100)**	
	or				
	Amburana cearensis kaempferol, Isokaempferide, Amburoside A	Cumaru	Bark	CEM (13.4)	Costa-Lotufo et al. (2003)
				MCF-7 (21.2)	
				CEM (2.6)	
				MCF-7 (5.5)	
				CEM (> 25)	
				MCF-7 (> 25)	
	Protocatechuic acid			CEM (> 25)	
				MCF-7 (> 25)	
Lamiaceae	*Hyptis salzmanni*		Leaves	KB (60)	Nascimento et al., 1984/85
Leguminoseae	*Tephrosia cinerea* Pers	Anil Bravo	Leaves	HL60 (> 100)	Not published
			Stem	HL60 (> 100)	
			Root	HL60 (16.0)	
				SW1573 (26.2)	
	Dipteryx lacunifera.				Pessoa et al. (2000)
	Vinhaticol		Root	CEM (13.0)	
	Copaifera langsdorffi	Copaiba		SW1573 (17.0)	
	Kaurenoic acidl		Oleo-resin	CEM-(78 μM)	Costa-Lotufo, et al. (2002)
	Cássia pudibunda			MCF-7-(78 μM)	
	Cássia tora Var.	Fedegoso	Leaves	KB (40)	Nascimento et al. (1984/1985)
			Stem	HL60 (82)	
	Mimosa hostilis Mart.	Jurema Preta	Leaves	HL60 (> 100)	
			Stem	HL60 (30.5)	
				HL60 (66.0)	
Moraceae	*Brosimum rubescens* Taubert	Muirapiranga			Pessoa et al. (2000)
	Xanthyletin		Wood	CEM (13.0)	
				SW1573 (17.0)	
Polygalaceae	*Bredemeyra floribunda*	Pau-Caixão			Leyva et al. (2000)
	1,7-Dihydroxy-3,4,8-trimethoxy-xanthone		Root	CEM (13.5)	
				SW1573 (17.0)	
Papilinoideae	*Dalbergia ecastophyllum* Taub		Root	KB (4)	Nascimento et al. (1984/1985)
	Desmodium canum		Stem	KB (2.0)	
Rutaceae	*Zanthoxylum gardneri* Engl.	Laranjinha			Pessoa et al. (2000)
	Sesselin		Root	CEM (15.0)	
				SW1573 (17.3)	

Table 3 (*continued*)

Family	Botanical name	Popular name	Plant part	Cell line (IC_{50} μg/mL)	References
Rhamnaceae	*Zizyphus joazeiro* Mart.	Juazeiro	Leaves	HL60 (63)	Not published
			Stem	HL60 (> 100)	
Solanaceae	*Physalis angulata* L.	Camapu	Leaves	HL60 (5.7)	Not published
				SW1573 (8.0)	
			Stem	HL60 (28.0)	
Scrofulariaceae	*Scoparia dulcis* L.	Vassourinha	Leaves	HL60 (10.2)	Not published
				SW1573 (13.6)	
Trigoniaceae	*Trigonia fasciculata* Griseb				Pessoa et al. (2000)
	Fasciculatin		Root	CEM (20.0)	
				SW1573 (20.0)	
Verbenaceae	*Lippia sidoides* Cham.	Alecrim Pimenta			
	Tectol		Stem		Costa et al. (2001)
Theaceae	Ternstroemia brasiliensis		Root	KB (9)	Nascimento et al. (1984/1985)

IC_{50}, concentration causing 50% inhibition of cell growth. Data are from 2 to 3 separate experiments.

for in vitro cytotoxicity. As an example, Costa et al. (2001) isolated 11 compounds and a new prenylated naphthoquinone, lippsidoquinone, from *Lippia sidoides*. The evaluation of the cytotoxic properties of these compounds against HL60, SW1573 and CEM cell lines showed that tectol and lippisidoquinone displayed significant activity. On the other hand, Costa-Lotufo et al. (2002b) studied the effects of kaurenoic acid, a diterpene isolated from the oleo-resin of *Copaifera langsdorffi* in developing sea urchin embryos, on tumor cell growth and on mouse and human erythrocytes. This diterpene displayed cytotoxic effects in four models suggesting a non-specific mechanism of action.

III. Concluding remarks

The rational use of the anticancer properties of northeastern Brazilian plants depends on several aspects. The establishment of a screening program with a multi-disciplinary team is extremely necessary to increase the efficiency of the development of new drugs. The narrow collaboration between chemists, pharmacologists and biologists is essential to elucidate the active principles of medicinal plants. The literature shows that less than 10% of biologically active extracts have been chemically investigated.

As pointed before, the great biodiversity and the high degree of endemism observed in caatinga justify the enormous potential of this vegetation in the isolation of new therapeutic compounds. However, natural ecosystems including caatinga and their respective wild biota have been destroyed for centuries, well before they can be better scientifically explored. The importance of utilizing this natural patrimony must be emphasized in the scientific community.

References

Almeida FRC, Matos ME, Rao VS. (1992) Antiinflammatory, antitumor and antifertility effects in rodents of two no-cucurbitacin glucosides from *Wilbrandia* species. Phytother Res 6:189–93.

Brito ARM, Brito AS. (1993) Forty years of Brazilian medicinal plant research. J Ethnopharmacol 39:53–67.

Costa SM, Lemos TLG, Pessoa ODL, Pessoa C, Montenegro RC, Braz-Filho RB. (2001) Chemical constituintes from *Lippia sidoides* and cytotoxic activity. J Nat Prod 64:792–5.

Costa-Lotufo LV, Cunha GM, Farias PM, Viana GS, Cunha KMA, Pessoa C, Moraes MO, Silveira E, Gramosa NV, Rao VSN. (2002b) The cytotoxic and embryotoxic effects of kaurenoic acid, a diterpene isolated from *Copaifera langsdorffi* óleo – resin. Toxicon 40:1231–4.

Costa-Lotufo LV, Ferreira MAD, Lemos TLG, Pessoa ODL, Viana GSB, Cunha GMA. (2002a) Toxicity to sea urchin egg development of the quinone fraction obtained from *Auxemma oncocalyx*. Braz J Med Biol Res 35:927–30.

Costa-Lotufo LV, Jimenez PC, Wilke DV, Leal LKAM, Cunha GMA, Silveira ER, Canuto KM, Viana GSB, Moraes MEA, Moraes MO, Pessoa C. (2003) Antiproliferative effects of several compounds isolated from *Amburana cearensis* A.C.Smith. Z Naturforsch 58c:675–80.

Cragg GM, Boyd M, Khama R, Newman D, Suasville EA. (1999) Natural product drug discovery and development. In: Romeo JT editor. The United States National Cancer

Institute Role. Phytochemicals in human health protection, nutrition, and plant defense. New York: Kluwer Academic, Plenum Publishers, pp. 1–29 Chapter 1..

Cragg GM, Newmann DJ. (1999) Discovery and development of antineoplastic agents from natural sources. Cancer Invest 17:153–63.

Cragg GM, Newmann DJ. (2005) Plants as a source of anti-cancer agents. J Ethnopharmacol 100:72–9.

Cragg GM, Newman DJ, Weiss RB. (1997) Coral reefs, forests, and thermal vents: the worldwide exploration of nature for novel antitumor agents. Semin Oncol 24:156–63.

Drews J. (2000) Drug discovery today – and tomorrow. DDT 5:2–3.

Elisabetsky E, Costa-Campos L. (1996) Medicinal plant genetic resources and international cooperation: the Brazilian perspective. J Ethnopharmacol 51:111–20.

Leyva A, Pessoa C, Boogaerdt F, Sokaroski R, Lemos TLG, Wetmore L, Huruta RR, Moraes MO. (2000) Oncocalyxones A and C 1,4 anthracenediones from *Auxemma oncocalyx*: comparison with anticancer 1,9 anthracenediones. Anticancer Res 20:1029–32.

Lima MA, Silveira ER, Marques MS, Santos RH, Gambardela MTP. (1996) Biologically active flavanoids and terpenoids from *Egletes viscosa*. Phytochemistry 41:217–23.

Lorenzi H, Matos FJA. (2002) Plantas medicinais no Brasil. São Paulo, Brazil: Instituto Plantarum de Estudos da Flora LTDA.

MINISTÉRIO DA SAÚDE/ INSTITUTO NACIONAL DE CÂNCER. Estimativas da incidência por câncer no Brasil para o ano 2000. Rio de Janeiro: INCA/ 1999–2000.

Moraes MO, Carvalho RC, Fonteneles MC. (1981–1982) Alterações hitopatológicas induzidas pelo *Cróton mucronifolius* Mull. Arg. Durante bloqueio do crescimento do carcinossarcoma 256 de Walker. Rev Méd Univ Fed Ceará XXI–XXII:3–17.

Moraes MO, Fonteneles MC, Matos FJ, Craveiro A. (1983) Efeitos de extratos e óleos essenciais de plantas do gênero Lippia do nordeste brasileiro. Ciência e Cultura 1:79–80.

Moraes MO, Fonteneles MC, Matos FJA, Moraes ME. (1980) Atividade antitumoral de plantas do Nordeste brasileiro. Ciência e Cultura 31:647.

Moraes MO, Fonteles MC, Moraes MEA. (1997) Screening for anticancer activity of plants from the Northeast of Brazil. Fitoterapia LXVIII(3):235–9.

Mors WB, Rizzini CT, Pereira NA. (2000) Medicinal plants of Brazil. Defilipps RA. Michigan-EUA: Reference Publications, Inc.

Muanza DN, Euler KL, Williams L, Newman DJ. (1995) Screening for antitumor and anti-HIV activities of medicinal plants from Zaire. Int. J Pharmacognosy 33:98–106.

Nakamura RM. (2000) Miracles of "Tahebo Extract" in cancer and other health problems. La Jolla, California-EUA: Mina – Helwig Company.

Nascimento SC, Chiappeta AA, Lima RMOC. (1990) Antimicrobial and cytotoxic activities in plants from Pernambuco, Brazil. Fitoterapia V.LXI(4):353–5.

Nascimeto SC, Mello JF, Chiappeta AA. (1984–1985) Agentes citotóxicos. Experimentos com células KB. Rev Inst Antibióticos 22:19–25.

Newman DJ, Cragg GM, Snader KM. (2000) The influence of natural products on drug discovery. Nat Prod Rep 17:215–34.

Newman DJ, Cragg GM, Snader KM. (2003) Natural products as sources of new drugs over the period 1981–2002. J Nat Prod 66:1022–37.

Pessoa C, Sant Anna E, Leyva A, Valle C, Moraes MO. (1999) A valiação da atividade antitumoral e citotoxicidade da Jatrofona, um diterpeno extraído da *Jatopha elliptica* Muell Arg. Rev Brás Farm 80(3/4):59–60.

Pessoa C, Silveira ER, Lemos TLG, Wetmore LA, Moraes MO, Leyva A. (2000) Antiproliferative effects of compounds derived from plants of Northeast Brazil. Phytother Res 14:187–91.

Pessoa ODL, Lemos TLG, Silveira ER, Braz-Filho R. (1993) Novel cordiachromes isolated from *Auxemma oncocalyx*. Nat Prod Lett 2:145–50.

Pezzuto JM. (1997) Plant-derived anticancer agents. Biochem Pharmacol 53:121–33.

Rao VS, Almeida FR, Moraes AP, Silva JV, Nascimento SC, Moraes MO. (1991) Evaluation of the purified fraction of *Wilbrandia (c.f.) verticicillata* for antitumoral activity. Mem Inst Oswaldo Cruz 86:43–5.

Rao VS, Figueiredo EG, Melo CL. (1994) Protective effect of ternatin, a flavanoid isolated from *Egletes viscosa* less in experimental liver injury. Pharmacology 48:392–7.

Rao VS, Santos FA, Sobreira TT, Souza MF, Melo CL, Silveira ER. (1997) Investigations on the gastroprotective and antidiarrhoeal properties of ternatin, a tetramethoxyflavone from *Egletes viscosa*. Planta Med 63:146–9.

Raskin I, Ribnicky D, Komarnytsky S, Llic N, Poulev A, Borisjuk N, Brinker A, Moreno D, Ripoll C, Yakoby N, O'Neal JM, Cornwell T, Pastor I, Fridlender B. (2002) Plants and human health in the twenty-first century. Trends in Biotechnol 20:522–30.

Rates SM. (2001) Plants source of drugs. Toxicon 39:603–13.

Santana CF, Lins LJP, Asfora JJ, Lima OG, Alburquerque IL, Gimino D. (1968) Observações sobre as propriedades antitumorais e toxicológicas do extrato do líber e de alguns componentes do cerne do Pau d'Arco (*Tabebuia avellandae*). Rev Inst Antibiót 8(1/2):89–94.

Santana CF, Lins LJP, Asfora JJ, Melo AM, Lima OG, Alburquerque IL. (1980–1981) Primeiras observações com emprego do Lapachol em pacientes humanos portadores de neoplasias malignas. Rev Inst Antibiót 20(1/2):61–8.

Sausville EA, Feigal E. (1999) Evolving approaches to cancer drug discovery and development at the National Cancer Institute. Ann Oncol 10:1287–92.

Schwartsmann G, Ratain MJ, Cragg GM, Wong JE, Saijo N, Fujiwara Y, Pazdur R, Newman DJ, Dagher R, Di Leone L. (2002) Anticancer drug discovery and development throughout the world. J Clin Oncol 20(18):47s–59s.

Silva MG. (1982) Câncer em Fortaleza 1978–1980. Fortaleza: Instituto e Registro de câncer do Ceará, p. 135.

Silveira ER, Falcão MF, Menezes Jr. A, Kingston D, Glass TE. (1995) Pentaoxygenated xanthones from *Bredemeyra floribunda*. Phytochemistry 39:1433–6.

Sinha BN, Dhasan B. (2002) Chemistry and pharmacology of *Jatropha* species (a review). Indian J Heterocyclic Chem 11:237–40.

Strohl W. (2000) The role of natural products in a modern drug discovery program. DDT 5:39–41.

Yunes RA, Calixto JB. (2001) Plantas Medicinais sob a ótica da química medicinal moderna. Santa Catarina, Brasil: Argos editora universitária.

Zhang JT. (2002) New drugs derived from medicinal plants. Therapie 57(2):137–50.

M.T.H. Khan and A. Ather (eds.)
Lead Molecules from Natural Products

Safety and efficacy of phytomedicines

Manoel Odorico de Moraes, Fernando Antônio Frota Bezerra, Letícia Veras Costa-Lotufo, Cláudia Pessoa, Maria Elisabete Amaral de Moraes

Abstract

In this chapter the safety and efficacy of the phytomedicines around the globe have been discussed. Medicinal plants and herbal medicines are much more economic to use, than those of synthetic modern medicines. Due to this reason and also due to easy availability many poor patients of developing countries are attracted to these, even in developed countries like Germany, USA, etc., interests of people are growing. People believe that these herbal medicines are always safe, but unfortunately this is not the case. There are number of cases reported in the scientific reports around the globe, about the toxicities of the plants, plant products and herbal medicines. This chapter emphasizes and summarizes some of the important points on these safety and regulation issues.

Keywords: *clinical trials, herbal drugs, phytomedicine, alternative medicine*

I. Introduction

It has been estimated that some 80% of the world's inhabitants, rely chiefly on traditional medicines for their primary health care needs, and it can safely be assumed that a major part of traditional therapy involves the use of plant extracts or their active principles (WHO, 1991). In many areas, especially in the tropics, an abundance of medicinal plants offer people access to products for use in the prevention and treatment of illness through self-medication (Costa et al., 1997; Matos et al., 2001; Lorenzi and Matos, 2002).

Despite the rich Brazilian flora, which represents more than 20% of the plant species known in the world (Tomlinson and Akerele, 1993), little has been done in Brazil to study their potential as a source of new drug molecules or as raw materials for pharmaceutical preparations (Ferreira, 1998).

In developing countries, like Brazil, the use of medicinal plants helps to reduce imports of drugs, thus boosting economic self-reliance (Ferreira, 1998). Furthermore, drug products derived from local plants tend to be more readily accepted, than those imported (WHO, 1999). There is little doubt that medicinal plants are important in developed countries as a source of prescription drugs in the form of their active principles (WHO, 1998). These drugs of known structure, number about 120 and are used globally. In developed countries, crude plant materials and extracts are of lesser importance as drugs, but there is a steady growth in their use.

Medicinal plants have long been used in the treatment of various diseases and this folk knowledge has been perpetuated along the generations. The Portuguese discoverers described the use of Brazilian medicinal plants by the indigenous peoples in 1500. In the middle of the XVI century, during the occupation of the Brazilian Northeast by the Dutch, the physician William Pies from the Occidental Indian Company, described the use of Ipecac, Jaborandi and Tobacco (Gottlieb and Mors, 1980).

Since the beginning of the industrialization era, the newborn original Brazilian industry started the manufacture of herbal drugs based on traditional information and formulations, and reached its production apex in the 1940s. This decade also saw the establishment of the foreign chemical and pharmaceutical industry in the country. From this moment on, synthetic drugs from international pharmaceutical laboratories have dominated the market (Petrovick, 1996).

Following the actual globalization tendency, Brazil has also sought for alternative drug sources from natural products, especially plants, classified by Brazilian Federal Regulation Agency (ANVISA) as 'Phytopharmaceutical', representing 20–25% of the local pharmaceutical market (Calixto, 2000). Despite their importance, almost none of these phytopharmaceutical products have been sufficiently clinically studied, in order to provide the necessary confirmation of their efficacy and safety, as every medicine actually demands. For many years, in fact, such products were marketed without clear legal registration rules by the Brazilian government (Petrovick et al., 1997).

There is a paradox in the current approach to the discovery and use of medications derived from natural products sources. On the one hand, there are many successful medications in current use that have been derived from plants (Tyler et al., 1998; Goldman, 2001). On the other, there is a suspicion of folk medicine by practitioners of scientific medicine and among the scientists whose responsibility is to discover drugs (Mills and Bone, 2000). The main reason for that is the non-confirmation of their efficacy and safety.

Insufficient data exist for most plants to guarantee their quality, efficacy and safety. The idea that herbal drugs are safe and free from side effects is false. Plants contain hundreds of constituents and some of them are very toxic, such as the most cytotoxic anti-cancer plant-derived drugs, digitalis and the pyrrolizidine alkaloids, etc. (Farnsworth, 1993; De Smet, 1995; Drew and Myers, 1997; Brinker, 1998; Cupp, 2000). However, the adverse effects of phytotherapeutic agents are less frequent compared with synthetic drugs, but well-controlled clinical trials have now confirmed that such effects really exist (Brown, 1992; D'Arcy, 1993).

Most medicinal plant preparations used in Brazil can be documented as having theoretical efficacy based on literature reports of scientific studies carried out experimentally, that is, either through *in vitro* and *in vivo* studies on extracts. Indeed, active principles have been identified in many of these preparations whose *in vitro* and/or *in vivo* pharmacological properties relate to the human uses for which the plants are intended. However, experimental *in vitro* and/or *in vivo* evidence does not insure that these preparations will be active when administered to humans. This presents a dilemma for Brazilian government that wish to approve plant products as phytomedicines for greater use among the poor population.

When it is considered that placebo preparations administered to humans will produce a desired therapeutic effect in about 30–35% of individuals, one might suggest that therapeutic efficacy would be achieved through the use of medicinal plants whether or not they contain biologically active constituents (Kienle and Kiene, 1996, 1997). Thus, until such time when it will be possible to evaluate medicinal plant preparations for efficacy in humans, the most important goal is to assure safety (Kinghron and Balandrin, 1993; WHO, 1998).

It is also important to stress that in Brazil it is not possible to establish safety for medicinal plant preparations based on epidemiological studies for several reasons. First, these types of studies would be costly. Second, little published data exist in countries where the major use of medicinal plants occurs; thus, general extrapolations based on a limited number of reports would have little meaning. Third, exact identification of products implicated in side effects claimed for medicinal plants is most often lacking.

II. Types of products from medicinal plants sold in Brazil marketplace

There are several categories of plant medicinal products in Brazil based on the nature of the manufacturing or manufacturing processes involved. Plants are used as:

1. Crude plant material (roots, leaves, bark, seeds, limbs, etc.) fresh or dried, sold in open markets or specialized shops all over the country, and used as infusions, cataplasms, teas, syrup, powdered material, tinctures, etc.
2. Alive pharmacy, a scientific orientated program adopted in certain poor parts of the country, especially in the Northeast region. It consists of spread herb gardens with well-known medicinal plants, among communities in the countryside, which have difficulty in accessing orthodox medicine (Matos, 1999). These herbs are used to prepare remedies as recommended by the program.
3. Dried plant material that is sold in bulk form or in the form of tea bags. The product may represent a single plant or a mixture of several plants. The intent is to add the product to hot water for a defined period of time and then either strain the mixture to remove the plant material or remove the tea bag. In either case, only the user consumes the hot water-soluble extractive.
4. Dried plant material that is reduced to a powdered form and is used as such, either in bulk form or placed into hard gelatin capsules. In the powdered form, the herbal material is used by adding it to hot water as indicated above. In the case of encapsulated materials, the capsule can be taken orally as such or the contents can be added to hot water and strained to produce herbal tea.
5. The manufacturer "advances" the herbal medicine. This is done by preparing a concentrated extract, usually by extraction with ethanol, removing the ethanol, adding an excipient to the extractive and producing either tablets or capsules. Hot water extracts can also be prepared, frozen and lyophilized. These result, however, in a product with greater bulk than an ethanol extract and, thus, is more difficult to formulate into capsules. In some cases, these types of extracts are made into liquid dosage forms, that is, as syrups or tinctures, in other cases, into ointments,

creams or gels. "Advanced" herbal medicines are preferentially adopted by the Brazilian phytomedicine modern industries (C.P.R.F.B., 1998).

Compared with well-defined synthetic drugs, phytomedicines in Brazil exhibit some marked differences. For example, the active principles are frequently unknown; standardization; stability and quality control are feasible but not easy; the availability and quality of raw materials are frequently problematic; well-controlled double-blind clinical and toxicological studies to prove their efficacy and safety are rare; empirical use in folk medicine is a very important characteristic; they have a wide range of therapeutic use and are suitable for chronic treatments; the occurrence of undesirable side effects seems to be less frequent with herbal medicines; but well-controlled randomized clinical trials have revealed that they also exist; and they usually cost less than synthetic drugs (Gillis, 2001).

III. Legal requirements for the use of phytomedicines in Brazil

The legal status for phytomedicines was defined by Brazilian Regulatory Agency for Medicines (ANVISA) on the RDC 48 of the 16th of May 2004, which established the legal requirements for the phytopharmaceutical drug registration in Brazil. They are finished, labeled medicinal products that contain as active ingredients aerial or underground parts of plants, or other plant material, or combinations thereof, whether in the crude state or as plant preparations (ANVISA/MS, 2004).

This act defines four different categories of phytomedicines considered by registration: traditional phytomedicines, pharmacopoeia phytomedicines, new phytomedicines and similar phytomedicines. By this Act, the phytomedicines in Brazil are regulated as drugs, used for diagnosis, cure, mitigation, treatment, prevention of diseases or improve quality of life. Therefore, they have to show quality, efficacy and safety.

The traditional phytomedicines include finished herbal products, which contain as active ingredients plants or parts of plants which refers to the long historical use of these medicines, their use is well established and widely acknowledged to be safe and effective.

The pharmacopeic phytomedicines are finished herbal products, whose formulations are described in Brazilian Pharmacopoeia and they have proved their safety and effectiveness.

The new phytomedicines consists of a finished herbal preparation made from one or more ingredients from new plants, with proved quality, efficacy and safety. The plants are considered as new if they have not been traditionally used by the populations or/and have not been described in "Brazilian Pharmacopoeia".

The similar phytomedicine is a finished plant preparation with the same ingredients, the same concentration of the active principle or markers and the same pharmaceutical formulation, used by the same route of administration and for the same therapeutic indication of a registered phytomedicine considered as a reference.

Phytomedicine agents registered by ANVISA are marketed as standardized preparations in the form of liquid, solid (powdered extract) or viscous preparations. They are prepared by maceration, percolation or distillation (volatile oils). Ethanol, water or mixtures of ethanol and water are used for the production of fluid extracts. Solid or powered extracts are prepared by evaporation of the solvents used in the process of extraction of the raw material. Some phytomedicines are greatly concentrated in order to improve their therapeutic efficacy.

IV. Evaluation of safety and efficacy

The phytomedicines have a long history and they are the sum total of the knowledge, skills and practices based on the theories, beliefs and experiences to different cultures, Amerindians, black slaves from Africa and Caucasians, whether explicable or not, used in the maintenance of health, as well as in the prevention, diagnosis, improvement or treatment of physical and mental illnesses (Wheelwright, 1994). However, insufficient data exist for most of these phytomedicines to guarantee their quality, efficacy and safety (Ferreira, 1998; Calixto, 2000). The idea that phytomedicines are safe and free from side effects is false. Plants contain hundreds of constituents and some of them are very toxic, such as the most cytotoxic anti-cancer plant-derived drugs, digitalis and the pyrrolizidine alkaloids, etc. However, the adverse effects of phytotherapeutic agents are less frequent compared with synthetic drugs, but well-controlled clinical trials have now confirmed that such effects really exist.

IV.A. Side effects and/or toxic reactions to phytomedicines

Side effects or toxic reactions associated with phytomedicines are rare. This could be due to the fact that phytomedicines are generally safe, that adverse reactions following their use are underreported, or because the side effects are of such a nature that they are not reported for example, minor allergic reactions. Most of the adverse effects reported for herbal medicines are associated with "crude drugs" or powdered forms of plant material, at least insofar as one can determine from the scientific literature. There is no way to completely eliminate the possibility that any substance, including phytomedicines, will produce allergic side effects in persons exposed to them.

The World Health Organization (WHO) Alma-Ata Declaration, in 1978, opened the door for a dialogue between traditional and modern health care, on the understanding that unsafe practices should be eliminated and that only what is both safe and effective should be promoted.

Safety should be the overriding criterion in the selection of phytomedicines. Screening, chemical analysis, clinical trials and regulatory measures should be undertaken in respect to phytomedicines (WHO, 1993). This chapter is not intended to represent an exhaustive review of the efficacy, safety, toxicity and/or quality control problems relating to phytomedicines. It is intended to highlight the major types of problems that need to be taken into consideration when testing them under clinical trials.

It would not be ethically acceptable or morally justifiable, prescribing phytomedicines without scientific evidence of its safety, efficacy and quality. Hence, scientific research is needed to provide additional information for the patients.

The comments given below are related to finished herbal preparation, which will be clinically evaluated using the allopathic system of medicine, applying all the general principles of clinical trials. However, there are special features regarding herbal remedies, which need to be kept in mind during their clinical evaluation.

The first feature is that at the time of clinical evaluation a lot may be known about the plant or its extract. There could be extensive literature about pre-clinical studies and use of those in the traditional medicine or, indeed, the phytomedicine may actually be in use by the population for many years. The scope and design of such studies with the substance to be clinically evaluated should be based on information on traditional use obtained from relevant literature, or by consultation with traditional medical practitioners. In conducting research and evaluating traditional medicine, knowledge and experience obtained through the long history of established practices should be respected. These phytomedicines are classified by ANVISA as traditional phytomedicines.

The second consideration is that it is possible that clinical evaluation of a plant extract has to be carried out which has never been in use before or has not ever been mentioned in scientific literature. These are considered as new phytomedicines by ANVISA.

The trails for these two types of substances cannot be the same and therefore, given below are specific guidelines for substances, which could fall into either group. The recommendations made earlier regarding informed consent, inducements for participation, information to be provided to the subject, withdraw from study and research involving children or persons not in full control of their senses all apply to these trials. These trials have also got to be approved by the appropriate scientific and ethical committees of the concerned institutions.

Clinical trials with herbal preparations should be carried out only after these have been standardized and markers identified to ensure that the substances being evaluated are always the same.

IV.B. Phytomedicines currently in use or mentioned in scientific literature of natural products (traditional phytomedicines)

It is important that the herbal preparation to be clinically evaluated has been described in scientific literature and prepared strictly in the same way, incorporating GMP norms or standardization as far as possible. It may not be necessary to undertake all its full range of toxicological tests of phase I studies, unless there are reports suggesting toxicity or the use is to be for more than three months. However, a clinical toxicology study may be needed before starting efficacy clinical trials (WHO, 1993). This study consists of an open study, non-randomized with 24 healthy volunteers of both sexes (12 men and 12 women), aged 18 to 50 years. The volunteers considered are qualified for inclusion in the study, after being submitted to clinical and laboratory investigations to evaluate if their records are compatible with standards for healthiness. They receive in an ambulatory manner, daily dose of the phytomedicine for 28 uninterrupted days (Meitnert, 1986). The laboratory tests included: hemoglobin, hematocrit, total and differential leukocyte counts, platelet count, sodium, potassium, creatinine, bilirubin total protein, albumin, glucose, alkaline phosphatase, glutamic oxalacetic transaminase (sGOT), glutamic pyruvic transaminase (sGPT), total cholesterol, triglycerides, uric acid, gamma GT and urinalysis (Fleiss, 1996; Pinheiro et al., 2001). Serological tests for hepatitis B, hepatitis C, HIV and βHCG (in females) were performed only prior to treatment. This clinical trial should be carried out in accordance with a written protocol agreed upon and signed by the investigator and the sponsor. Any change(s) subsequently required must be similarly agreed on and signed by the investigator and sponsor and appended to the protocol as amendments.

The greatest challenge in validating efficacy of phytomedicines in Brazil is to systemically assess and evaluate the therapeutic outcome of these therapies by means of contemporary research methodologies. The clinical trial adopted to study therapeutic efficacy of these phytomedicines investigates the effectiveness for treating a particular medical condition or achieving a specific health goal. It is a double-blind, randomized controlled trial and each trial follows a protocol written, detailed plan that explains why there is a need for the study, what it is intended to do, and how it will be conducted. This type of research seeks to determine if a phytotherapic agent is clinically active and for that it must compare the effect of treatment with the effect of no treatment, ensure that people who receive treatment are randomly selected into that group (control for placebo effect), control for any confounding factors, identify a clearly diagnosed illness and a standardized treatment, provide for a noticeable change, and use measures which are precise, valid and reliable.

In general, traditional procedure-based therapies are relatively safe, if they are performed properly by well-trained physicians (clinical pharmacologists). But, accidents do occasionally occur, most probably when physicians are not fully trained. Therapies should be performed within accepted parameters, and the indications for a therapy should be evidence based when possible. Serious adverse effects of therapies are rare, but supportive facilities should be readily available. Accordingly, the evaluation of adverse effects should be considered a priority area for systematic evaluation of safety of these therapies.

Clinical research aimed at evaluating traditional medicine should incorporate the conventional concepts of research design, such as randomized controlled trials or other types of clinical studies, such as observational studies (Bulpitt, 1993). The level of evidence on efficacy of traditional phytomedicines can be significantly increased by well-designed clinical trails (Good and Clark, 1996). It is essential that the outcome measures chosen be appropriate to the research question (Friedman et al., 1996). The number of patients in a study needs to be adequate, in order to be able to determine any clinically important differences between the study groups. With respect to the study design, the statistical methods used should be appropriate to the proposed analysis of the study's outcome (Chow and Liu, 1998). Appropriate outcomes may include quantitative and qualitative outcomes, primary and/or secondary outcomes and generic and/or highly specific outcomes.

It is essential that the sample represent the target population of patients to which the results would be generalized (Hamilton, 1994). The reliability of the categorization/diagnostic criteria used in the study should be considered and stated. The source of the patients under study should be comprehensively described along with details of the recruitment process (Harris and Fitzerald, 1990). The inclusion and exclusion criteria should be completely described and rationalized (Spilker, 1991). Any potential bias in patient selection, recruitment and enrolment should be excluded (Cohen and Posner, 2000). It is preferable to compare a phytomedicine with both a well-established treatment and another control group (from the list of control groups) to determine whether the phytomedicine is useful in the context of current best practice (Moraes and Moraes, 2000).

One specific problem in clinical research of traditional medicine is the simultaneous conventional treatment of patients (e.g., cancer patients), in a study it may not be ethically possible to withdraw the conventional treatment. Therefore, in such cases, the focus of research may be on the additional or supportive effects of traditional medicine. Research on combinations of traditional and conventional medicine should always consider potential therapeutic interactions and side effects (Brown, 1992; D'Arcy, 1993; De Smet, 1995; Brinker, 1998).

Traditional phytomedicines are used not only to prevent, diagnose, improve and treat illness, but also to maintain health and improve the quality of life (Spilker, 1993; Högel and Gaus, 1995; Keller, 1996).

IV.C. For new phytomedicines not in use or mentioned in literature

The other situation would be when a phytomedicine has to be clinically evaluated for its therapeutic effect. This is, for all purposes, a new herbal medicine never been tested before, a new indication for an existing herbal medicine, or a significantly different dosage form or route of administration, the general principles and requirements for a clinical trial should be very similar to those which apply to conventional drugs (Iber et al., 1987; Glenny and Nelmes, 1996).

Before a new phytomedicine is tested in a clinical trial, there must be adequate data from *in vivo* and/or *in vitro* studies to validate its claimed therapeutic efficacy and safety (Moss, 1998). Although, for both safety and efficacy, a pharmacological effect observed *in vitro* or in animal models is not necessarily applicable to humans, all the pre-clinical tests such as acute, sub-acute and chronic toxicity, cytotoxicity, carcinogenicity, genotoxicity and teratogenicity, which need to be carried out with any new synthetic compound would need to be carried out with this preparation before clinical evaluation (Baguley and Kerr, 2002; Anonymous-FDA, 2000). *In vitro* data usually serve to verify the reported mechanism of action in animals or humans. Such data have to be confirmed by clinical studies. Well-documented reports of pharmacological activity in animals or humans may be viewed as having scientific rationale.

When the pre-clinical tests have been completed, and there is no evidence of undue toxicity, a decision has to be taken whether the substance should be clinically evaluated. If the decision is made to carry out a clinical evaluation, this has to be carried out by a physician of the modern system of conventional medicine i.e., allopathic.

It should be tested just as any new synthetic drug would be tested (trails of phase I, II, III and IV) and all the observations made earlier would also apply to this compound or extract (Shapiro and Louis, 1983; Spilker, 1987).

Well-established, randomized controlled clinical trials provide the highest level of evidence for efficacy. Such studies facilitate the acceptance of phytomedicines in conventional medicine (Tygstrup et al., 1982; Moraes et al., 1998; Pinheiro et al., 2001). However, methods such as randomization and use of a placebo may not always be possible as they may involve ethical issues as well as technical problems.

For example, it may be not possible to have a placebo control if the herbal medicine has a strong or prominent smell or taste, as is the case for products containing certain essential oils. In addition, patients who have been treated previously with the phytomedicine under investigation that has a characteristic organoleptic property cannot be randomized into control groups. In the case of herbal medicines with a strong flavor, placebo substances with the same flavor may have a similar function. In such cases, it may be advisable to use a low dosage of the same herbal medicine as a control (WHO, 1993). Alternatively, a positive control, such as well-established treatment, can be used.

IV.C.1. The phase I trials of phytomedicines

The phase III trials are studies in healthy volunteers, or sometimes patients, to determine the safety, tolerability and sometimes the pharmacodynamic effects and pharmacokinetic profile of markers (absorption, distribution, metabolism and excretion) of the phytomedicines (Spilker and Schoenfelder, 1990). Evidence of efficacy may be gained if patients, disease models or biomarkers are used.

To be submitted to phase I trials, the phytomedicines must contain sufficient information to demonstrate that the product is safe for testing in humans and that the clinical protocol is properly designed for its intended objectives (Friedman et al., 1996).

Some traditional phytomedicines and all new phytomedicines are required to be submitted to this phase of clinical trials.

IV.C.2. The phase II trials of phytomedicines

These trials are performed in a limited number of subjects and are often, at a later stage, of a comparative (e.g. placebo-controlled) design. Their purpose is to demonstrate therapeutic activity and to assess short-term safety of the phytomedicines in patients suffering from a disease or condition for which the active ingredient is intended. This phase also aims at the determination of appropriate dose ranges or regimens and (if possible) clarification of dose–response relationships in order to provide an optimal background for the design of extensive therapeutic trials.

IV.C.3. The phase III trials of phytomedicines

The phase III trials are done in larger (and possibly varied) patient groups with the purpose of determining the short-term and long-term safety/efficacy balance of formulation(s) of the phytomedicines, and of assessing its overall and relative therapeutic value (Pocock, 1987). The pattern and profile of any frequent adverse reactions must be investigated and special features of the product must be explored (e.g. clinically relevant drug interactions, factors leading to differences in effect such as age). These trials should preferably be of a randomized double-blind design, but other designs may be acceptable, e.g. long-term safety studies. Generally, the conditions under which these trials are carried out should be as close as possible to normal conditions of use.

IV.C.4. The phase IV trials of phytomedicines

Studies performed after marketing of the pharmaceutical product. Trials in phase IV are carried out on the basis of the product characteristics on which the marketing authorization was granted and are normally in the form of post-marketing surveillance, or assessment of therapeutic value or treatment strategies. Although methods may differ, these studies should use the same scientific and ethical standards as applied in pre-marketing studies. After a product has been placed on the market, clinical trials designed to explore new indications, new methods of administration or new combinations, etc., are normally considered as trials for new pharmaceutical products (WHO, 1998).

V. Conclusion

To conclude, it should be stressed that there is no "alternative medicine". There is only scientifically proven, evidence-based medicine supported by solid data or unproven medicine for which scientific evidence is lacking. Herbal medicines proven to be useful according to this standard should be accepted. Those failing should be rejected. In either case, only scientifically valid data will support a convincing decision.

References

Anonymous – FDA. (2000) Guidance for industry botanical drug products. Washington, US: Government Printing Office, Food and Drug Administration.

ANVISA/MS (Agência Nacional de Vigilância Sanitária – Ministério da Saúde) Resolução RDC n. 48 de 16 de maio 2004.

Baguley BC, Kerr DJ. (2002) Anticancer drug development. San Diego: Academic Press.

Brinker F. (1998) Herb contraindications and drug interactions, 2nd edition. Sandy, OR: Eclectic Medical Publications.

Brown RG. (1992) Toxicity of Chinese herbal remedies. Lancet 340:673.

Bulpitt J. (1993) Randomised controlled clinical trials, 1ª Edition. Den Haag: Martinus Nijhoff.

Calixto JB. (2000) Efficacy, safety, quality control, marketing and regulatory guidelines for herbal medicines (phytotherapeutic agents). Braz J Med Biol Res 33:179–89.

Chow S, Liu J. (1998) Design and analysis of clinical trials concepts and methologies, 1ª edition. New York: Wiley.

Cohen A, Posner J editors. A guide to clinical drug research. 2nd edition. Boston: Kluwer Academic Publishers.

Costa AFE, Moraes ME, Frota FA, Moraes MO. (1997) Plantas Medicinais Usadas pelos Pacientes Atendidos nos Ambulatórios do Hospital Walter Cantídio da Universidade Federal do Ceará. Pesq Med 1:20–5.

C.P.R.F.B. (1998) (Comissão Permanente de Revisão da Farmacopeia Brasileira) Farmacopéia Brasileira, 4ª Edição. São Paulo: Atheneu Editora.

Cupp MJ editor. Toxicology and clinical pharmacology of herbal products. Totowa: Humana Press; 2000.

D'Arcy PF. (1993) Adverse reactions and interactions with herbal medicines. Part 2. Drug interactions. Adv Drug React Toxicol Rev 12:147–62.

De Smet PAGM. (1995) Health risks of herbal remedies. Drug Safety 13:81–93.

Drew A, Myers SP. (1997) Safety issues in herbal medicine: implications for the health professions. Med J Aust 166:538–41.

Farnsworth NR. (1993) Relative safety of herbal medicines. Herbal Gram 29:36A–H.

Ferreira SH editor. Medicamentos a partir de plantas medicinais no Brasil. Rio de Janeiro: Academia Brasileira de Ciências.

Fleiss J. (1996) The design and analysis of clinical experiments, 1ª edition. New York: Wiley.

Friedman M, Furberg D, DeMets L. (1996) Fundamentals of clinical trials, 3ª edition. St Louis: Mosby.

Gillis CN. (2001) Biomedical science and herbal medicine: a reluctant but necessary alliance. FASEB Newsletter 34:123–34.

Glenny H, Nelmes P. (1996) Handbook of clinical drug research, 1ª edition. Oxford: Blackwell.

Goldman P. (2001) Herbal medicines today and the roots of modern pharmacology. Ann Intern Med 135:594–600.

Good S, Clark C. (1996) The principles and practice of clinical trials, 1ª edition. Edinburg: Churchill Livingstone.

Gottlieb OR, Mors WB. (1980) Potential utilization of Brazilian wood extractives. J Agric Food Chem 28:196–215.

Hamilton M. (1994) Lectures on the methodology of clinical research, 2ª edition. Edinburgh: Churchill Livingstone.

Harris E, Fitzerald JD. (1990) The principles and practice of clinical trials, 1ª edition. Edinburgh: Churchill Livingstone.

Högel J, Gaus W. (1995) Studies on the efficacy of unconventional therapies. Arzneimittel-Forschung/Drug Res 45:88–92.

Iber F, Riley W, Murray P. (1987) Conducting clinical trials, 1ª edition. New York: Plenum.

Keller K. (1996) Herbal medicinal products in Germany and Europe: experiences with national and European assessment. Drug Inform J 30:933–48.

Kienle GS, Kiene H. (1996) Placebo effect and placebo concept: a critical methodological and conceptual analysis of reports on the magnitude of the placebo effect. Altern Ther Health Med 2:39–54.

Kienle GS, Kiene H. (1997) The powerful placebo effect: fact or fiction? J Clin Epidemiol 50:1311–8.

Kinghron AD, Balandrin MF editors. Human medicinal agents from plants. Washington (DC): American Chemical Society.

Lorenzi H, Matos FJA. (2002) Plantas medicinais no Brasil: nativas e exóticas. São Paulo: Instituto Plantarum.

Matos FA, Viana GSB, Bandeira MM. (2001) Guia fitoterápico. Ceará: Governo do estado do Ceará.

Matos FJA. (1999) FARMÁCIAS VIVAS, Editora UFC, Fortaleza.

Meitnert C. (1986) Clinical trials – design, conduct and analysis, 1ª edition. Oxford: Oxford University Press.

Mills S, Bone K. (2000) Principles and practice of phytotherapy: modern herbal medicine. Toronto: Churchill Livingstone.

Moraes MEA, Moraes Filho MO. (2000) Ensaios Clínicos de Medicamentos no Brazil. Farmacos e Medicamentos 6:36–40.

Moraes MO, Castillo DLC, Batista TAB, Gomes PRP, Melo GS, Bezerra FAR, Nucci G. (1998) Estudo Toxicológico e Laboratorial do Fitoterápico "Essência de Vida Olina. Anais do XIII Congresso da FESB. Minas Gerais: Caxambu, pp. 133.

Moss RW. (1998) Herbs against cancer: history and controversy. New York: Equinox Press.

Petrovick P.R. (1996) Normatização da Indústria Fitofarmacêutica. In: 1ª Reunión de Coordinación Internacional. Programa Iberoamericano de Ciencia y Tecnología para el Desarrollo (CYTED). Red Iberoamericana de Productos Fitofarmacéuticos (RIPROFITO), Sep. 28–Oct. 01, Antigua-Guatemala, 91–93.

Petrovick PR, González Ortega G, Bassani VL. (1997) From a medicinal plant to a pharmaceutical dosage form. A (still) long way for the Brazilian medicinal plants. Ciência e Cultura 49:364–9.

Pinheiro MNA, Ramos DV, Santana GSM, Soares AKA, Uchoa CRA, Bezerra FAF, Moraes Filho MO, Moraes ME. (2001) Estudo da Toxicologia Clínica do Rhemoflora (Fitoterápico). Anais do XVI Congresso da FESB. Minas Gerais: Caxambu, pp. 441.

Pocock J. (1987) Clinical trials: a practical approach, 4ª edition. New York: Wiley.
Shapiro H, Louis A. (1983) Clinical trials: issues and approaches, 1ª edition. New York: Marcel Dekker.
Spilker B. (1987) Guide to planning and mananging multiple clinical studies, 1^a edition. New York: Raven Press.
Spilker B, Schoenfelder J editors. Presentation of clinical. 1^a edition. New York: Raven Press.
Spilker B editor. Guide to clinical trials. 1^a edition. Philadelphia: Lippincott Williams & Wilkins.
Spilker B editor. Quality of life assessments in clinical trials. 1^a edition. Philadelphia: Lippincott Williams & Wilkins.
Tomlinson TR, Akerele O editors. Medicinal plants: their role in health & biodiversity. Philadelphia: University of Pennsylvania Press.
Tygstrup N, Lachin J, Juhl E editors. The randomized clinical trial and therapeutic decisions. 1^a edition. New York: Marcel Dekker.
Tyler VE, Brady LR, Robbers JE editors. Phamacognosy. 9th edition. Philadelphia: Lea & Febiger.
Wheelwright EG editor. Medicinal plants and their history. New York: Dover Publications, Inc.
WHO. (1991) Guidelines for the assessment of herbal medicines. Geneva: World Health Organization.
WHO. (1993) Research guidelines for evaluating the safety and efficacy of herbal medicine. Geneva: World Health Organization, pp. 1–86.
WHO. (1998) Regulatory situation of herbal medicines. A worldwide review. Geneva: World Health Organization, pp. 1–43.
WHO. (1999) WHO monographs on selected medicinal plants, Vol. I. Geneva: World Health Organization.

M.T.H. Khan and A. Ather (eds.)
Lead Molecules from Natural Products

Pharmacological and biochemical profiling of lead compounds from traditional remedies: the case of *Croton cajucara*

Maria Aparecida M. Maciel, Tereza Neuma C. Dantas, Janaína Keila P. Câmara, Angelo C. Pinto, Valdir F. Veiga Jr., Carlos R. Kaiser, Nuno A. Pereira, Cristina M.T.S. Carneiro, Frederico A. Vanderlinde, Antônio J. Lapa, Aniele R. Agner, Ilce M.S. Cólus, Juliana Echevarria-Lima, Noema F. Grynberg, Andressa Esteves-Souza, Kenia Pissinate, Aurea Echevarria

Abstract

This present phytomedicinal study aims at both identifying the variety of the *Croton cajucara* lead molecule *trans*-dehydrocrotonin (DCTN), in all parts of $1\frac{1}{2}$ to 6 year old plants, and perform a general evaluation of its biological efficacy. In addition, the biochemical evaluation of *trans*-crotonin (CTN), *trans*-cajucarin B (*t*-CJC-B), *cis*-cajucarin B (*c*-CJC-B), acetyl aleuritolic acid (AAA), cajucarinolide (CJCR) and isocajucarinolide (ICJCR) were performed. The pharmacological action of bioactive extracts and semi-purified fractions of this *Croton* are discussed herein. The phytochemical investigation proved that only the stem bark of the mature plants (MP) is a rich source of DCTN (1.4% of dry bark), while 3-year-old plants contained only 0.26%. In young plants (YP) the triterpene AAA was the major component and in these the diterpene DCTN was not present. Phytochemical investigation of the *C. cajucara* stem bark yields AAA, CTN, DCTN, *t*-CJC-B, *c*-CJC-B, CJCR, ICJCR, *trans*-cajucarin A (*t*-CJC-A) and isosacacarin. In addition, the common metabolites 4-hydroxy-3-methoxy-benzoic acid (vanilic acid), 4-hydroxy-benzoic acid and 2-methylamino-3-(4-hydroxyphenyl)propanoic acid (*N*-methyltyrosine) were isolated. The structure elucidation of DCTN and CTN was revised using 600 MHz NMR (nuclear magnetic resonance) spectrum, confirming their original structure. Synthetic transformation of DCTN afforded great amount of semi-synthetic CTN, CJCR and ICJCR derivatives, which were submitted to pharmacological experiments. Among the pharmacological experiments, DCTN and CTN were studied for their *in vivo* antitumor activities on Ehrlich carcinoma and Sarcoma 180 (S-180), and *in vivo* release was evaluated. The *in vitro* antiproliferative effects of DCTN and CTN were determined on the cultured Ehrlich carcinoma cells. The cytotoxic effects of CJCR, ICJCR, *t*-CJC-B, *c*-CJC-B, a tea preparation, and a MeOH extract were evaluated for human K562 leukemia and Ehrlich carcinoma cells. The tea preparation was also studied for its hepatotoxic effects on the weight variations and glutamic pyruvic transaminase (GPT) dosage in mice, in which GPT value variations revealed a dosage–response relation. The genotoxic action of DCTN was examined in Swiss mouse bone marrow cells *in vivo*, submitted to acute intraperitoneal treatment, by micronucleus (MN) and chromosomal aberration (CA) tests. Statistical analysis indicated that DCTN doses (50% and 75% of the LD_{50} via intraperitoneal or gavage treatments) were antimutagenic with regard to cyclophosphamide. The oral administration of DCTN and AAA

(100 mg/kg) reduced gastrointestinal transit of mice and index of gastric mucosa damage (GMD) induced by cold stress. After pylorus ligature in rats, similar dose of DCTN decreased acid secretion and volume. These compounds (1 μM) also reduced the ^{14}C-AP uptake induced by histamine (10^{-5} M); however, uniquely DCTN decreased the uptake induced by bethanechol.

Keywords: *Croton cajucara, clerodanes, trans-dehydrocrotonin, phenolic acids, vanillic acid, N-methyl-tyrosine, cytotoxic effects, hepatotoxic effects, gastrointestinal effects*

Abbreviations: DCTN, *trans-dehydrocrotonin*; CTN, *trans*-crotonin; *t*-CJC-B, *trans*-cajucarin B; *c*-CJC-B, *cis*-cajucarin B; AAA, acetyl aleuritolic acid; AAA–ME or AAA–Me, acetyl aleuritolic acid methyl ester; CJCR, cajucarinolide; ICJCR, isocajucarinolide; MP, mature plants; YP, young plants; CE-MP, chloroform extract of mature plants; $CHCl_3$, chloroform; CE-YP, chloroform extract of young plants; TF, terpenoidic fractions; ME, methanol extract; AE, aqueous extract or infusion preparation, or tea preparation; TF-AE-MP, terpenoidic fractions of an aqueous extract of mature plants; HRGC, high-resolution gas chromatography; HRGC–MS, high-resolution gas chromatography–mass spectrometry; TLC, thin layer chromatography; NMR, nuclear magnetic resonance; COSY, correlation spectroscopy; HSQC, heteronuclear single quantum coherence; HETCOR, heteronuclear single quantum coherence; HMBC, heteronuclear multiple bond correlation; HMQC, heteronuclear multiple quantum correlation; COLOC, correlation spectroscopy for long-range couplings; DEPT, distortionless enhancement by polarization transfer; NOESY, nuclear overhauser and exchange spectroscopy; NOE, nuclear overhauser enhancement; IMD, index of mucosal damage; GMD, gastric mucosa damage; Cim, cimetidine; GPT, glutamic pyruvic transaminase; MN, micronucleus; CA chromosomal aberration; S-180, Sarcoma 180; 5-FU, fluorouracil; MTT, 3-(4, 5-dimethylthiazol-2-yl)-2, 5-diphenyltetrazolium bromide; PBS, phosphate-buffered saline; TNF-α, tumor necrosis factor-α.

I. Introduction

It seems paradoxical that at a time when modern scientific medicine appears to be making such giant strides and enjoying unparalleled prestige, interest should be taken in traditional medicine in both developed and developing countries (Bannerman, 1982). This statement is valid when one considers *Croton cajucara*, an Amazon plant largely used in the traditional health care in the Amazon region of northern Brazil. *C. cajucara* Benth. (Euphorbiaceae) occurs widely in this region, where it is popularly known as 'sacaca' that among the Tupi Indians language means 'witchery'. The stem bark is popularly used as tea or capsules for the treatment of several diseases, specially for the treatment of liver, stomach, kidney disorders and also to lower blood cholesterol (Van den Berg, 1993; Di Stasi et al., 1989; Martins, 1989). This plant has also been used to treat diabetes, diarrhea, fever, jaundice, hepatitis and malaria (Van den Berg, 1982, 1993; Di Stasi et al., 1994). Furthermore, the leaf is used by over-weight people for weight loss, but frequently toxic hepatitis appears as side effect (Maciel et al., 2000, 2002; Veiga Jr. et al., 2005). The tea concentration is determined by the mode of preparation. For example, a soupspoon of stem bark in 100 mL of boiling water is used to prepare tea, to be taken at a dose of a cup twice a day for a period of 2–3 weeks. Similar preparations could be performed using 25 g of stem bark in 1000 mL of boiling water, in this case each cup of the tea must be warmed before ingestion. The capsule prepared with 250 mg of the dry stem bark is to be taken once a day (Maciel et al., 2000). Popular preference for the capsule over the tea is explained by the stem bark bitterness. The leaves toxicity reported by folk medicinal information could be correlated with both strong doses and prolonged use required for losing weight or slimming, because this toxicological effect was not observed by Farias et al. (1996) who found an absence of acute toxicity of a hydroalcoholic extract of *C. cajucara* leaves. Concerning to this *Croton* the cure process is not only lead by the pharmacological principle, but also by the history of the safe use of this plant. In fact, its medicinal belief presented in the Amazon culture survived through generation, providing health among the users (Maciel et al., 2000). On the other hand, during the 1990s several cases of toxic hepatitis correlated to this plant were notified in public hospitals of Belém city (state capital of Pará) located at Amazon region. This disease originated from abusive use (strong doses and prolonged treatments) of both leaves and stem bark. In spite of the warning of this toxic hepatitis effect on diet treatments, many Brazilian people are still using this species to lose weight or to prevent disease (Maciel et al., 2002a, b). The *C. cajucara* toxic warning first came in the 1990s from the traditional medicine of the Amazon region, and was reinforced in the early 21st century by a Brazilian TV program called 'Globo Reporter,' in which its cytotoxicity was focused (Veiga Jr. et al., 2005).

In other regions of Brazil the use of extracts from Brazilian medicinal plants in the treatment of human disease is a common practice, which has increased greatly in recent years. Meanwhile, many vegetal extracts are used by the people without any knowledge of the side effects they can have on their health. Commercial prescription drugs containing medicinal plants including *C. cajucara* are common in Belém. However, in many cases medicinal plants are used without any medical recommendation or control. The street market called *Ver-o-peso* in Belém is part of the centuries-old folk medicine culture of the Brazilian Amazon region. According to Prance (1992), Van den Berg correlated 1200 different medicinal plants available for sale at *Ver-o-peso* and the widespread use of *C. cajucara* is a typical example of the strong cultural belief by the Amazonians. The Amazon forest is well known for its great diversity of species of plants. In its Brazilian part, several of them have been used as medicine according to popular tradition, although, at present, there is still lack of knowledge concerning their chemical composition. In this work a brief review on such an important Amazon medicinal plant *C. cajucara* and new phytopharmacological experiments including cytotoxic assays, isolated constituents and DCTN semi-synthetic derivatives obtainment are discussed herein, cytotoxicity of this plant being the target point of this search.

II. *Croton cajucara* general pharmaco-chemical profiling

In order to validate the traditional use of *C. cajucara* as a therapeutic safe plant and also to improve its scientific studies, we have been undertaking an extensive phytopharmacological search oriented by its traditional medicine. Among the approaches we are performing, cytotoxic model experiments on the major isolated diterpene *trans*-dehydrocrotonin (DCTN). Regarding this compound, only the stem bark of the mature plants (MP; 4–6 years old) is a rich source of this clerodane-type diterpene (Maciel et al., 1998). Advances in our search with DCTN proved that it was neither genotoxic nor cytotoxic to bone marrow cells (i.p. treated mice) (Agner et al., 1999, 2001). Its antigenotoxicity will be briefly discussed herein. Meanwhile,

Rodríguez and Haun (1999) observed that the cytotoxicity of DCTN on V79 cells line and on rat hepatocytes was time–exposure dependent.

Our previous pharmacological studies performed with the isolated terpenoids, e.g., DCTN, *trans*-crotonin (CTN) and acetyl aleuritolic acid (AAA), proved striking correlation among these compounds with the folk traditional therapeutic use of *C. cajucara*; DCTN being the lead compound. Specifically, DCTN was insect antifeedant (Kubo et al., 1991) and showed anti-inflammatory and antinociceptive (Carvalho et al., 1996; Maciel et al., 2002a) effects. It was proved to be a hypoglycemic (Farias et al., 1997), antifungic (Souza et al., 2000), antiatherogenic and hypolipidemic (Silva et al., 2001a, b) agent. Additionally, DCTN demonstrated hypotensive and bradycardic effects *in vivo* that were mostly related to the endothelium-dependent and -independent vasorelaxant effects on aortic rings, and a direct negative chronotropic effect on right atria of rats *in vitro* (Silva et al., 2005).

As part of our ongoing phytochemical studies, we here report the isolation of two phenolic acids (vanillic acid and 4-hydroxy-benzoic acid) and an amino acid (*N*-methyltyrosine). Such phenols have not been reported, however, from *C. cajucara* prior to this study. Recent work on phenolics was related to either their benefits to human health or to their use in phytogenetic studies (Simmonds, 2003). Shibamoto & Nishimura (1982) reported the presence of some phenols in coffee beans. Related to the phenolics isolated from *C. cajucara* they showed remarkable antioxidant activity in other species (Hung & Yen, 2002; Ohsugi et al., 1999). Reinforcing this subject, different kinds of kempferol metabolites proved to be antioxidant agents (Bonina et al., 2002; Jonson et al., 2003; Marfak et al., 2003; Murota and Terao, 2003). *C. cajucara* has two of them, e.g., kempferol 3,4′,7-trimethyl ether and 3,7-dimethyl ether (Maciel et al., 2000). Based on these results this *Croton* may possess antioxidant property and could attract special attention to its possible role in the prevention of oxidative human process.

II.A. Seasonal variety of DCTN and its pharmacological importance

Extensive classical phytochemical studies with the stem bark of *C. cajucara* led to the discovery of clerodane-type diterpenes, e.g., DCTN, CTN, *cis*-dehydrocrotonin, *trans*-cajucarin A (*t*-CJC-A), *cis*-cajucarin B (*c*-CJC-B), *trans*-cajucarin B (*t*-CJC-B), cajucarinolide (CJCR), isocajucarinolide (ICJC) and isosacacarin (Simões et al., 1979; Itokawa et al., 1989, 1990; Kubo et al., 1991; Ichihara et al., 1992; Maciel et al., 1998, 2000, 2002a, 2003). The triterpene AAA was also obtained (Maciel et al., 1998). The presence of the bioactive terpenoids DCTN and AAA was investigated in the stem bark of 18 months to 6-year-old plants. Through this study it was proved that their concentration depends on the plant age. The largest amount (1.4%) of DCTN was detected in MP, while 3-year-old plants contained only 0.3% of this compound. DCTN was found to be absent in the stem bark of young plants (YP; 18-month-old plants). Meanwhile, the triterpene AAA was preponderant (0.2%) in YP (Figure 1). DCTN was not detected in the leaves and branches, and its content in roots (0.002%) and stem (0.2%) were much lower when compared to the stem bark amounts (Maciel et al., 2000).

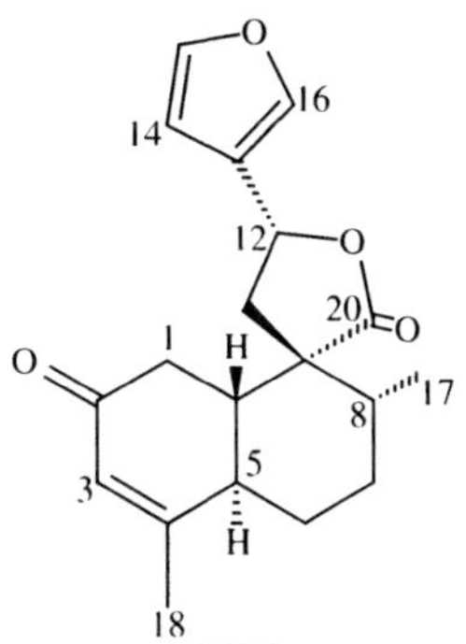

DCTN

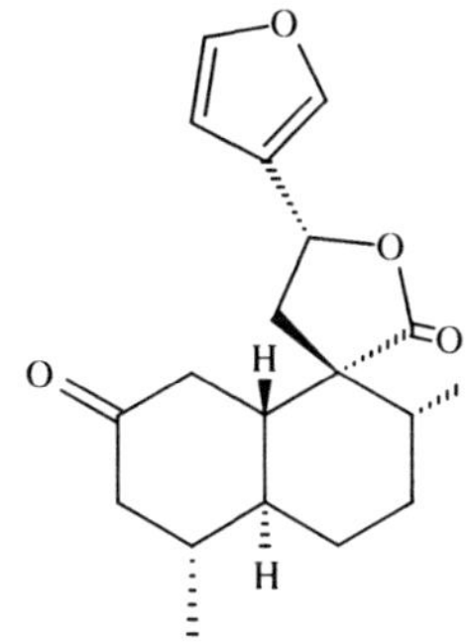

CTN

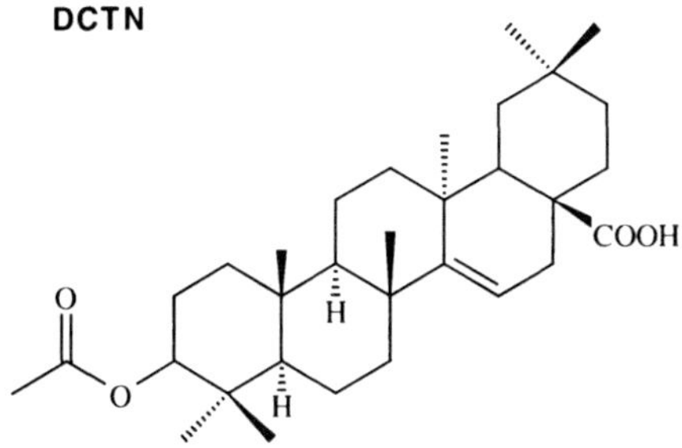

AAA

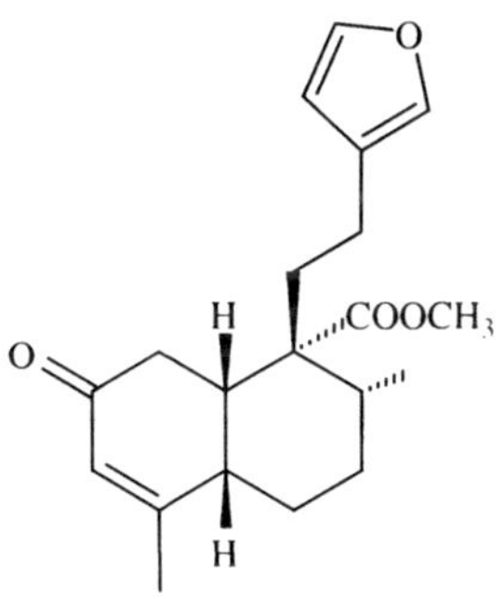

c-CJC-B

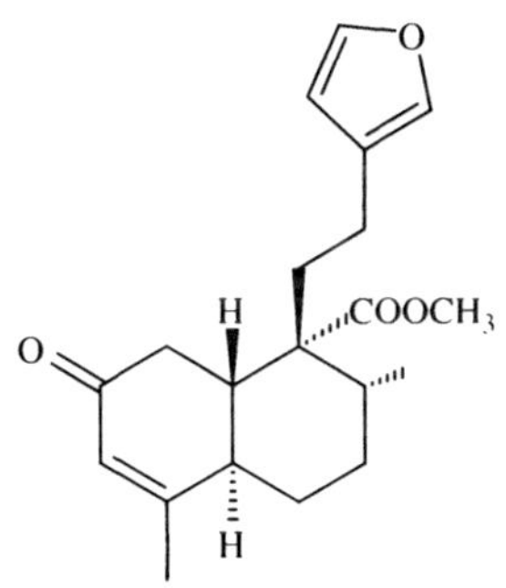

t-CJC-B

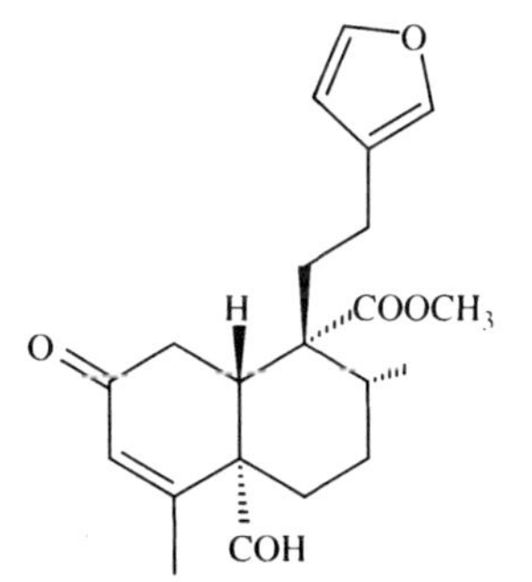

t-CJC-A

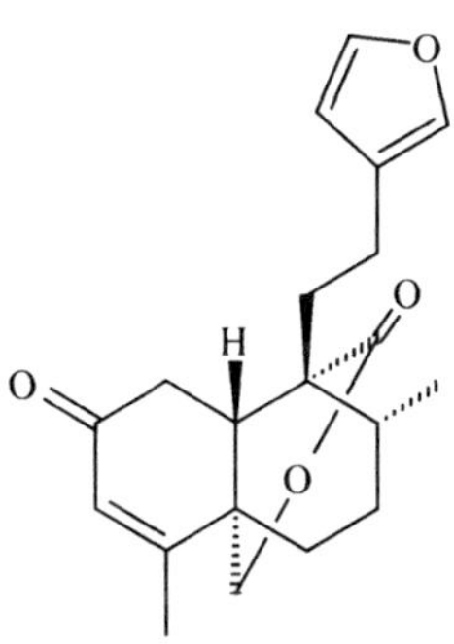

isosacacarin

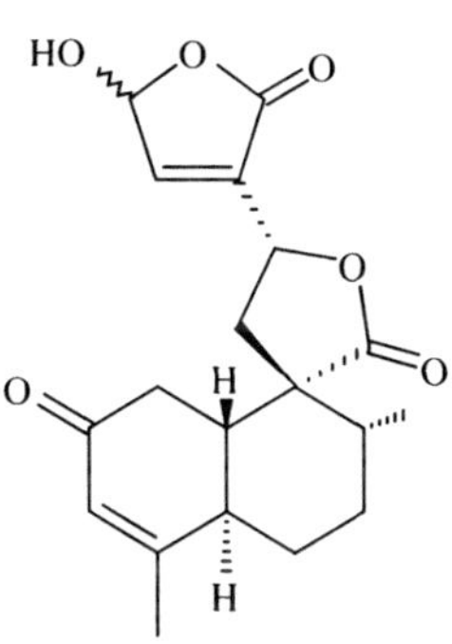

CJCR

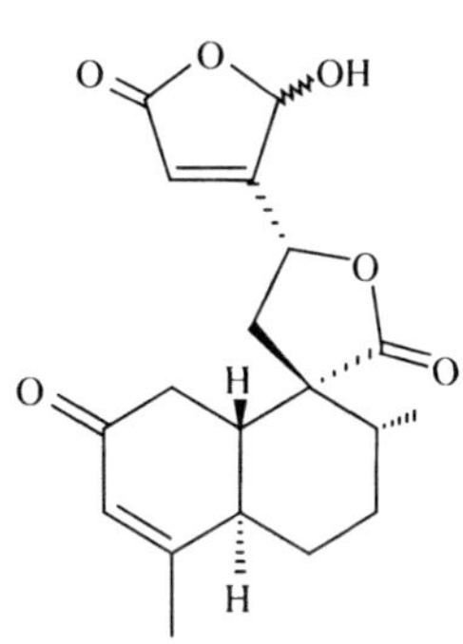

ICJCR

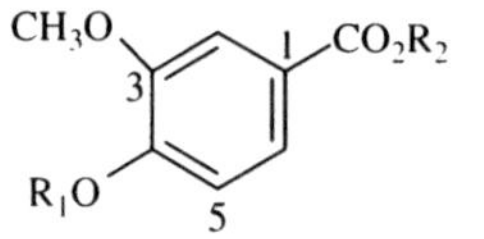

vanillic acid (R_1, R_2 = H)

R_1 = H R_2 = Me **(derivative I)**

R_1, R_2 = Me **(derivative II)**

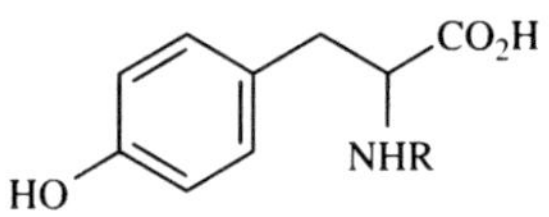

N-methyltyrosine
(R = CH3)

tyrosine
(R = H)

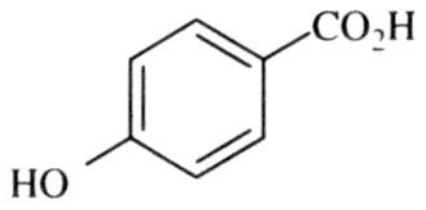

4-hydroxy-benzoic acid

Since the main pharmacological activity of *C. cajucara* was correlated to DCTN (Maciel et al., 2000, 2002b), we performed high-resolution gas chromatography (HRGC) investigations in terpenoidic fractions (TF) obtained from the stem bark of MeOH extracts of MP [TF-ME-MP, eluted with CH_2Cl_2/MeOH (9:1)], and also from 3-year-old plant, youngest plants (TF-ME-YP) and from an infusion preparation obtained from 50 g of powdered material of MP (AE-MP, extracted with 200 mL of distilled water at 100 °C during 5 min, and then filtered under vacuum). This infusion was assumed to correspond to an aqueous extract (AE-MP). The TF of AE-MP (TF-AE-MP) was obtained by extraction with $CHCl_3$ (10×50 mL).

The whole obtained TFs were esterified with diazomethane and then submitted to HRGC analysis. When a peak was found near to DCTN elution region and to AAA–ME (acetyl aleuritolic acid methyl ester) elution region, the presence of these compounds were confirmed by HRGC–mass spectrometry (HRGC–MS) and the peak compared with calibration curves and quantified by HRGC. TF-AE-MP afforded 59.9% of DCTN and 1.3% of AAA-ME. As expected, TF-ME-MP showed DCTN as the major component (78.7%) and TF of 3-year-old plants contained only 37.5%. In addition, DCTN was not detected in TF-ME-YP and the triterpene AAA-ME being preponderant (9.9%). These results showed that the isolated amount of DCTN and AAA by classical chromatography fractionation of MeOH extracts (Maciel et al., 2000) were in agreement with those found by HRGC–MS investigations.

The relative low hot-water solubility of DCTN may account to the observed difference in the amount of DCTN in TF-AE-MP and TF-ME-MP (59.9% and 78.7%, respectively). These significant contents indicate DCTN as an important component of the tea (correlated to TF-AE-MP) and capsules (correlated to TF-ME-MP) of *C. cajucara*.

Plant extracts are usually complex mixtures, which contain several molecules of different sizes with varied functional groups. Such extracts are a challenge to the chemist of natural products. Ion exchange chromatography in non-aqueous medium, used for the separation of basic or acidic fractions from plant extracts, is an important unit operation in preparative scale separations. Recently we successfully

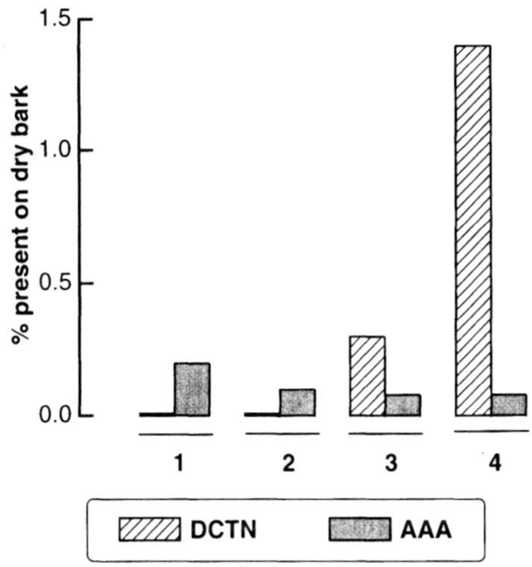

Fig. 1. Presence of the terpenoids DCTN and AAA in both hexane and MeOH extract of the stem bark of plants with ages ranging from $1\frac{1}{2}$ to 6 years old.

1. Plants of $1\frac{1}{2}$ year that came from poor conditions of quantity of solar radiation. DCTN was absent and AAA was found to be present (0.2% of the original material).
2. Plant of $1\frac{1}{2}$ year that came from good quality and quantity of solar radiation. DCTN was absent and AAA was found to be present (0.1% of the original material).
3. Plant of 3 years that came from good quality and quantity of solar radiation. DCTN and AAA were found to be present (0.3% DCTN and 0.08% AAA of the original material).
4. Plant of 4–6 years that came from poor conditions of quantity of solar radiation (native area of the Amazon region). DCTN and AAA were found to be present (1.4% DCTN and 0.08% AAA of the original material).

applied ion exchange chromatography (an anionic macroporous resin in non-aqueous medium) for separation of the AAA on methanolic extract from the stem bark of *C. cajucara* (Barreto Jr. et al., 2005). This technique showed to be effective in the purification of those extracts in which DCTN were the target drug to be isolated.

Concerning AAA biological importance, this compound proved to possess antibiotic (Addae-Mensah et al., 1992) and antibacterial (Peres et al., 1997) activities and also presented anti-inflammatory and gastrointestinal effects in mice (Maciel et al., 2000), as briefly described below. AAA showed lack of efficacy on both antitumoral *in vivo* assays (Grynberg et al., 1999) and gastric mucosa damage (GMD). The latter was caused by ulcers induced by cold stress (Maciel et al., 2000). The chloroform extract of young plants (CE-YP) did not produce any effect on GMD. In fact, this result shows correlation with the higher amount of AAA in YP. As expected, DCTN and chloroform extract of mature plants (CE-MP) reduced GMD index. CE-MP showed great amount of DCTN (Maciel et al., 2000). Compared to normal control CE-YP, AAA and DCTN demonstrated significant decrease in the charcoal grastrointestinal transit. Unexpectedly CE-MP, which is a rich source of AAA and DCTN, was ineffective (Maciel et al., 2000).

The biological importance of the natural and semi-synthetic CTN has been reported (Grynberg et al., 1999; Maciel et al., 2000; Hiruma-Lima et al., 2002; Anazetti et al., 2003). Natural CTN was found to be present only in the stem bark of MP as minor metabolites, this phytochemical limitation led us to obtain the semi-synthetic CTN derivative. Within the program of chemical transformation of the target molecule DCTN, which means preparing DCTN derivative compounds for screening as pharmacological agents, CTN, CJCR and ICJCR have been already obtained. CTN was obtained in great amount by hydrogenation of DCTN in 95% ethanol over prereduced 10% charcoal over palladium, followed by recrystallization from hexane/Me_2CO. The adequate amount (9.6 g) of semi-synthetic CTN allowed us to perform some pharmacological studies (Maciel et al., 2000). Among them its anti-inflammatory, antinociceptive and antiulcerogenic action are briefly described below. Related to CJCRs derivatives (ICJCR derivatives), they were obtained by regio-specific oxidation of substituted furans, which gives the corresponding hydroxybutenolides. Using this method, which was previously described by Kernan and Faulkner (1988), the ICJCR derivatives were synthesized in good yields by synthetic transformation of the DCTN furan moiety, with singlet oxygen, generated from molecular oxygen by irradiating a polymer-bound rose bengal catalyst in dichloromethane solution at −78 °C, in the presence of a hindered base, such as diisopropylethylamine. In the absence of a hindered base, this reaction yielded an isomeric mixture of CJCR and ICJCR, which was purified by chromatography fractionation (Maciel et al., 2002). The structure elucidation of CJCR and ICJCR derivatives were determined by spectroscopy, including 2D-NMR experiments (correlation spectroscopy, COSY 45; heteronuclear single quantum coherence, HSQC; and heteronuclear multiple bond correlation, HMBC) and the observed data corresponded to those described by Ichihara et al. (1992) who found those clerodanes in early chemical studies of *C. cajucara*.

The anti-inflammatory effect of *C. cajucara* was recently associated to the natural AAA and DCTN, the semi-synthetic CTN, and also to AE obtained from the stem bark of this *Croton*. The antinociceptive effect was only correlated to DCTN and AE (Vanderlinde et al., 1998; Maciel et al., 2002b). The CTN derivative and the natural AAA did not induce nociception (Maciel et al., 2002). The relative low anti-inflammatory activity observed for DCTN and CTN compared to the significant effects of AE suggest a probable synergism among DCTN, CTN, *t*-CJC-B, *c*-CJC-B, cajucarin A, CJCR, ICJCR and isosacacarin, which were detected to be present in AE by thin layer chromatography (TLC) analysis (compared with authentic samples) and confirmed by HRGC–MS study.

Reinforcing this chemical-pharmacological result of AE, a significant anti-inflammatory activity was previously correlated to the hydroxybutenolide-clerodanes CJCR and ICJCR (Ichihara et al., 1992). In addition, a polar fraction AE-F3 (eluted with MeOH/H_2O) obtained from AE showed to possess a strong analgesic effect, in such fraction the aminoacid 2-methylamino-3-(4-hydroxyphenyl)propanoic acid (*N*-methyltyrosine) and a probable quinoline alkaloid (detected by Dragendorff reagent in TLC analysis and ^1H NMR spectrum) were found to be present (Vanderlinde et al., 2000; Maciel et al., 2002b).

Reinforcing the anti-inflammatory and antinociceptive effects of *C. cajucara*, Bighetti et al. (1999) report these effects in rodents testing the essential oil obtained from its stem bark.

Other *C. cajucara* leaf constituents such as *β*-sitosterol, stigmasterol and sitosterol-3,*O*-*β*-glucoside proved to be bioactive. These compounds showed biological effects against snake venom (Mors, 1991; Bilia et al., 1996; Peres et al., 1997). The bioactivy clerodane CJCR was also isolated from this part of *C. cajucara* (Maciel et al., 2000).

Among the popular use of this *Croton*, the gastroprotective widespread use was correlated to its essential oil (Hiruma-Lima et al., 1998, 2000) and also to AAA, CTN and DCTN (Brito et al., 1998; Hiruma-Lima et al., 1999; Maciel et al., 2000). According to our results, DCTN, CTN and AAA reduced the index of gastric lesions induced by restraint-in-cold. Dose related to DCTN and CTN inhibited *in vivo* the basal acid secretion in pylorus-ligature rats and oxyntic glands isolated from rabbit gastric mucosa. DCTN, CTN and AAA decreased *in vitro* uptake basal acid secretion induced by histamine and measured with the ^{14}C-aminopyrine uptake method. Uniquely, DCTN inhibited ^{14}C-AP uptake induced by bethanechol. The effect of DCTN and CTN on gastric secretion was obtained by pylorus ligature in rats ($n = 6$). The volume of gastric juice secreted by control animals (DMSO 0.5%), over 4 h, was 2.2 ± 0.4 mL with a pH 1.9 ± 0.1 and total acidity of 11.3 ± 2.9 mEq[H$^+$]/L^{-1}. DCTN (10100 mg/kg) injected into the duodenal lumen (i.d.) produced a dose-related effect of acid secretion, volume and pH. The highest dose of DCTN determined significant decrease of acid secretion and volume by 75% (2.8 ± 1.1 mEq[H$^+$]/L^{-1}/4 h) and 55% (1 ± 0.2 mL), respectively. The pH of the gastric juice was raised to 3.6 ± 0.6. In oxyntic glands isolated from rabbit gastric mucosa, the effect of the tested compounds of acid formation was estimated with the ^{14}C-aminopyrine (^{14}C-AP) uptake method. DCTN, CTN and AAA (1 mM) reduced significantly by 37.4 ± 1.7%, 44.4 ± 1.1% and 41.7 ± 0.2%, respectively, and the ^{14}C-AP uptake induced by histamine (10^{-5} M) (Figure 2). Uniquely DCTN decreased 57.8 ± 1.5% uptakes induced by bethanechol (10^{-5} M). These compounds (10^{-6} to 10^{-4} M) did not change the basal acid secretion. The gastrointestinal transit effects of DCTN, CTN and AAA were determined by charcoal meal method (Stickney and

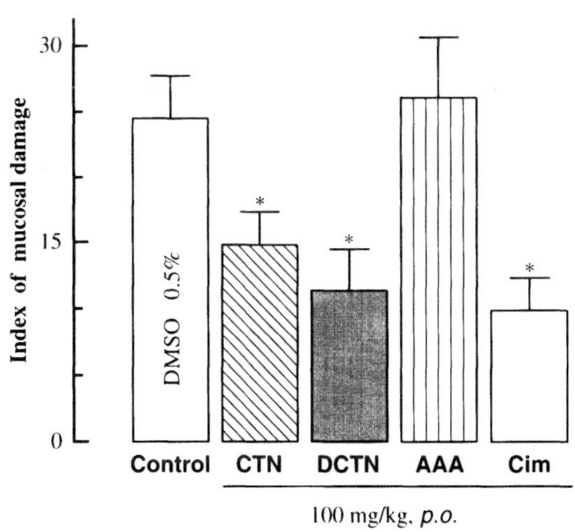

Fig. 2. IMD in rats by stress-in-cold after mobilized (4 °C for 2 h). The animals were treated orally (p.o.) with 100 g/kg of each compounds DCTN, CTN, and acetil aleuritolic acid (AAA) obtained from the bark of *C. cajucara*, or positive control Cim 60 min before restraint. The columns and vertical bars are means ± Sem, $n = 6$ for each group; * different of control group $p < 0.05$.

Northrup, 1959). Oral administration of mice with DCTN or AAA (100 mg/kg) decreased the charcoal gastrointestinal transit by 25%, 30% and 40%, respectively. Similar treatment with CTN was ineffective (Table 1).

These results together with those anti-inflammatory and antinociceptive properties observed by Campos et al. (2002) for leaf extracts of *C. cajucara* may account for the wide popular use of this plant as an analgesic, anti-inflammatory, antidiarrheal and antiulcerogenic agent. Reinforcing the leaf biochemical importance, Rosa et al. (2003) report the *in vitro* leishmanicidal effects of a linalool-rich essential oil from the leaves of this *Croton* against *Leishmania amazonensis*.

II.B. Hepatic toxicological evaluation of the aqueous extract

The folk medicine *C. cajucara* in the Amazon region is also used for the treatment of liver diseases (Van den Berg, 1982, 1993; Kubo et al., 1991). However, this plant can cause toxic hepatitis when used over the safe control. In order to evaluate a probable toxic hepatitis as a side effect of its stem bark infusions (corresponding to a tea preparation, AE), this ingredient was administrated to Swiss albino mice with initial weight between 28 and 35 g. The treated animals were divided into three groups of ten, one of them being the control group, and the other two groups were treated with 12.5 and 25 g/L of AE, in the wells exclusively. Their weight variation was average once a week during 42 days of the whole treatment. Table 2 shows the animals weight and Table 3 the effect of the glutamic pyruvic transaminase (GPT), in which the average dosages were strictly correlated to hepatic investigations, based on the biologic action of GPT, an enzyme that has a predominant hepatic origin and possess remarkable alterations in acute hepatitis cases (Reitman and Frankel, 1957). The

Table 1
Effect of the terpenoids DCTN, CTN and AAA in gastrointestinal transit in mice

Treatment	Dose (g/kg, p.o.)	% of control length traveled by charcoal meal	p^*
DCTN	0.1	68.5 ± 3.9	<0.01
CTN	0.1	93.9 ± 6.2	>0.05
AAA	0.1	59.8 ± 5.9	<0.01
Atropine	5×10^{-3}	20.1 ± 6.3	<0.01

Note: Each group represents the mean $\pm$ Sem of results from 10 animals.
$^*p<0.01$ significant (compared with control group).

Table 2
Animals average weight variation (g) during the test period (days)

Group control	0	7	14	21	28	35	42
	28.3	29.1	29.8	30.8	30.4	30.8	31.3
Treated with 12.5 g/L AE	31.9	32.2	31.9	32.1	32.7	32.6	33.6
Treated with 25 g/L AE	30.6	28.3	28.5	31.0	31.7	33.8	34.2

Table 3
Average GPT values variation

Groups	Control	12.5 g/L tea	25 g/L tea
GPT u/L	61.4	64.3	69.5

obtained results proved that the control group animals as well as the animals treated with 12.5 g/L AE gained weight regularly, while the animals treated with 25 g/L AE showed initially remarked weight loss being regular in the rest of the period of treatment. Although the results were found in intervals considered normal for male mice and the dosage–response relation observed, we suggest other investigations such as an anatomopathological evaluation of mice liver, in order to confirm the absence of hepatic toxicity of AE.

This pharmacological result jointly with the significant contents of DCTN in AE proved that drinking *C. cajucara* tea instead of taking capsules is an unambiguous way to be protected against DCTN over doses side effects.

II.C. Antiproliferative effects of the isolated terpenoids and stem bark extracts

The minor compounds such as CTN, CJCR and ICJCR mentioned above, can be obtained from the major component DCTN, and then from the derivatives ingredients utilized on *in vivo* and *in vitro* experiments, the case of CTN (Maciel, 2000, 2002b). In order to identify the biological antiproliferative action of the stem bark of *C. cajucara*, we studied the effects of the natural AAA, DCTN, CTN, *t*-CJC-B and *c*-CJC-B against the growth of ascitic Ehrlich carcinoma cells and human K562 leukemia. In addition, the semi-synthetic CTN, CJCR and ICJCR derivatives were tested. The experiments utilized MTT assay (Mosmann, 1983) and the positive controls were quercetin (for Ehrlich carcinoma) and vincristine (for human K562 leukemia).

II.D. In vitro *cytotoxic effects*

The results of inhibitory effects on Ehrlich carcinoma cells by terpenoid derivatives DCTN, CTN, CJCR and ICJCR showed a significant cytotoxicity in dependent dose response, for culture of 48 h, with IC_{50} values of 52.2 μg/mL (166 μM, DCTN); 51.8 μg/mL (164 μM, CTN); 22.4 μg/mL (65 μM, CJCR); and 3.5 μg/mL (10 μM, ICJCR), while quercetin showed IC_{50} of 44 μM (Table 4). The tested CTN, CJCR and ICJCR derivatives against Ehrlich carcinoma also indicated a similar antiproliferative activity. The triterpene AAA and its methylated derivative, AAA-Me, did not present cytotoxic activity. According to Wattenberg (1995), compounds containing an α,β-unsaturated carbonyl moiety have been shown to bind to receptors that induce increased activities of phase II enzymes responsible for metabolizing xenobiotic agents. In fact, AAA does not have this specific moiety. Meanwhile, DCTN, CJCR and ICJCR, which contain this reactive electrophilic moiety, proved to be cytotoxic to Ehrlich cells. On the one hand, exception could be observed in the case of CTN which does not have the α,β-unsaturated carbonyl unit, but was

Table 4
Values of IC_{50} of metabolites of *C. cajucara* against Ehrlich carcinoma and human K562 leukemia cells

	Ehrlich carcinoma		Human K562 leukemia cells	
Metabolites	IC_{50} (μM)	IC_{50} (μg/mL)	IC_{50} (μM)	IC_{50} (μg/mL)
DCTN	166	52.2	N.A.	N.A.
CTN	164	51.8	N.A.	N.A.
AAA	N.A.	N.A.	N.A.	N.A.
AAA-Me	N.A.	N.A.	N.A.	N.A.
CJCR	65	22.4	36	12.6
ICJCR	10	3.5	43	14.9
t-CJC-B	N.A.	N.A.	38	12.6
c-CJC-B	N.A.	N.A.	N.A.	N.A.

N.A.: not active.

effective. On the other, the clerodanes *c*-CJC-B and *t*-CJC-B present an α,β-unsaturated carbonyl moiety and showed to be ineffective. The conformations of these clerodanes may account to these unexpected results.

Considering the cytotoxic effect of ICJCR (10 μM) we suggest that its higher antiproliferative action on growth cell cancer is correlated to the stereoposition of the hydroxyl group in the hydroxybutenolide moiety. In fact, ICJCR (OH at position 16) is isomeric with CJCR (OH at position 15). It is important to observe that DCTN which has only one α,β-unsaturated carbonyl moiety proved to be highly antiproliferative when compared to CJCR which has this moiety and an additional one in the hydroxybutenolide unit.

The natural CJCR and ICJCR were studied for their effects against human K562 leukemia cells in culture of 96 h, in that vincristine, in 60 nM, was used as positive control. The inhibitory effects on cell growth were dose–response dependent, with IC_{50} values of 36 μM for CJCR, and 43 μM for ICJCR (Table 4). In addition, *t*-CJCR-B presented inhibitory effect with IC_{50} value of 38 μM. Meanwhile, the lead molecule DCTN and also *c*-CJC-B, CTN (natural and semi-synthetic), AAA and its derivative (AAA-Me) lack efficacy against K562 cell culture. The phenolic acids (vanilic acid, as well as 4-hydroxy-benzoic acid) were also assayed in the same experimental conditions, being inactive against both Ehrlich and human K562 leukemia cells. Figure 3 shows the comparative cytotoxic effects of the tested clerodanes.

The cytotoxic effects of the stem bark of *C. cajucara* was extended to both MeOH extract and tea preparation of MP (ME-MP and AE-MP, respectively). These ingredients showed significant antiproliferative effects. The AE afforded three polar fractions AE-F1, AE-F2 and AE-F3 after chromatography (fractionation on a silica gel column). The terpenoide AAA and the clerodanes DCTN, CTN, *c*-CJCR-B and *t*-CJC-B were found to be present in AE-F1 (by TLC analysis with authentic samples). As expected in such fraction, the concentration was highest in DCTN (detected by HRGC–MS investigations). As mentioned earlier, the amino acid (*N*-methyltyrosine) and a probable quinoline alkaloid revealed to be present in AE-F3.

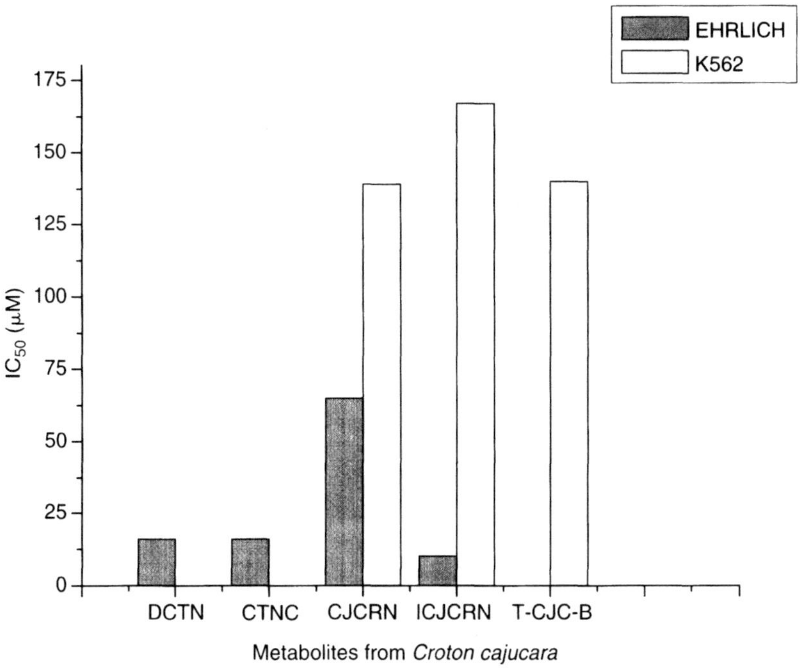

Fig. 3. Graphic of the IC_{50} values of *C. cajucara* metabolites against Ehrlich carcinoma and human K562 leukemia cells.

The MeOH extract yielded fractions ME-F1 (corresponding to the essential oil), ME-F2 (a mixture of essential oil and compound AAA) and ME-F7-22 (a mixture of DCTN, CTN, *c*-CJC-B and *t*-CJC-B). The chemical composition of these fractions was detected by TLC analysis with authentic samples and then confirmed by HRGC–MS analysis.

The cytotoxic effects of ME-F1, ME-F2 and ME-F7-22, and also AE-F1 and AE-F3 against K562 and Ehrlich cells were determined. The results indicated a significant antiproliferative effect on human K562 leukemia cells, with values of the IC_{50} being 16 μg/mL for ME-F1; 5 μg/mL (ME-F2); 6 μg/mL (ME-F7-22); 18 μg/mL (AE-F1); and 26 μg/mL (AE-F3). Moreover, the results of Ehrlich carcinoma cells did not indicate significant cytotoxicity for these tested fractions. The antiproliferative effects of methanolic fractions indicated ME-F7-22 as the more cytotoxic fraction suggesting a synergic effect among DCTN, CTN, *c*-CJC-B and *t*-CJC-B. On the other hand, ME-F1 compared to ME-F2 showed lower antiproliferative effect due to the presence of the ineffective triterpene AAA. Figure 4 shows the results for methanolic and AE fractions on K562 cell culture.

II.E. In vivo *antitumor effects*

The clerodanes DCTN and CTN derivative were used on *in vivo* experiments against Ehrlich carcinoma and ascitic murine tumors Sarcoma 180 (S-180) (Grynberg et al., 1999). Their results are shown in Table 5, in that 5-FU was used as a positive

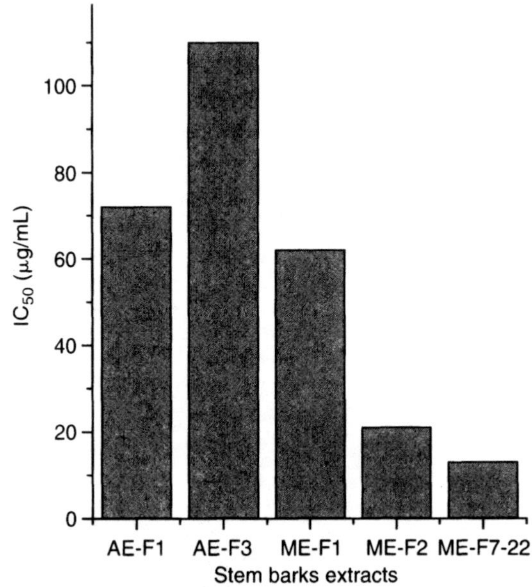

Fig. 4. Graphic of the IC_{50} values to stem bark fractions of *C. cajucara* against human K562 leukemia cells.

Table 5
In vivo effects of DCTN, CTN and 5-FU on ascitic S-180 and Ehrlich tumor growth

Metabolite	Tumor	Treatment (days)	Total dose (mg/kg)	(mol/kg)	% T/C[a]
5-FU	S180	1 and 2	38	0.292	140
DCTN	S180	1 and 2	84	0.257	137
		1 and 6	33.3	0.102	91
CTN	S180	1 and 2	80	0.244	121
DCTN	Ehrlich	1, 2 and 3	120	0.368	128
CTN	Ehrlich	1, 2 and 3	120	0.365	110
5-FU	Ehrlich	1, 2 and 3	80	0.615	144

[a]% T/C≥125 for significant antitumor activity.

control. The diterpene DCTN presented a significant activity against S-180 and Ehrlich at doses of 84 and 120 mg/kg, %T/C at 137 and 128, respectively. A borderline antitumor activity was observed in the maximum soluble dose in the schedule of treatment against S-180 with CTN derivative, while there was no activity on Ehrlich carcinoma in the schedule of the used treatment. The antitumor activity was dose–response dependent for both tumors, since doses of DCTN lower than 84 mg/kg were not effective. Furthermore, there was no weight loss or evidence of toxicity by macroscopic examination of the organs in non-bearing tumor mice treated with those clerodanes.

These results led us to assay for *in vivo* TNF-α induction in Ehrlich tumor bearing mice treated with DCTN and CTN. A significant activity, $p<0.05$, of TNF-α was obtained in ascitic supernatant after treatment with DCTN at the dose of 120 mg/kg. The prolongation in survival time on the treatment of mice with DCTN increased TNF-α secretion at the area of tumor cell growth (Grynberg et al., 1999). Also, other diterpenes like taxol-enhanced TNF-α production.

II.F. Genotoxic activity

The genotoxic action of DCTN was examined in Swiss mouse bone marrow cells *in vivo*, submitted to acute intraperitoneal treatment, by micronucleus (MN) and chromosomal aberration (CA) tests. The statistical tests (Anova and Tukey) made to compare the results obtained in each of the three doses of DCTN (138.75, 277.50 and 416.25 mg/kg b.w.) with the negative control showed that the frequencies of MN and mitotic index were equal to the negative control and that the frequencies of CA were lower than that observed in the negative control. Both the MN and CA tests did not show large differences in relation to the treatments and the utilized protocols, which may have indicated which one of tests is the most adequate for this type of analysis. Thus, we can say that both tests showed a similar sensitivity. In relation to the averages of micronucleated polynucleated erythrocytes (PCEs) and mitotic indices obtained after treatment with the three doses of DCTN was not statistically lower than the ones obtained for the negative control group. Therefore, the tested doses did not show cytotoxic and genotoxic effect in the bone marrow cells of mice. This result is in accordance with the hepatic toxicological evaluation of AE, therefore does not agree with the ones of Kubo et al. (1991) who found *in vitro* cytotoxic effect of the methanol extract (ME) of the stem bark of *C. cajucara* in hepatocytes of mice.

Our results (Agner et al., 1999) showed that DCTN is not a mutagenic substance and revealed the need to evaluate its antimutagenic potential.

II.G. Antigenotoxic activity

The antigenotoxic effects of three doses of DCTN were assessed in Swiss mice bone marrow cells. The DCTN was administered intraperitoneally and by gavage in acute treatments to mice in doses equivalent to 138.75, 277.5 and 416.25 mg/kg b.w. The bone marrow cells were analyzed by MN and CA assays, which have commonly been utilized as a predictor of the genotoxicity or antigenotoxicity of different substances. Comparisons were performed between the three doses of DCTN used and the negative control group. Statistical analysis (ANOVA and Tukey) indicated that doses of 50% and 75% of the LD_{50} via intraperitoneal injection or gavage treatment were antigenotoxic with regard to cyclophosphamide (positive control). However, the dose of 25% of LD_{50} was only antimutagenic when administered by gavage (Agner et al., 2001). These results proved that the DCTN reduced the damage caused by cyclophosphamide, an alkylant agent, in Swiss mice bone marrow cells. Although the evaluation of toxicity (Maciel et al., 2000) and mutagenicity (Agner et al., 1999) of DCTN showed negative results. We suggest further studies, in order to confirm scientifically the safe therapeutic properties of *C. cajucara*.

II.H. Croton cajucara phytochemical studies

The hexane and MEs of the stem bark of MP were submitted to chromatography fractionation on a silica gel column affording AAA (0.08%) and DCTN (1.4%) as major compounds and CTN (0.002%), *t*-CJC-B (0.005%), *c*-CJC-B (0.001%), *t*-CJC-A (0.001%), CJCR (0.0003%), ICJCR (0.0001%) and isosacacarin (0.002%) as minor components (Maciel, 1998, 2000, 2003). In addition, two aromatic acids 4-hydroxy-3-methoxy-benzoic acid (commonly known as vanillic acid) and 4-hydroxy-benzoic acid [0.005% in the ratio 87:13 (RMN), respectively], an amino acid 2-methylamino-3-(4-hydroxyphenyl) propanoic acid (0.014%, commonly known as *N*-methyltyrosine) and an uncharacterized quinoline alkaloid were isolated.

During our experimental work we observed that clerodane-type diterpenes possessing α,β-unsaturated ketone (DCTN, *c*-CJC-B, *t*-CJC-B, *t*-CJC-A, CJCR, ICJCR and isosacacarin) and/or lactone ring (CTN, DCTN, *c*-CJC-B, *t*-CJC-B, *t*-CJC-A, CJCR, ICJCR and isosacacarin) were detected by Dragendorff reagent largely used to detect alkaloid compounds.

II.I. NMR structural study of trans-dehydrocrotonin

The most complete NMR analysis of the lead compound DCTN was previously performed by Itokawa et al. (1989) using a 400 MHz instrument, but some hydrogens, e.g., H-1, H-3, H-5, H-6, H-7, H-11, H-15 and H-16 were not precisely assigned due to overlapping in the ^{1}H spectrum. The present work shows that using 600 MHz instrument, the above-cited missing signals appeared enough separated allowing unambiguous assignments.

The C-11 methylene hydrogens (δ_H 2.42 and 2.37) were easily recognized due to mutual cross signals with H-12 (δ_H 5.43) in the COSY spectrum. With the aid of the heteronuclear multiple quantum correlation (HMQC) spectrum, it was feasible to distinguish the remaining C-1 (δ_H 2.19 1α and 2.54 1β), C-6 (δ_H 2.27 6α and 1.18 6β) and C-7 (δ_H 1.66 7α and 1.88 7β) methylene groups (δ_C 39.76, 28.23 and 30.13, respectively). The first methylene δ_C corresponds to C-1 in view of the deshielding caused by the neighbor carbonyl group and also due to the cross signal with H-3 (δ_H 5.89) in the HMBC spectrum. The C-1 hydrogen at δ_H 2.54 must be at β position since it shows only one great coupling constant (due to the *geminal* H-1α, J_{HH} 15.61 Hz) and due to the correlations with H-11 (δ_H 2.37) and with H-10 (δ_H 1.81) in the nuclear overhauser and exchange spectroscopy (NOESY) spectrum. The H-10 shows $^3J_{H,H}$ coupling constants, e.g., 13.80 Hz (H-10, H-1α), 10.70 Hz (H-10, H-5α) and 2.80 Hz (H-10, H-1β). The stereochemistry of the decalin and lactone units of this compound was determined by the correlations of H-8 with H-6β and with H-10 in the NOESY spectrum, which also showed correlation between H-17 methyl group (δ_C 17.57, δ_H 1.160, *d*, 6.80 Hz) and H-12.

The hydrogens from the furan moiety were assigned as follows: in the COSY spectrum the H-16 shows correlation with H-12 due to a long-range coupling ($^4J_{HH}$ 0.72 Hz), C-14 and C-16 show correlations with H-12 in the HMBC spectrum, and H-14 and H-16 show correlations with H-1β, H-11 and H-12 in the NOESY spectrum. The fact that only H-16 shows coupling with H-12 and that the H-14/H-12 nuclear overhauser enhancement (NOE) cross signal is much less intense than the H-16/H-12 suggests a hindered rotation around the C-12/C-13 bond.

II.J. NMR structural study of trans-crotonin

The spectral data of compound CTN was quite similar to those of DCTN. The main difference remains at ring A due to the absence of $\Delta^{3,4}$ unsaturation which in DCTN structure showed to be at δ_{C-3} 126.73, δ_{C-4} 165.70, δ_{H-3} 5.89 and δ_{H-18} 1.97. The presence of the methylene (vicinal to ketone C-2) and methine groups (vicinal to methyl C-18) was confirmed by the new assignments at δ_{C-3} 50.79, δ_{C-4} 39.45, δ_{C-18} 20.20, δ_{H-3a} 2.16, δ_{H-3e} 2.35, δ_{H-4} 1.42 and δ_{H-18} 1.06. The methyl group at C-4 was assigned to be *equatorial* and the H-4 assigned as an *axial* one since the J value between H-4 and H-5a was 11.28 Hz and also due to correlation with H-6a (δ_H 0.94) in addition to the correlation of H-18 (attached to C-4) with H-5a (δ_H 2.05) in the NOESY spectrum.

The diaxial couplings between H-3a and H-5a ($^3J_{HH}$ 12.81 and 11.48 Hz) reinforce the H-5 as an *axial* one and confirm the diaxial $J_{H4a,H5a}$ value. The ^{1}H coupling constants and NOESY correlations from hydrogens of the ring A revealed that this ring is closed to a chair conformation, on the contrary to DCTN in which the C-1–C-5 bonds are on the same plane. The ^{1}H chemical shifts and coupling constants also improve the previous work performed by Itokawa et al. (1989) with addition of these new data, mainly to the hydrogens of the positions C-1, C-3, C-6, C-15 and C-16.

The remaining assignments of DCTN, CTN as well as new NMR structural study of *c*-CJC-B, *t*-CJC-B and isosacacarin were recently reported (Maciel et al., 2003).

II.K. NMR structural study of aromatic acids

The structures of vanillic acid, 4-hydroxy-benzoic acid and *N*-methyltyrosine were identified by spectroscopic analysis (including 300 MHz ^{1}H and ^{13}C NMR and MS experiments) and by chemical transformation of the vanillic acid, with diazomethane giving the two methylated derivative esters 4-hydroxy-3-methoxy-methyl benzoate (derivative I) [δ7.56 (H-2, *d*, $J = 1.8$ Hz), 6.94 (H-5, *d*, $J = 8.2$ Hz), 7.64 (H-6, *dd*, $J = 8.2, 1.8$ Hz), 3.96 (OMe, *s*), 3.90 (CO_2Me)] and 3,4-dimethoxy-methyl benzoate (derivative II) [δ7.55 (H-2, *d*, $J = 2.0$ Hz), 6.89 (H-5, *d*, $J = 8.4$ Hz), 7.68 (H-6, *dd*, $J = 8.4, 2.0$ Hz), 3.93 (C3-OMe, *s*), 3.94 (C4-OMe, *s*), 3.90 (CO_2Me)] utilized for structure confirmation. The measured values in the ^{1}H and ^{13}C NMR ($CDCl_3$/MeOH, 4:1) data of *N*-methyltyrosine showed peaks at δ7.84 (H-2,6, *d*, $J = 8.7$ Hz), 6.77 (H-3,5, *d*, $J = 8.7$ Hz) and δ121.7 (C-1), 132.6 (C-2,6), 115.3 (C-3,5), 161.9 (C-4), 169.7 (CO_2H). ^{1}H and ^{13}C NMR (DMSO/DCl) data of *N*-methyltyrosine were in accordance with the authentic sample of 2-amino-3-(4-hydroxyphenyl) propanoic acid (commonly known as tyrosine) obtained from commercial material. The different observed peaks were assigned to the N-Me group of *N*-methyltyrosine [δ_H 2.5 (Me, *s*) and δ_C 32.2 (Me)].

III. Experimental procedures

III.A. Chemical procedures

III.A.1. General experimental procedures

Melting points were determined with a Kofler (Jasco DIP-370) apparatus and they have not been corrected. Optical activities were measured in $CHCl_3$ on a Perkin-Elmer 341 polarimeter. IR spectra (KBr and $CHCl_3$) were taken on a Perkin-Elmer FT-16PC spectrophotometer and UV spectra (MeOH) on a GBC UV/VIS 911A CG instrument. The $^{13}C\{^1H\}$/DEPT135 spectra were recorded on Varian-Gemini and Bruker-Avance spectrometers (300 MHz for ^1H and 75 MHz for ^{13}C), and the ^1H and all 2D experiments were recorded on a Bruker-Avance spectrometer (600 MHz for ^1H and 150 MHz for ^{13}C). Mass spectra were run on a CG/MS Finnigan-4000 and VG Auto Spec Q at 70 eV. TLC was carried out on 0.25 mm layers of silica gel PF 254 (Merck). Classical column chromatography was performed with silica gel 60 (70–230 mesh).

III.A.2. Chromatographic analysis

Standard solutions at five different concentrations of pure samples of DCTN and AAA–ME in ethyl acetate and the integrated HRGC peak areas from 1.0 μL injections were used to prepare calibration curves for quantification. GC peaks from TF were compared to calibration curves, and the concentration of each metabolite was calculated as percentage. Standards and TF were subjected to HRGC–MS to confirm the identity of GC peaks. The MS scan range was from 40 to 600 a.m.u.

III.A.3. Plant material

The leaves and stem bark of *C. cajucara* were collected in Jacundá, State of Pará (Amazon region, Brazil) and were identified by Nelson A. Rosa. A voucher specimen (no. 247) has been deposited in Herbarium of the Museu Paraense Emílio Goeldi (Belém, Brazil).

Extraction and isolation of compounds. The extraction of the powdered stem bark (6 kg) of native MP was carried out with hexane and then MeOH in a Soxhlet apparatus for 48 h. After solvent evaporation, the hexane extract (471.8 g) was filtered over a silica gel (900 g) column with hexane (2000 mL, fr A), CH_2Cl_2 (8000 mL, fr B) and MeOH (5000 mL, fr C). Fraction B was chromatographed with mixtures of hexane/CH_2Cl_2 (500 mL for each eluted fraction) affording fractions 1–28 [frs 1 and 2 eluted with hexane/CH_2Cl_2 (9:1); frs 3, 4 (8:2); frs 5, 6 (7:3); frs 7–16 (6:4); frs 17–20 (5:5); frs 21–24 (1:4); and frs 25–28 (0:1)] and then with hexane/EtOAc affording fractions 29–40 [frs 29, 30 (9:1); frs 31, 32 (7:3), frs 33–36 (5:5); and frs 37–40 (0:1)]. Fractions 3–6 gave 4.5 g of AAA [white needles; mp 302–303 °C, lit (Addae-Mensah et al., 1992) 299–301 °C; $[\alpha]_D$ +21 ($CHCl_3$, *c* 0.1); crystallization from $CHCl_3$; TLC analysis of AAA were eluted with hexane/EtOAc (8:2); detection with H_2SO_4/MeOH (1:1) reagent, R_f 0.5]. Fractions 7–36 (17,000 mL) showed a solid material which was crystallized from hexane/Me_2CO (1:1) affording 37.2 g of DCTN [colorless crystals; mp 139–140 °C, lit (Kubo et al., 1991) 138.5–140.5 °C; $[\alpha]_D$ +10.6° ($CHCl_3$, *c* 0.6); TLC analysis with hexane/EtOAc (6:4); detection with H_2SO_4/MeOH (1:1) and

Dragenforff reagents, R_f 0.5]. The mother liquor residue from frs 9–24 was filtered over silica gel [hexane/EtOAc (7:3, 1000 mL) (5:5, 2000 mL)] to yield an amorphous solid material and an oil residue. The solid material was crystallized from hexane/Me_2CO (1:1) affording 0.151 g of CTN [colorless crystals; mp 130–132 °C, lit (Itokawa et al., 1990) 131–132 °C; $[\alpha]_D$ + 1.5 ($CHCl_3$, *c* 0.8); TLC analysis with hexane/EtOAc (7:3); detection with H_2SO_4/MeOH (1:1); and Dragenforff reagents, R_f 0.4]. The oil residue was subjected to preparative TLC using hexane/EtOAc (8:2) as solvent (eluted four times; R_f 0.4) to yield 0.308 g of *t*-CJC-B [colorless oil; $[\alpha]_D$ −10.2 ($CHCl_3$, *c* 1.6)] and 0.064 g of *c*-CJC-B [colorless oil; $[\alpha]_D$ −13.1 ($CHCl_3$, *c* 1.7)].

Fraction C was chromatographed with mixtures of hexane/EtOAc (500 mL for each eluted fractions] affording fractions FC_1–FC_5. From fractions FC_1 (8:2) and FC_2 (7:3) 0.101 g of AAA was obtained. Fractions FC_3 (1:1), FC_4 (1:3) and FC_5 (0:1) yielded 22.3 g of DCTN. Fractions C_4 and C_5 remaining material showed a solid material which after crystallization [$CHCl_3$/Me_2CO (9:2)] afforded 0.020 g of CJCR [colorless needles; mp 202–204 °C, lit 202–203 °C (Ichihara et al., 1992); its TLC analysis were performed with CH_2Cl_2/EtOAc/MeOH (7:3:0.2); detection with H_2SO_4/MeOH (1:1) and Dragendorff reagents, R_f 0.5] and 0.006 g of ICJCR [colorless needles; mp 204–205 °C, 205–206 °C, lit (Ichihara et al., 1992); TLC analysis with CH_2Cl_2/EtOAc/MeOH (7:3:0.2), R_f 0.5].

The MeOH extract (202.0 g) was chromatographed with mixtures of hexane/EtOAc affording fr_1 (7:3), fr_2 (6:4), fr_3 (5:5), fr_4 (1:3) and fr_5 (0:1) and then with EtOAc/MeOH [fr_6 (1:1) and fr_7 (0:1); 1000 mL for each eluted fraction]. Fractions fr_1–fr_7 (7000 mL) were worked out using fractions B and C with similar technique to give 0.290 g of AAA, which was obtained from fr_1 and 26.3 g of DCTN (fr_2–fr_7). Fractions fr_1 and fr_2 remaining material afforded 0.051 g of a mixture of the phenolic acids, vanillic acid and 4-hydroxy-benzoic acid. This mixture was subjected to preparative TLC using hexane/EtOAc (6:4) as solvent (eluted five times, detection with $FeCl_3$ reagent, R_f 0.5) to yield 0.037 g of vanillic acid (white needles, mp 209–210 °C) and 0.005 g of 4-hydroxy-benzoic acid (white needles, mp 214–215 °C). Fractions fr_4 and fr_5 remaining material was filtered over a silica gel with hexane/EtOAc (1:1, 1:3 and 0:1) giving a crystalline material, which was crystallized from hexane/Me_2CO (1:1) to afford 0.062 g of CJCR [colorless oil; $[\alpha]_D$ −65.6 ($CHCl_3$, *c* 0.03); TLC analysis with hexane/EtOAc/MeOH (8:2:0.5), R_f 0.4]. Fractions fr_6 and fr_7 remaining material was filtered over a silica gel (MeOH as solvent) to give 0.143 g of *N*-methyltyrosine [white powder; mp 240–241 °C; crystallization from MeOH/H_2O/HCl (8:1.5:0.5); TLC analysis were performed using *n*-BuOH/Me_2CO/AcOH/H_2O (2.0:3.5:3.5:1.0), detection with Ninhydrin reagent, R_f 0.4].

III.A.4. Oxidation of DCTN using singlet oxygen

A solution of DCTN (0.3 g) in dichloromethane (30 mL) containing polystyrene-bound rose bengal catalyst (2 mL) was stirred at −78 °C under an atmosphere of oxygen and irradiated with a 500 W tungsten incandescent lamp. Aliquots of the reaction mixture were periodically analyzed by TLC until the reaction appeared complete (6 h). The reaction mixture was allowed to warm to room temperature and filtered through a pad of cotton, and the solvent was evaporated under a vacuum. The synthetic material after evaporation was submitted to chromatography on a silica gel column eluted with a mixture of hexane/EtOAc (1:1) and then MeOH to obtain the semi-synthetic derivatives CJCR and ICJCR. The methanolic fraction afforded 0.2 g (65%) of colorless needles, mp 204–205 °C corresponding to ICJCR.

III.A.5. Complementary spectroscopic data

Aromatic acids (mixture of vanillic acid and 4-hydroxy-benzoic acid): IR ν_{max} cm^{-1} (KBr): 3412, 2920, 2867, 1669, 1594, 1548, 1429, 1286, 1203, 1110, 1026, 912, 762. UV λ max nm (MeOH): 290, 256 and 215, (MeOH/NaOH): 282 and 217; (MeOH/$AlCl_3$): 223, 268 and 295; (MeOH/$AlCl_3$/HCl): 289, 259 and 217. MS *m/z* (rel. int. %): 168 (100), 153 (60), 138 (30), 121 (44), 97 (28), 93 (12), 65 (18).

^1H NMR, $CDCl_3$, 300 MHz (vanillic acid): 7.47 (H-2, *d*, *J* = 1.9 Hz), 6.82 (H-5, *d*, *J* = 8.2 Hz), 7.54 (H-6, *dd*, *J* = 8.2, 1.9 Hz), 3.83 (OMe, *s*). ^1H NMR, $CDCl_3$/MeOH (4:1), 300 MHz: 7.56 (H-2, *d*, *J* = 1.9 Hz), 6.89 (H-5, *d*, *J* = 8.2 Hz), 7.61 (H-6, *dd*, *J* = 8.2, 1.9 Hz), 3.92 (OMe, *s*). ^{13}C NMR, $CDCl_3$/MeOH (4:1), 75.4 MHz: 122.1 (C-1), 112.8 (C-2), 147.3 (C-3), 151.1 (C-4), 114.8 (C-5), 124.6 (C-6), 169.3 (CO_2H), 56.0 (OMe).

Aromatic acid derivative I: ^1H NMR, $CDCl_3$, 300 MHz: 7.56 (H-2, *d*, *J* = 1.8 Hz), 6.94 (H-5, *d*, *J* = 8.2 Hz), 7.64 (H-6, *dd*, *J* = 8.2, 1.8 Hz), 3.96 (OMe, *s*), 3.90 (CO_2Me). ^{13}C NMR, $CDCl_3$, 75.4 MHz: 123.5 (C-1), 111.2 (C-2), 146.5 (C-3), 150.5 (C-4), 114.2 (C-5), 124.3 (C-6), 166.7 (COOMe), 52.7 (COOMe), 56.2 (OMe).

Aromatic acid derivative II: ^1H NMR, $CDCl_3$, 300 MHz: 7.55 (H-2, *d*, *J* = 2.0 Hz), 6.89 (H-5, *d*, *J* = 8.4 Hz), 7.68 (H-6, *dd*, *J* = 8.4, 2.0 Hz), 3.93 (C3-OMe, *s*), 3.94 (C4-OMe, *s*), 3.90 (CO_2Me). ^{13}C NMR, $CDCl_3$, 75.4 MHz: 123.0 (C-1), 110.2 (C-2), 146.8 (C-3), 152.7 (C-4), 111.9 (C-5), 123.5 (C-6), 166.7 (COOMe), 52.0 (COOMe), 56.0 (C3-OMe), 56.0 (C4-OMe).

4-Hydroxy-benzoic acid: ^1H NMR, $CDCl_3$, 300 MHz: 7.84 (H-2,6, *d*, *J* = 8.7 Hz), 6.77 (H-3,5, *d*, *J* = 8.7 Hz). ^1H NMR, $CDCl_3$/MeOH (4:1), 300 MHz: 7.92 (H-2,6, *d*, *J* = 8.7 Hz), 6.84 (H-3,5, *d*, *J* = 8.7 Hz). ^{13}C NMR, $CDCl_3$/MeOH (4:1), 75.4 MHz: 121.7 (C-1), 132.6 (C-2,6), 115.3 (C-3,5), 161.9 (C-4), 169.7 (CO_2H).

N-methyltyrosine: IR ν_{max} cm^{-1} (KBr): 3410, 3205, 3010, 2958, 2922, 1605, 1588, 1515, 1449, 1387, 1319, 1253, 1104, 832, 645, 539. MS *m/z* (rel. int. %): 195 $[M]^+$ (7), 88 (100), 107 (85), 150 (30). ^1H NMR, DMSO/DCl, 300 MHz: 7.01 (H-2,6, *d*, *J* = 9.6 Hz), 6.67 (H-3,5, *d*, *J* = 9.6 Hz), 3.05 (CH_2, *m*), 4.04 (CH, *t*, *J* = 6.6 Hz), 2.5 (Me, *s*) . ^{13}C NMR, DMSO/DCl, 75.4 MHz: 124.8 (C-1), 131.0 (C-2,6), 115.9 (C-3,5), 156.9 (C-4), 34.3 (CH_2), 61.7 (CH), 32.2 (Me), 169.8 (CO_2H).

Tyrosine (authentic sample from commercial material): colorless crystals, mp 247–248 °C. IR ν_{max} cm^{-1} (KBr): 3203, 3119, 3017, 2958, 2929, 1609, 1589, 1512, 1452, 1363, 1330, 1244, 1214, 1099, 1042, 877, 841, 798, 739, 649, 575, 528. ^1H NMR, DMSO/DCl, 300 MHz: 6.9 (H-2,6, *d*, *J* = 8.4 Hz), 6.67 (H-3,5, *d*, *J* = 8.4 Hz), 2.98 (CH_2, *m*), 3.98 (CH, *t*, *J* = 5.2 Hz). ^{13}C NMR, DMSO/DCl, 75.4 MHz: 125.3 (C-1), 131.3 (C-2,6), 116.2 (C-3,5), 156.9 (C-4), 35.4 (CH_2), 53.9 (CH), 170.7 (CO_2H).

III.A.6. Remaining compounds

The isolation and structure elucidation of the remaining compounds was previously achieved by classical chromatography and then spectroscopic measurements including

2D (HETCOR, HMQC, HMBC, COLOC and COSY 45) NMR experiments (Itokawa et al., 1989, 1990; Kubo et al., 1991; Ichihara et al., 1992; Addae-Mensah et al., 1992; Maciel et al., 1998, 2000, 2002, 2003).

III.B. Pharmacological procedures

III.B.1. Toxicological GPT methodology

The stem bark pieces of *C. cajucara* were ground and the powder added to boiling water, resulting in infusions with 12.5 and 25 g/L concentrations. The tea was given to the animals in the wells exclusively for the whole test period. Swiss albino mice with initial weight between 28 and 35 g were kept under standard environmental conditions and fed with pressed ration normally.

The animals were divided into three groups of ten, one of them being the control group and the other two groups being treated with 12.5 and 25 g/L tea. All the animals had their weight measured once a week, on the same schedule. The GPT dosage was administered in the mice eyes. After a 21-day period of the tea ingestion, the blood from the eyes of the mice was taken out in capillaries and centrifugated (3000 rpm) posteriorly for 5 min, and the plasma (off hemolysis) was submitted to the colorimetric system to dosage the transaminases (Reitman and Frankel, 1957). The observations were performed through the spectrophotometer (505 mm) or green filter, marking the zero with distilled water. The GPT unit values were found using the calibration curve.

III.B.2. Antiproliferative effects methodology

Plant material. The extraction of the powdered bark was done with MeOH in a Soxhlet apparatus for 24 h and an AE of the stem bark was also obtained by boiling in water. The MeOH extract was chromatograped on a silica gel column affording ME-F1 (eluted with hexane); ME-F2 (hexane/CH_2Cl_2, 7:3); F7–22 (hexane/CH_2Cl_2, 1:1); ME-F23-38 (CH_2Cl_2); and ME-F39-54 (CH_2Cl_2/EtOAc, 7:3–1:1). The AE after lyophilization procedure was submitted to chromatography on a silica gel column affording three polar fractions AE-F1 (eluted with EtOAc), AE-F2 (MeOH) and AE-F3 (MeOH/H_2O, 7:3). The residue obtained from AE-F3 was dissolved in water and then submitted to liofilization procedure. The cytotoxicity of the essential oil (ME-F1 and ME-F2), as well as all obtained fractions from both MeOH and AE extracts, and also pure DCTN, CJCR and ICJCR were evaluated against K562 (human leukemia) and Ehrlich carcinoma (mouse breast carcinoma) in three independent experiments. TLC analysis of AE-F2 showed that compounds DCTN and also *t*-CJC-A, CJCR, ICJCR and sacacarin were present in this fraction, in spite of that, AE-F2 has not yet been studied for its biological properties, nor has ME-F23-38 and ME-F39-54.

In vivo *experiments*

Animals and tumors. Female inbred DBA/2 mice weighing 20–25 g from FIOCRUZ animal house (RJ) were used. Groups of five mice were housed in plastic cages under standard laboratory conditions. S-180 and Ehrlich carcinoma ascitic tumors were maintained by weekly i.p. passages in DBA/2 mice. Each one received 5×10^5 cells i.p. harvested from the mouse bearing 7-day-old tumors.

Administration of drugs. The administration of drugs was initiated i.p. after 24 h of the tumor inoculation in two or three doses (in 0.04 mL of DMSO/saline). The control animals received the same vehicle.

Antitumor evaluation. Antitumor activity was determined on the basis of the increase in the survival time of treated group (T) as compared to that of control group (C) using the formula: %T/C = 100 × [mean survival time (days) of treated mice/mean survival time (days) of control mice]. The values of %T/C$\geqslant$125 indicates significant antitumor activity. The experiments were repeated at least once (Geran et al., 1972).

In vitro *experiments.* The assays against K562 (human leukemia) and murine Ehrlich carcinoma (breast carcinoma mice) were performed in three independent experiments. The cells were grown on RPMI-1640 medium supplemented with 5% fetal bovine serum, 2 mM glutamine, 100 μg/mL streptomycin and 100 U/mL penicillin, and were incubated at 37 °C under 5% CO_2 atmosphere. In these experiments, cells were plated in 96 well plates (2×10^4 cell/mL for K562 and 5×10^5 cell/mL for Ehrlich) with the fractions and were incubated for 96 h (for K562) and 48 h (for Ehrlich). Tumor cell growths were quantified by the ability of living cells to reduce the yellow dye MTT to a purple formazan product (Mosmann, 1983). Upon incubation, the MTT was added and after 3 h the formazan was dissolved in 100 μL-acidified isopropanol. The absorbance was measured using an Elisa microplate reader and the cytotoxic effect was quantified as the percentage of control absorbance to reduce dye at 550 nm.

III.B.3. Genotoxic and antigenotoxic activity

Drugs. Different amounts of DCTN were dissolved in a mixture of DMSO (0.1 mL) and distilled water (0.2 mL) and were administered intraperitoneally by gavage to the mice in doses equivalent to 25% (138.75 mg/kg b.w.), 50% (277.5 mg/kg b.w.) and 75% (416.25 mg/kg b.w) of the LD_{50}, as previously determined by Carvalho et al. on mice (555 mg/kg b.w) treated p.o. In the i.p treatments, the same LD_{50} calculated for mice p.o was used due to restricted amounts of DCTN. Cyclophosphamide (Sigma, CAS: 6055-19-2) was utilized as a positive control because of its strong clastogenic action. Each animal received an intraperitoneal injection of 10 mg/kg b.w. of this drug diluted in distilled water. The negative control group received an intraperitoneal injection or gavage of a solution of DMSO–H_2O (1:2), the vehicle used for the DCTN dilution.

Protocols. The genotoxic and antigenotoxic potentials of DCTN were assessed by the MN and CAs tests on mice bone marrow cells under acute treatment, in order to evaluate the genetic risks derived from the consumption of *C. cajucara*.

After each animal had been weighed, they were divided into groups of ten, with each group comprising five males and five females per treatment; subsequently each animal received by gavage or intraperitoneal injection, 0.1 mL of solution per each 10 g b.w. of solvent, cyclophosphamide or sub-lethal doses of DCTN.

To assay the genotoxicty of DCTN, the MN test consisted of treatment with DCTN (i.p. or gavage) at 0 h and sacrifice at 24 h. The CA test consisted of treatment (i.p. or gavage) at 0 h, colchicine at 22 h and sacrifice at 24 h.

Micronucleus assay. The used methodology was previously described by Schmid (1975), with modifications. The animals were killed by cervical dislocation, one femur was dissected, and the bone marrow was flushed out with 2 mL fetal bovine serum (Cultilab, Brazil). The cell suspension was centrifuged and the cell pellet was resuspended in fetal bovine serum and smeared on a clean glass slide. The preparations were dried, then fixed with methanol and stained with 5% Giemsa diluted in phosphate buffer (0.06 M Na_2HPO_4 and 0.06 M KH_2PO_4, pH 6.8).

Chromosomal aberration assay. The methodology employed was described by Ford and Hamerton (1956), with modifications. Two hours before sacrifice, the animals received a single intraperitoneal injection of 1% colchicine (Sigma, CAS: 64-86-8), 0.3 mL for each 30 g of b.w. After the animals were sacrificed by cervical dislocation, one femur was dissected and the bone marrow cells were flushed out with 2 mL of NaCl (0.9%) solution and then centrifuged. The bone marrow suspension was washed with 0.075 M KCl cold hypotonic solution and then fixed with 1:3 acetic acid:methanol. Chromosome preparations were air dried and stained with 5% Giemsa diluted in phosphate buffer (0.06 M Na_2HPO_4 and 0.06 M KH_2PO_4, pH 6.8).

Cytological analysis. Two thousand PCEs were analyzed per animal for the MN assay. Two hundred metaphase cells per animal were used for the CA assay, and for the mitotic index (MI) calculation 1000 cells were analyzed using an Olympus biocular microscope.

The criteria utilized for the CA and MN analysis were previously described by Preston et al. (1987), while those for MN analysis were described by Huber et al. (1983) and Titenko-Holland et al. (1997).

Statistical analysis. For the statistical analysis of the MN frequencies, CA, cells with CA and mitotic index, the data were transformed relative to the cytogenetic parameter analyzed through the formula: $y = \sqrt{(x + 1/2)}$. The statistical tests utilized were Anova and Tukey at the 95% level of confidence ("System Analysis Statistical").

III.B.4. Antiulcer activity

Animals. F1 mice (25–30 g), a hybrid from a cross between inbred C57Bl/6 female and Balb/C male, Wistar rats weighing 200–300 g and male albino rabbits weighing 1.5–2.5 kg of either sex were used in this study. All animals were kept under a controlled light/dark cycle and temperature (22 ± 2 °C), with free access to food and water. When necessary, animals were deprived from food 15–24 h before the experiments.

Antiulcer evaluation. Rats were treated with DCTN, CTN and AAA (100 mg/kg, p.o.). After 1 h, gastric lesions were induced by restraint-in-cold as detailed elsewhere (Fischman et al., 1991). The animals were killed under ether anesthesia after 2 h restraint at 4 °C. The stomach was dissected out, the mucosal side was gently washed to remove remaining food and inspected under magnification to determine the number of ulcers and to score the index of GMD. This was independently done by two different colleagues upon the same preparation, in that considering the color, edema and hemorrhage of the gastric folds, the number of pethechyae, the number and size of ulcers (Gamberini et al., 1991).

Determination of gastric acid secretion. A pylorus ligature was carefully done in rats under ether anesthesia and the DCTN (10–100 mg/kg) was injected into the duodenal lumen (i.d.). After 4 h the animals were killed by deep ether anesthesia and the gastric secretion collected with a pipette. The final volume and pH were directly determined after washing the mucosal side of the stomach with 2 mL of distilled water. Total acidity of the gastric juice was titrated with 0.1 N NaOH using 2% phenolphthalein as indicator (Berglindh et al., 1976).

Oxyntic gland preparation. The glands were prepared from New Zealand rabbit gastric mucosa (1–2 kg) by collagenase digestion (Berglindh et al., 1976). The animals were not starved. They were sacrificed by cervical dislocation and abdomen were opened, both thoracic and abdominal aorta were clamped. A warm (37 °C) oxygenated PBS was then pumped into the aorta and perfusion was done. When the stomach appeared totally exsanguined after perfusion of some 80 mL of PBS, it was rapidly removed, cut along the lesser curvature, emptied and rinsed several times with PBS. The cardial and antral regions were discarded. By blund dissection the mucosa could easily be separated from the muscular and sub-muscular layers. The corpus mucosa was stripped off and minced into small pieces. These pieces were washed twice in oxygenated PBS and then incubated in 50 mL of collagenase enzyme solution (0.2 mg/mL), continuously gassed with 95% O_2 to 5% CO_2 for 1 h at 37 °C. The resulting suspension was filtered through nylon mesh and then washed with PBS for three times before adjusting the volume to give a glandular concentration of approximately 15 mg dry weight/mL (Berglindh et al., 1976).

Gland viability. Gland viability was made/evaluated at room temperature with the dye exclusion technique using trypan blue (4 mg/mL) dissolved in PBS. Equal volumes of dye solution and gland suspension were mixed and immediately observed under the microscope. Normally the cells were able to exclude the dyes and only two to three cells in a gland showed a loss of viability and became stained.

^{14}C-Aminopyrine accumulation/uptake. Acid formation was estimated from ^{14}C-AP uptake in the glandular pellet. The terpenoids DCTN, CTN and AAA (10^{-6} to 10^{-3} M) were pre-incubated for 15 min at 37 °C in 1 mL of gland suspension (15 mg dry weigth/mL) in polypropylene tubes. At this time 10 nCi/mL of ^{14}C-AP solution was added to the corresponding secretagogues histamine or bethanechol (10 μM) and the incubation continued for 15 min more. The reaction was stopped by addition of 1.5 mL of cold PBS and 500 μL of nitric acid 95%. Samples of 100 μL from remnant solution were passed to vials with 3.5 mL of scintillation liquid and counted in scintillator B. The experiments were performed in triplicates and the results from 5 to 6 different preparations were expressed as percentage of ^{14}C-AP uptake relative to the vehicle group (Berglindh et al., 1976).

Gastrointestinal transit. To test the possible effects of DCTN, CTN and AAA in the gastrointestinal motility, the mice were fasted for 24 h after the gastrointestinal transit was analyzed, as described previously (Arbos et al., 1993). The samples were washed three times with 2.5 mL of ice PBS and the pellet digested by boiling with DCTN and AAA for 1 h after being given a standard charcoal meal (10%, 100 μL/10 g, p.o.). Control animals received the same volume of DMSO 0.5% (100 μL/10 g, p.o.) or positive control atropine (5 mg/kg, p.o.) 60 min before being tested. The mice were sacrificed 30 min after administration of the charcoal meal and the distance the charcoal meal had traveled was measured. Data were expressed as the percentage of the gut the charcoal meal traveled compared to the control group (100%).

Drugs. Bethanechol chloride, histamine dihydrochloride (Sigma), colagenase enzyme Type I (Sigma), dimethylamino ^{14}C-aminopyrine with specific activity of 109 mCi/nmol (Amersham Sciences), activated charcoal (Merck), ranitidine (Antak-Glaxo), cimetidine (Cim), atropine. All other chemicals were of analytical grade.

Statistics. Data were expressed as means $\pm$ Sem. Statistical significance of the results was determined using a one-way analysis of variance followed by Tukey Kramer test. Data were considered different at a significance level of $p < 0.05$.

Conclusions

The improvement of the *C. cajucara* phytopharmacological studies shows that the therapeutic potential of this plant is undeniable. New phytochemical investigation yields two aromatic acids (vanillic acid and 4-hydroxy-benzoic acid) and an amino acid (*N*-methyltyrosine). Since the aromatic acids proved to be antioxidant agents, *C. cajucara* could be cited now as an antioxidant plant. The amount of DCTN and AAA obtained by classical chromatography fractionation was in agreement with those found by GC–MS investigation. The DCTN highest concentration was 1.4% of dry stem bark of MP. It was proved that its concentration depends on the plant age. DCTN was found to be absent in the stem bark of YP, in that AAA being preponderant. DCTN synthetic transformation afforded in great amount semi-synthetic ingredients, e.g., CTN, CJCR and ICJC which can be isolated as minor natural constituents of *C. cajucara*. The tested bioactive terpenoids AAA, *t*-CJC-B, *c*-CJC-B, CTN, DCTN, CJCR and ICJCR had their purity confirmed by NMR analysis. Taking as an example, 1D and 2D-NMR experiments with DCTN indicated only DCTN peaks.

The clerodanes DCTN, CTN, *t*-CJC-B, CJCR and ICJCR and also bioactive extracts (including a tea preparation, AE) and semi-purified fractions showed cytotoxic effect against human K562 leukemia and/or S-180 and ascitic Ehrlich carcinoma cells. The inhibitory effects in cell growth were dose-dependent. In that, DCTN treatment of mice has the same antitumor effect as 5-FU a recognized anticancer drug on S-180 and Ehrlich tumors.

Using the genotoxic approach to test DCTN antimutagenic effect, it was found that DCTN doses of 50% and 75% of the LD_{50} via intraperitoneal or gavage treatments were antimutagenic with regard to cyclophosphamide. However dose of 25% of the LD_{50} was only antimutagenic when administered by gavage.

The results of the hepatotoxic effects of the AE proved that *C. cajucara* was not cytotoxic, in that the treated mices gained weight regularly during entire period of treatment.

DCTN and AAA reduced gastrointestinal transit effects by 30% and 40%. CTN lacks efficacy. These compounds reduced the index of gastric mucosa damage (IMD) induced by cold stress comparatively to the control vehicle (0.5% DMSO). The clerodanes CTN and DCTN reduced the IMD by 40% and 54%, respectively. AAA was ineffective. The effect on gastric secretion was observed by highest dose of DCTN. The terpenoids DCTN, CTN and AAA reduced significantly the ^{14}C-AP uptake induced by histamine. Uniquely DCTN decreased uptake induced by bethanechol. These compounds did not change the basal acid secretion.

The cytotoxicity of this plant was not confirmed by the experiments performed in this work. Further investigations on its cytotoxicity should be done to improve the preclinical experiments.

Acknowledgments

The authors wish to thank financial support from CNPq and CAPES-PRODOC program and CNRNM for the facilities in the use of the 600 MHz NMR instrument.

References

Addae-Mensah I, Achenbach H, Thoithi GN, Waibel R, Mwangi JW. (1992) Epoxychiromodine and other constituents of *Croton megalocarpus*. Phytochemistry 31:2055–8.

Agner AR, Maciel MAM, Pinto AC, Cólus ISM. (2001) Antigenotoxicity of *trans*-dehydrocrotonin, a clerodane diterpene from *Croton cajucara*. Planta Med 67:815–9.

Agner AR, Maciel MAM, Pinto AC, Pamplona GRS, Cólus IMS. (1999) Investigation of genotoxic activity of *trans*-dehydrocrotonin, a clerodane diterpene from *Croton cajucara*. Teratogen Carcin Mut 19:377–84.

Anazetti MC, Melo PS, Duran N, Haun M. (2003) Comparative cytotoxicity of dimethylamide-crotonin in the promyelocytic leukemia cell line (HL60) and human peripheral blood mononuclear cells. Toxicology 188(2–3):261–74.

Arbos ML, Zegri A, Lopez-Soriano FRJ, Argiles JM. (1993) A simple method for determining the rate of gastrointestinal transit in the rat. Arch Int Physiol Biochim Biophys 101:113–5.

Bannerman RH. (1982) Traditional medicine in modern health care. World Health Forum 3(1):8–26.

Barreto Jr AG, Biscaia Jr EC, Veiga Jr VF, Pinto AC, Carvalhaes SF, Maciel MAM. (2005) Ion exchange chromatography applied to the fractionation of the Copaíba Oil (*Copaifera multijuga*) and Sacaca (*Croton cajucara*) extracts. Quim Nova 28(4), 719–22.

Berglindh T, Helander HF, Obrink KJ. (1976) Effects of secretagogues on oxygen consumption, aminopyrine accumulation and morphology in isolated gastric glands. Acta Physiol Scand 97:401–14.

Bighetti EJ, Hiruma-Lima CA, Gracioso JS, Brito AR. (1999) Antiinflammatory and antinociceptive effects in rodents of the essential oil of *Croton cajucara* Benth. J Pharm Pharmacol 51:1447–53.

Bilia AR, Mendez J, Morelli I. (1996) Phytochemical investigations of the *Licania* genus, Flavonoids and triterpenoids from *Licinia carii*. Pharm Acta Helv 71:191–7.

Bonina F, Puglia C, Ventura D. (2002) In vitro antioxidant and in vivo photoprotective effects of a lyophilized extract of *Capparis spinosa* L. buds. J Cosmet Sci 53:321–35.

Brito AR, Rodriguez JA, Hiruma-Lima CA, Haun M, Nunes DS. (1998) Antiulcerogenic activity of *trans*-dehydrocrotonin from *Croton cajucara*. Planta Med 64:126–9.

Campos AR, Albuquerque FAA, Rao VSN, Maciel MAM, Pinto AC. (2002) Investigations on the antinociceptive activity of crude extracts from *Croton cajucara* leaves in mice. Fitoterapia 73:116–20.

Carvalho JCT, Silva MFC, Maciel MAM, Pinto AC, Nunes DS, Lima RML, Bastos JK, Sarti SJ. (1996) Investigation of anti-inflammatory and antinociceptive activities of *trans*-dehydrocrotonin, a 19-*nor*-clerodane diterpene from *Croton cajucara*. Part 1. Planta Med 62:402–4.

Di Stasi LC, Hiruma CA, Guimarães EM, Santos CM. (1994) Medicinal plants popularly used in Brazilian Amazon. Fitoterapia LXV(6):529–40.

Di Stasi LC, Santos EMG, Santos CM, Hiruma CA. (1989) Plantas Medicinais da Amazônia. Editora UNESP, Brazil: São Paulo, pp. 127–8.

Farias RAF, Neto MFO, Viana GSB, Rao VSN. (1996) Effects of *Croton cajucara* extract on serum lipids of rats fed a high fat diet. Phytother Res 10:697–9.

Farias RAF, Rao VSN, Viana GSB, Silveira ER, Maciel MAM, Pinto AC. (1997) Hypoglycemic effect of *trans*-dehydrocrotonin, a *nor*-clerodane diterpene from *Croton cajucara*. Planta Med 66:558–60.

Fischman LA, Skorupa LA, Souccar C, Lapa AJ. (1991) The water extract of *Coleus barbatus* Benth. decreases gastric secretion in rats. Mem Inst Oswaldo Cruz 86(Suppl II):141–3.

Ford CE, Hamerton JL. (1956) A colchicine hypotonic citrate squash sequence for mammalian chromosome. Stain Technol 3:247–51.

Gamberini MT, Skorupa LA, Souccar C, Lapa AJ. (1991) Inhibition of gastric secretion by a water extract from *Braccharis triptera*. Mart Mem Inst Oswaldo Cruz 86(Suppl II):137–9.

Geran RI, Greenberg NF, MacDonald MM, Schumacher AM, Abott BJ. (1972) Protocols for screening chemical agents and natural products against animals and other biological systems. Cancer Chemother Rep 3:1–103.

Grynberg NF, Echevarria A, Lima JE, Pamplona SGSR, Pinto AC, Maciel MAM. (1999) Anti-tumor activity of two 19-nor-clerodane diterpenes: *trans*-dehydrocrotonin and *trans*-crotonin from *Croton cajucara*. Planta Med 65:687–9.

Hiruma-Lima CA, Gracioso JS, Nunes DS, Brito AR. (1998) Effects of an essential oil from the bark of *Croton cajucara* Benth. on experimental gastric ulcer models in rats and mice. J Pharm Pharmacol 51:341–6.

Hiruma-Lima CA, Gracioso JS, Rodríguez JA, Nunes DS, Brito AR. (2000) Gastroprotective effects of essential oil from the bark of *Croton cajucara* Benth. (Euphorbiaceae). J Ethnopharmacol 69:229–34.

Hiruma-Lima CA, Spadari-Bratfisch RC, Grassi-Kassisse DM, Brito AR. (1999) Antiulcerogenic mechanisms of dehydrocrotonin, a diterpene lactone obtained from *Croton cajucara*. Planta Med 65:325–30.

Hiruma-Lima CA, Toma W, Gracioso JS, Almeira AB, Batista LM, Magri L, de Paula AC, Soares FR, Nunes DS, Brito ARS. (2002) Natural *trans*-crotonin: the antiulcerogenic effect of another diterpene isolated from the bark of *Croton cajucara* Benth. Biol Pharm Bull 25:452–6.

Huber R, Streng S, Bauchinger M. (1983) The suitability of the human lymphocyte micronucleus assay system for biological dosimetry. Mutat Res 111:185–93.

Hung C, Yen G. (2002) Antioxidant activity of phenolic compounds isolated from *Mesona procumbens* Hemsl. J Agric Food Chem 50:2993–7.

Ichihara Y, Takeya K, Hitotsuyanagi Y, Morita H, Okuyama S, Suganuma M, Fujiki H, Motidome M, Itokawa H. (1992) Cajucarinolide and isocajucarinolide: anti-inflammatory diterpenes from *Croton cajucara*. Planta Med 58:549–51.

Itokawa H, Ichihara Y, Kogima H, Watanabe K, Takeya K. (1989) *Nor*-clerodane diterpenes from *Croton cajucara*. Phytochemistry. 28:1667–9.

Itokawa H, Ichihara Y, Shimizu M, Takeya K, Motidome M. (1990) Cajucarins A and B, new clerodane diterpenes from *Croton cajucara*, and their conformations. Chem Pharm Bull 38:701–5.

Jonson EL, Schmidt WF, Emche SD. (2003) Kaempferol (rhamnosyl) glucoside, a new flavonol from *Erythroxylum coca* var. *ipadu*. Biochem Syst Ecol 31:59–67.

Kernan RM, Faulkner DJ. (1988) Regioselective oxidation of 3-alkylfurans to 3-alkyl-4-hydroxybutenolides. J Org Chem 53:2773–6.

Kubo I, Asaka Y, Shibata K. (1991) Insect growth inhibitory *nor*-diterpenes: *cis*-dehydrocrotonin and *trans*-dehydrocrotonin from *Croton cajucara*. Phytochemistry 30:2545–6.

Maciel MAM, Pinto ACP, Arruda AC, Pamplona SGSR, Vanderlinde FA, Lapa AJ, Cólus IMS, Echevarria A, Grynberg NF, Farias RAF, Luna Costa AM, Rao VSN. (2000) Ethnopharmacology, phytochemistry and pharmacology: a successful combination in the study of Croton cajucara. J Ethnopharmacol 70:41–55.
Maciel MAM, Pinto AC, Brabo SN, Silva MN. (1998) Terpenoids from Croton cajucara. Phytochemistry 49:823–8.
Maciel MAM, Pinto AC, Kaiser CR. (2003) NMR and structure review of some natural furoclerodanes. Magn Reson Chem 41:278–82.
Maciel MAM, Pinto AC, Veiga Jr. VF, Grynberg NF, Echevarria A. (2002a) Medicinal plants: the need for multidisciplinary scientific studies. Quim Nova 25:429–38.
Maciel MAM, Pinto AC, Veiga Jr. VF, Martins JR, Grynberg NF, Echevarria A, Lapa AJ, Vanderlinde FA. (2002b) *Croton cajucara* as an alternative to traditional medicine in a modern health system. Phytochem Pharmacol II Ser. Recent Prog Med Plants 8:502–17.
Marfak A, Trouillas P, Allais DP, Champavier Y, Calliste CA, Duroux JL. (2003) Radiolysis of kaempferol in water/methanol mixtures. Evaluation of antioxidant activity of kaempferol and products formed. J Agr Food Chem 51:1270–7.
Martins JEC. (1989) Plantas medicinais de uso na amazônia. Belém, Brazil: Centros de Estudos Jurídicos do Pará, pp. 92.
Mors WB. (1991) Plants active against snake bite. In: Wagner H, Farnsworth NR, editors. Economic and medicinal plant research. New York: Academic Press, Vol. 5, pp. 353–73.
Mosmann TJ. (1983) Rapid colorimetric assay for cellular growth and survival: application to proliferation and cytotoxicity assays. J Immunol Methods 65:55–63.
Murota K, Terao J. (2003) Antioxidative flavonoid quercetin: implication of its intestinal absorption and metabolism. Arch Biochem Biophys 417:12–7.
Ohsugi M, Fan W, Hase K, Xiong Q, Tezuka Y, Komatsu K, Namba T, Saitoh T, Tazawa K, Kadota S. (1999) Active-oxygen scavenging activity of traditional nourishing-tonic herbal medicines and active constituents of *Rhodiola sacra*. J Ethnopharmacol 67:111–9.
Peres MT, Monache FD, Cruz AB, Pizzolatti MG, Yunes RA. (1997) Chemical composition and antimicrobial activity of *Croton urucurana* Baillon. J Ethnopharmacol 56:223–6.
Prance GT. (1992) Out of the Amazon. London: HMSO Publications, pp. 87.
Preston RJ, Dean BJ, Galloway S, Holden H, Mc Fee AF, Shelby M. (1987) Analysis of chromosome aberration in bone marrow cells. Mutat Res 189:157–65.
Reitman S, Frankel S. (1957) A colorimetric method for the determination of serum glutamic oxalacetic and glutamic pyruvic transaminase. Am J Clin Path 28:56.
Rodríguez JA, Haun M. (1999) Cytotoxicity of *trans*-dehydrocrotoning from *Croton cajucara* on V79 cells and rat hepatocytes. Planta Med 65:522–6.
Rosa MDS, Mendonca RR, Bizzo HR, Rodrigues ID, Soares RMA, Souto-Padron T, Alviano CS, Lopes AHCS. (2003) Antileishmanial activity of a linalool-rich essential oil from *Croton cajucara*. Antimicrob Agents Chemother 47(6):1895–901.
Schmid W. (1975) The micronucleus test. Mutat Res 31:9–15.
Shibamoto T, Nishimura O. (1982) Isolation and identification of phenols in oil of vetiver. Phytochemistry 21:793.
Silva RM, Oliveira FA, Cunha KMA, Maia JL, Maciel MAM, Pinto AC, Nascimento NRF, Santos FA, Rao VSN. (2005) Cardiovascular effects of trans-dehydrocrotonin, a diterpene from *Croton cajucara* in rats. Vasc Pharmacol 43:11–8.
Silva RM, Santos FA, Maciel MAM, Rao SN, Pinto AC. (2001a) The lipid-lowering effect of *trans*-dehydrocrotonin from *Croton cajucara* Benth. In mice fed on high-fat diet. J Pharm Pharmacol 53:535–9.
Silva RM, Santos FA, Maciel MAM, Pinto AC, Rao VS. (2001b) Effect of *trans*-dehydrocrotonin, a 19-*nor*-clerodane diterpene from *Croton cajucara* on experimental hypertriglyceridaemia and hypercholesterolaemis induced by Triton WR 1339 (tyloxapol) in mice. Planta Med 67:763–5.
Simmonds MSJ. (2003) Flavonoid-insect interactions: recent advances in our knowledge. Phytochemistry 64:21–30.
Simões JC, Silva AJR, Serruya H, Bentes MHS. (1979) Desidrocrotonina, um norditerpeno de *Croton cajucara* Benth. (Euphorbiaceae). Ciência e Cultura 31:1140–1.

Souza MAA, Maciel MAM, Souza SR. (2000) Determinação das propriedades fungicidas do óleo essencial de sacaca (*Croton cajucara*) em ensaio in vitro com fungos de solo, XXIV Reunião Brasileira de Fertilidade do Solo e Nutrição de plantas, Fertbio, Brasil, 5–927.

Stickney JC, Northrup DW. (1959) Effect of gastric emptying upon propulsive motility of small intestine in rat. Proc Soc Exp Biol Med 101:582.

Titenko-Holland N, Windham G, Kolachana P, Reinisch F, Parvatham S, Osorio AM, Smith MT. (1997) Genotoxicity of malathion in human lymphocytes assessed using the micronucleus assay in vitro and in vivo: a study of malathion-exposed workers. Mutat Res 338:85–95.

Van den Berg ME. (1982) Plantas Medicinais na Amazônia – Contribuição ao seu Conhecimento Sistemático. Belém, Brazil: Gráfica Editora Falangola, pp. 58–59.

Van den Berg ME. (1993) Plantas Medicinais na Amazônia – Contribuição ao seu Conhecimento Sistemático, 2nd edition, Belém, Brazil: Museu Paraense Emílio Goeldi, CNPq-MCT (Coleção Adolpho Ducke), pp. 124, 125, 168.

Vanderlinde FA, Maciel MAM, Pinto AC, Torres LMB, Souccar C, Castro MSA, Lapa AJ. (2000) Peripheral opioid component involved in the antinociceptive activity of fraction F3 obtained from barks aqueous extracts of *Croton cajucara* Benth. (sacaca), XVI Latin-american Congress of Pharmacology, Águas de Lindóia, SP, Brazil, Setor 10, abstract 10.079, p. 198.

Vanderlinde FA, Pinto AC, Maciel MAM, Kaiser CR, Castro MSA, Torres LB, Santos AL, Souccar C, Lapa AJ. (1998) Atividade Antiedematogênica e Antinociceptiva do Extrato Aquoso e Fração das cascas do *Croton cajucara* Benth. (Euphorbiaceae), XV Simpósio de Plantas Medicinais do Brasil, Águas de Lindóia, SP, Brazil, 01.098, p. 67.

Veiga VF, Pinto AC, Maciel MAM. (2005) Medicinal plants: safe cure? Quim Nova 28:519–28.

Wattenberg LJ. (1995) Chalcones, myo-inositol and other novel inhibitors of pulmonary carcinogenesis. J Cell Biochem 22:162–8.

M.T.H. Khan and A. Ather (eds.)
Lead Molecules from Natural Products

Antihypertensive peptides from natural resources

Toshiro Matsui, Kiyoshi Matsumoto

Abstract

Blood pressure is regulated by many physiological systems in humans. In the vasopressor system, for example the renin-angiotensin system is thought to play an important role, where angiotensin II, which is produced from the cleavage of angiotensin I by angiotensin I-converting enzyme (ACE) shows a potent vasocontractive activity. In this review, preparation of bioactive peptides from natural resources as well as their underlying antihypertensive mechanism will be discussed.

Keywords: *peptide, angiotensin I-converting enzyme, hypertension, functional food*

Abbreviations: ACE, angiotensin I-converting enzyme; RAS, renin–angiotensin-aldosterone system; SHR, spontaneously hypertensive rat; THM, Tsukuba-hypertensive mouse.

I. Background

The importance of biologically active substances contained in foods has so far received much attention. There is an important link between diet and disease prevention, thus the effects of food on diseases such as hypertension, diabetes, obesity, osteoporosis and cardiovascular is a growing field of study. In addition, the increase in aging population in many developed countries and the growing pressure on public health spending has led to a greater emphasis on prevention of disease and on more individual responsibility for health care provision. In Japan, the Ministry of Health, Labor and Welfare proposed a new legislation in 1991, concerning food functionality to design an appropriate prophylaxis of lifestyle-related diseases. Presently, many functional food products with a health claim (FOSHU: foods for specified health uses) are available in the local market. An example of FOSHU product with successful body-modulating effects is the antihypertensive food for borderline hypertensives, which were evidentially developed on the basis of extensive intervention trials. It is becoming increasingly evident at present that some peptides in the foods are responsible for the appearance of antihypertensive effect although their physiological mechanisms have not been fully clarified.

Peptides, which are essentially composed of L-amino acid residues, are critical to providing physicochemical and functional properties as well as nutritional properties. For nutritional aspects, small peptides such as di- and tri-peptides are recognized as a suitable nitrogen resource (Silk et al., 1980) because of their high absorption rather than their free amino acid form (Adibi and Morse, 1971). In addition to their nutritional prevalence, some latent physiological functions of peptides have been reported: promotion of Ca absorption (Tsuchihata et al., 2001), stimulation of phagocytosis (Kawamura, 1990), antioxidant action against unsaturated fatty acids (Amarowicz and Shahidi, 1997) and lowering effect of blood pressure (Cheung et al., 1980). Thus, bioactive effects induced by specific food peptides are becoming of much interest in food industries. Hence, in this chapter, we focus on the latent disease-prevention property of peptides against hypertension disease.

II. Renin–angiotensin–aldosterone system

Hypertension is known to be closely correlated with onsets of cardiovascular and apoplexy diseases, of which over 90% is classified as essential hypertension, and lifestyle or food style is assumed to be involved in the pathogenesis. Promotion of systemic blood pressure (BP), i.e., hypertension, is a result of disruption of Na^+-K^+ balance or increase in fluid volume as well as of increase in vascular resistance. Arterial BP regulation is mainly achieved by diverse metabolic systems; pressor and depressor hormonal systems, and nerve systems (Sealey and Laragh, 1990). Among these, the renin–angiotensin–aldosterone system (RAS) is one of the predominant pressor systems, being widely distributed in not only circulatory blood system but also in diverse organs such as brain, lung, abdominal aorta and kidney. In particular, circulatory RAS is well demonstrated as a main BP-promoting system (Figure 1). In this system, angiotensinogen from liver is primarily cleaved by renal renin to produce angiotensin (Ang) I, which is a decapeptide, and shows no pressor effect. Ang I is then converted by the action of angiotensin I-converting enzyme (ACE; EC 3.4.15.1, dicarboxypeptidase) to potent vasoconstrictor Ang II by cleaving the dipeptide His-Leu from the C-terminal of Ang I. The major physiological role of Ang II is to exert a vasoconstrictive effect at the vessel wall via the binding of Ang II to Ang II receptor (AT1) (Johnston and Burrell, 1995). Ang II also participates in the increment of extracellular fluid volume through the stimulation of adrenal aldosterone release, by which renal tubular sodium reabsorption is enhanced. Thus, it has been recognized that circulatory RAS plays an important role in the development and maintenance of hypertension, although other circulatory complaints such as diabetes and hyperlipemia followed by arteriosclerosis or dysfunctions of kidney or heart cooperatively elicit the hypertension.

III. Target for RAS regulation

There is also a close relationship between RAS and depressor hormonal (kinin–kallikrein) system, in which ACE is concerned in both the hydrolysis of

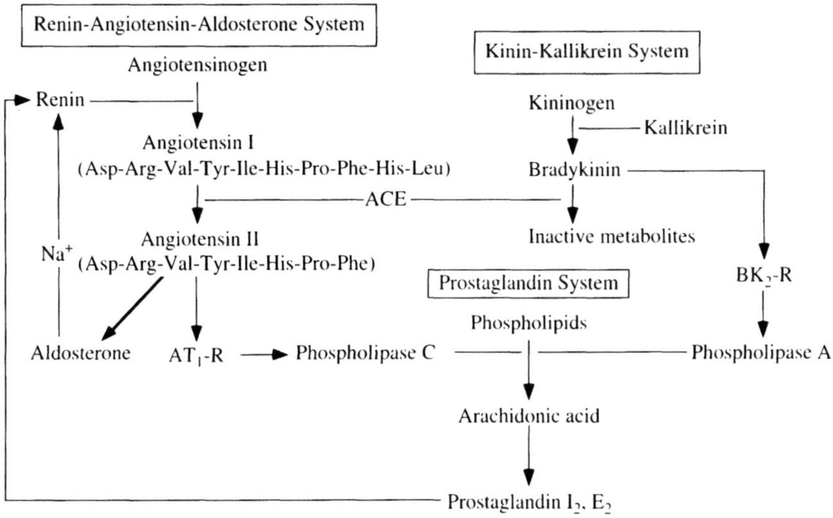

Fig. 1. Circulatory rennin-angiotensin system and its related blood pressure regulatory systems.

bradykinin with potent vasodilating action and the production of Ang II (Figure 1). Therefore, in order to prevent the pathogenesis of hypertension or to treat an elevated BP, suppression of Ang II production via the retardation of ACE action may be greatly effective (Johnston and Burrell, 1995). From this point of view, a pharmacologic ACE blocker, captopril, had been established in 1977, following the development of other blockers including AT1 antagonists and renin inhibitors (Natoff, 1987). As illustrated in Figure 2, captopril (IC_{50}; 35 nM) is a peptidic drug, and is designed on the basis of a basal structure of Ala-Pro. Similarly, enalapril (IC_{50}; 5.2 nM) has been successfully synthesized on the structural basis of Phe-Ala-Pro. Then, why are these ACE inhibitory drugs derived from any peptide backbones? As described above, ACE is a dicarboxypeptidase with two zinc metals, preferentially cleaved by Ang I decapeptide to Ang II octapeptide. Considering that it has four functional amino acid residues of Tyr, Arg, Glu, and Lys at the active site, and three hydrophobic binding subsites (Ehler and Riordan, 1990), a favorable blockade of ACE action would thus be achieved by hydrophobic peptidic inhibitors that show high affinity with active sites. The article by Cheung (Cheung et al., 1980) has supported this prevalence that small dipeptides with hydrophobic and aromatic amino acid residues at the C-terminal have a potential to inhibit ACE activity, although their inhibitory activities were less than 1/500-fold lower than those of therapeutic drugs.

Some controversies exist concerning the depressor mechanisms by which peptides regulate the systemic BP. In the sections below, we focus mainly on the in vitro ACE inhibitory effects of bioactive peptides derived from natural resources by physicochemical or enzymatic treatments. We will also discuss their physiological effects in rats and humans as a basis for understanding their roles in preventing hypertension.

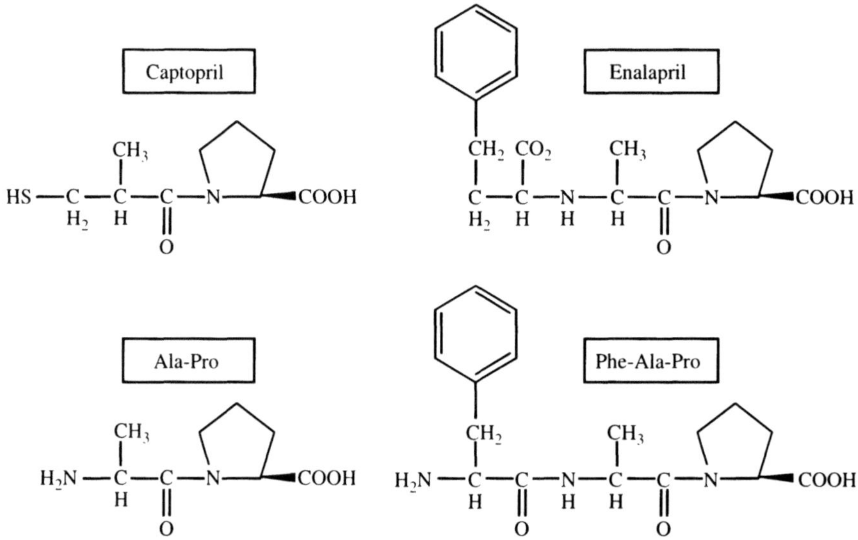

Fig. 2. Structures of ACE inhibitory drugs and their related peptides.

IV. Preparation and identification of ACE inhibitory peptides from natural proteins

The study of the preparation and identification of natural ACE inhibitors derived from foodstuffs has become an active area of research to prevent the pathogenesis of hypertension. IC_{50} value is usually used as a measure of ACE inhibitory activity of inhibitors, indicating the concentration of inhibitor required for inhibits 50% of ACE activity. The assay is carried out by using a synthetic substrate of Hip-His-Leu analogous to Ang I. Since Maruyama and Suzuki (1982) reporting of the production of some ACE inhibitory peptides by tryptic hydrolysis of casein, a number of active peptides have already been identified as in vitro ACE inhibitor. Table 1 summarizes the lists of ACE inhibitory peptides so far reported in the 45 papers (all the cited references can be seen in our web: http://www.agr.kyushu-u.ac.jp/biosci-biotech/bunseki/) and about 400 species of peptides that exert, to some extent, in vitro ACE inhibitory activity. It is apparent that aromatic amino acid residues including Pro at the C-terminal are essential in inhibiting ACE powerfully. Peptide length is not resposible for revealing the action. The Table also demonstrates that all the identified peptides are poor inhibitors, compared to drugs; even one of the strongest ACE inhibitor, e.g., Phe-Thr-His-Ile-Ala-Trp (IC_{50}; 0.18 μM), is weak and still not comparable to drugs (captopril; 0.035 μM). However, daily intake of bioactive peptides or antihypertensive food must be of much benefit for gradual improvement of BP in borderline hypertensives because of their expected mild BP lowering effect.

In addition to the identification of active peptides, their preparation from potential bioresources is also important in providing a physiologically functional food. A number of natural foodstuffs such as sardine (Sugiyama et al., 1991; Matsui et al., 1993), tuna (Kohama et al., 1988), bonito bowels (Matsumura et al., 1993), casein (Maruyama and Suzuki, 1982), lactoglobulin (Mullally et al., 1997), wheat

Table 1
ACE inhibitory peptides

Amino acid sequence (one-letter symbol)	IC_{50} (µmol/l)	Amino acid sequence (one-letter symbol)	IC_{50} (µmol/l)
AF	190	AVK	16
AP	230	AIM	3
AY	88	AKK	3.1
AW	10	ALA	71
DY	100	ALP	240
DW	13	AQK	1800
FP	749	AVL	7.1
FY	25	AVM	8
GF	630	AYV	17
GI	1200	DYG	2700
GP	450	EGQ	69
GY	210	FFY	13
GW	30	FIL	19
HY	26	FLM	57
IF	930	FMG	52
IP	130	FQF	6.9
IR	830	FQP	12
IW	2.0	FIL	19
IY	3.7	FSP	101
KF	116	FVA	6
LF	349	FWN	18.3
LY	38.5	GFG	75
MF	44.7	GFI	73
MY	193	GGF	21
QK	885	GHF	1100
RF	230	GIG	30
RP	180	GIY	97
RL	2439	GLY	8.8
RW	16	GKP	352
RY	51	GPL	2.55
TF	17.8	GPV	4.67
VY	22	GRP	20
VF	53	GVL	88
VK	13	GVY	400
VP	420	HHL	2
VQ	1300	HIK	> 100
WI	82	IGS	55
WL	51	IKP	1.7
YG	1100	ILP	270
YL	122	IMY	1.8
YP	2440	IPA	141
YQ	628	IPP	5
VW	1.6	IRA	6.4
AAL	93	IRP	1.8
AAF	92	ITF	49
AAV	25	IVY	0.48
ADY	38	IYP	61
AFP	610	LAA	13
AIP	670	LAF	53
AIM	3	LAH	63
AKK	3.1	LAV	48
ALA	71	LAY	3.9
ALP	240	LEK	800

Table 1 (*continued*)

Amino acid sequence (one-letter symbol)	IC_{50} (μmol/l)	Amino acid sequence (one-letter symbol)	IC_{50} (μmol/l)
LEL	13	ALPP	280
LGI	29	EVLP	> 1000
LIF	77	FVAP	10
LIY	0.82	GLYP	190
LKL	188	GRPR	470
LKP	0.32	GVYP	140
LLP	57	GWAP	3.9
LNP	43	HHTF	84
LNY	81	HIKW	> 100
LPP	9.6	HIRL	1153
LQL	81	IAIP	470
LQP	1.9	INSQ	36
LQQ	100	IRPV	31
LRP	0.27	KAIP	330
LSP	1.7	KLEK	> 2500
LTF	2.7	KTAP	37
LVR	14	LIYP	10
LVV	65	LHLP	210
LYP	6.6	LLNP	> 1000
LWW	56	LPHA	560
MGI	56	LPLP	720
MVV	31	LPPP	> 1000
PAP	87	LVYP	170
PGT	53	NILP	560
PLP	430	PAQK	22
PLP	430	QAFP	> 1000
PSY	16	QPIP	860
QGE	69	RHQG	330
QVY	81	RPVQ	300
RFH	330	SVAK	1000
SHP	280	TAPY	13.6
SVA	> 2500	TVPY	2
TAP	3.5	VFPS	0.46
TGP	53	VPQP	> 1000
TKY	2.3	VVRP	8
VAA	13	WHHT	110
VAP	2	YGGY	16.2
VAV	260	YGLF	733
VAY	42	YLLF	172
VLP	320	YPER	133
VMP	29	YPHK	1000
VPP	9	YRPY	320
VRP	2.2	AKLEK	500
VVL	28	ALPHA	10
VYP	44	AVVRP	74
WAP	71	DIGYY	3.4
WWL	51	DYVGN	0.72
WWL	51	FFVAP	6
YAV	46	GPFPI	177
YDA	47	GRVMP	250
YQY	4	GVYPH	2500
YVP	200	HLPLP	41
AIPP	900	HQIYP	110
ALPH	1800	IAIPP	600

Table 1 (*continued*)

Amino acid sequence (one-letter symbol)	IC_{50} (μmol/l)	Amino acid sequence (one-letter symbol)	IC_{50} (μmol/l)
IKWGD	> 100	KVLPVP	5
IPAQK	260	LHLPLP	2.9
IRAQQ	160	LPPPVH	> 1000
IRPVQ	1.4	MIPAQK	300
IWHHT	5.1	PAHIAW	> 100
IYPRY	4.1	PAVVLP	45
LALPP	790	PAVVRP	18
LKPNM	17	PDHIAW	> 100
LYPVK	5	PHQIYP	30
NLHLP	420	PPPVHL	110
PKAIP	185	PQEVLP	910
PLIYP	4.4	PQNILP	440
PNSHP	> 1000	PTFIAW	> 100
PQAFP	300	PTHDAW	> 100
PQPIP	340	PTHGAW	> 100
PRHQG	55	PTHIAW	0.39
PSFQP	73	PTHIDW	13
PTHIK	> 100	PTHIKW	1.8
PTHIW	> 100	PTHKAW	> 100
PTPAP	33	PTHVAW	1.5
PVPQP	110	PVRAVP	170
QEVLP	> 1000	QPLIYP	3.6
QKTAP	30	QPQAFP	610
QNILP	> 1000	QSLVYP	41
SLVYP	40	RPRHQG	22
SVAKL	900	SVAKLE	2200
TVVPG	2.2	TPVVVP	749
TYLGS	0.86	TTMPLW	16
VHIPP	10	VAKLEK	100
VHLAP	4.5	VHLPPP	200
VHLPP	18	VAKLEK	100
VLPIP	31	VLPIPQ	5300
VLPYP	36	VLPYPV	420
VKAGF	83	VPQPIP	290
VYPHK	7.6	VYPFPG	221
WHHTF	46	YKVPQL	22
YGFLP	260	ALPMHIR	42.6
YGLYP	260	APGAGVY	1.7
ATHIAW	3.5	AVPYPQR	15
AVNPIR	14	DLIPAQK	77
DYGLYP	62	DLMPAQK	56
ENLHLP	> 1000	DMIPAQK	45
EPKAIP	380	DRVYIHP	5
EPLIYP	7.1	GRPRHQG	34
FVEPIP	> 1000	IPQEVLP	690
GGVIPN	0.74	KTTMPLW	28.7
GVYPHK	1.6	KVLPVPQ	> 1000
GGVIPN	0.74	LENLHLP	> 1000
HIKWGD	50	LPQNILP	46
HQAAGW	60	NLHLPLP	51
IKPLNY	43	NNVMLQW	41
IWHHTF	2.5	NPAVVRP	19
IVGRPR	300	PPHQIYP	22
KVLAGM	30	PQPLIYP	3

Table 1 (*continued*)

Amino acid sequence (one-letter symbol)	IC_{50} (μmol/l)	Amino acid sequence (one-letter symbol)	IC_{50} (μmol/l)
PQTLALP	110	SIQSQPQAFP	560
PTHIKWD	38	YANPAVVRP	7.8
PTHIKWG	7.6	DRVYIHPFHL	75
QVPQPIP	660	FAQTQSLVYP	25
SEPKAIP	430	GIQSEPKAIP	430
SQPQAFP	710	GKMVKVVSWY	6
SVAKLEK	82	LLNPPHQIYP	9.6
THIKWGD	50	LTDLENLHLP	> 1000
TQSLVYP	64	SFQPQPLIYP	1.4
VLIPAQK	114	TDQHQDKIYP	380
YLYEIAR	16	TVYTKGRVMP	38
ANPAVVRP	25	PSFQPQPLIYP	2.7
DRVYIHPF	35	SGIQSEPKAIP	650
ENLHLPLP	155	SSIQSQPQAFP	> 1000
EPQPLIYP	3.8	HPFAQTQSLVYP	26
NPPHQIYP	37	HSGIQSEPKAIP	> 1000
PFFDPQIP	410	HSSIQSQPQAFP	> 1000
PTHIKWGD	0.9	SLVYPFPGPIHN	38.5
QPQPLIYP	4	YPSFQPQPLIYP	4.8
QSEPKAIP	750	FHSGIQSEPKAIP	520
QSQPQAFP	700	IHPFAQTQSLVYP	19
QTQSLVYP	73	IYPSFQPQPLIYP	7.6
RDMPIQAF	209	VHSSIQSQPQAFP	> 1000
SKVLPVPE	39	KFHSGIQSEPKAIP	> 1000
VGRPRHQG	5.4	KIHPFAQTQSLVYP	39
YLYEIARR	86	KIYPSFQPQPLIYP	8.6
YYPQIMQY	24.8	KVHSSIQSQPQAFP	> 1000
AFKAWAVAR	1.7	DKIYPSFQPQPLIYP	107
ALKAWSVAR	3.4	KYPVQPFTESQSLTL	93
AQTQSLVYP	76	YQQPVLGPVRGPFPIIV	101
DRVYIHPFH	5	PPQSVLSLSESKVLPVPE	25
EPIPYGFLP	180	LLYQQPVLGPVRGPFPIIV	21
FQPQPLIYP	1.8	DELQDKIHPFAQTQSLVYPF-PGPIHNS	4
IQSEPKAIP	570		
IQSQPQAFP	630	LPQNIPPLTQTPVVVPPFL-QPEVMGVSK	144
IVGRPRHQG	6.2		
LENLHLPLP	86	QTQYTDAPSFSDIP-NPIGSENSEKTTMPLW	346
LNPPHQIYP	25		
RVYIHPFHL	7.5		

(Matsui et al., 1999a) and buckwheat (Li et al., 2002), royal jelly (Matsui et al., 2002c) and so on have already been utilized as latent food material capable of inhibiting the ACE. Enzymatic hydrolysis of natural proteins is a convenient and effective treatment in this field; except for fermented foods such as soybean and milk products as soy source (Kinoshita et al., 1993) or fermented milk (Isono, 1996). Gastrointestinal proteases such as pepsin, trypsin and chymotrypsin, and food processing proteases such as alkaline protease, papain, thermolysin, Denazyme AP, etc., are the useful ones. Optimal enzymatic hydrolysis treatment or its selection may be greatly influenced by food matrices. In the case of enzymatic hydrolysis of wheat germ (Matsui et al., 1999a), which has high contents of lipids (9.7%) and sugars

(47%), defat treatment was needed to achieve a susceptible protein hydrolysis. Defat treatment is also suitable for elevating ACE inhibitory activity of fish protein (Sugiyama et al., 1991). Namely, wheat germ was defatted by hexane at 80 °C for 5 h, followed by the hydrolysis with 1.0 wt% α-amylase for 5 h. The obtained hydrolysate was subjected to a subsequent *Bacillus licheniformis* alkaline protease hydrolysis. As summarized in Table 2, 0.5 wt% alkaline protease hydrolysis for 8 h was the best condition for wheat germ hydrolysis to yield strong ACE inhibitory activity (IC_{50}; 0.67 mg-protein/ml). Heat treatment of food prior to hydrolysis is also beneficial for enhancing ACE inhibitory activity of hydrolysate. This is because of easy susceptibility of enzymes to heat. On heating protein is denatured, in which the inner hydrophobic region is exposed by unfolding. The efficiency of heat treatment of food protein has been proven in sardine muscle hydrolysis (Ukeda et al., 1991). In any case, however, the smaller the peptide length, the bitter the taste of protein hydrolysate. In general, hydrophobic peptides with bitterness may exert a potent ACE inhibitory activity (Table 1). Thus, it is of great importance in developing a functional "food" in which this discrepancy induced by peptides is resolved.

One successful processing is subsequent purification after enzymatic hydrolysis of a given food protein. In order to obtain more strong ACE inhibitory food material with good taste, a reversed-phase column chromatography (YMC ODS-AQ 120-S50 column (3.5 cm × 13 cm)) with ethanol was applied for further fractionation of the hydrolysate of sardine muscle (Matsui et al., 1993). Ethanol fractionation has some

Table 2
Preparation of ACE inhibitory foods from wheat germ under various hydrolysis conditions

Pretreatment[a]	Protease[b]	IC50 (mg protein ml^{-1})
Intact		
	Pepsin	2.38
	Chymotrypsin	2.68
	Trypsin	3.03
	Alkaline protease	1.69
	Protease YP-SS	2.56
Defatted		
	Pepsin	1.50
	Chymotrypsin	2.73
	Trypsin	2.93
	Alkaline protease	0.98
	Protease YP-SS	2.44
Defatted α-Amylase		
	Pepsin	1.78
	Chymotrypsin	2.40
	Trypsin	2.49
	Alkaline protease	0.67
	Protease YP-SS	1.77

[a]Pretreatment conditions of wheat germ were as follows: defatting by *n*-hexane; 80 °C for 5 h, α-amylase; 1.0 wt% addition for 5 h at 37 °C.
[b]Hydrolysis of wheat germ by a given protease was done under the conditions of 1.0 wt% addition for 5 h.

Table 3
Successive separation of sardine muscle hydrolysate by ODS column-chromatography

Fraction	Protein-g(yield)	IC50(mg protein/ml)	Bitterness
Hydrolysate	13.6 (100)	0.26	±
0% Ethanol	6.1 (45.6)	1.222	
10%	3.9 (29.8)	0.015	±
25%	2.3 (17.6)	0.078	+ +
50%	0.3 (2.3)	0.041	+ + +
99.5%	0.1 (0.8)		±

Sardine muscle hydrolysate by alkaline protease (13.6 g) was put into a YMC ODS-AQ 120-S50 column (3.5 × 13 cm) and successively eluted with 500 ml of 0, 10, 25, 50 and 99.5% ethanol.

advantages in that the obtained fraction can be directly applied to food and that ethanol as an eluent is more appropriate to the fractionation of hydrolysate because of its high degree of hydrophobicity than methanol. By successive elution with ethanol, efficient fractionation was completed as shown in Table 3. Among them, 10% ethanol fraction (the higher polar elution fraction) had the most potent activity with the IC_{50} value of 0.015 mg protein/ml. This indicates that many polar peptides responsible for potent ACE inhibition is present in it. The higher polar elution fraction is useful in preparing physiologically functional food of better taste with little bitterness and high solubility in water. The 10% ethanol fraction also gave low-sensory scores concerning umami and bitterness. Kato et al. (1989) have reported that the umami taste was elicited by peptides having Glu at the N-terminal, while peptides having Glu and/or Asp at another position, especially at the C-terminal, elicited the sour taste. Therefore, the low scores of the 10% ethanol fraction are presumed to be owing to masking effect by tasty peptides.

V. Antihypertensive effect of natural peptides with *in vitro* ACE inhibitory activity

In this section, toxicity, antihypertensive effect and metabolism of ACE inhibitory peptides are discussed by using one of the predominant ACE inhibitory peptides, Val-Tyr (VY) and Ile-Val-Tyr (IVY), derived from sardine muscle (Seki et al., 1995a) and wheat germ hydrolysates (Matsui et al., 1999a), respectively. It is undoubtedly of importance to know whether bioactive peptides exert antihypertensive effect in vivo (in rat or human), and whether it has toxicity if used as food, even though they have strong ACE inhibitory activity. An essential evaluation step to speculate their effects is Rat or mouse experiment. In a toxicity report of peptide by Hartleb and Leuschner (1997), a mixture of low-molecular weight peptides from porcine spleen has showed a median lethal concentration (LD_{50}) of 0.112 mg/kg in rats for a single dose. On the other hand, when 10 mg/kg dose of IVY was injected to 10 *ddy* mice no remarkable change in the body weight was observedthe phase between administration (33.4 ± 1.4 g) and the end of the trial (35.8 ± 2.5 g). No unusual motility after injection in of any of the mice was observed (Matsui et al., 2000). Furthermore, no mouse was dead during the 1 week study, indicating that the (1 week LD_{50}) was more than 10 mg/kg in IVY. Immunological study of sardine hydrolysate also revealed that natural ACE inhibitory hydrolysate would not give any antigenicity, and be effective for food material, in which the hydrolysate did not exert active systemic anaphylaxis as well as homologous passive cutaneous anaphylaxis in male Harrley guinea-pig (Seki et al., 1995b). Hence, it is assumed that smaller peptides derived from proteins would have less toxicity or antigenicity.

Spontaneously hypertensive rats (SHR) that are derived from normotensive Wistar rats and are designated as optimal experimental model of essential hypertensive subject have been primarily used for clarifying antihypertensive effect of active peptides with ACE inhibitory activity. As a result of intravenous administration test of VY (Matsufuji et al., 1995) and IVY (Matsui et al., 2000) in SHR, both inhibitors reduced the mean arterial blood pressure (MAP), and the reduction of MAP increased with dose (Table 4). The reduction ratio of IVY was almost consistent with that of the 50 mg/kg dose of VY (9.9%). In addition, single oral administration test of VY to SHR, in which 10 mg/kg dose of VY caused a significant systolic BP (SBP) reduction of 12 mmHg after 2 h (Seki et al., 1999), also provided evidence that small peptides have an ability to exert antihypertensive effect. However, in most cases of active hydrolysate, at least less than 0.1 g dosage seems to be necessary for obtaining sufficient depressor effect in rat experiments. For instance, Saito et al (1994) reported that the antihypertensive effect of sake lee hydrolysate, in which 1 g/kg of oral administration resulted in a 15 mmHg of SBP lowering in SHR after 4 h-administration. An exception was the in vivo SHR study of sardine hydrolysate (Seki et al., 1999), in which 10 mg dosage was enough to lower the SBP in SHR, and long-lasting effect induced by it was observed during 8 h-protocol period (Figure 3). However, it was found that natural peptide has still weaker antihypertensive potency by a factor of > 1/100 than captopril (Matsufuji et al., 1995).

Antihypertensive effect of some functional foods has also been demonstrated in mild hypertensive subjects (> 140/ > 90 mmHg) and/or high normal BP subjects (130–139/85–89 mmHg). (Medium and heavy hypertensive subjects are not our object of study in the prophylaxis of hypertension disease by health-improving "food").

Table 4
Intravenous administration of Val-Tyr and Ile-Val-Tyr into SHRs

Inhibitor	Dose (mg/kg)	BP (mmHg) Before	BP (mmHg) After	Reduction ratio(%)
VY	50	182.2 ± 4.0	164.2 ± 3.8*	9.9
IVY	5	184.7 ± 8.49	165.5 ± 14.5*	10.4
Captopril	0.03	182.2 ± 10.2	168.0 ± 6.1*	7.8

BP is shown as mean ± S.D., $n = 5$. Significant differences from before administration by paired Student's t test:
*$p < 0.05$.

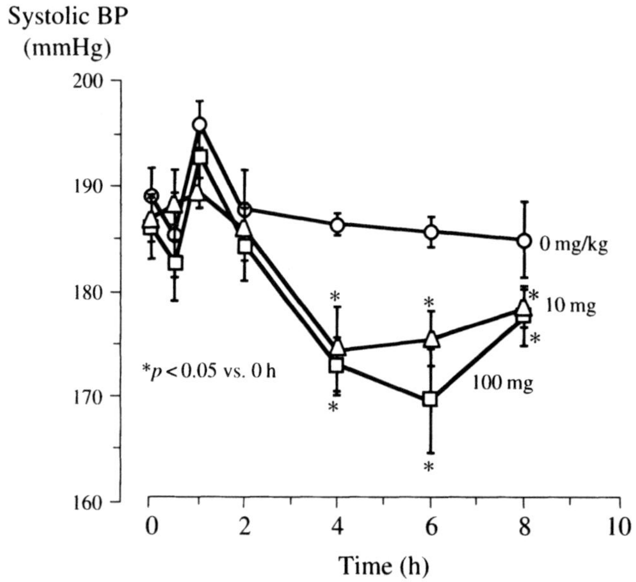

Fig. 3. Effect of single oral administration of sardine muscle hydrolysate on systolic blood pressure of 15-week SHRs.

For instance, intake of tryptic hydrolysate of casein (20 g/day) resulted in a mild BP reduction behavior (ΔSBP/ΔDBP = 4.6/6.6 mmHg) after 4-week administration ($n = 18$; 141 mmHg/99 mmHg) (Sekiya et al., 1992). Sour fermented milk containing ACE inhibitors, Ile-Pro-Pro and Val-Pro-Pro, also reduced SBP by 15 mmHg (3.4 mg in 95 ml/day) (Hata et al., 1996). Similar positive result was obtained in the sardine hydrolysate containing 3 mg of VY in a 100 ml-drink, where a significant BP reduction (ΔSBP/ΔDBP = 9.7/5.6 mmHg after 4-week administration) was demonstrated by a randomized double-blind placebo-controlled human study ($n = 17$, 147/91 mmHg) (Figure 4) (Kawasaki et al., 2000). These human-volunteer studies provide some useful information on BP lowering effect induced by functional food, by which BP was reduced moderately and any adverse effects such as cough were not observed. Therefore, these findings evidentially reveal that intake of antihypertensive foods prepared from natural food resources are of great benefit for regulating or improving mild hypertension disease.

VI. Absorption and antihypertensive mechanism of natural ACE inhibitory peptides

The findings described above strongly lead us to investigate how active peptide(s) elicits antihypertensive effect in human systemic system. Before elucidating the mechanism, it is of importance to know their pharmacokinetics. Studies on peptide transport have demonstrated that peptide can be absorbed faster than amino acid itself (Gardner, 1982) and the transport is carried out through PepT1 and T2 transporters at the intestinal membrane (Yang et al., 1999). Within this restriction, di- and

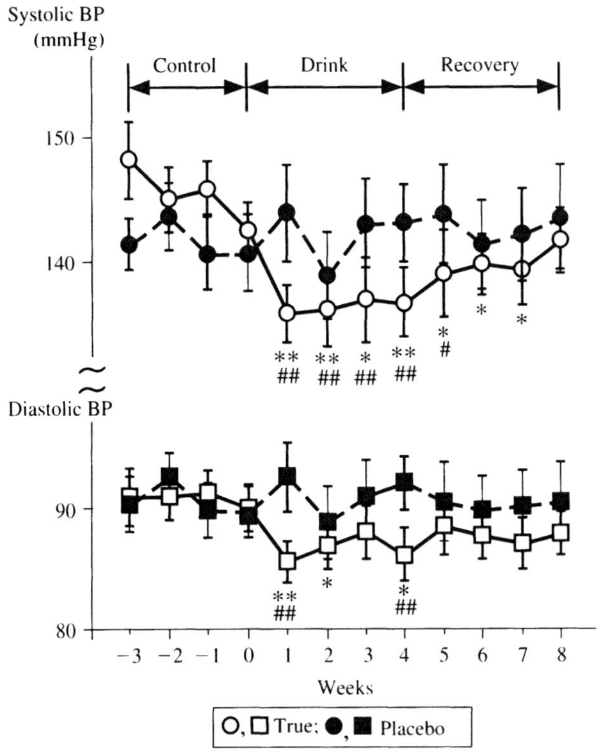

Fig. 4. Change in blood pressure of mild hypertensive volunteers during the double-blind placebo-controlled study-period after the intake of peptide-drink (3 mg of Val-Tyr/100 ml-drink; twice a day). $^{\#}p<0.01$; $^{\#\#}p<0.001$ vs. placebo, $^{*}p<0.01$; $^{**}p<0.001$ vs. 0 h

tri-peptides possessing a free terminal carboxyl group and at least one peptide bond are found to be a preferable penetrant. These findings lead us to investigate how much amount of antihypertensive peptide(s) can be absorbed in human subjects. However, detailed investigations on absorption of bioactive peptides are yet to be done owing to the lack of sufficient assay methods. Recently, a high-sensitive detection system of small peptides has been proposed, in which dipeptide (VY) with ACE inhibitory activity was successfully detected by using column-switching HPLC method (Matsui et al., 1999b). Fluorimetric derivatization of VY by naphthalene 2, 3-dialdehyde allowed its high-sensitive detection of > 10 fmol/ml of plasma. By using this assay method, VY absorption behavior into normotensive and mild hypertensive human circulatory blood systems was clarified for the first time (Matsui et al., 2002b, a). As shown in Figure 5, VY absorption curve continued maximally over the second hour postprandially, and the increment continued 8 h after the intake. The pharmacokinetics of VY in normotensive human subjects are as follows; $t1/2$: 3.1 h, t_{max}: 2 h, C_{max}: 1934 ± 145 fmol/ml-plasma, AUC0-24 h dosed at 12 mg of VY: 9,185 ± 688 fmol h/ml-plasma. Although absorption behavior of small peptides has not been clarified thoroughly, it is speculated that they could be absorbed into the body intact with slow absorption rate.

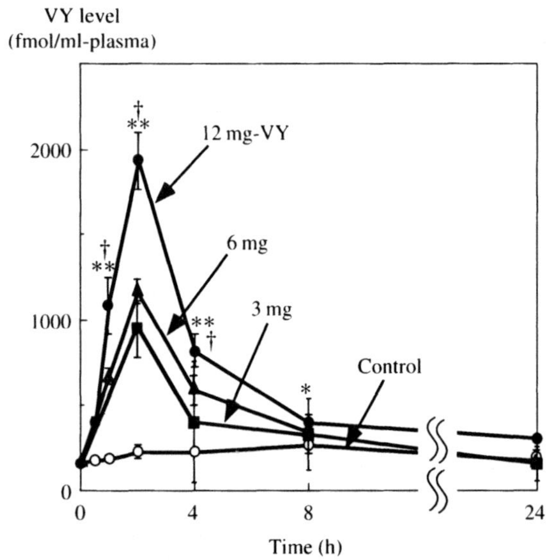

Fig. 5. Val-Tyr absorption profiles into normotensive human plasma after the intake of Va-Tyr drink. *$p<0.05$; **$p<0.01$ vs. 0 h, †$p<0.01$ vs. control

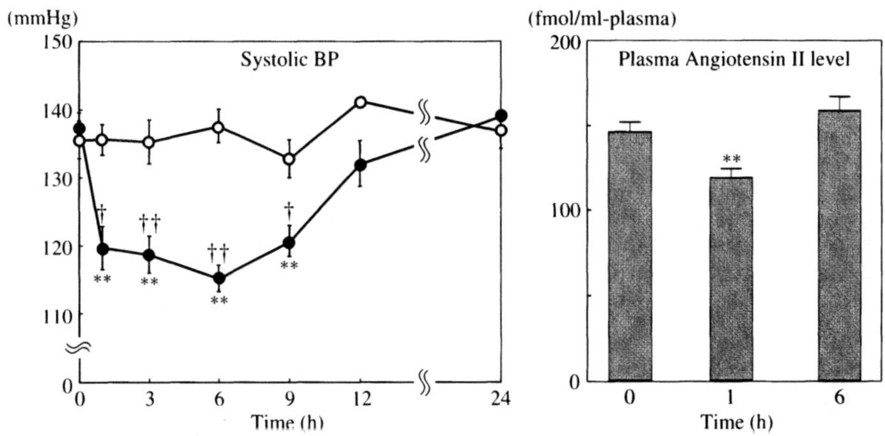

Fig. 6. Change in systolic blood pressure and plasma angiotensin II level of 11-week THM after oral administration of Val-Tyr given a dose of 0.1 mg/g (closed circle) and vehicle (open circle). Each value is expressed as mean ± SEM ($n = 7$). Significant difference against 0 h; **$p<0.01$, against control group; †$p<0.05$, ††$p<0.01$.

The next step is the elucidation of depressor mechanism of small peptides. Useful information on antihypertensive effect induced by peptide-intake was gathered from an experiment on performed by using hypertensive mice (Matsui et al., 2003). According to this study when VY was administered orally into Tsukuba Hypertensive Mouse (THM) having both human renin gene and human angiotensinogen gene, a prolonged BP reduction up to 9 h was observed. The effect clearly demonstrated that the absorbed VY acts on the enhanced human RAS so as to lower the BP. On the other hand, reduction of plasma Ang II levels was observed only at 1 h after the administration (Figure 6), suggesting that the suppression of human RAS by the absorbed VY was a transient effect. Consequently, a long-lasting BP lowering effect induced by VY-intake must be closely associated with local RAS, and not with the circulatory RAS. Some reports provide useful information that one possible explanation of antihypertensive mechanism may be a because of retardation of local RAS such as abdominal aorta or kidney (Okunishi et al., 1991). However, further studies is needed to clarify the mechanism of antihypertensive peptides, because no direct and informative results on their stimulation and regulation of local RAS in diverse organs have been obtained yet.

References

Adibi SA, Morse EL. (1971) Intestinal transport of dipeptides in man. J Clin Invest 50:2266–75.

Amarowicz R, Shahidi F. (1997) Antioxidant activity of peptide fractions of capelin protein hydrolysates. Food Chem 58:355–9.

Cheung H-S, Wang F-L, Ondetti MA, Sabo EF, Cushman DW. (1980) Binding of peptide substrates and inhibitors of angiotensin-converting enzyme. J Biol Chem 255:401–7.

Ehler MRW, Riordan JF. (1990). In: Laraph JH, Brenner BM editors. Hypertension: pathophysiology, diagnosis, and management. New York: Raven Press, p. 1217.

Gardner MLG. (1982) Absorption of intact peptides: studies on transport of protein digests and dipeptides across rat small intestine in vitro. Quarterly J Exp Physiol 67:629–37.

Hartleb M, Leuschner J. (1997) Toxicological profile of a low molecular weight spleen peptide formulation used in supportive cancer therapy. Drug Res 47:1047–51.

Hata Y, Yamamoto M, Ohni M, Nakajima K, Nakamura Y, Takano T. (1996) A placebo-controlled study of the effect of sour milk on blood pressure in hypertensive subjects. Am J Clin Nutr 64:767–71.

Isono Y. (1996) Peptide inhibitors for angiotensin I-converting enzyme from Masai fermented milk. Food Sci Technol Int 2:213–6.

Johnston CI, Burrell LM. (1995) Evolution of blockade of the renin–angiotensin system. J Human Hypertens 9:375–80.

Kato H, Rhue MR, Nishimura T. (1989). In: Teranishi R, Buttery RG, Shahidi F editors. Flavor chemistry – trends and developments. Washington, DC: ACS Symposium Series 388, p. 158.

Kawamura Y. (1990) Prospects and basic problems of physiologically active peptides derived from food proteins. Shokuhin to Kaihatsu (in Japanese) 26:28–32.

Kawasaki T, Seki E, Osajima K, Yoshida M, Asada K, Matsui T, Osajima Y. (2000) Antihypertensive effect of Valyl-Tyrosine, a short chain peptide derived from sardine muscle hydrolysate. J Human Hypertens 14:519–23.

Kinoshita E, Yamakoshi J, Kikuchi M. (1993) Purification and identification of an angiotensin I-converting enzyme inhibitor from soy source. Biosci Biotechnol Biochem 57:1107–10.

Kohama Y, Matsumoto S, Oka H, Teramoto T, Okabe M, Mimura T. (1988) Isolation of angiotensin converting enzyme inhibitor from tuna muscle. Biochem Biophys Res Commun 155:332–7.

Li C-H, Matsui T, Osajima Y. (2002) Latent production of angiotensin I-converting enzyme inhibitors from buckwheat protein. J Peptide Sci 8:267–74.

Maruyama S, Suzuki H. (1982) A peptide inhibitor of angiotensin I converting enzyme in the tryptic hydrolysate of casein. Agric Biol Chem 46:1393–4.

Matsufuji H, Matsui T, Ohshige S, Kawasaki T, Osajima K, Osajima Y. (1995) Antihypertensive effects of angiotensin fragments in SHR. Biosci Biotechnol Biochem 59:1398–401.
Matsui T, Li C-H, Osajima Y. (1999a) Preparation and characterization of novel bioactive peptides responsible for angiotensin I-converting enzyme inhibition from wheat germ. J Peptide Sci 5:289–97.
Matsui T, Matsufuji H, Seki E, Osajima K, Nakashima M, Osajima Y. (1993) Inhibition of angiotensin I-converting enzyme by *Bacillus licheniformis* alkaline protease hydrolysates derived from sardine muscle. Biosci Biotechnol Biochem 57:922–5.
Matsui T, Tamaya K, Kawasaki T, Osajima Y. (1999b) Determination of angiotensin metabolites in human plasma by fluorimetric high-performance liquid chromatography using a heart-cut column-switching technique. J Chromatogr B 729:89–95.
Matsui T, Li C-H, Tanaka T, Maki T, Osajima Y, Matsumoto K. (2000) Depressor effect of wheat germ hydrolysate and its novel angiotensin I-converting enzyme inhibitory peptide, Ile-Val-Tyr, and the metabolism in rats and human plasma. Biol Pharm Bull 23:427–31.
Matsui T, Tamaya K, Seki E, Osajima K, Matsumoto K, Kawasaki T. (2002b) Val-Tyr as a natural antihypertensive dipeptide can be absorbed into the human circulatory blood system. Clin Exp Pharm Physiol 29:204–8.
Matsui T, Tamaya K, Seki E, Osajima K, Matsumoto K, Kawasaki T. (2002a) Absorption of Val-Tyr with in vitro angiotensin I-converting enzyme inhibitory activity into the circulating blood system of mild hypertensive subjects. Biol Pharm Bull 25:1228–30.
Matsui T, Yukiyoshi A, Doi S, Sugimoto H, Yamada H, Matsumoto K. (2002c) Gastrointestinal enzyme production of bioactive peptides from royal jelly protein and their antihypertensive ability in SHR. J Nutr Biochem 13:80–6.
Matsui T, Hayashi A, Tamaya K, Matsumoto K, Kawasaki T, Murakami K, Kimoto K. (2003) Depressor effect induced by dipeptide, Val-Tyr, in hypertensive transgenic mice is, in part, due to the suppression of human circulating renin-angiotensin system. Clin Exp Pharm Physiol 30:262–5.
Matsumura N, Fujii M, Takeda Y, Sugita K, Shimizu T. (1993) Angiotensin I-converting enzyme inhibitory peptides derived from bonito bowels autolysate. Biosci Biotechnol Biochem 57:695–7.
Mullally MM, Meisel H, FitzGerald RJ. (1997) Identification of a novel angiotensin I-converting enzyme inhibitory peptide corresponding to a tryptic fragment of bovine β-lactoglobulin. FEBS Lett 402:99–101.
Natoff IL. (1987) Preclinical studies on angiotensin converting enzyme inhibitors. Cardiovasc Drugs and Therapy 1:15–27.
Okunishi T, Kawamoto Y, Kurobe Y, Oka K, Ishii K, Tanaka M, Miyazaki M. (1991) Pathogenic role of vascular angiotensin-converting enzyme in the spontaneously hypertensive rat. Clin Exp Pharm Physiol 18:649–59.
Saito Y, Wanezaki K, Kawano A, Imayasu S. (1994) Antihypertensive effects of peptide in Sake and its by-products on spontaneously hypertensive rats. Biosci Biotechnol Biochem 58:812–6.
Sealey JE, Laragh JH. (1990). In: Laraph JH, Brenner BM editors. Hypertension: pathophysiology, diagnosis, and management. New York: Raven Press, p. 1287.
Seki E, Osajima K, Matsufuji H, Matsui T, Osajima Y. (1995a) Val-Tyr, an angiotensin I-converting enzyme inhibitor from sardines that have resistance to gastrointestinal proteases. Nippon Nogeikagaku Kaishi (in Japanese) 69:1013–20.
Seki E, Osajima K, Matsufuji H, Matsui T, Osajima Y. (1995b) Immunological applicability for foods of short chain peptides derived from sardines. Nippon Nogeikagaku Kaishi (in Japanese) 69:1171–4.
Seki E, Kawasaki T, Yoshida M, Osajima K, Tamaya K, Matsui T, Osajima Y. (1999) Antihypertensive effect of sardine peptide and Val-Tyr in SHR. Nippon Eiyo Shokuryo Gakkaishi (in Japanese) 52:271.
Sekiya S, Kobayashi Y, Kita E, Imamura Y, Toyama S. (1992) Antihypertensive effects of tryptic hydrolysate of casein on normotensive and hypertensive volunteers. J Jpn Soc Nutr Food Sci 45:513–7.

Silk DBA, Fairclough PD, Clark ML, Hegarty JE, Marrs TC, Addison JM, Burston D, Clegg KM, Matthews DM. (1980) Use of a peptide rather than free amino acid nitrogen source in chemically defined elemental diets. J Parenter Enteral Nutr 4:548–53.

Sugiyama K, Takada K, Egawa M, Yamamoto I, Onzuka H, Oba K. (1991) Hypotensive effect of fish protein hydrolysate. Nippon Nogeikagaku Kaishi (in Japanese) 65:35–43.

Tsuchihata H, Suzuki T, Kuwata T. (2001) The effect of casein phosphopeptides on calcium absorption from calcium-fortified milk in growing rats. Br J Nutr 85:5–10.

Ukeda H, Matsuda H, Kuroda H, Osajima K, Matsufuji H, Osajima Y. (1991) Preparation and separation of angiotensin I-converting enzyme inhibitory peptides. Nippon Nogeikagaku Kaishi (in Japanese) 65:1223–8.

Yang CY, Dantzig AH, Pidgeon C. (1999) Intestinal peptide transport systems and oral drug availability. Pharm Res 16:1331–43.

M.T.H. Khan and A. Ather (eds.)
Lead Molecules from Natural Products

Xanthones as therapeutic agents: chemistry and pharmacology

Noungoue Tchamo Diderot, Ngouela Silvere, Tsamo Etienne

Abstract

Xanthones are a class of heterocyclic compounds containing oxygen, with a yellow coloration and all of them have dibenzo-γ-pyrone as the basic skeleton. A number of this class of compounds has been isolated from a variety of natural sources (higher plants, fungi and lichens). They have gradually risen to great importance because of their many interesting pharmacological and biological properties.

Keywords: *xanthones, natural source, taxonomy, classification, structures, pharmacological properties, synthesis*

Abbreviations: DNA, deoxyribonucleic acid; IC_{50}, inhibitory concentration at 50%; ED_{50}, effective dose at 50%; MAOI, monoamine oxidase inhibitors.

I. Introduction

The term xanthone derives from the Greek ξαντηου (xanthos), which means yellow (Roberts, 1961). Xanthones are a class of heterocyclic compounds containing oxygen, with a yellow coloration and all of them have dibenzo-γ-pyrone (**1**) as the basic skeleton. (**1**) does not, so far as is known, occur in nature, but a number of its oxygenated derivatives, which are yellow in colour, have been isolated from a variety of natural sources (higher plants, fungi and lichens) (see Figure 1).

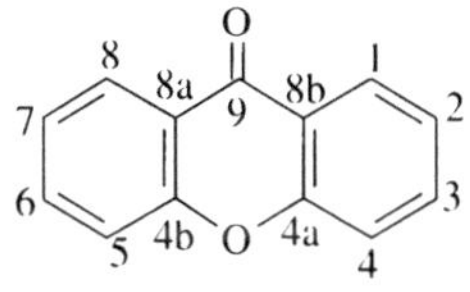

Dibenzo-γ-pyrone (**1**)

Fig. 1. Dibenzo-γ-pyrone (**1**).

They have gradually risen to great interest because of their taxonomic importance in some families (Mandal et al., 1992) as well as because of their many interesting pharmacological properties.

II. Distribution in nature

In 1961, Roberts noticed that xanthones had been isolated from lower fungi, lichens and from only three families (Gentianaceae, Guttiferae and Anacardiaceae) of flowering plants.

Mandal et al. (1992), however, in their review, listed plants belonging to 20 families that have been found to produce xanthones. They noticed that Gentianaceae and Guttiferae families were, in this order, the principal sources of xanthone derivatives.

In order to have an in-depth picture, the different plants that have been found to produce all kinds of xanthones during the period from 1992 to mid-2003 are recorded in Table 1, which shows that xanthones were still principally obtained from Guttiferae and Gentianaceae families. Since 1992, many new families and genera of higher plants, fungi and lichens have been found to be sources of xanthones.

III. Obtaining of xanthones

III.A. Classification

The xanthones already isolated may be classified into five groups (Mandal et al., 1992).

III.A.1. Simple oxygenated xanthenes

They can be further subdivided into six groups depending on the degree of oxygenation pattern of the basic skeleton:

(a) Mono-oxygenated (**2**) (Gunatilaka et al., 1982): they are unusual, and only three compounds were isolated from Hypericaceae and Guttiferae families.
(b) Di-oxygenated (**3**) (Monache et al., 1983): 12 derivatives of this class were identified.
(c) Tri-oxygenated (**4**) (Henry and Caventou, 1821): the gentisine (**4**) was the first isolated xanthone.
(d) Tetra-oxygenated (**5**) (Noungoue et al., 2001): they were more numerous and majority were isolated from the Gentianaceae family.
(e) Penta-oxygenated (**6**) (Liu and Huang, 1982): only a small number was encountered in nature.
(f) Hexa-oxygenated (**7**) (Rodriguez et al., 1995): it has the highest degree of oxygenation observed so far.

Table 1
List of families, genera and species known to produce xanthone (1992–2003)

Family Amaranthaceae
Genus *Amaranthus*
A. paniculatus

Family Annonaceae
Genus *Anascagorea*
A. luzonensis

Family Asteraceae
Genus *Senecio*
S. mikanioide

Family Clusiaceae
Genus *Clusia*
C. grandiflora
C. insignis
C nemerosa
C. renggerioides
C. rosea

Family Eriocaulaceae
Genus *Leiothri*
L. curvifolia
L. flavexens

Family Fabaceae
Genus *Erythrina*
E. variegata

Family Filicineae
Genus *Dovallia*
D. solida

Family Gentianaceae
Genus *Canscora*
C. lucidissima

Genus *Centausium*
C. erythraea
C. littorale

Genus *Gentiana*
G. acuta
G. barbata
G. karelinii
G. lutea
G. olgida

Genus *Gentianella*
G. florida
G. nitida
G. tristicha
G. wevadensis

Genus *Gentianopsis*
G. barbata

Genus *Holenia*
H. Campanulata
H. cerniculata

Genus *Hoppe*
H. fastigiata

Genus *Schultsia*
S. Guianensis
S. lisianthoide

Genus *Swertia*
S. Alata
S. Calycina
S. Chirata
S. ciliata
S. cuneata
S. cordota
S. decora
S. erythrosticta
S. franchetiana
S. moxin
S. pryenalskii
S. punicea
S. speciosa
S. verticillifolia

Genus *Tripterospermum*
T. japonicum
T. lanceolatum

Genus *Veratrilla*
V. baillonii

Family Guttiferae
Genus *Calophyllun*
C. austroindicum
C. bracteatum
C. brasilienses
C. caledonicum
C. colaba
C. inophyllum
C. moonii
C. opetalum
C. panciflorum
C. teysmannii
C. thwaitessi
C. tomentosum

Genus *Cratoxylum*
C. cochinchinense
C. formosanum
C. moingayi
C. sumatranum

Genus *Garcinia*
G. altroviridis

Table 1 (*continued*)

G. bracteata
G. camborgia
G. cowa
G. dioica
G. dulcis
G. Forbesii
G. Hanburyi
G. kola
G. latissima
G. livingstonei
G. mangostana
G. opaca
G parvifolia
G. scortechinii
G. subelliptica
G. vilersiana

Genus *Hoploclathra*
H. paniculata

Genus *Hypericum*
H. androsaemum
H. annulatum
H. axyron
H. balearicum
H. brasiliense
H. geminiflorum
H. henryi
H. inodorum
H. japonicum
H. patulum
H. perforatum
H. roeperanum

Genus *Kielmeyera*
K. coriacea
K. lathrophyton

Genus *Mamnea*
M. acuminata
M. siumensis

Genus *Martla*
M. laxiflora

Genus *Mesua*
M. ferrea

Genus *Montrouziera*
M. sphaeroidus

Genus *Poeciloneuron*
P. pauciflorum

Genus *Psorospermum*
P febrifugum

Genus *Symphonia*
S. globulifera

Genus *Tovomita*
T. brasiliensis
T. brevistaminea

Genus *Vismia*
V. guineensis
V. latifolia
V. magnoliaefolia
V. parviflora

Family Leguminoceae
Genus *Cyclopia*
C. intermedia

Family Moraceae
Genus *Artocorpus*
A. teysmanii
A. communis

Genus *Cudrania*
C. cochinchinensis

Genus *Maclura*
M. Tinctoria
M. chochinchinensis

Genus *Morus*
M. insignis

Family Nyetayinaceae
Genus *Boerhaovia*
B. diffusa

Family Podostemaceae
Genus *Podostemum*
P. marathrum
P. oserya

Family Polygalaceae
Genus *Bredemeyera*
B. brevifolia
B. floribunde

Genus Monnina
M. sylvatica

Genus *Polygala*
P. tenuifolia
P. virgata
P. wattersii
P. sibirica
P. caudata
P. cyparissias
P. nyikensis

Genus *Securidaca*
S. inappendiculata

Table 1 (*continued*)

Family Ulmaceae	Genus *Lecanora*
Genus *Trema*	*L. Elisein*
T. orientalis	*L. Epibryon*
	L. moyrhoferi
Fungi	*L. parmelinoides*
Cercospora beticola kescins	*L. subfusca*
Neosartorya fischeri	
Penicillium vermiculatum	Genus *Pyrenula*
Phomopsis longicolla	*P. japonica (lichen mycobints)*
Lichens	Genus *Rinodina*
Genus *Dimelaena*	*R. thiomela*
D. australiensis	
D. lecamora	Genus *Sporopodium*
D. lecidella	*S. vezdeaniem*
D. isabellina	*S. xantholeucun*
D. micarea	
D. pertusania	

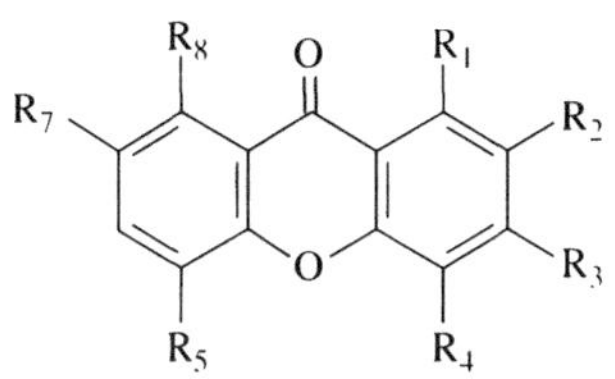

2-Hydroxyxanthone (R_1=R_3=R_4=R_5=R_7=R_8=H ; R_2=OH) (**2**)
3-Hydroxy-2-methoxyxanthone (R_1=R_4=R_5=R_7=R_8=H ; R_3=OH ; R_2=OMe) (**3**)
1,7-Dihydroxy-3-methoxyxanthone (R_2=R_4=R_5=R_8=H ; R_1=R_7=OH ; R_3=OMe) (**4**)
1-Hydroxy-3, 7, 8-trimethoxyxanthone (R_2=R_4=R_5=H ; R_1=OH ; R_3=R_7=R_8=OMe) (**5**)
1-Hydroxy-2, 3, 4, 5-tetramethoxyxanthone (R_7=R_8=H ; R_1=OH ; R_2=R_3=R_4=R_5=OMe) (**6**)
1-Hydroxy-2, 3, 4, 5, 7-pentamethoxyxanthone (R_8=H ; R_1=OH ;R_2=R_3=R_4=R_5=R_7=OMe) (**7**)

Substituents on the xanthone nuclei are mainly OH and OMe groups, and CH_3 or Cl groups are rarely observed on lichens or on fungi xanthones.

III.A.2. Glycoside xanthenes

They may be divided into *O*-glycosides and *C*-glycosides according to the nature of the glycosidic linkage:

(a) *O*-glycosides xanthones (**8**) (Agrawal and Singh, 1988): the most common oxygenation being 1,3,5,8-(tetra-oxygenated) and also 1,2, 3,7; 1,3,6,8; 1,3,4,7; 1,2,5,6,8; 1,3,4,7,8; and 2,3,5,7,8. The glycosyl moiety at either 1,4,5,7 or 8 position has been encountered in the isolated *O*-glycoside xanthones. They are easily hydrolysed in enzymatic or acid environment (Hostettmann and Wagner, 1977). Some diglycosides involving the positions 6:8; 2:7 and 2:6 have been encountered. Most *O*-glycoside xanthones have the sugar moiety attached to position 1 of the xanthone nucleus, which is difficult to explain considering the vicinity of the carbonyl function, since this might create a strain.

(b) *C*-glycoside xanthones (**9**) (Mandal et al., 1992): they are more resistant to hydrolysis compared with *O*-glycoside xanthones, but their occurrence is very much limited. Mangiferin was the first glycoside xanthone isolated in 1908 from *Mangifera indica* (Anacardiaceae) and has also been reported in a number of ferns.

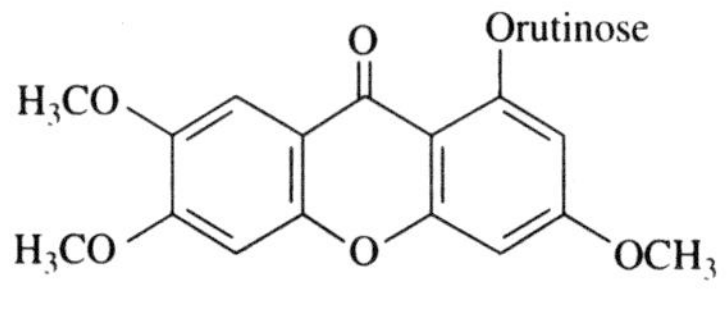

□-L-Rhamnopyrannosyl (1,6) - □-D-glucopyrannose (**8**)

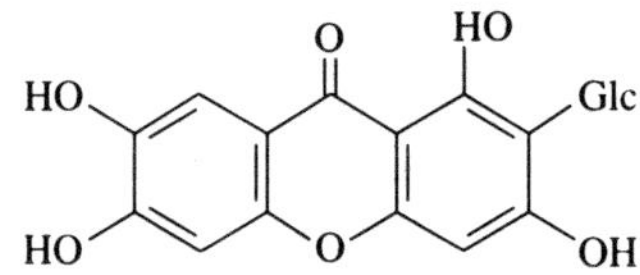

Mangiferin (**9**)

III.A.3. Prenylated xanthones

The family Guttiferae appears to produce a large number of xanthones with isopentenyl and geranyl substituents. Prenylated and the corresponding pyranoxanthones have been reported to occur in Moraceae; (**10**) Iinuma et al. (1994); (**11**) Gunasekera et al. (1981).

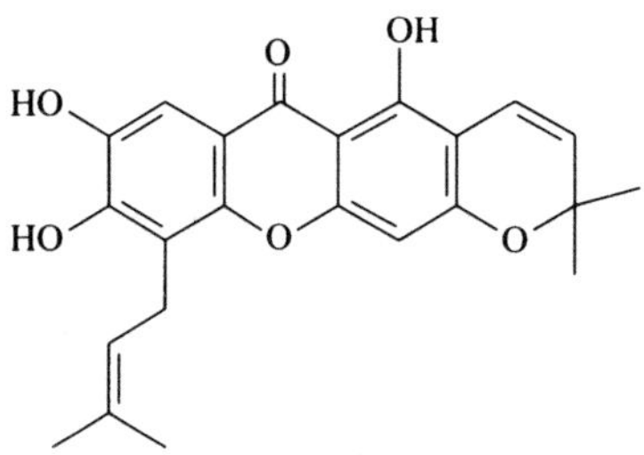

Caloxanthone (**10**)

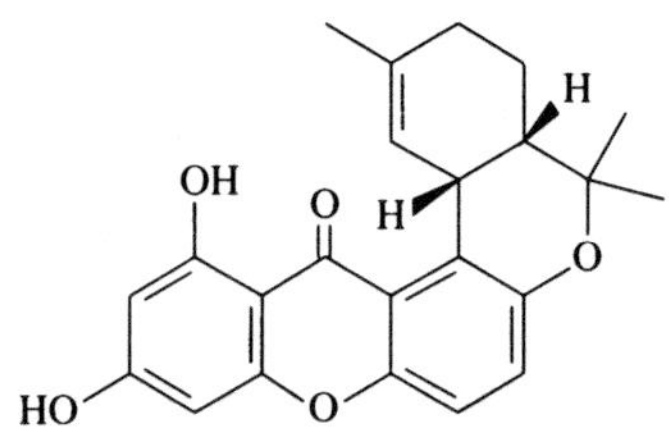

Calozeyloxanthone (**11**)

III.A.4. Xanthonolignoids

They are a relatively rare group of natural products and principally occur in some genera of Guttiferae family: *Kielmeyera*, *Caraipa* (Castelao et al., 1977), *Psorospermum* (Abou-Shoer et al., 1989). These compounds are very close in the skeletal patterns formed from the association of the xanthone nucleus and the lignoid pattern (coniferyl alcohol or syringenin). The most representative is candesin D (**12**) (Cardona et al., 1986).

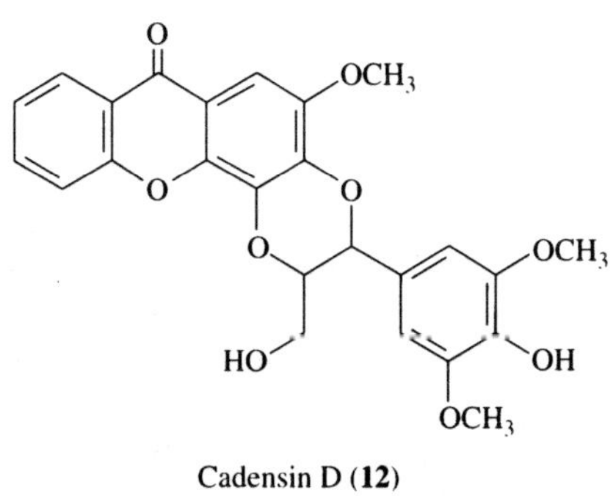

Cadensin D (**12**)

III.A.5. Miscellaneous

Besides these groups, some xanthones with unusual substitutions have been isolated from different plant sources including lichens, which could not be classified in the usual manner. These compounds have been grouped as Miscellaneous. We can name chloride compounds (**13**) and (**14**) from *Dimelaena lichen* (Elix and Bennett, 1990; Elix et al., 1991) as examples of this group.

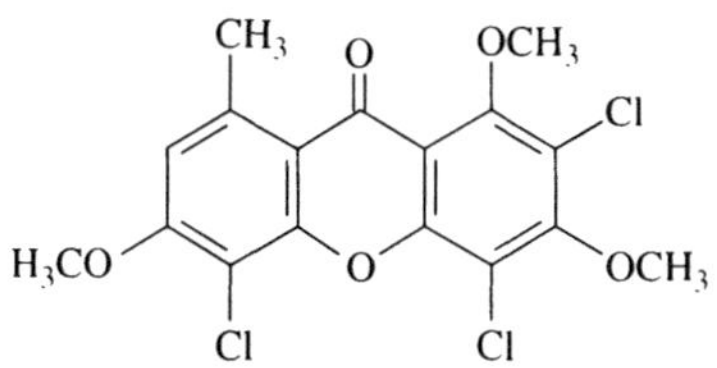

1,3,6-Tri-*O*-methylarthotheline (**13**)

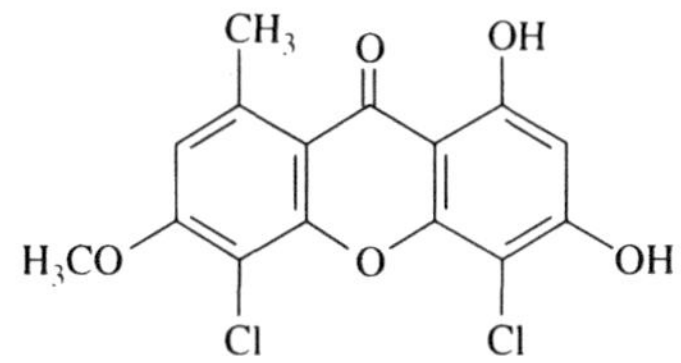

4,5-Dichloro-6-*O*-methylnorlichexanthone (**14**)

III.B. Methods of isolation and structural investigation

Xanthones can be found in all parts of the plant and can be extracted with an increasing polarity of organic solvents (Petroleum ether, methylene chloride, methanol ...) according to the type of xanthone (prenylated, aglycon or not, glycosides ...). But in fungi or lichens the extraction is different. They are cultured in a specific environment and then extracted with appropriate solvent (Davies et al., 1960; Ahmed et al., 1992). Purification of xanthones from extracts can be made by various classical chromatographic techniques using different solvent mixtures and silica gel as absorbent, according to the nature of the oxygenation pattern (Arends, 1970; Hostettmann et al., 1986).

These secondary metabolites that are xanthones, can be readily detected by their colour in UV light by using a general phenolic spray (Harborne, 1994).

The structural investigation of this class of compounds is done with ease using spectroscopic means.

Nowadays, the use of IR and UV spectroscopy in xanthone chemistry is marginalised.

However, UV spectral assignments can easily be made with the availability of a considerable amount of data. A lot of information on the position of hydroxyl groups can be obtained by the use of $AlCl_3$ shifts for chelated hydroxyls, as well as sodium acetate, sodium hydroxide and boric acid shifts (Mesquita et al., 1968).

In IR spectroscopy, the effect of chelation on the IR carbonyl frequency of hydroxy xanthones may be a useful feature for analysing the spectra of xanthones, as well as for detecting unchelated hydroxyl and methyl groups (Roberts, 1961).

Since it allows access to the molecular weight, mass spectrum is very valuable for preliminary examination. No systematic investigation of the electron-impact-induced fragmentation of xanthones appears to have been made, except that made by Arends et al. (1973) in their study of electron-impact-induced fragmentation of monohydroxy and monomethoxy xanthones.

^{1}H-NMR and ^{13}C-NMR spectroscopic analyses are the most useful tool in the structure elucidation of xanthones. The data obtained from ^{1}H-NMR spectroscopic analysis are of great value for locating aromatic protons, through comparison with reference data and analysis of spin–spin coupling; these enable easy prediction of the oxygenation pattern for this class of compounds (Figure 2, Barraclough et al., 1970).

Arends & Helboe (1972) and Helboe & Arends (1973) have conduted a study on the influence of OH, OCH_3 and allyl substituents in the chemical shifts of the proton of the phenolic hydroxyl (Table 2). The hydroxyl protons of the aromatic nuclei resonate in the range 9.25–13.35 ppm, the variation of these values depends on the position of the hydroxyl and methoxyl substituents with respect to carbonyl function. This influence is less for OH in positions 2/7 and 4/5, but it increases in position 3/6 because of conjugation with the carbonyl. When the OH is in positions 1/8, the chemical shift is higher (11.45–13.35 ppm) owing to the chelation with the carbonyl.

The ^{13}C-NMR data of a large number of all classes of xanthones have been reported with chemical shifts assigned (Castelao et al., 1977; Westerman et al., 1977; Frahm and Chaudhuri, 1979). The calculated and observed chemical shifts carried out by Mandal et al. (1992) were close enough in most cases to allow unambiguous assignments of carbon signals. Thus, the xanthones carbonyl group resonates in the range 174–182 ppm. Within this range, the presence of a hydrogen bonded 1-hydroxyl group at C1 showed a signal at 179–182 ppm, whereas the loss of this intramolecular hydrogen bonding on O-alkylation caused the carbonyl resonance to move upfield by about 5 ppm.

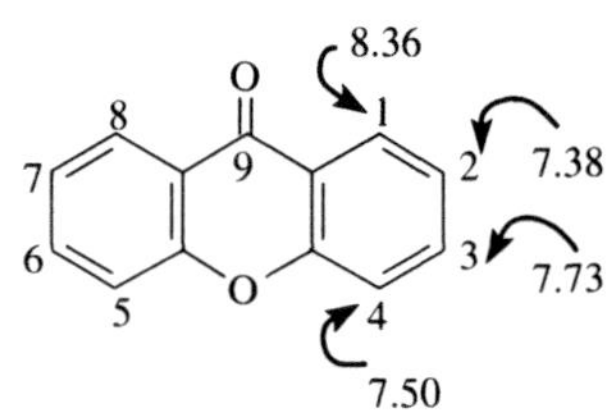

Fig. 2. Protons attribution in xanthones (Barraclough et al., 1970).

Table 2
Chemical shifts of the proton of the phenolic hydroxyl

Chemical shift δ (ppm)	Position
9.25–9.45	2 or 7 – OH with OR in 1 or 8, respectively
9.35–9.60	4 or 5 – OH with OR in *ortho* or *para*
9.70–10.05	2 or 7 – OH
10.35–10.55	4 or 5 – OH
	3 or 6 – OH with OR in 4 or 5, respectively
10.80–11.10	3 or 6 – OH
11.45–12.00	1 and 8 – OH
	1 or 8 – OH with OR en 4 or 5, respectively
12.50–12.90	1 or 8 – OH
12.90–13.35	1 or 8 – OH with OR in 3 and 6
	1 or 8 – OH with OR in 8 or 1, respectively

Source: Arends (1972).

Table 3
Assignment of aryl carbon signals (DMSO-d_6)

Carbon relatively to OH	Position of carbon in the ring	Increment (ppm)
C-*ipso*	1, 8	$+35.5 \pm 0.5$
	2, 3, 4, 5, 6, 7	$+29.3 \pm 1.0$
C-*ortho*	1, 4, 5, 8	-16.0 ± 1.5
	8b, 4a, 4b, 8a	-11.5 ± 1.5
	2, 3, 6, 7 move with OH-*ortho*, respectively in C-1, 4, 5, 8 in C-3, 2, 7, 6	-14.5 ± 1.0 -10.5 ± 1.0
C-*meta*	1, 8	$+1.0 \pm 1.0$
	8a, 8b	$+1.0 \pm 1.0$
	4a, 4b	$+0.5 \pm 1.0$
C-*para*	1, 4, 5, 8	-10.5 ± 1.0
	8b, 4a, 4b, 8	-7.0 ± 1.0

Source: Frahm and Chaudhuri (1979).

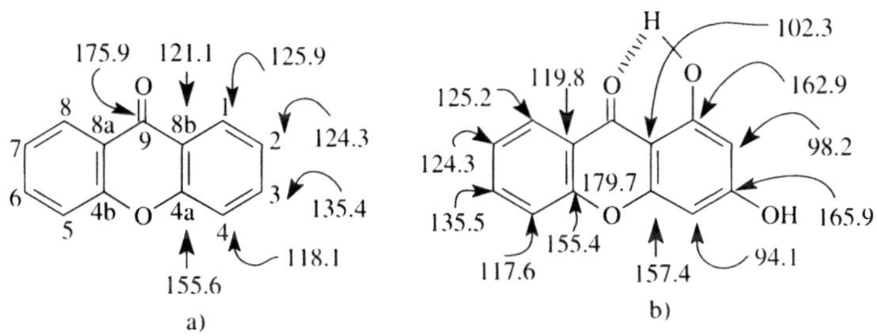

Fig. 3. Carbons attribution: (a) non substituted xanthone; (b) 1, 3-dihydroxysubstituted xanthone (DMSO-d_6) (Frahm and Chaudhuri, 1979).

On the other hand, Frahm and Chaudhuri (1979) studied ^{13}C-NMR spectra of some polyhydroxyxanthones and established a rule that allows undoubtful assignment of aryl carbon signals (Table 3). Chemical shifts of carbon that wear hydroxyl group (C-*ipso*) in position 1 or 8 are higher (up to about 36 ppm) because of the intermolecular hydrogen bonding between the O-atom of carbonyl and OH in 1 and/or in 8, whereas, the other C-*ipso* chemical shifts (6 ppm) are smaller. Double chelation (1,8-dihydroxy) induces the carbonyl carbon to move downfield by about 6 ppm, while monochelation (1 or 8-hydroxy) shifts in the same way by 4.5 ppm. Supplementary oxygenation in 3 or 6 causes a shield of carbonyl carbon by 1.5 ppm (see Figure 3).

IV. Use of some plant sources containing xanthones

It has now been observed that a growing number of plant species containing xanthones exhibit various biological properties and are used as chemotherapeutic agents in indigenous medicine for the treatment of many diseases. Typical examples are the popular use of *Swertia chirata* and *Garcinia mangostana* as febrifuge (Nakatani et al., 2002), *Hypericum perforatum* (St John's Wort) for managing different depressive state (Rocha et al., 1994) and *Trema orientalis* for anti-cardiovascular disease (Noungoue, 1998).

- Total extract of *S. chirata* has also been used in the traditional medicine to improve the liver function and as an anti-malarial (Mandal et al., 1992);
- The total extract of *Canscora decussata* has been found to be useful in the treatment of certain mental disorders (Ghosal and Chaudhuri, 1975);
- *S. franchetiana* possesses the ability to reduce fever, and it is employed for the treatment of hepatogenous jaundice and cholecystitis (Wang et al., 1994);
- The fruit hull of *G. mangostana* has been used in indigenous medicine for the treatment of skin infection, wounds and diarrhoea (Mahabusarakam et al., 1987). It also has anti-inflamatory property (Likhitwitayawuid et al., 1998);
- The whole dried flowering aerial part of *Centaurium erythraea* is used as stomachic, cholagogue or bitter tonic (Schimmer and Mauthner, 1996);
- *Calophyllum caledonicum*: the heartwood of this tropical tree is used by local populations against fungi and termites (Morel et al., 2002);
- The leaves of *M. indica* are prepared as tea and the juice of the leaf is considered to be useful in bleeding dysentery; its bark and seeds are used as astringent. This plant is also used for melancholia and nervous debility (Yoshimi et al., 2001);
- Bark of *G. subelliptica* had been utilised as a source of a yellow-coloured dye (Minami et al., 1996);
- The stem bark of *Anthocleista vogelli* is used by local populations for the treatment of fever, stomach ache and as purgative (Okorie, 1976);
- The balsam from bark of *C. inophyllum* is used as a cicatrising, while an infusion or decoction of the leaves has been used as an eye cure (Basnet et al., 1995);
- The roots of *C. decussata* are used as a laxative, diuretic and for liver troubles, and also as nerve tonic and in tuberculosis and fevers, whereas the aerial portions are used for the treatment of insanity, epilepsy and nervous debility (Ghosal et al., 1973);
- *Cudrania cochin* is used as a remedy to cure neuralgia, rheumatics, hepatitis and contusions (Chang et al., 1994);
- Extracts obtained from *Eustoma grandiflorum* are used to treat constipation, nervous debility, tuberculosis, anorescia and fever (Sullivan et al., 1977);
- Seeds and leaves of *G. dulcis* have been used in folk medicine against lymphatitis, parotitis and struma (Iinuma et al., 1996);
- *Hoppea dichotama* is used in the treatment of haemorrhoids, in cardiac disorders and in certain mental disorders (Ghosal et al., 1978);
- Extracts of *H. perforatum* are widely used as a drug for depression treatment (Rath et al., 1996);
- *H. roeperanum* is employed to cure female sterility (Rath et al., 1996);
- Leaves of *Lawsonia inernis* are used for treating some skin diseases including tinea of the legs (Mahmoud et al., 1980);
- *Monnina obtusifolia* is used in the folk medicine as an anti-fungal, anti-tumoural, antiseptic and as skin cleanser (Pinto et al., 1994);
- The roots of *Polygala tenuifolia* are used as expectorant, tonic and sedative agent (Ikeya et al., 1991);

- *Psorospemum febrifugum* and *S. macrosperma* have been used as febrifuge, for the treatment of leprosy treatment, as a poison antidote and a purgative (Kupchan et al., 1980; Zhou et al., 1989);
- *Saponaria vaccaria* is reputed to possess sudorific, emetic and laxative properties and is also used in indigenous system of medicine for the treatment of jaundice, rheumatism, hepatic eruption and venereal ulcers (Kazmi et al., 1989);
- *S. chirayita* is used as a traditional remedy for chronic fever, anaemia, asthma and liver disorders (Asthana et al., 1991);
- The extract of *S. hookeri* is used by local population for the treatment of microbial infection in man, in hypertension and as a mood enhancer (Ghosal et al., 1980);
- The whole plant of *S. japonica* has been used for stomach complaints (Basnet et al., 1994);
- The extracts of the *S. chirata* is used as a bitter tonic, stomachic, as remedy for a febrifuge, and as an anthelmintic. It is also used as a remedy for scanty urine, for epilepsy, for certain types of mental disorders and as blood purifier. Its decoctions are commonly used for constipation and weak liver function in children (Mandal et al., 1992);
- *S. mileensis* and *S. mussoti* are especially efficacious for treating acute viral hepatitis (Zhou et al., 1989);
- The leaves decoction of *T. orientalis* are used against fever, hypertension, epilepsy, diarrhoea and gout. Its stem bark is used by the local population to cure bronchitis, asthma and cough and also as antidote of poison (Adjanohoun et al., 1984; Iwu, 1993; Noungoue, 1998)

V. Biosynthesis of xanthones

The biosynthetic pathway to xanthones in plants has been abundantly studied by many authors *in vivo* (Fujita and Inoue, 1980) and *in vitro* with markers (Gröger et al., 1968; Locksley and Murray, 1970; Gupta and Lewis, 1971). They attempted to inter-relate the observed oxygen patterns of natural xanthones and correlate them with recognised oxygenation patterns. They showed that two processes are involved in their synthesis.

V.A. Acetate polymalonic route

It has been shown for some xanthones in lower plants (micro-organisms and lichens) that their synthesis was totally acetate-derived from seven acetate units (McMaster et al., 1960; Birch et al., 1976). The biosynthetic mechanism of ravelenin (**15**) from *Helminthosporium ravenelii* proposed by Birch et al. (1976) gives an illustration of the acetate polymalonic route. Benzophenone is involved as an intermediate (see Figure 4).

V.B. Mixed shikimate acetate pathway

In general it is accepted that ring A with its CO group attached are from the shikimic acid pathway, and ring B arises from acetate-malonate polyketide route

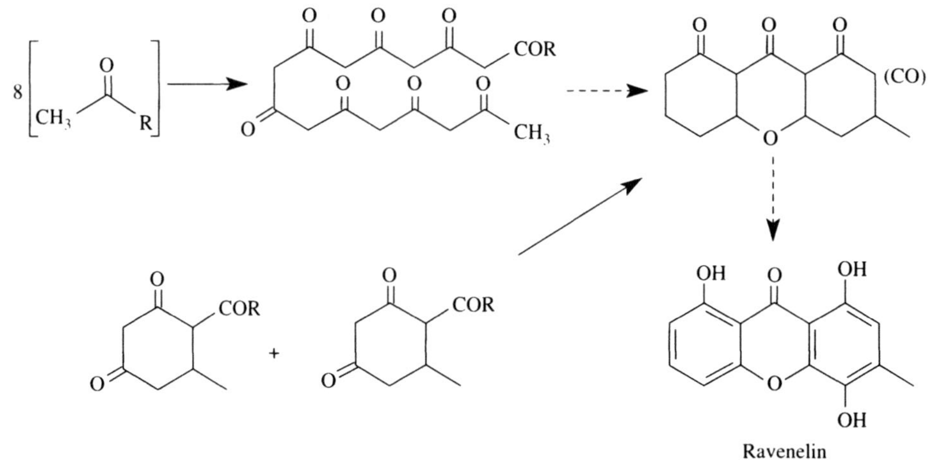

Fig. 4. Biosynthesis of xanthones lower plants: the case of ravenelin (**15**) (Birch et al., 1976).

(Locksley and Muray, 1970; Afzal and Al-Hassan, 1980; Sultanbawa, 1980). Hence, the oxygenation patterns of all xanthones in higher plants are formed by a mixed shikimate-acetate pathway. The two moieties condense to form benzophenone or benzophenone-like intermediates, which react intramolecularly to give xanthones. The occurrence of both xanthones and their precursor benzophenones has been observed in *Symphonia globulifera*, Guttiferae (Locksley et al., 1967) and *Gentiana lutea*, Gentianaceae (Atkinson et al., 1969). The mechanisms of this reaction involves either phenol oxidative coupling (Lewis, 1963), quinone addition (Ellis et al., 1967), dehydration between hydroxyl groups on the acetate and shikimate-derived rings (Markham, 1965a, b) or spirodienone formation and subsequent rearrangement to form the xanthone (Gottlieb, 1968; Carpenter et al., 1969).

Fujita and Inoue (1980) showed that biosynthesis of mangiferin, a *C*-glycosylxanthone isolated from *Anemarrhena asphodeloïdes* (Liliaceae) involves a condensation of one *p*-coumarate and two malonates. For prenylated xanthones, prenylation can occur at the benzophenone or at the xanthone stage (Bennett and Lee, 1989).

VI. Synthesis of xanthones

Six general methods have been reported for synthesizing xanthone derivatives.

1. *Michael–Kostanecki method.* The first xanthone synthesis was carried out by Kostanecki (1892). In this method, an equimolar mixture of a polyphenol and an *O*-hydroxybenzoic acid are heated in the presence of a dehydrating agent such as acetic anhydride or zinc chloride (see Figure 5).

   Grover et al. (1955) using a mixture of phosphorus oxychloride and zinc chloride as condensing agent obtained good results. This modified method has the advantage that the reaction temperature is lower. When hydroquinone, resorcinol or

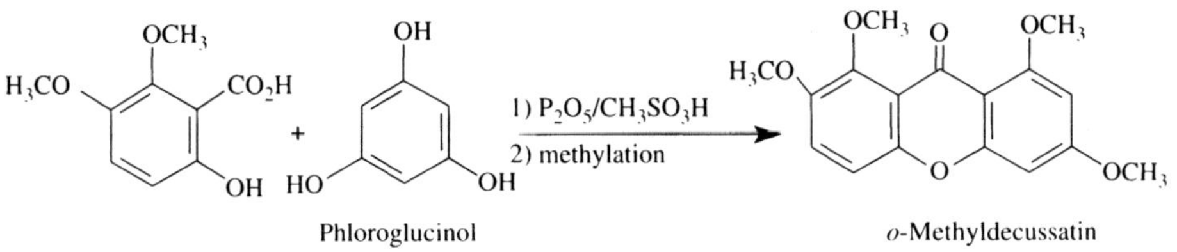

Fig. 5. Synthesis of *O*-methyldecussatin (Pillai et al., 1986).

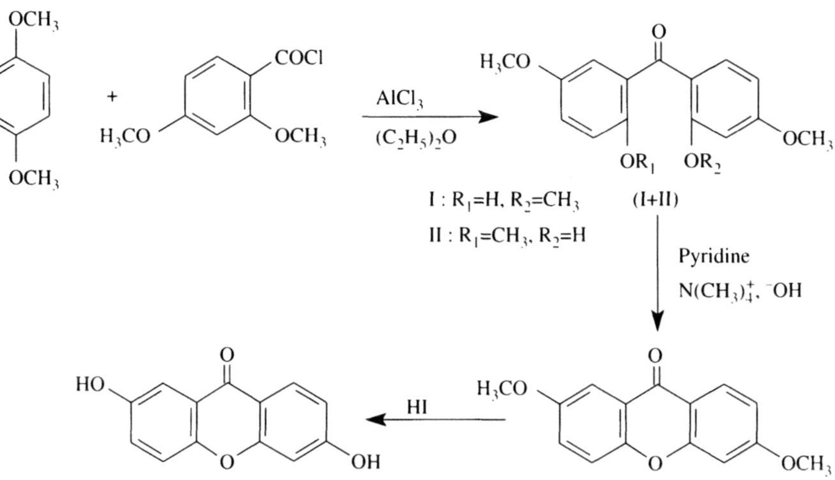

Fig. 6. Preparation of 3,7-dihydroxyxanthone (Lin et al., 1993).

pyrogallol are used as polyphenol, poor yields are obtained; the benzophenone intermediate being required in this case.

2. *Friedel–Crafts method.* The Friedel–Crafts method involves a benzophenone intermediate (Quillinan and Scheinmann, 1973; Gil et al., 1987, 1988; Ravi et al., 1994) as shown in the synthesis of 3,7-dihydroxyxanthone (Lin et al., 1993) (see Figure 6).
3. *Robinson–Nishikawa method.* This is an ingenious and good method (a variant of Hoesch synthesis), but the yield is low. It has been modified by Atkinson and Heilbron (1926) and proceeds via a ketimino intermediate as shown for the synthesis of 3,6-dihydroxyxanthone (see Figure 7).
4. *Asahina–Tanase method* (Asahina & Tanase, 1940). This is an interesting method for the synthesis of some methoxylated xanthones or xanthones with acid sensitive substituents (Granoth and Pownall, 1975). Recently Vitale et al. (1994) modified the procedure as shown in Figure 8.
5. *Tanase method.* The Tanase method enables the synthesis of polyhydroxyxanthones. It has been used for the preparation of partially methylated polyhydroxyxanthones with pre-established orientation of some substituents, for example the synthesis of 3,8-dihydroxy-1-methoxyxanthone (Pillai et al., 1986) (see Figure 9).

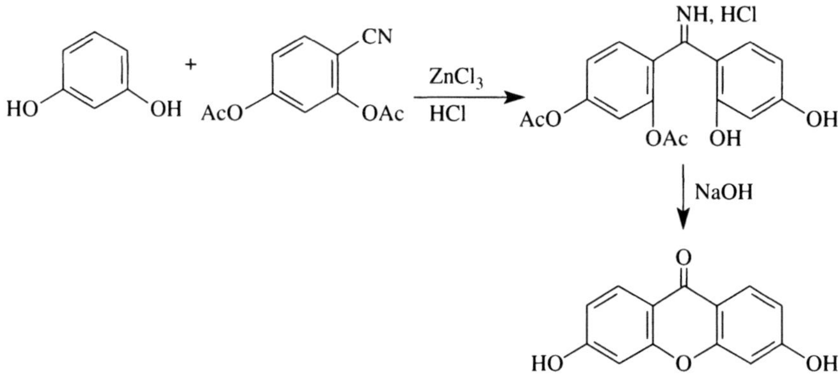

Fig. 7. Robinson–Nishikawa method (Atkinson and Heilbron, 1926).

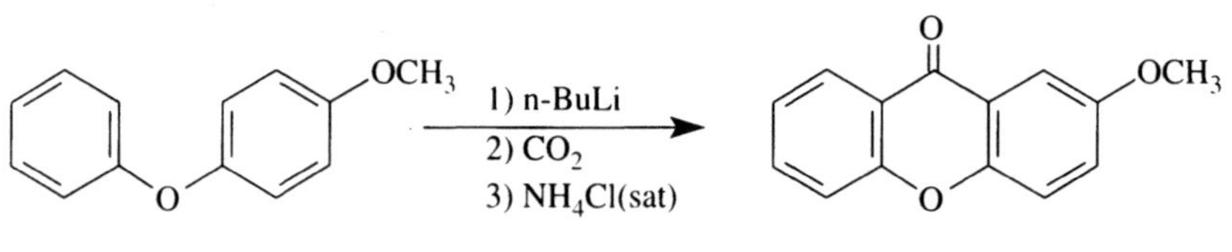

2-Methoxyxanthone (52%)

Fig. 8. Asahina–Tanase method (Vitale et al., 1994).

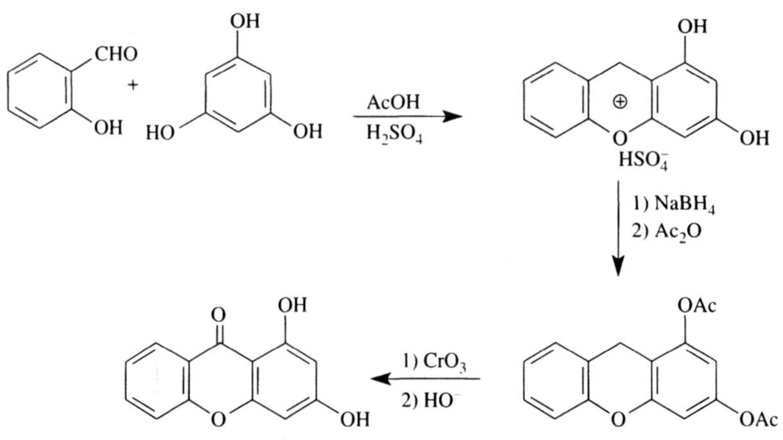

1,3-Dihydroxyxanthone

Fig. 9. Synthesis of 1,3-dihydroxyxanthone (Pillai et al., 1986).

6. *Ullman method.* In the Ullman method, a phenol and an *O*-chlorobenzoic acid are condensed and the diphenylether obtained cyclises to give the xanthone. This method has been successfully applied to the synthesis of euxanthone (Ullmann and Pauchaud, 1906) (see Figure 10).

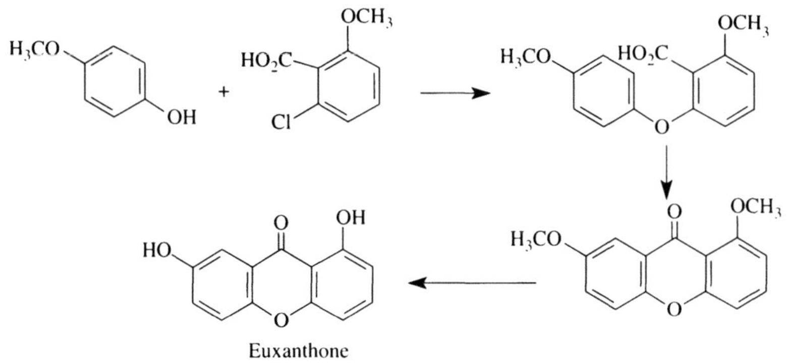

Fig. 10. The Ullman method for the preparation of euxanthone.

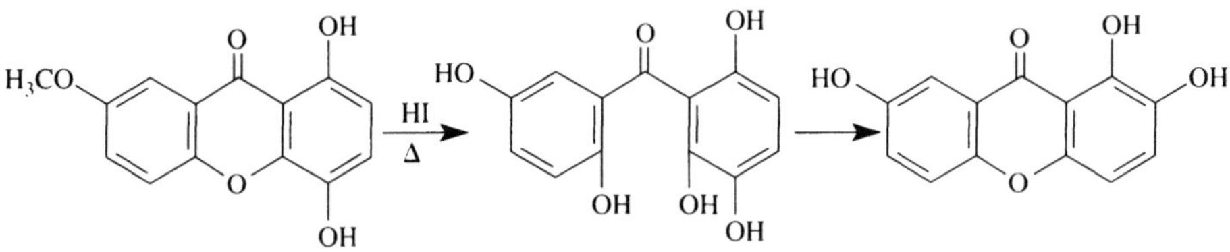

Fig. 11. Demethylation of xanthone with HI (Philbin et al., 1956).

VI.A. Demethylation of xanthones

Most naturally occurring xanthones contain methoxyl groups and can serve as starting material for obtaining other xanthones by demethylation. For this propose, hydriodic acid or aluminium chloride or morpholine are often used (Chaudhuri et al., 1978b). It should be noted that this demethylation reaction is often accompanied by the Wessely–Moser rearrangement. For example, the demethylation of 1,4-dihydroxy-7-methoxyxanthone, which also led to the isolation of 1,2,7-trihydroxyxanthone (see Figure 11).

VI.B. Synthesis of prenylated xanthones

VI.B.1. O-prenylated xanthones

They have been obtained by O-alkylation of an hydroxyxanthone with prenylbromide in the presence of potassium carbonate (Burling et al., 1965; Patel and Trivedi, 1988; Noungoue et al., 2000), but very little work has been done in this area. The alkylation generally takes place on the aromatic ring, when a prenylated derivative reacts with a naturally occurring or synthetic xanthones.

Three methods of C-prenylation are known. They are as follows.

1. C-prenylation with 2-methylbut-3-en-2-ol in the presence of boron trifluoride in ether. Only one case has been reported in the literature with 1,3-dihydroxyxanthone

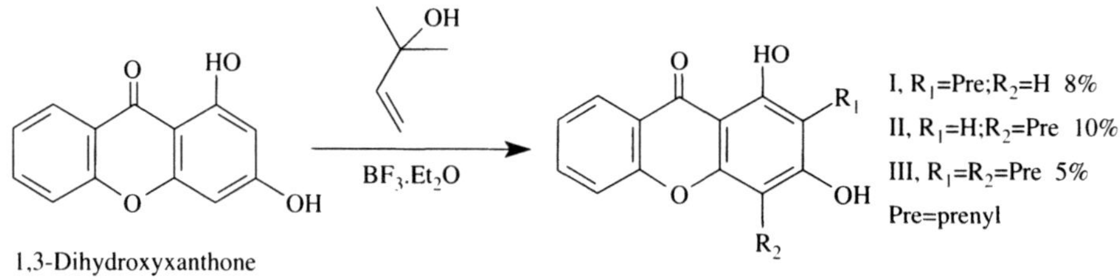

Fig. 12. *C*-prenylation with 2-methylbut-3-en-2-ol (Anand and Jain, 1973).

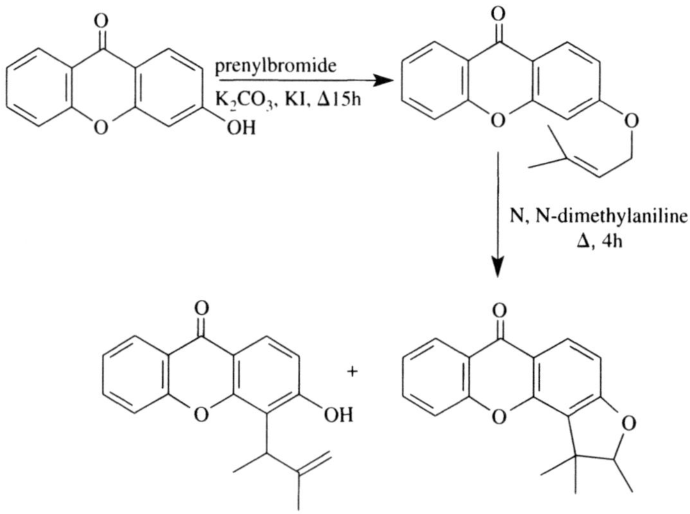

Fig. 13. *C*-prenylation with prenylbromide in the presence of a strong base (Anand and Jain, 1974).

leading to prenylated xanthones I, II and III (Anand and Jain, 1973) (see Figure 12).

2. The reaction of 1,3-dihydroxy-5,8-methyoxyxanthone with prenylbromide in the presence of sodium methoxide yields a mixture of O- and C-prenylated xanthones I, II and III (Anand and Jain, 1974) (see Figure 13).
3. Another method of prenylation of aromatic ring starts with an O-prenylation followed by a C-prenylation by Claisen rearrangement. Patel and Trivedi (1988) made the C-prenylation of 3-hydroxyxanthone using this method (see Figure 14).

VII. Bioactivities of xanthones

Numerous reports have appeared in the literature concerning the interesting pharmacological properties of naturally occurring and synthetic xanthones.

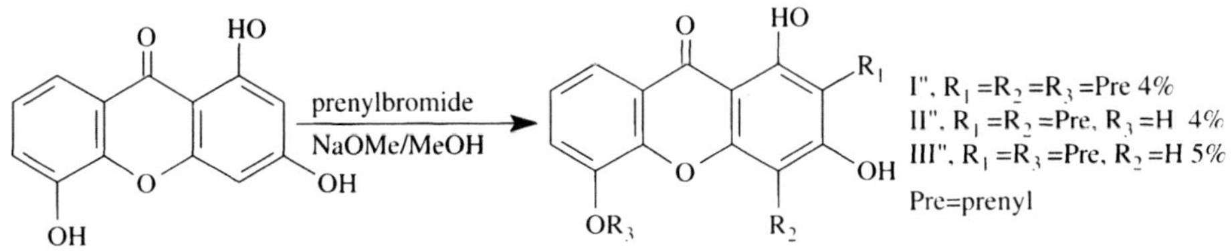

Fig. 14. *C*-prenylation through claisen rearrangement (Patel and Trivedi, 1988).

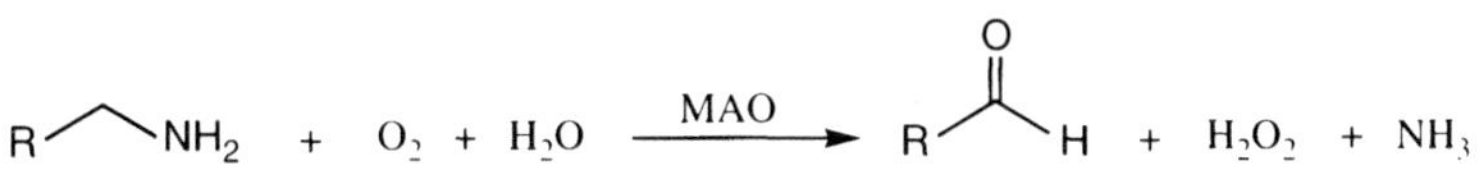

Fig. 15. Mechanism of desamination of monoamine by MAO (Fowler and Ross, 1984).

VII.A. Action on the central nervous system (CNS)

Among all the pharmacological properties of xanthones, the most studied one is the stimulation of CNS by the inhibition of MAO. The two isoenzymes of MAO, MAO-A and MAO-B play a key role in the regulation of some physiological amines and act by the desamination of some neurotransmitter such as catecholamin serotonin or tyromin as shown below(Fowler and Ross, 1984) (See Figure 15).

Inhibitors of MAO are used as anti-depressant drugs. The inhibitions *in vitro* and *in vivo* of MAO by the plant extracts rich in xanthones or by the pure xanthones have been reported in many works (Fowler and Ross, 1984).

Almost all the xanthones tested against MAO came from Gentianaceae or Guttiferae families.

The study of glycosylxanthones showed that aglycons 1,3-dihydroxyled inhibited both MAO-A and MAO-B, whereas bellidifolin (1,5,8-trihydroxy-3-methoxyxanthone) (**15**) was found to be a higher selective inhibitor of MAO-A ($IC_{50} = 0.31\,\mu M$ for MAO-A and $IC_{50} = 180\,\mu M$ for MAO-B) compared with the reference compound pargylin ($IC_{50} = 0.60\,\mu M$ for MAO-A and $IC_{50} = 0.028\,\mu M$ for MAO-B) (Schaufelberger & Hostettmann, 1988). Except for mangiferin (**9**), glycoside xanthones possess very weak MAOI activity and sometimes exhibit depressive activity, especially *O*-glucoside xanthones (Bhattacharya, 1972).

VII.B. Cardiovascular activity

Some studies carried out *in vitro* and *in vivo* showed that inhibition of the angiotensin-I converting-enzyme induced a hypotensive effect (Cushman and Cheung, 1971; Hof et al., 1987; Bernstein et al., 1989). Important inhibition properties of that enzyme have been observed with tetrahydroxyled xanthone isolated from *Tripterospermum lanceolatum*. To inhibit the enzyme, it is necessary for the hydroxyl group in position 4 of the xanthone skeleton to chelate Zn^{2+} in the right way, the ions of the angiotensin-I-converting enzyme (Chen et al., 1992). Norathyriol

(1,3,6,7-tetrahydroxyxanthone) (**16**) was reported to be a good vasorelaxing agent (Ko et al., 1991).

VII.C. Anti-fungal and anti-bacterial activities

A series of natural prenyled xanthones have been shown to display great inhibition on three phytopathogen fungi (*Fusarium oxysporum vasinfectum*, *Alternaria tenuis* and *Dreschlera oryza*), particularly γ-mangostine (1,3,6,7-tetrahydroxy-2,8-diprenyl-xanthone) (**17**) that was active on these three fungi at 1000 ppm (Gopalakrishnan et al., 1997). Also, 1,7-dihydroxy-4-methoxy (**18**) and 1,7-dihydroxy-3,5,6-trimethoxyxanthone (**19**) have shown good activity against *Cladosporium cucumerium* (Marston et al., 1993).

In general, anti-microbial activities have been observed on prenyled xanthones (Bennet et al., 1989). Synthetic heptacyclic xanthones (Cervinomycine A_1 and A_2) exhibited antibiotic activity against anaerobic bacteria, mycoplasma and some gram-positive bacteria (Mehta et al., 1994). Recently (Nkengfack et al., 2002), three prenylated xanthones (1,7,8-trihydroxy-2,2-dimethylpyrano[5′,6′: 3,4]xanthone (globulixanthone C) (**20**), 1,6-dihydroxy-5-methoxy-7-(3-methylbu-2-enyl)xanthone (globulixanthone D) (**21**) and globulixanthone E (**22**)) isolated from the root bark of *S. globulifera* were tested for their anti-microbial potential against *Staphylococcus aureus*, *Bacillus subtilis* and *Vibrio anguillarium* bacteria. The activities of the three compounds were equivalent to those demonstrated by streptomycin. Anti-tubercular activity of polyhydroxylated xanthones in positions 1,3,7 was noticed against *Mycobacterium tuberculosis* H37 Ra (ICM: 0.5 mol/nl) (Hambloch et al., 1985). Also xanthones from *C. decussata* have shown potent tuberculostatic activity (Ghosal and Chaudhuri, 1975).

VII.D. Cytotoxic and anti-tumoral properties

Furanoxanthone derivatives have been shown to be the most potent inhibitors of tumour cell growth among xanthones. Psorespermine (**23**) (a dihydrofuranoxanthone) exhibited significant anti-tumour activity in the cells culture derived from a human carcinoma of the nasopharinx (K.B) *in vitro* system and also on leukaemia P388 of mice *in vivo* system (Kupchan et al., 1980). Abou-Shoer et al. (1988) also reported that many furanoxanthones show interesting cytotoxic properties on cells cultured from human adenocarnoma in the colon.

Xantholignoids from Guttiferae were responsible for the cytotoxity of mammary cancers (Abou-Shoer et al., 1989). Prenylated xanthones from *Garcinia* (Guttiferae) especially 4-(3′,7′-dimethylocta-2′,6′-dienyl)-1,3,5-trihydroxyxanthone (**24**) have been shown *in vitro* to have great inhibition activity with $IC_{50} = 0.6\,\mu g/ml$ compared with 5-fluorouracil ($IC_{50} = 0.4\,\mu g/ml$) on colon cancer cells culture (Sordat et al., 1992). Chemoprevention is a very promising means to control cancer (Wattenberg, 1985). Mangiferin (**9**) was tested in bowel carcinogenesis of a male F344 rat and showed great inhibitory effects, and can be classified as a potential chemopreventive agent (Yoshimi et al., 2001). Polymethoxylated xanthone derivatives from *C. erythraea* showed strong anti-mutagenic property in *solvanella*

(Shimmer and Mauthner, 2001). Anti-mutagenic compounds may be useful in preventing tumors from being induced or promoted (Ramel et al., 1986).

VII.E. Anti-diabetic property

Bellidifolin (1,5,8-trihydroxy-3-methoxyxanthone) (**15**) obtained from *S. japonica* was found to be a potent hypoglycemic agent in induced diabetic rats (Basnet et al., 1994, 1995).

VII.F. Hepatoprotective activity

From the studies carried out by Fernandez et al. (1995) , synthetic xanthones and xanthonolignoids were shown to exhibit good *in vitro* hepatoprotective property. The tested compounds prevented tert-butylhydroperoxide-induced lipid peroxidation and cell death in freshly isolated rat hepatocytes. All compounds were also effective in preventing perturbation of cell glutathione homeostasis. The two xanthones (3,4-dihydroxy-2-methoxyxanthone(**25**) and 2,3-dihydroxy-4-methoxyxanthone(**26**)) were more effective than the two xanthonolignoids (transkielcorin (**27**) and trans-isokielcorin B(**28**)).

VII.G. Anti-parasitic activity

Some prenylated xanthones isolated from *G. cola* have been shown to some extent to possess good anti-malarial property against *Plasmodium falciparum* in the concentration range of $IC_{50} = 1.50–3.00\,\mu g/ml$ compared with chloroquine ($IC_{50} = 0.03$ $\mu g/ml$) and pyrimethamine ($IC_{50} = 2.80\,\mu g/ml$) taken as reference compounds (Likhitwitayawuid et al., 1998). The two xanthone dimers, phomoxanthones A (**29**) and B (**30**), from fungus *phomopsis* spp., also exhibited significant activity against *P. falciparum* ($IC_{50} = 0.11$ and $0.33\,\mu g/ml$, respectively) compared with chloroquine diphosphate ($IC_{50} = 0.16\,\mu g/ml$) and artemisin ($IC_{50} = 0.0011\,\mu g/ml$) (Isaka et al., 2001).

VII.H. Anti-inflammatory effects

Gopalakrishnan et al. (1980) reported a prominent anti-inflammatory property of prenylated xanthones from *C. inophyllum* and *Mesua ferrea*. More recently, Lin et al. (1996) tested 18 synthetic xanthones and reported structure-activity relationships of these various oxygenated xanthones. 1,3- and 3,5-dihydroxyxanthones showed strong inhibitory effects on the release of β-glucuronidase and histamine ($10\,\mu g/ml$ each) from rat peritonial mast cells. Whereas 1,6-dihydroxyxanthone and 1,3,8-trihydroxyxanthone showed very good inhibitory activity on the release of β-glucoronidose and lysozyme (each at $10\,\mu g/ml$), respectively, from rat neutrophils. 1,3- and 1,6-dihydroxyxanthones, 1,3,7-trihydroxyxanthone and 1,3,5,6-; 2,3,6,7- and 3,4,5,6-tetrahydroxyxanthones showed potent inhibitory effects on superoxide formation of rat neutrophils. In this study, 1,6- and 3,5-dihydroxyxanthones significantly reduced the paw swelling in polymyxin B-induced oedema of normal as well as adrenalectomised mice.

VII.I. HIV-inhibitory activity

Wang et al. (1994) isolated the first flavonexanthone *C*-glucoside swertifrancheside [1,5,8-trihydroxy-3-methoxy-7-(5′,7′,3″,4″-tetrahydroxy-6′-*C*-*β*-D-glucopyranoxyl-4′-oxy-8′-flavyl)-xanthone] (**31**). This compound was found to be a potent inhibitor of DNA polymerase activity of HIV-1 reverse transcriptase by 99.8% at 200 μg/ml (ED_{50} = 30.9 μg/ml), while the dimer xanthone swertipunicoside (**32**) displayed an ED_{50} of 3.0 μg/ml.

Two related xanthones namely mangostin (**33**) and *γ*-mangostin have also been found to be active against HIV-1 protease. The type of inhibition by both compounds is non-competitive (Vlietinck et al., 1998).

Prenylated xanthones of the plant *Maclura tinctoria* have been reported to exhibit moderate anti-HIV activity (Groweiss et al., 2000).

VII.J. P-glycoprotein inhibitory property

New prenylated xanthones [(**34**), (**35**), (**36**) and (**37**)] obtained by hemisynthesis from naturally occurring xanthone (Decussatin) were found to be potent modulators of P-glycoprotein, a protein involved in multi-drug resistance phenomenon (Noungoue et al., 2000).

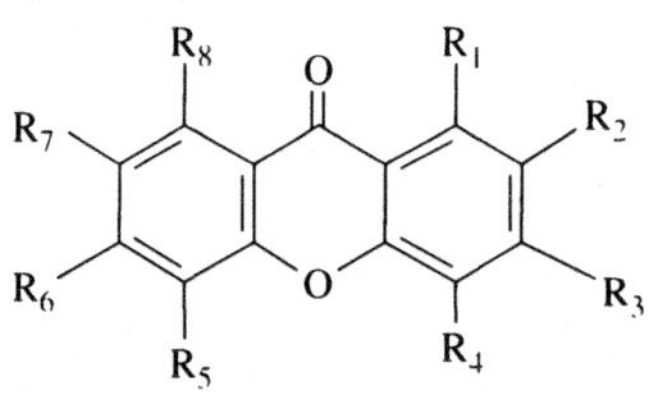

1, 5,8-Trihydroxy-3-methoxyxanthone : R_1=R_5=R_8=OH, R_3=MeOH, R_2=R_4=R_6=R_7=H (**15**)
1, 3, 6, 7-Tetrahydroxyxanthone : R_1=R_3=R_6=R_7=OH, R_2=R_4=R_5=R_8=H (**16**)
1, 3, 6, 7-Tetrahydroxy-2, 8-diprenyl-xanthone : R_1=R_3=R_6=R_7=OH, R_4=R_5=H, R_2=R_8=Pre (**17**)
1, 7-Dihydroxy-4-methoxy : R_1=R_7=OH, R_4=MeOH, R_2=R_3=R_5=R_8=H (**18**)
1, 7-Dihydroxy-3, 5, 6-trimethoxyxanthone : R_1=R_7=OH, R_3=R_5=R_6=MeOH, R_2=R_5=R_8=H (**19**)
Globulixanthone D R_1=R_6=OH, R_5=MeOH, R_2=R_3=R_4=R_8=H, R_7=Pre (**21**)
3,4-Dihydroxy-2-methoxyxanthone R_3=R_4=OH, R_2=MeOH, R_1=R_5=R_6=R_7=R_8=H (**25**)
2,3-Dihydroxy-4-methoxyxanthone R_2=R_3=OH, R_4=MeOH, R_1=R_5=R_6=R_7=R_8=H (**26**)
1, 3, 6-Trihydroxy-7-methoxy-2, 8-diprenyl-xanthone : R_1=R_3=R_6=OH, R_7=MeOH, R_4=R_5=H, R_2=R_8=Pre (**33**)

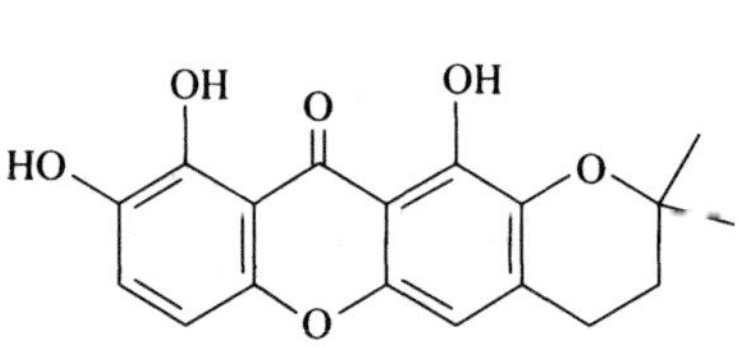

Globulixanthone C (**20**)

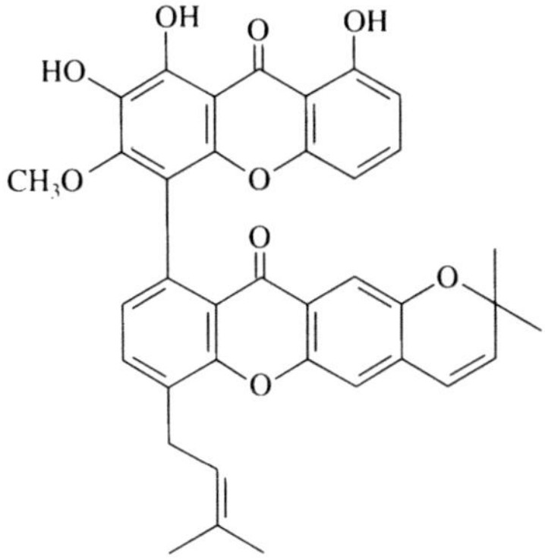

Globulixanthone E (**22**)

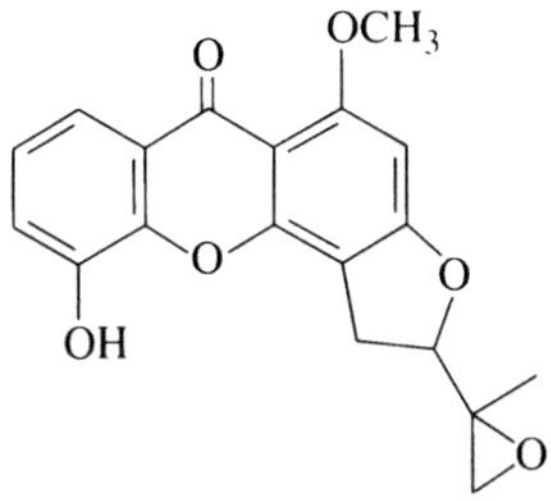

Psorespermine (**23**)

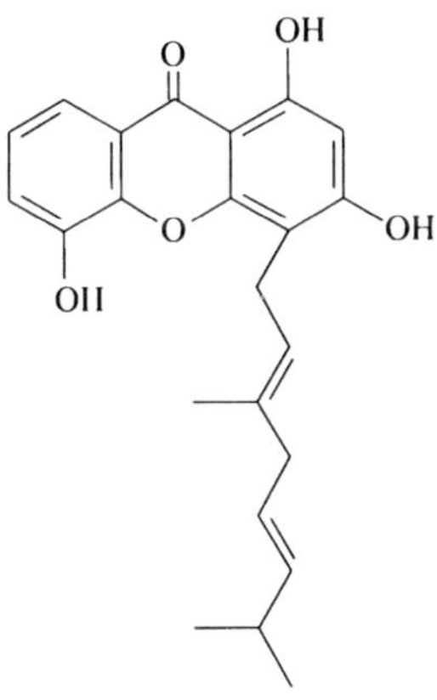

(**24**)

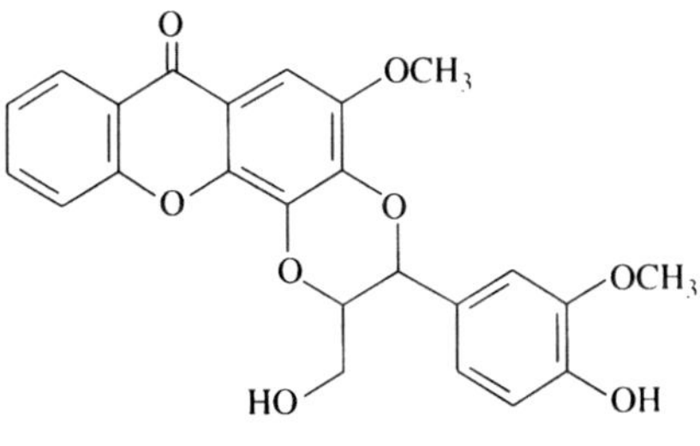

Transkielcorin (**27**)

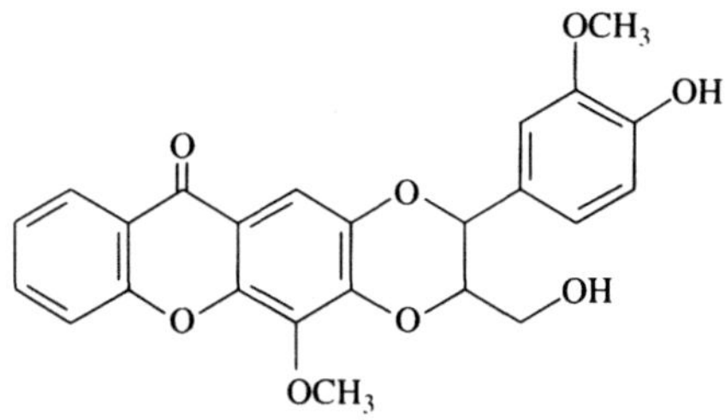

Trans-isokielcorin B (**28**)

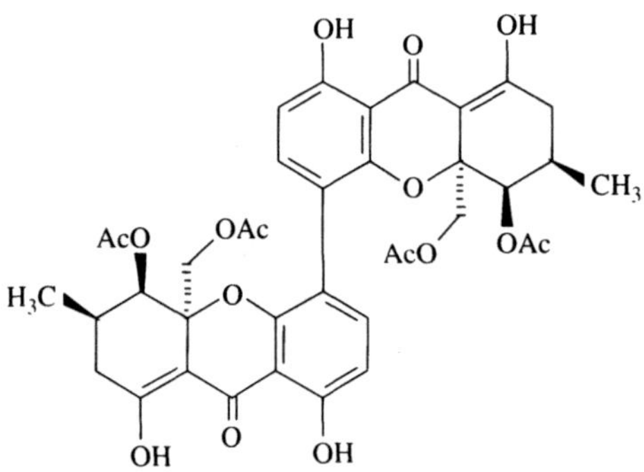

Phomoxanthone A (**29**)

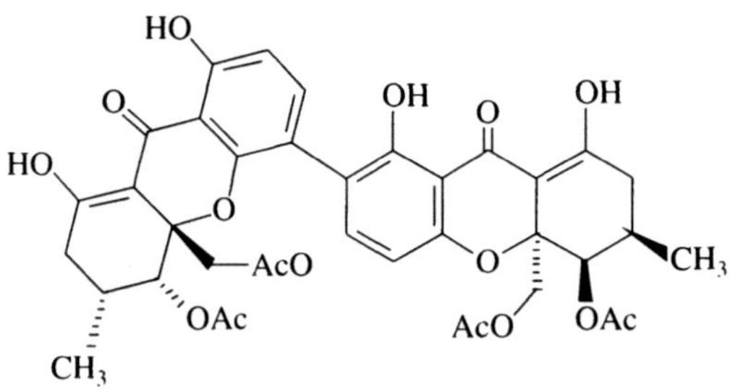

Phomoxanthone B (**30**)

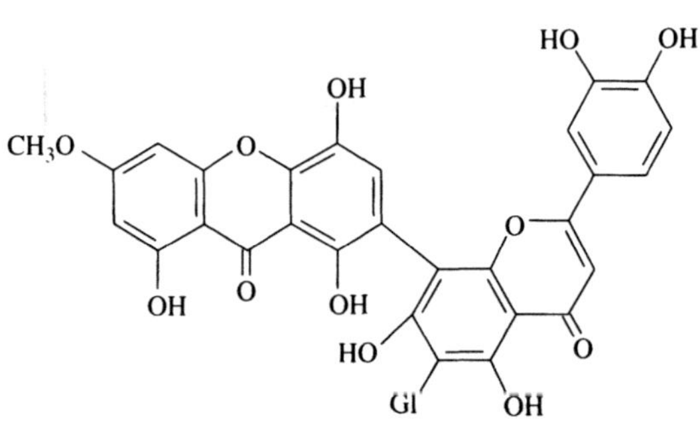

Swertifrancheside (**31**)

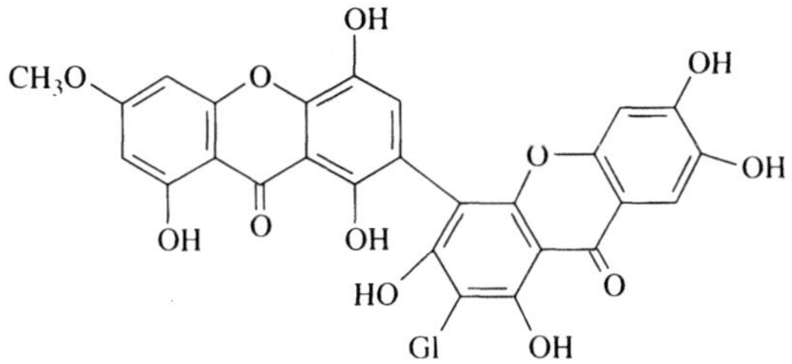

Swertipunicoside (**32**)

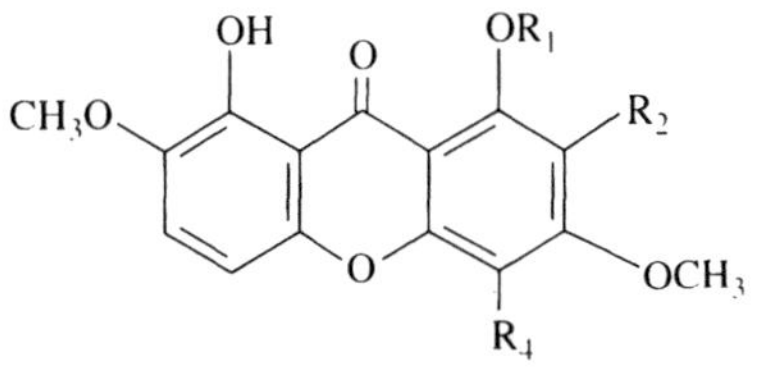

3,7,8-Trimethoxy-1-(3,3-dimethylallyl)xanthone R_2=R_4=H, R_1=Pre (**34**)
1-Hydroxy-3,7,8-trimethoxy-2-(1,1-dimethylallyl)xanthone R_1=R_4=H, R_2=Pre (**35**)
1-Hydroxy-3,7,8-trimethoxy-4-(3,3-dimethylallyl)xanthone R_1=R_2=H, R_4=Pre (**36**)
1-Hydroxy-3,7,8-trimethoxy-2-(3,3-dimethylallyl)xanthone R_1=R_4=H, R_2=Pre (**37**)

VIII. Conclusion

Xanthones belong to the an oxygen heterocycles group, which have gradually great regard because of their many interesting biological and pharmacological activities. The main natural source of this class of compounds are plants from many families whose number has progressively increased. Diversification of biological tests concerning other pathologies might contribute in raising the interest in xanthones.

References

Abou-Shoer M, Boettner FE, Chang CJ, Cassady JM. (1988) Phytochemistry 27:2795–800.
Abou-Shoer M, Habib AA, Chang CJ, Cassady JM. (1989) Phytochemistry 28:2483–7.
Adjanohoun EJ, Assi LA, Chibon P, De Vecchy H, Duboze E, Eymé J, Gassita JN, Goudote E, Guinko S, Keita A, Koudogbo B, Le Bras M, Mourambou I, Mve-Mengome E, Nguéma MG, Ollome JB, Posso P, Sita P. (1984) Agence de Coopération Culturelle et Technique, Paris.
Afzal M, Al-Hassan JM. (1980) Heterocycles 14:1173–7.
Agrawal A, Singh J. (1988) Phytochemistry 27:3692–4.
Ahmed SA, Bardshiri E, McIntyre CR, Simpson TJ. (1992) Aust J Chem 45:249–74.
Anand SM, Jain AC. (1973) Indian J Chem 11:1237–40.
Anand SM, Jain AC. (1974) Aust J Chem 27:1515–20.
Arends P. (1970) J Chromatogr 47:550–1.
Arends P, Helboe P. (1972) Acta Chem Scand 26:4180–2.
Arends P, Helboe P, Moller J. (1973) Org Mass Spectrom 7:667–81.
Asahina Y, Tanase Y. (1940) New synthesis of xanthones. Proc Imp Acad 16:297–298. (CA 34:7906).

Asthana RK, Sharma NK, Kulshreshtha DK, Chatterjee SK. (1991) Phytochemistry 30:1037–40.
Atkinson H, Heilbron IM. (1926) J Chem Soc, 2688–2691.
Atkinson JE, Gupta P, Lewis JR. (1969) Tetrahedron 24:1507–11.
Barraclough D, Locksley HD, Scheinmann F, Magalhaes MT, Gottlieb OR. (1970) J Chem Soc (B), no. 1, 603–612.
Basnet P, Kadota S, Shimizu M, Namba T. (1994) Planta Med 60:507–11.
Basnet P, Kadota S, Shimizu M, Takata Y, Kobayashi M, Namba T. (1995) Planta Med 61:402–5.
Bennett GJ, Lee HH. (1989) Phytochemistry 28:967–98.
Bernstein KE, Martin BM, Edwards AS, Bernstein EA. (1989) J Biol Chem 264:11945–51.
Birch AJ, Baldas J, Hlubucek JR, Simpson TJ, Westerman PW. (1976) J Chem Soc Perkin Trans I no. 8, 898–904.
Burling ED, Jefferson A, Scheinmann F. (1965) Tetrahedron 21:2653–69.
Cardona ML, Fernandez MI, Pedro JR, Seoane E, Vidal R. (1986) J Nat Prod 49:95–7.
Carpenter I, Locksley HD, Scheinmann F. (1969) Phytochemistry 8:2013–26.
Castelao JF, Gottlieb OR, De Lima RA, Mesquita AAL. (1977) Phytochemistry 16:735–40.
Chang CH, Lin CC, Hattori M, Namba T. (1994) J Ethnopharmacol 44:79–85.
Chaudhuri RK, Zymalkowski F, Ghosal S. (1978b) J Pharm Sci 67:1321–2.
Chen C-H, Lin J-Y, Lin C-N, Hsu S-Y. (1992) J Nat Prod 55:691–5.
Cushman DW, Cheung HS. (1971) Biochem Pharmacol 20:1637–48.
Davies JE, Kirkaldy D, Roberts JC. (1960) J Chem Soc 3(437):2169–2178.
Elix JA, Bennett SA. (1990) Aust J Chem 43:1587–90.
Elix JA, Bennett SA, Jiang H. (1991) Aust J Chem 43:1157–62.
Ellis RC, Whalley WB, Ball K. (1967) Chem Comm no. 1, 803.
Fernandez ER, Carvalho FL, Remiao FG, Bastos ML, Pinto MM, Gottlieb OR. (1995) Pharm Res 12:1756–60.
Fowler CJ, Ross SB. (1984) Med Res Rev 4:323–58.
Frahm AW, Chaudhuri RK. (1979) Tetrahedron 35:2035–8.
Fujita M, Inoue T. (1980) Chem Pharm Bull 28:2476–81.
Ghosal S, Sharma PV, Chaudhuri RK, Bhattacharya SK. (1973) J Pharm Sci 62:926–30.
Ghosal S, Chaudhuri RK. (1975) J Pharm Sci 64:888–9.
Ghosal S, Sharma PV, Chaudhuri RK, Bhattacharya SK. (1978) J Pharm Sci 67:55–60.
Ghosal S, Biswas K, Jaiswal D. (1980) Phytochemistry 9:123–8.
Gil S, Sanz V, Tortajada A. (1987) J Nat Prod 50:301–4.
Gil S, Parra M, Sanz V, Tortajada A. (1988) J Nat Prod 51:339–42.
Gopalakrishnan C, Shankaranarayanan D, Nazimudeen SK, Viswanathan S, Kameswaran L. (1980) Ind J Pharmac 12:181–91.
Gopalakrishnan G, Banumathi B, Suresh G. (1997) J Nat Prod 60:519–24.
Gottlieb OR. (1968) Phytochemistry 7:411–21.
Granoth I, Pownall HJ. (1975) J Org Chem 40:2088–91.
Gröger D, Erge D, Franck B, Ohnsorge U, Flasch H, Hüper F. (1968) Chem Ber 101:1970–8.
Grover PK, Shah GD, Shah RC. (1955) J Chem Soc no. 4, 3982–3985.
Groweiss A, Cardellina JH, Boyd MR. (2000) J Nat Prod 63:1537–9.
Gunasekera SP, Sotheeswaran S, Sultanbawa MUS. (1981) J Chem Soc, Perkin Trans 1(7):1831–5.
Gunatilaka AA, Leslee S, Jasmin AMY, Subramanian S. (1982) Phytochemistry 21:1751–3.
Gupta P, Lewis JR. (1971) J Chem Soc (C), no. 4, 629–631.
Hambloch H, Frahm AW, Wiedemann B. (1985) Eur J Med Chem-Chim Ther 20:71–7.
Harborne JB. (1994) The flavonoids advances in research since 1986. London: Chapman & Hall.
Helboe P, Arends P. (1973) Acta Chem Scand 27:2237–8.
Henry and Caventou (1981) J Pharm Chim 7:178–179.
Hof RP, Salzmann R, Siegl H. (1987) Am J Cardiol 59:30B–6B.
Hostettmann K, Wagner H. (1986) Preparative chromatography techniques. Application in natural product isolation. Phytochemistry 16:821–9.

Hostettmann K, Hostettmann M, Marston A. (1986) Berlin: Springer.
Iinuma M, Tosa H, Tanaka T, Yonemori S. (1994) Phytochemistry 35:527–32.
Iinuma M, Ito T, Tosa H, Tanaka T. (1996) Five new xanthones from *Garcinia dulcis*. J Nat Prod 59:472–5.
Ikeya y, Sugama K, Okada M, Mitsuhashi H. (1991) Phytochemistry 35:2061–4.
Isaka M, Jaturapat A, Rukseree K, Danwisetkanjana K, Tanticharoen M, Thebtaranonth Y. (2001) J Nat Prod 64:1015–8.
Iwu MM. (1993) Handbook of African medicinal plants. Florida: CRC Press.
Kazmi SNUH, Ahmed Z, Malik A. (1989) Phytochemistry 28:3572–5.
Ko FN, Lin CN, Liou SS, Huang TF, Teng CM. (1991) Eur J Pharmacol 192:133–9.
Kostanecki V. (1892) Ber D Chem 25:1649–51.
Kupchan SM, Streelman DR, Sneden AT. (1980) J Nat Prod 43:296–301.
Lewis JR. (1963) Proc Chem Soc 363:373–9.
Likhitwitayawuid K, Phadungcharoen T, Krungkrai J. (1998) Planta Med 64:70–2.
Lin CN, Chung M-I, Liou S-J, Lee T-H, Wang J-P. (1996) J Pharm Pharmacol 48:532–8.
Lin CN, Liou SS, Ko FN, Teng CM. (1993) J Pharm Sci 82:11–6.
Liu J, Huang M. (1982) Zhongeaoyao 13: 433–435. (1983 Chem Abstr 98: 140530)
Locksley HD, Moore I, Scheinmann F. (1967) Tetrahedron 23:2229–34.
Locksley HD, Murray IG. (1970) J Chem Soc (C), no. 2, 392–398.
Mahabusarakam W, Wiriyachitra P, Taylor W. (1987) J Nat. Prod 50:474–80.
Mahmoud ZF, Salam NAA, Khafagy SM. (1980) Fitoterapia 51:153–6.
Mandal S, Das PC, Joshi PC. (1992) J Indian Chem Soc 69:611–36.
Markham KR. (1965a) Tetrahedron 21:1449–52.
Markham KR. (1965b) Tetrahedron 21:3687–95.
Marston A, Hamburger M, Sordat-Diserens I, Msonthi JD, Hostettmann K. (1993) Phytochemistry 33:809–12.
McMaster WJ, Scott AI, Trippett S. (1960) J Chem Soc 5(894):4628–31.
Mehta G, Shah SR, Venkateswarlu Y. (1994) Tetrahedron 50:11729–42.
Mesquita AAL, Corrêa D, de B, Gottlieb OR, Magalhaes MT. (1968) Anothers Chim Acta 42:311–23.
Minami H, Takahashi E, Kodama M, Fukuyama Y. (1996) Phytochemistry 41:629–33.
Monache FD, Mac-Quhal MM, Monache GD, Bettolo GBM, De Lima RA. (1983) Phytochemistry 22:227–9.
Morel C, Seraphin D, Teyrouz A, Larcher G, Bouchara J-P, Litaudon M, Richomme P, Bruneton J. (2002) Planta Med 68:41–4.
Nakatani K, Nakahata N, Arakawa T, Yasuda H, Ohizumi Y. (2002) Biochemical Pharmacology 63:73–9.
Nkengfack AE, Mkounga P, Meyer M, Fomum ZT, Bodo B. (2002) Phytochemistry 61:181–7.
Noungoue TD. (1998) Thesis, Faculty of Pharmacy, University of Grenoble I, France.
Noungoue TD, Cartier G, Dijoux-Franca M-G, Mariotte A-M, Tsamo E. (2001) Pharma Biol 39:202–5.
Noungoue TD, Dijoux-Franca M-G, Mariotte A-M, Tsamo E, Daskiewiez JB, Bayet C, Barron D, Conseil G, Dipietro A. (2000) Bioorg Med Chem 10:1343–5.
Okorie DA. (1976) Phytochemistry 15:1799–801.
Patel GN, Trivedi KN. (1988) J Indian Chem Soc 65:192–3.
Philbin EM, Swirski J, Wheeler TS. (1956) J Chem Soc 4(857):4455–8.
Pillai RKM, Naiksatam P, Johnson F, Rajagopalan R, Watts PC, Cricchio R, Borras S. (1986) J Org Chem 51:717–23.
Pinto DCG, Fuzzati N, Pazmino XC, Hostettmann K. (1994) Phytochemistry 7:875–7.
Quillinan AJ, Scheinmann F. (1973) J Chem Soc Perkin Trans I no. 13, 1329–1337.
Ramel C, Alekperov UK, Ames BN, Kada T, Wattenberg LW. (1986) Mutat Res 168:47–65.
Rath G, Potterat O, Mavi S, Hostettmann K. (1996) Phytochemistry 43:513–7.
Ravi P, Vathani P, Reddy GC. (1994) Indian J Heterocyc Chem 3:209–10.
Roberts JC. (1961) Chem Rev 61:591–605.
Rocha L, Marston A, Kaplan MAC, Evans HS, Thull U, Testa B, Hostettman K. (1994) Phytochemistry 36:1381–5.

Rodriguez S, Wolfender J-L, Odontuya G, Purev O, Hostettmann K. (1995) Phytochemistry 40:1265–72.
Schaufelberger D, Hostettmann K. (1988) Planta Med 54:219–21.
Schimmer O, Mauthner H. (1996) Planta Med 62:561–4.
Sullivan G, Stiles FD, Rosler KHA. (1977) J Pharma Sci 66:828–35.
Sultanbawa MUS. (1980) Tetrahedron 36:1465–506.
Vitale AA, Romanelli GP, Autino JC, Pomilio AB. (1994) J Chem Res (S) 82–83.
Ullmann F, Pauchaud L. (1906) Ann 350:108–15.
Vlietinck AJ, Bruyne TD, Apers S, Pieters LA. (1998) Planta Med 64:97–109.
Wang J-N, Hou C-Y, Liu Y-L, Lin L-Z, Gil RR, Cordell GA. (1994) J Nat Prod 57:211–7.
Wattenberg LW. (1985) Cancer Res 45:1–8.
Westerman PW, Gunasekera SP, Sultanbawa MUS, Kazlauskas R. (1977) Org Magn Reson 9:631–6.
Yoshimi N, Matsunaga K, Katayama M, Yamada Y, Kuno T, Quoi Z, Hara A, Yamahara J, Mori H. (2001) Cancer Lett 163:163–70.
Zhou HM, Liu YL, Blasko G, Cordell GA. (1989) Phytochemistry 28:3569–71.

M.T.H. Khan and A. Ather (eds.)
Lead Molecules from Natural Products

Inhibition of immunodeficiency type-1 virus (HIV-1) life cycle by medicinal plant extracts and plant-derived compounds

ROBERTO GAMBARI, ILARIA LAMPRONTI

Abstract

Identification of molecules inhibiting the different steps of the life cycle of the human immunodeficiency virus type 1 (HIV-1) is central for the development of an efficient therapeutic treatment of AIDS. In this respect, this research takes great advantage from the fact that the molecular biology of the HIV-1 life cycle is well known. The present review summarizes and discusses strong evidences demonstrating that several medicinal plant extracts display anti-HIV-1 activities. Moreover, several compounds isolated from bio-active extracts were characterized in detail and found to exhibit inhibitory effects against different steps of the HIV-1 life cycle, including virus–cell fusion and virus absorption, reverse transcription, integration and proteolytic cleavage. Clinical trials on limited numbers of patients suffering from AIDS have been already reported, supporting the concept that some medicinal plant extracts or single molecules derived from them display anti-HIV effects not only *in vitro* but also *in vivo*. Therefore, medicinal plant extracts should be considered as an excellent source of clinically relevant anti-HIV-1 molecules.

Keywords: *medicinal plants, HIV-1, AIDS, ethnopharmacology*

Abbreviations: AIDS, acquired immunodeficiency syndrome; HIV-1, human immunodeficiency virus, type 1; RT, reverse transcriptase; PR, protease; PI, protease inhibitor; LTR, long terminal repeat; NF-kappaB, nuclear factor kappaB; HAART, highly active antiretroviral therapy; AZT (zidovudine), EMSA, electrophoretic mobility shift assay.

I. Introduction

The life cycle of the human immunodeficiency type-1 virus (HIV-1) is one of the major targets for the development of pharmaceutical compounds of great interest in biomedicine, considering the fact that HIV-1 infection causes AIDS, one of the major causes of mortality in the developed world. Accordingly, significant efforts have been made in the recent past to identify molecules inhibiting the different biological steps of the HIV-1 life cycle. In this respect, this research field takes great advantage from the fact that the molecular biology of the HIV-1 life cycle is well known and has been the object of several excellent research papers and reviews (Wang et al., 2000; Pani et al., 2002; Bukrinskaya, 2004; Bannwarth and Gatignol, 2005; Nielsen et al., 2005).

The important steps in HIV-1 infection are virus–cell attachment, gp120-CD4 binding, gp120-coreceptor binding, viral fusion, viral assembly and disassembly, reverse transcription, nuclear import of the pre-integration complex, proviral integration, viral transcription, processing of viral transcripts and nuclear export, assembly of new virions (Figure 1). In addition to HIV-1 proteins, several cellular factors are involved in HIV replication (Bukrinskaya, 2004; Nielsen et al., 2005).

As far as anti-HIV-1 drugs, the current therapeutic approach is based on the combined use of different molecules, such as AZT (zidovudine), enfuvirtide (the first

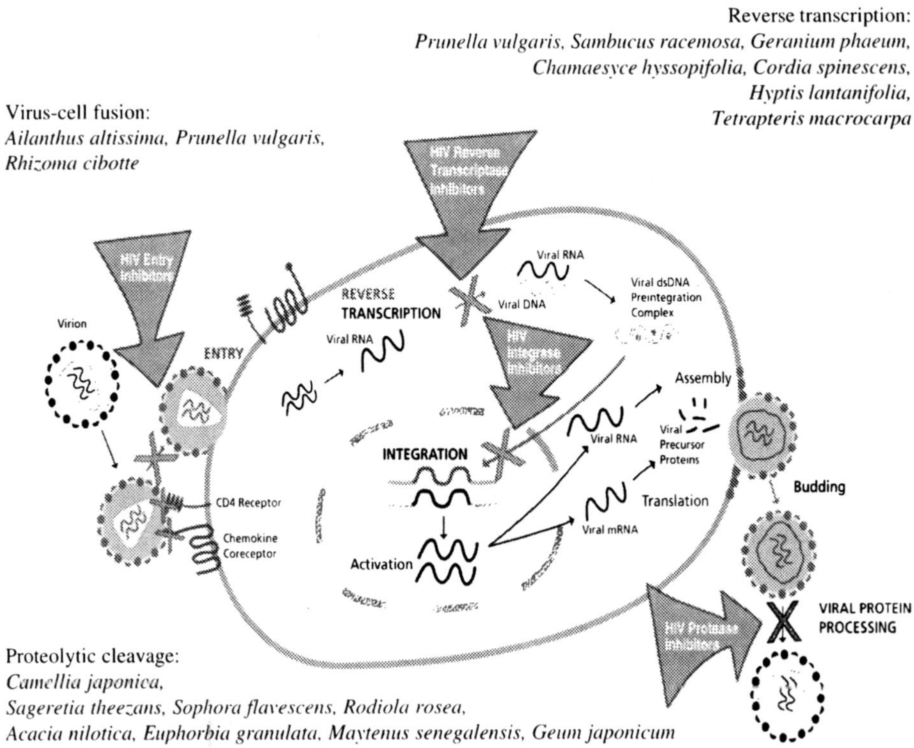

Fig. 1. Steps of the HIV-1 life cycle affected by medicinal plant extracts. The effects of *Ailanthus altissima* on virus–cell fusion was reported by Chang and Woo (2003); the effects on reverse transcription were found using extracts from *Prunella vulgaris* (Kageyama et al., 2000), *Sambucus racemosa* (Mlinaric et al., 2000), *Geranium phaeum* (Mlinaric et al., 2000), *Chamaesyce hyssopifolia* (Matsuse et al., 1999), *Cordia spinescens* (Matsuse et al., 1999), *Hyptis lantanifolia* (Matsuse et al., 1999), *Tetrapteris macrocarpa* (Matsuse et al., 1999); inhibition of proteolytic cleavage was obtained with extracts from *Camellia japonica* (Park et al., 2002), *Sageretia theezans* (Park et al., 2002), *Sophora flavescens* (Park et al., 2002), *Rodiola rosea* (Min et al., 1999), *Acacia nilotica* (Hussein et al., 1999), *Euphorbia granulata* (Hussein et al., 1999), *Maytenus senegalensis* (Hussein et al., 1999), *Geum japonicum* (Xu et al.,1996); Liu et al. (2002) studied the effects of *Prunella vulgaris* and *Rhizoma cibotte* on gp41.

fusion inhibitor), tenofovir (a reverse transcriptase inhibitor), atazanavir (a protease inhibitor), tipranavir (another protease inhibitor) (Barbaro et al., 2005).

Reviews on the alternative approaches aimed at the treatment of AIDS patients have been published (Pande and Ramos, 2003; Dionisio et al., 2004; Pereira and Paridaen, 2004; Barbaro et al., 2005; Boffito et al., 2005). The most promising approach is the highly active antiretroviral therapy (HAART) employing the combined use of drugs exhibiting different mechanisms of action. This approach, that has markedly decreased mortality and morbidity of AIDS patients, consists of a combination of three or more of the following classes of antiretroviral drugs: reverse transcriptase inhibitors, protease inhibitors and a recently approved fusion inhibitor. However, HAART cannot completely eradicate HIV-1 from the body, results in long-term toxicity and eventually leads to the emergence of drug-resistant HIV-1 strains. Therefore, more anti-HIV-1 compounds are of great interest.

As far as interest of medicinal plants and compounds isolated from them is concerned, it is relevant to note that by simply looking to the recent literature, several reports have been published in which plant extracts have been claimed to exhibit anti-HIV-1 activity (Atta-ur-Rahman, 1996; Vlietinck et al., 1998; Matthee et al., 1999; Bedoya et al., 2001).

For instance, in a first study, el-Mekkawy et al. (1995) reported the inhibitory effects of Egyptian folk medicines on HIV-1 reverse transcriptase. These authors examined the activity of extracts of 41 medicinal plants and found that the extracts of fruits of *Phyllanthus emblica*, *Quercus pedunculata*, *Rumex cyprius*, *Terminalia bellerica*, *Terminalia chebula* and *Terminalia horrida* showed significant inhibitory activity on HIV-1 reverse transcriptase. Interestingly, this study led, through a bioassay guided-fractionation of the methanol extract of the fruit of *P. emblica*, to the isolation of putranjivain A as a potent inhibitory substance, together with 1,6-di-*O*-galloyl-beta-D-glucose, 1-*O*-galloyl-beta-D-glucose, kaempferol-3-*O*-beta-D-glucoside, quercetin-3-*O*-beta-D-glucoside and digallic acid (el-Mekkawy et al., 1995).

In another study, Bedoya et al. (2002), as part of a screening project aimed at the identification of novel anti-AIDS agents from natural sources, evaluated extracts of 15 medicinal plants widely used in the folk medicines of North America and Europe. Most of the extracts tested were relatively nontoxic to human lymphocytic MT-2 cells, but only extracts of *Hysopp officinalis* and *Dittrichia viscosa* exhibited anti-HIV activity in an *in vitro* MTT assay. This study indicates that anti-HIV-1 activity is restricted to a minority of medicinal plants, and it is not a simply unspecific characteristic of plant extracts.

The observation that only a minority of plant extracts is active against HIV-1 was confirmed in other studies in which comparison with activity on life cycle of different disease-related viruses was performed. For instance, Ruffa et al. (2004) studied 15 Argentine medicinal plants for their *in vitro* antiviral activity against herpes simplex virus types 1 and 2 (HSV-1 and 2), bovine viral diarrhoea virus type 1 (BVDV-1), influenza virus type A (Inf A) and HIV-1. In this case, antiviral activity was evaluated by a reduction in cytopathic effect, plaque-forming units and p24 HIV-1 antigen. While they found activity of the analyzed plant extracts on BVDV-1, HSV-1 and Inf A, none of the plants assayed against HIV-1 displayed any antiviral activity.

Taken together, the results available in the literature are consistent with the hypothesis that large number of plant extracts should be screened in order to find those active against HIV-1 (Li et al., 1993, Wu et al., 1996; Lin et al., 1997; Chang et al., 2003).

Accordingly, Park et al. (2002), in order to identify substances with anti-HIV activity in traditional medicines, screened 101 extracts of Korean medicinal plants for their inhibitory effects on HIV type-1 protease. Of the extracts tested, strong inhibitory effects were observed in the acetone extracts of the pericarp and leaves of *Camellia japonica*, in the water extract of the leaves of *Sageretia theezans* and in the methanol extract of the aerial part of *Sophora flavescens*. Camelliatannin H, isolated from the pericarp of *C. japonica*, showed a potent inhibitory activity on HIV-1 PR.

Tables 1 and 2 show a partial list of the medicinal plant extracts found to be active against HIV-1.

It should be underlined that some studies indicated also the possible target of the anti-HIV-1 plant extracts used. For instance, stem bark of *Ailanthus altissima* inhibits virus–cell fusion (Chang et al., 2003). Other reports were focused on the activity against HIV-1 reverse transcriptase, showing that extracts from *Prunella vulgaris* (Kageyama et al., 2000), *Sambucus racemosa* (Mlinaric et al., 2000), *Geranium phaeum* (Mlinaric et al., 2000), *Chamaesyce hyssopifolia*, *Cordia spinescens*, *Hyptis lantanifolia*, *Tetrapteris macrocarpa* (Matsuse et al., 1999) are potent inhibitors of this crucial enzyme. Plant extracts targeting gp41 (Liu et al., 2002) and the HIV-1 protease proteolytic cleavage (Xu et al., 1996; Lin et al., 1997; Hussein et al., 1999; Min et al., 1999) were also reported. Of course, these analyses display the obvious difficulty related to the fact that plant extracts are very much heterogeneous with respect to bioactive compounds responsible for the found biological effects.

Figure 1 shows a partial list of medicinal plant extracts targeting specific biological steps of the HIV-1 life cycle.

II. Comparative *in vitro* effects of extracts from medicinal plants and AZT

Several authors tried to understand whether the activity of anti-HIV plants extracts is comparable or even better than commonly used anti-HIV drugs, including AZT. An example of such studies is that reported by Ayisi and Nyadedzor (2003) on the comparative *in vitro* effects of AZT and extracts of *Ocimum gratissimum*, *Ficus polita*, *Clausena anisata*, *Alchornea cordifolia*, *Elaeophorbia drupifera* against HIV-1 and HIV-2 infections. Interestingly, they found that some plant extracts were more active than AZT in inhibiting HIV-1 life cycle. For instance, in Molt-4/HIV cultures early cytopathic effect of cell fusion was unaffected by AZT but was completely inhibited by all the employed plant extracts at noncytotoxic concentrations. Other studies in which the activity of plant extracts was studied together with that of AZT are those published by Lopez et al. (2003) and Piras et al. (1997).

Table 1
Medicinal plants exhibiting anti-HIV-1 activities (families in alphabetical order, from Anacardiaceae to Guttiferae)

Anacardiaceae	*Rhus succedanea* (Lin et al., 1997)
Annonaceae	*Annona squamosa* (Wu et al., 1996), *Polyalthia suberosa* (Li et al., 1993)
Araucariaceae	*Ailanthus altissima* (Chang et al., 2003)
Aspidiaceae	*Dryopteris crassirhizoma* (Min et al., 2001a)
Asteraceae	*Achyrocline flaccida* (Hnatyszyn et al., 1999), *Aspilia pluriseta* (Cos et al., 2002), *Aster scaber* (Kwon et al., 2000), *Calendula officinalis* (Kalvatchev et al., 1997), *Chrysanthemum morifolium* (Hu et al., 1994)
Boraginaceae	*Cordia spinescens* (Matsuse et al., 1999)
Caprifoliaceae	*Sambucus racemosa* (Mlinaric et al., 2000)
Celastraceae	*Maytenus senegalensis* (Hussein et al., 1999), *Tripterygium hypoglaucum* (Duan et al., 2000), *Tripterygium wilfordii* (Chen et al., 1992; Duan et al., 2000)
Cistaceae	*Tuberaria lignosa* (Bedoya et al., 2001)
Clusiaceae	*Garcinia mangostana* (Chen et al., 1996), *Garcinia multiflora* (Lin et al., 1997)
Combretaceae	*Buchenavia capitata* (Beutler et al., 1992), *Combretum paniculatum* (Asres et al., 2001), *Terminalia bellerica* (el-Mekkawy et al., 1995), *Terminalia chebula* (el-Mekkawy et al., 1995), *Terminalia horrida* (el-Mekkawy et al., 1995)
Cornaceae	*Cornus kousa* (Vlietinck et al., 1998)
Crassulaceae	*Rodiola rosea* (Min et al., 1999)
Cucurbitaceae	*Momordica charantia* (Lee-Huang et al., 1990), *Trichosanthes kirilowii* (Lee-Huang et al., 1991)
Cynomoriaceae	*Cynomorium songaricum* (Ma et al., 1999)
Dipterocarpaceae	*Vatica cinerea* (Zhang et al., 2003)
Euphorbiaceae	*Alchornea cordifolia* (Ayisi et al., 2003), *Chamaesyce hyssopifolia* (Matsuse et al., 1999), *Croton tiglium* (el-Mekkawy et al., 1999; el-Mekkawy et al., 2000), *Elaeophorbia drupifera* (Ayisi et al., 2003), *Euphorbia granulata* (Hussein et al., 1999), *Gelonium multiflorum* (Lee-Huang et al., 1995), *Homalanthus nutans* (Gustafson et al., 1992), *Jatropha curcas* (Matsuse et al., 1999), *Mallotus japonicus* (Min et al., 2001a), *Maprounea africana* (Pengsuparp et al., 1994, 1995), *Phyllanthus amarus* (Notka et al., 2003, 2004), *Phyllanthus emblica* (el-Mekkawy et al., 1995), *Phyllanthus niruri* (Qian-Cutrone, 1996), *Phyllanthus sellowianus* (Asres et al., 2001)
Fabaceaea	*Acacia nilotica* (Hussein et al., 1999), *Gleditsia japonica* (Konoshima et al., 1995)
Fagaceae	*Quercus pedunculata* (el-Mekkawy et al., 1995)
Ganodermaceae	*Ganoderma lucidum* (el-Mekkawy et al., 1998)
Gentianaceae	*Swertia franchetiana* (Wang et al., 1994; Pengsuparp et al., 1995)
Geraniaceae	*Geranium phaeum* (Mlinaric et al., 2000)
Guttiferae	*Calophyllum lanigerum* (Matthee et al., 1999)

III. From unfractionated plant extracts to pure compounds and from the laboratory to the clinic

Several classes of molecules isolated from medicinal plants have been found to exert anti-HIV-1 activities, including low molecular weight molecules, peptides and proteins.

Table 2
Medicinal plants exhibiting anti-HIV-1 activities (families in alphabetical order, from Hippocastanaceae to Zygophyllaceae)

Hippocastanaceae	*Aesculus chinensis* (Yang et al., 1999)
Lamiaceae (Labiatae)	*Anisomeles indica* (Shahidul et al. 2000), *Hyptis capitata* (Kashiwada et al., 1998), *Hyptis lantanifolia* (Matsuse et al., 1999), *Mentha piperita* (Yamasaki et al., 1998), *Ocimum basilicum* (Yamasaki et al., 1998), *Ocimum gratissimum* (Ayisi et al., 2003), *Melissa officinalis* (Yamasaki et al., 1998), *Perilla frutescens* (Yamasaki et al., 1998), *Prunella vulgaris* (Yamasaki et al., 1998; Liu et al., 2002; Kim et al., 2000), *Rosmarinus officinalis* (Paris et al., 1993), *Satureja montana* (Yamasaki et al., 1998)
Lauraceae	*Cinnamomum cassia* (Premanathan et al., 2000)
Leguminosae	*Gymnocladus chinensis* (Konoshima et al., 1995)
Malpighiaceae	*Tetrapteris macrocarpa* (Matsuse et al., 1999)
Melianthaceae	*Bersama abyssinica* (Asres et al., 2001)
Mimosaceae	*Prosopis glandulosa* (Kashiwada et al., 1998)
Moraceae	*Ficus polita* (Ayisi et al., 2003)
Olacaceae	*Ximenia americana* (Asres et al., 2001)
Orchidaceae	*Cymbidium hybrid* (Balzarini et al., 1992), *Epipactis helleborine* (Balzarini et al., 1992)
Parmeliaceae	*Cetraria islandica* (Pengsuparp et al., 1995)
Plumbaginaceae	*Limonium tetragonum* (Min et al., 2001a)
Polygonaceae	*Rumex bequaertii* (Cos et al., 2002), *Rumex cyprius* (el-Mekkawy et al., 1995)
Rosaceae	*Agrimonia pilosa* (Min et al., 2001a), *Eriobotrya japonica* (De Tommasi et al., 1992), *Geum japonicum* (Xu et al., 1996), *Rosa damascena* (Mahmood et al., 1996), *Sanguisorba minor magnolii* (Bedoya et al., 2001), *Waldsteinia fragarioides* (Abou-Karam and Shier, 1992)
Rutaceae	*Clausena anisata* (Ayisi et al., 2003)
Sapindaceae	*Dodonaea angustifolia* (Asres et al., 2001), *Xanthoceras sorbifolia* (Ma et al., 2000)
Theaceae	*Camellia japonica* (Park et al., 2002)
Thymelaeceae	*Wikstroemia indica* (Hu et al., 2000)
Urticaceae	*Urtica dioica* (Balzarini et al., 1992)
Verbenaceae	*Clerodendron trichotomum* (Kim et al., 2001)
Viscaceae	*Phoradendron juniperinum* (Kashiwada et al., 1998)
Zygophyllaceae	*Larrea tridentata* (Gnabre et al., 1996)

Despite the fact that most of the reported studies have been done *in vitro*, *in vivo* effects of a limited number of molecules have also been reported. For instance, Notka et al. (2004) administered orally to volunteers plant material from *Phyllanthus amarus*, being able to demonstrate a potent anti-HIV activity in blood. Sera at a final concentration of 5% reduced HIV replication by more than 30%. These results support the conclusion that *P. amarus* has inhibitory effects on HIV not only *in vitro* but also *in vivo*.

Among the very few plant-derived anti-HIV products used in clinical trials, glycyrrhizin, papaverine, trichosanthin, castanospermine, *N*-butyl-1-deoxynojirimycin and acemannan have been administered to a limited number of patients suffering from AIDS (Vlietinck et al., 1998).

IV. The HIV-1 life cycle and the mechanism of action of anti-HIV compounds from medicinal plants

Despite the fact that the molecular target(s) of the biological action of several anti-HIV substances, including alkaloids (*O*-demethyl-buchenavianine, papaverine), polysaccharides (acemannan), lignans (intheriotherins, schisantherin), phenolics (gossypol, lignins, catechol dimers such as peltatols, naphthoquinones such as conocurvone) and saponins (celasdin B, Gleditsia and Gymnocladus saponins), has not been fully elucidated, the molecular targets of several isolated compounds from medicinal plants have been identified (Vlietinck et al., 1998).

As already pointed out, the HIV-1 life cycle has been deeply investigated and several reviews are available on this particular issue (Wang et al., 2000; Pani et al., 2002; Bukrinskaya, 2004; Bannwarth and Gatignol, 2005; Nielsen et al., 2005). The first step is of course virus–cell fusion and virus adsorption; other critical molecular events are reverse transcription and integration. After integration, transcription of HIV-1 is of great importance, as well as TAR–Tat and RRE–Rev interactions. Finally, translation, proteolytic cleavage, glycosylation and assembly/release are biological steps that can be considered as molecular targets of anti-HIV activity (see Figure 1).

Therefore, several research groups tried to link the anti-HIV-1 activity of compounds isolated from medicinal plants with specific steps of the HIV-1 life cycle.

Table 3 reports a partial list of molecules isolated from plant extracts, divided for their effects on specific HIV-1 life cycle steps.

It should be observed that some molecules inhibit several HIV-1 life cycle steps. For instance, triterpenes inhibit virus absorption, but also virus–cell fusion and reverse transcription (Pengsuparp et al., 1995; Vlietinck et al., 1998); curmarins inhibit virus absorption, reverse transcription, integration and proteolytic cleavage (Vlietinck et al, 1998). On the other hand, some class of anti-HIV-1 compounds exhibit restricted activity, such as lectins (inhibiting virus–cell fusion) (Table 3).

While in most cases a good correlation is found between activity on a specific HIV-1 life cycle step and inhibition of HIV-1 infection in cellular models, this statement cannot be generalized. One example is the study by Qian-Cutrone et al., who found that niruriside, isolated from the MeOH extract of the dried leaf of *Phyllanthus niruri* L., is a new and strong HIV REV/RRE binding inhibitor (Qian-Cutrone et al., 2005). In this case, the screening of natural products was designed for their ability to inhibit the binding of HIV-REV protein to ^{33}P-labeled RRE RNA. Unfortunately, niruriside did not protect CEM-SS cells from acute HIV infection (Qian-Cutrone et al., 2005).

V. Inhibition of transcription factors involved on HIV-1 transactivation

While several cellular transcription factors play a crucial role in the HIV-1 cell cycle, acting on the HIV-1 long terminal repeat (LTR) (Wang et al., 2000; Pani et al., 2002; Bukrinskaya, 2004; Bannwarth and Gatignol, 2005; Nielsen et al., 2005), few examples have been reported on selective effects of medicinal plant derived products on pathways related to these transcription factors.

Table 3
Steps of the HIV-1 life cycle affected by molecules isolated from plant extracts

1. Virus adsorption	Chromone alkaloids (schumannificine), isoquinoline alkaloids (michellamines), sulphated polysaccharides and polyphenolics, flavonoids, coumarins (glycocoumarin, licopyranocoumarin), phenolics (caffeic acid derivatives, galloyl acid derivatives, catechinic acid derivatives), tannins, triterpenes (glycyrrhizin and analogues, soyasaponin and analogues)
2. Virus–cell fusion	Lectins (mannose- and *N*-acetylglucosamine-specific) and triterpenes (betulinic acid and analogues) (Fujioka et al., 1994)
3. Reverse transcription	Alkaloids (benzophenanthridines, protoberberines, isoquinolines, quinolines) (McCormick et al., 1996), coumarins (calanolides and analogues) (Matthee et al., 1999), flavonoids (robustaflavone, hinokiflavone, kaempferol acetylrhamnosides) (Min et al., 2001b; Lin et al., 1997)
	Phloroglucinols, lactones (protolichesterinic acid) (Pengsuparp et al., 1995), tannins, iridoids (fulvoplumierin), triterpenes (1 *β*-hydroxyaleuritolic acid 3-p-hydroxybenzoate) (Pengsuparp et al., 1995; Sun et al., 1996)
	Flavonone–xanthone glucoside (swertifrancheside) (Pengsuparp et al., 1995)
4. Integration	Coumarins (3-substituted-4-hydroxycoumarins), lignans (arctigenin and analogues), phenolics (curcumin, *O*-caffeoyl derivatives) (Kim et al., 2000; Kwon et al., 2000)
	Phenylpropanoid glycosides (acteoside, acteoside isomer, leucosceptoside A, plantainoside C, jionoside D, martynoside, isomartynoside) (Kim et al., 2001)
	Flavonoids (quercetin 3-*O*-(2″-galloyl)-alpha-*L*-arabinopyranoside) (Kim et al., 1998)
5. Translation	Single chain ribosome inactivating proteins (SCRIP's)
6. Proteolytic cleavage (protease inhibition)	Saponins (ursolic and maslinic acids), xanthones (mangostin and analogues), cumarins, tannins (camelliatannin H) (Park et al., 2002)
	Epigallocatechin-(4beta→8, 2beta→*O*-7)-epicatechin (Ma et al., 2000), 3-oxotirucalla-7, 24-dien-21-oic acid (Ma et al., 2000)
	Triterpene acids (oleanolic acid, ursolic acid and its hydrogen malonate) (Ma et al., 2000; Ma et al., 1999)
	Diterpenes (carnosic acid) (Paris et al., 1993)
7. Glycosylation	Alkaloids, including indolizidines (castanospermine and analogues), piperidines (1-deoxynojirimicin and analogues), pyrrolizidines (australine and analogues)
8. Assembly/release	Naphthodianthrones (hypericin and pseudohypericin), photosensitisers (terthiophenes and furoisocoumarins), phospholipids (Hudson et al., 1993)
9. Rev/RRE interactions	Niruriside (Qian-Cutrone et al., 1996)

Source: Unless otherwise indicated, informations are from Vlietinck et al. (1998).

In this respect, Sancho et al. (2004) studied the activity of imperatorin. Interestingly, imperatorin did not inhibit the reverse transcription nor the integration steps in the HIV-1 viral cell cycle. These authors found that the transcription factor Sp1 is critical for the inhibitory activity of imperatorin. These authors found that imperatorin strongly inhibits other Sp1-related genes, such as cyclin D1 expression, leading to a deep alteration of cell cycle, characterized by an arrest of the cells at the G(1) phase of the cell cycle. These results highlight the potential of Sp1 transcription factor as a target for natural anti-HIV-1 compounds, such as furanocoumarins, that might have a potential therapeutic role in the management of AIDS.

In a recent report, we demonstrated that several Bangladeshi medicinal plant extracts are active in inhibiting molecular interactions between DNA and nuclear factor kappaB (NF-kB), another transcription factor deeply involved in HIV-1 transcriptional activation (Lampronti et al., 2005). We employed the electrophoretic mobility shift assay (EMSA) as a suitable technique for the identification of plant extracts altering the binding between transcription factors and the specific DNA elements. In this paper, we have analyzed the activity of extracts from the medicinal plants *Hemidesmus indicus, Polyalthia longifolia, Aphanamixis polystachya, Moringa oleifera, Lagerstroemia speciosa, Paederia foetida, Cassia sophera, Hygrophila auriculata* and *Ocimum sanctum*. We found that low concentrations of *Hemidesmus indicus, Polyalthia longifolia, Moringa oleifera* and *Lagerstroemia speciosa*, and very low concentrations of *Aphanamixis polystachya* extracts inhibit the interactions between nuclear factors and target DNA elements mimicking sequences recognized by NF-kB. On the contrary, high amount of extracts from *Paederia foetida, Cassia sophera, Hygrophila auriculata* or *Ocimum sanctum* were unable to inhibit NF-kB/DNA interactions. Extracts inhibiting NF-kB binding activity might be a source for the development of anti-HIV-1 drugs.

VI. Conclusions

In this review, we have described studies demonstrating that several medicinal plant extracts display anti-HIV-1 activity. In addition, several compounds isolated from bioactive extracts were found to exhibit inhibitory activities against different steps of the HIV-1 life cycle, including virus–cell fusion and virus absorption, reverse transcription, integration and proteolytic cleavage (Pengsuparp et al., 1995; Vlietinck et al, 1998).

Accordingly, clinical trials on limited numbers of patients suffering from AIDS have been already reported (Vlietinck et al., 1998; Notka et al., 2004), supporting the concept that some medicinal plant extracts or single molecules derived from them display anti-HIV effects not only *in vitro* but also *in vivo*.

Therefore, medicinal plant extracts should be considered as an excellent source of clinically relevant anti-HIV-1 molecules.

Acknowledgments

Work supported by Fondazione Italiana Ricerca sulla Fibrosi Cistica, by AIRC, by Associazione Veneta per la Lotta alla Talassemia, Rovigo and by UE (Interventi Strutturali Obiettivo 2).

References

Abou-Karam M, Shier WT. (1992) Isolation and characterization of an antiviral flavonoid from Waldsteinia fragarioides. J Nat Prod 55:1525–7.

Asres K, Bucar F, Kartnig T, Witvrouw M, Pannecouque C, De Clercq E. (2001) Antiviral activity against human immunodeficiency virus type 1 (HIV-1) and type 2 (HIV-2) of ethnobotanically selected Ethiopian medicinal plants. Phytother Res 15:62–9.

Atta-ur-Rahman. (1996) Plants and AIDS. Nature 383:294.

Ayisi NK, Nyadedzor C. (2003) Comparative *in vitro* effects of AZT and extracts of *Ocimum gratissimum*, *Ficus polita*, *Clausena anisata*, *Alchornea cordifolia*, and *Elaeophorbia drupifera* against HIV-1 and HIV-2 infections. Antiviral Res 58:25–33.

Balzarini J, Neyts J, Schols D, Hosoya M, Van Damme E, Peumans W, De Clercq E. (1992) The mannose-specific plant lectins from Cymbidium hybrid and Epipactis helleborine and the (N-acetylglucosamine) n-specific plant lectin from Urtica dioica are potent and selective inhibitors of human immunodeficiency virus and cytomegalovirus replication *in vitro*. Antiviral Res 18:191–207.

Bannwarth S, Gatignol A. (2005) HIV-1 TAR RNA: the target of molecular interactions between the virus and its host. Curr HIV Res 3:61–71.

Barbaro G, Scozzafava A, Mastrolorenzo A, Supuran CT. (2005) Highly active antiretroviral therapy: current state of the art, new agents and their pharmacological interactions useful for improving therapeutic outcome. Curr Pharm Des 11:1805–43.

Bedoya LM, Palomino SS, Abad MJ, Bermejo P, Alcami J. (2002) Screening of selected plant extracts for *in vitro* inhibitory activity on human immunodeficiency virus. Phytother Res 16:550–4.

Bedoya LM, Sanchez-Palomino S, Abad MJ, Bermejo P, Alcami J. (2001) Anti-HIV activity of medicinal plant extracts. J Ethnopharmacol 77:113–6.

Beutler JA, Cardellina 2nd JH, McMahon JB, Boyd MR, Cragg GM. (1992) Anti-HIV and cytotoxic alkaloids from *Buchenavia capitata*. J Nat Prod 55:207–13.

Boffito M, Acosta E, Burger D, Fletcher CV, Flexner C, Garaffo R, Gatti G, Kurowski M, Perno CF, Peytavin G, Regazzi M, Back D. (2005) Current status. Antivir Ther 10:375–92.

Bukrinskaya AG. (2004) HIV-1 assembly and maturation. Arch Virol 149:1067–82.

Chang YS, Woo ER. (2003) Korean medicinal plants inhibiting to human immunodeficiency virus type 1 (HIV-1) fusion. Phytother Res 17:426–9.

Chen K, Kashiwada Y, Zhang DC, Hu CQ, Jin JQ, Nozaki H, Kilkuskie RE, Tramontano E, Cheng YC, McPhail DR, McPhail AT, Lee KH, Shi Q. (1992) Anti-aids agents, 6. Salaspermic acid, an anti-HIV principle from Tripterygium wilfordii, and the structure-activity correlation with its related compounds. J Nat Prod 55:340–6.

Chen SX, Wan M, Loh BN. (1996) Active constituents against HIV-1 protease from *Garcinia mangostana*. Planta Med 62:381–2.

Cos P, Hermans N, De BT, Apers S, Sindambiwe JB, Witvrouw M, De CE, Vanden BD, Pieters L, Vlietinck AJ. (2002) Antiviral activity of Rwandan medicinal plants against human immunodeficiency virus type-1 (HIV-1). Phytomedicine 9:62–8.

De Tommasi N, De Simone F, Pizza C, Mahmood N, Moore PS, Conti C, Orsi N, Stein ML. (1992) Constituents of *Eriobotrya japonica*. A study of their antiviral properties. J Nat Prod 55:1067–73.

Dionisio D, Esperti F, Messeri D, Vivarelli A. (2004) Priority strategies for sustainable fight against HIV/AIDS in low-income countries. Curr HIV Res 2:377–93.

Duan H, Takaishi Y, Imakura Y, Jia Y, Li D, Cosentino LM, Lee KH. (2000) Sesquiterpene alkaloids from *Tripterygium hypoglaucum* and *Tripterygium wilfordii*: a new class of potent anti-HIV agents. J Nat Prod 63:357–61.

el-Mekkawy S, Meselhy MR, Kusumoto IT, Kadota S, Hattori M, Namba T. (1995) Inhibitory effects of Egyptian folk medicines on human immunodeficiency virus (HIV) reverse transcriptase. Chem Pharm Bull (Tokyo) 43:641–8.

el-Mekkawy S, Meselhy MR, Nakamura N, Hattori M, Kawahata T, Otake T. (1999) 12-*O*-acetylphorbol-13-decanoate potently inhibits cytopathic effects of human immunodeficiency virus type 1 (HIV-1), without activation of protein kinase C. Chem Pharm Bull (Tokyo) 47:1346–7.

el-Mekkawy S, Meselhy MR, Nakamura N, Hattori M, Kawahata T, Otake T. (2000) Anti-HIV-1 phorbol esters from the seeds of *Croton tiglium*. Phytochemistry 53:457–64.

el-Mekkawy S, Meselhy MR, Nakamura N, Tezuka Y, Hattori M, Kakiuchi N, Shimotohno K, Kawahata T, Otake T. (1998) Anti-HIV-1 and anti-HIV-1-protease substances from *Ganoderma lucidum*. Phytochemistry 49:1651–7.

Fujioka T, Kashiwada Y, Kilkuskie RE, Cosentino LM, Ballas LM, Jiang JB, Janzen WP, Chen IS, Lee KH. (1994) Anti-AIDS agents, 11. Betulinic acid and platanic acid as anti-HIV principles from *Syzigium claviflorum*, and the anti-HIV activity of structurally related triterpenoids. J Nat Prod 57:243–7.

Gnabre JN, Ito Y, Ma Y, Huang RC. (1996) Isolation of anti-HIV-1 lignans from *Larrea tridentata* by counter-current chromatography. J Chromatogr A 719:353–64.

Gustafson KR, Cardellina JH, McMahon JB, Gulakowski RJ, Ishitoya J, Szallasi Z, Lewin NE, Blumberg PM, Weislow OS, Beutler JA, Buckheit RW, Gagg GM, Cox PA, Bader JP, Boyd MR. (1992) A nonpromoting phorbol from the samoan medicinal plant Homalanthus nutans inhibits cell killing by HIV-1. J Med Chem 35:1978–86.

Hnatyszyn O, Broussalis A, Herrera G, Muschietti L, Coussio J, Martino V, Ferraro G, Font M, Monge A, Martinez-Irujo JJ, Sanroman M, Cuevas MT, Santiago E, Lasarte JJ. (1999) Argentine plant extracts active against polymerase and ribonuclease H activities of HIV-1 reverse transcriptase. Phytother Res 13:206–9.

Hu CQ, Chen K, Shi Q, Kilkuskie RE, Cheng YC, Lee KH. (1994) Anti-AIDS agents, 10. Acacetin-7-O-beta-D-galactopyranoside, an anti-HIV principle from *Chrysanthemum morifolium* and a structure-activity correlation with some related flavonoids. J Nat Prod 57:42–51.

Hu K, Kobayashi H, Dong A, Iwasaki S, Yao X. (2000) Antifungal, antimitotic and anti-HIV-1 agents from the roots of *Wikstroemia indica*. Planta Med 66:564–7.

Hudson JB, Harris L, Towers GH. (1993) The importance of light in the anti-HIV effect of hypericin. Antiviral Res 20:173–8.

Hussein G, Miyashiro H, Nakamura N, Hattori M, Kawahata T, Otake T, Kakiuchi N, Shimotohno K. (1999) Inhibitory effects of Sudanese plant extracts on HIV-1 replication and HIV-1 protease. Phytother Res 13:31–6.

Kageyama S, Kurokawa M, Shiraki K. (2000) Extract of *Prunella vulgaris* spikes inhibits HIV replication at reverse transcription *in vitro* and can be absorbed from intestine *in vivo*. Antivir Chem Chemother 11:157–64.

Kalvatchev Z, Walder R, Garzaro D. (1997) Anti-HIV activity of extracts from *Calendula officinalis* flowers. Biomed Pharmacother 51:176–80.

Kashiwada Y, Wang HK, Nagao T, Kitanaka S, Yasuda I, Fujioka T, Yamagishi T, Cosentino LM, Kozuka M, Okabe H, Ikeshiro Y, Hu CQ, Yeh E, Lee KH. (1998) Anti-AIDS agents, 30. Anti-HIV activity of oleanolic acid, pomolic acid, and structurally related triterpenoids. J Nat Prod 61:1090–5.

Kim HJ, Woo ER, Shin CG, Hwang DJ, Park H, Lee YS. (2001) HIV-1 integrase inhibitory phenylpropanoid glycosides from *Clerodendron trichotomum*. Arch Pharm Res 24:286–91.

Kim HJ, Woo ER, Shin CG, Park H. (1998) A new flavonol glycoside gallate ester from *Acer okamotoanum* and its inhibitory activity against human immunodeficiency virus-1 (HIV-1) integrase. J Nat Prod 61:145–8.

Kim SN, Lee JY, Kim HJ, Shin CG, Park H, Lee YS. (2000) Synthesis and HIV-1 integrase inhibitory activities of caffeoylglucosides. Bioorg Med Chem Lett 10:1879–82.

Konoshima T, Yasuda I, Kashiwada Y, Cosentino LM, Lee KH. (1995) Anti-AIDS agents, 21. *Triterpenoid saponins* as anti-HIV principles from fruits of *Gleditsia japonica* and *Gymnocladus chinensis*, and a structure-activity correlation. J Nat Prod 58:1372–7.

Kwon HC, Jung CM, Shin CG, Lee JK, Choi SU, Kim SY, Lee KR. (2000) A new caffeoyl quinic acid from aster scaber and its inhibitory activity against human immunodeficiency virus-1 (HIV-1) integrase. Chem Pharm Bull (Tokyo) 48:1796–8.

Lampronti I, Khan MTH, Bianchi N, Ather A, Borgatti M, Vizziello A, Fabbri E, Gambari R. (2005) Bangladeshi medicinal plants extracts inhibiting molecular interactions between nuclear factors and target DNA sequences mimicking NF-kB binding sites. Med Chem 1:327–33.

Lee-Huang S, Huang PL, Huang PL, Bourinbaiar AS, Chen HC, Kung HF. (1995) Inhibition of the integrase of human immunodeficiency virus (HIV) type 1 by anti-HIV plant proteins MAP30 and GAP31. Proc Natl Acad Sci USA 92:8818–22.

Lee-Huang S, Huang PL, Kung HF, Li BQ, Huang PL, Huang P, Huang HI, Chen HC. (1991) TAP 29: an anti-human immunodeficiency virus protein from *Trichosanthes kirilowii* that is nontoxic to intact cells. Proc Natl Acad Sci USA 88:6570–4.

Lee-Huang S, Huang PL, Nara PL, Chen HC, Kung HF, Huang P, Huang HI, Huang PL. (1990) MAP 30: a new inhibitor of HIV-1 infection and replication. FEBS Lett 272:12–8.

Li HY, Sun NJ, Kashiwada Y, Sun L, Snider JV, Cosentino LM, Lee KH. (1993) Anti-AIDS agents, 9. Suberosol, a new C31 lanostane-type triterpene and anti-HIV principle from *Polyalthia suberosa*. J Nat Prod 56:1130–3.

Lin YM, Anderson H, Flavin MT, Pai YH, Mata-Greenwood E, Pengsuparp T, Pezzuto JM, Schinazi RF, Hughes SH, Chen FC. (1997) *In vitro* anti-HIV activity of biflavonoids isolated from *Rhus succedanea* and *Garcinia multiflora*. J Nat Prod 60:884–8.

Liu S, Jiang S, Wu Z, Lv L, Zhang J, Zhu Z, Wu S. (2002) Identification of inhibitors of the HIV-1 gp41 six-helix bundle formation from extracts of Chinese medicinal herbs *Prunella vulgaris* and *Rhizoma cibotte*. Life Sci 71:1779–91.

Lopez S, Armand-Ugon M, Bastida J, Viladomat F, Este JA, Stewart D, Codina C. (2003) Anti-human immunodeficiency virus type 1 (HIV-1) activity of lectins from Narcissus species. Planta Med 69:109–12.

Ma C, Nakamura N, Hattori M, Kakuda H, Qiao J, Yu H. (2000) Inhibitory effects on HIV-1 protease of constituents from the wood of *Xanthoceras sorbifolia*. J Nat Prod 63:238–42.

Ma C, Nakamura N, Miyashiro H, Hattori M, Shimotohno K. (1999) Inhibitory effects of constituents from *Cynomorium songaricum* and related triterpene derivatives on HIV-1 protease. Chem Pharm Bull (Tokyo) 47:141–5.

Mahmood N, Piacente S, Pizza C, Burke A, Khan AI, Hay AJ. (1996) The anti-HIV activity and mechanisms of action of pure compounds isolated from *Rosa damascena*. Biochem Biophys Res Commun 229:73–9.

Matsuse IT, Lim YA, Hattori M, Correa M, Gupta MP. (1999) A search for anti-viral properties in Panamanian medicinal plants. The effects on HIV and its essential enzymes. J Ethnopharmacol 64:15–22.

Matthee G, Wright AD, Konig GM. (1999) HIV reverse transcriptase inhibitors of natural origin. Planta Med 65:493–506.

McCormick JL, McKee TC, Cardellina JH, Boyd MR. (1996) HIV inhibitory natural products, 26. Quinoline alkaloids from *Euodia roxburghiana*. J Nat Prod 59:469–71.

Min BS, Bae KH, Kim YH, Miyashiro H, Hattori M, Shimotohno K. (1999) Screening of Korean plants against human immunodeficiency virus type 1 protease. Phytother Res 13:680–2.

Min BS, Kim YH, Tomiyama M, Nakamura N, Miyashiro H, Otake T, Hattori M. (2001a) Inhibitory effects of Korean plants on HIV-1 activities. Phytother Res 15:481–6.

Min BS, Tomiyama M, Ma CM, Nakamura N, Hattori M. (2001b) Kaempferol acetylrhamnosides from the rhizome of *Dryopteris crassirhizoma* and their inhibitory effects on three different activities of human immunodeficiency virus-1 reverse transcriptase. Chem Pharm Bull (Tokyo) 49:546–50.

Mlinaric A, Kreft S, Umek A, Strukelj B. (2000) Screening of selected plant extracts for *in vitro* inhibitory activity on HIV-1 reverse transcriptase (HIV-1 RT). Pharmazie 55:75–7.

Nielsen MH, Pedersen FS, Kjems J. (2005) Molecular strategies to inhibit HIV-1 replication. Retrovirology 2:10.

Notka F, Meier GR, Wagner R. (2003) Inhibition of wild-type human immunodeficiency virus and reverse transcriptase inhibitor-resistant variants by *Phyllanthus amarus*. Antiviral Res 58:175–86.

Notka F, Meier G, Wagner R. (2004) Concerted inhibitory activities of *Phyllanthus amarus* on HIV replication *in vitro* and *ex vivo*. Antiviral Res 64:93–102.

Pande V, Ramos MJ. (2003) Nuclear factor kappa B: a potential target for anti-HIV chemotherapy. Curr Med Chem 10:1603–15.
Pani A, Loi AG, Mura M, Marceddu T, La Colla P, Marongiu ME. (2002) Targeting HIV: old and new players. Curr Drug Targets Infect Disord 2:17–32.
Paris A, Strukelj B, Renko M, Turk V, Pukl M, Umek A, Korant BD. (1993) Inhibitory effect of carnosic acid on HIV-1 protease in cell-free assays. J Nat Prod 56:1426–30.
Park JC, Hur JM, Park JG, Hatano T, Yoshida T, Miyashiro H, Min BS, Hattori M. (2002) Inhibitory effects of Korean medicinal plants and camelliatannin H from *Camellia japonica* on human immunodeficiency virus type 1 protease. Phytother Res 16:422–6.
Pengsuparp T, Cai L, Constant H, Fong HH, Lin LZ, Kinghorn AD, Pezzuto JM, Cordell GA, Ingolfsdottir K, Wagner H, et al. (1995) Mechanistic evaluation of new plant-derived compounds that inhibit HIV-1 reverse transcriptase. J Nat Prod 58:1024–31.
Pengsuparp T, Cai L, Fong HH, Kinghorn AD, Pezzuto JM, Wani MC, Wall ME. (1994) Pentacyclic triterpenes derived from *Maprounea africana* are potent inhibitors of HIV-1 reverse transcriptase. J Nat Prod 57:415–8.
Pereira CF, Paridaen JT. (2004) Anti-HIV drug development – an overview. Curr Pharm Des 10:4005–37.
Piras G, Makino M, Baba M. (1997) Sho-saiko-to, a traditional Kampo medicine, enhances the anti-HIV-1 activity of lamivudine (3TC) *in vitro*. Microbiol Immunol 41:835–9.
Premanathan M, Rajendran S, Ramanathan T, Kathiresan K, Nakashima H, Yamamoto N. (2000) A survey of some Indian medicinal plants for anti-human immunodeficiency virus (HIV) activity. Indian J Med Res 112:73–7.
Qian-Cutrone J, Huang S, Trimble J, Li H, Lin PF, Alam M, Klohr SE, Kadow KF. (1996) Niruriside, a new HIV REV/RRE binding inhibitor from *Phyllanthus niruri*. J Nat Prod 59:169–96.
Ruffa MJ, Wagner ML, Suriano M, Vicente C, Nadinic J, Pampuro S, Salomon H, Campos RH, Cavallaro L. (2004) Inhibitory effect of medicinal herbs against RNA and DNA viruses. Antivir Chem Chemother 15:153–9.
Sancho R, Marquez N, Gomez-Gonzalo M, Calzado MA, Bettoni G, Coiras MT, Alcami J, Lopez-Cabrera M, Appendino G, Munoz E. (2004) Imperatorin inhibits HIV-1 replication through an Sp1-dependent pathway. J Biol Chem 279:37349–59.
Shahidul Alam M, Quader MA, Rashid MA. (2000) HIV-inhibitory diterpenoid from *Anisomeles indica*. Fitoterapia 71:574–6.
Sun HD, Qiu SX, Lin LZ, Wang ZY, Lin ZW, Pengsuparp T, Pezzuto JM, Fong HH, Cordell GA, Farnsworth NR. (1996) Nigranoic acid, a triterpenoid from *Schisandra sphaerandra* that inhibits HIV-1 reverse transcriptase. J Nat Prod 59:525–7.
Vlietinck AJ, De Bruyne T, Apers S, Pieters LA. (1998) Plant-derived leading compounds for chemotherapy of human immunodeficiency virus (HIV) infection. Planta Med 64:97–109.
Wang JN, Hou CY, Liu YL, Lin LZ, Gil RR, Cordell GA. (1994) Swertifrancheside, an HIV-reverse transcriptase inhibitor and the first flavone-xanthone dimer, from Swertia franchetiana. J Nat Prod 57:211–7.
Wang WK, Chen MY, Chuang CY, Jeang KT, Huang LM. (2000) Molecular biology of human immunodeficiency virus type 1. J Microbiol Immunol Infect 33:131–40.
Wu YC, Hung YC, Chang FR, Cosentino M, Wang HK, Lee KH. (1996) Identification of ent-16 beta, 17-dihydroxykauran-19-oic acid as an anti-HIV principle and isolation of the new diterpenoids annosquamosins A and B from *Annona squamosa*. J Nat Prod 59:635–7.
Xu HX, Zeng FQ, Wan M, Sim KY. (1996) Anti-HIV triterpene acids from *Geum japonicum*. J Nat Prod 59:643–5.
Yamasaki K, Nakano M, Kawahata T, Mori H, Otake T, Ueba N, Oishi I, Inami R, Yamane M, Nakamura M, Murata H, Nakanishi T. (1998) Anti-HIV-1 activity of herbs in Labiatae. Biol Pharm Bull 21:829–33.
Yang XW, Zhao J, Cui YX, Liu XH, Ma CM, Hattori M, Zhang LH. (1999) Anti-HIV-1 protease triterpenoid saponins from the seeds of *Aesculus chinensis*. J Nat Prod 62:1510–3.
Zhang HJ, Tan GT, Hoang VD, Hung NV, Cuong NM, Soejarto DD, Pezzuto JM, Fong HH. (2003) Natural anti-HIV agents. Part IV. Anti-HIV constituents from *Vatica cinerea*. J Nat Prod 66:263–8.

M.T.H. Khan and A. Ather (eds.)
Lead Molecules from Natural Products
© 2006 Published by Elsevier B.V.

Anticancer properties of saffron, *Crocus sativus* Linn.

JOSÉ-ANTONIO FERNÁNDEZ

Abstract

Crocus sativus **L.**, commonly known as **saffron**, is the raw material for one of the most expensive spices in the world and it has been used in folk medicine for centuries. This chapter provides a review of the recent literature about the analysis and production of antitumour agents present in this plant and its potential application in cancer biotherapy.

Keywords: *antioxidant, antitumour activity, apoptosis, carotenoids, chemoprevention, corm, crocetin, crocin, crocus sativus, cytotoxicity, macrophage activation, saffron, tissue culture*

Abbreviations: ID50, dose producing 50% cell growth inhibition; NO, nitric oxide; PKC, protein kinase C.

I. Introduction

In the continued search for new antitumour agents, investigators dedicate many efforts to the research on natural compounds and their effects in modifying cancer risks, delaying carcinogenesis, or inhibiting tumour formation. From purple foxglove, *Digitalis purpurea* (Plantaginaceae), which produce digitalins, to the Pacific yew, *Taxus brevifolia* (Taxaceae), from which taxanes were isolated (paclitaxel and docetaxel), plants have been a source of research material for useful drugs (da Rocha et al., 2001; Lindholm et al., 2002; Raskin et al., 2002). Recent research suggests that many edible fruits, vegetables, herbs and spices contain chemicals that may reduce the incidence of cancer (Aruna and Sivaramakrishnan, 1990; Unnikrishnan and Kuttan, 1990; Dragsted et al., 1993; Rogers et al., 1993; Willett, 1994, 1999; Havas, 1997; Kelloff et al., 2000; Croce, 2001; Cohen, 2002a; Milner, 2002; Furst, 2002).

Commercial **saffron**, one of the most expensive spices in the world, is composed of the dry stigmas of the **saffron** flower (***Crocus sativus*** Linn.), a perennial herb of the Iridaceae family cultivated in Iran, India, Greece, Spain, Morocco, China, Azerbaijan,

Italy, France, Turkey, Israel, Egypt, United Arab Emirates, Mexico, Switzerland, Algeria, Australia and New Zealand. **Saffron** is mostly used as spice and food colourant and, less extensively, as textile dye or perfume, but folk herbal medicines have used **saffron** for the treatment of numerous illnesses due to its analgesic and sedative properties (Basker and Negbi, 1983; Locock, 1995; Robinson, 1995).

Chemical analysis of **saffron** extracts revealed that the main constituents are the **carotenoids crocetin** (also called α-**crocetin** or **crocetin**-I) and its glycosidic forms digentiobioside (**crocin**), gentiobioside, glucoside, gentioglucoside and diglucoside; β-**crocetin** (monomethyl ester), γ-**crocetin** (dimethylester), *trans*-**crocetin** isomer, 13-*cis*-**crocetin** isomer; α-carotene, β-carotene, lycopene, zeaxanthin and mangi**crocin**, a xanthone-**carotenoid** glycosidic conjugate. The monoterpene aldehydes picro**crocin** and its deglycosylated derivative safranal (dehydro-β-cyclocitral) are important components of **saffron** responsible of its bitter flavour and aroma, respectively (Figure 1). Antocianins, flavonoids, vitamins (especially riboflavin and thiamine), amino acids, proteins, starch, mineral matter, gums and other chemical compounds have been described also in **saffron** (see reviews of Ríos et al., 1996; Winterhalter and Straubinger, 2000; Abdullaev, 2002). Interestingly, the **carotenoids crocetin** and **crocin**, the characteristic pigments of **saffron** stigma, are also major components of the cape jasmine fruit, *Gardenia jasminoides* Elliss (Rubiaceae), widely used as ornamental, natural food colourant source and as a Chinese herbal medicine (Watanabe and Terabe, 2000). Studies carried out with **crocin** and **crocetin** extracted from *Gardenia* are reviewed also in the present article.

There are several reviews published in the past years about the phytochemical and biomedical uses of the **saffron** (***Crocus sativus*** Linn.). In a recent review report Abdullaev and Espinosa-Aguirre discussed on the overview of experimental in vitro and in vivo investigations focused on the anticancer activity of saffron and its principal ingredients (Abdullaev and Espinosa-Aguirre, 2004). The authors also discussed the potential uses

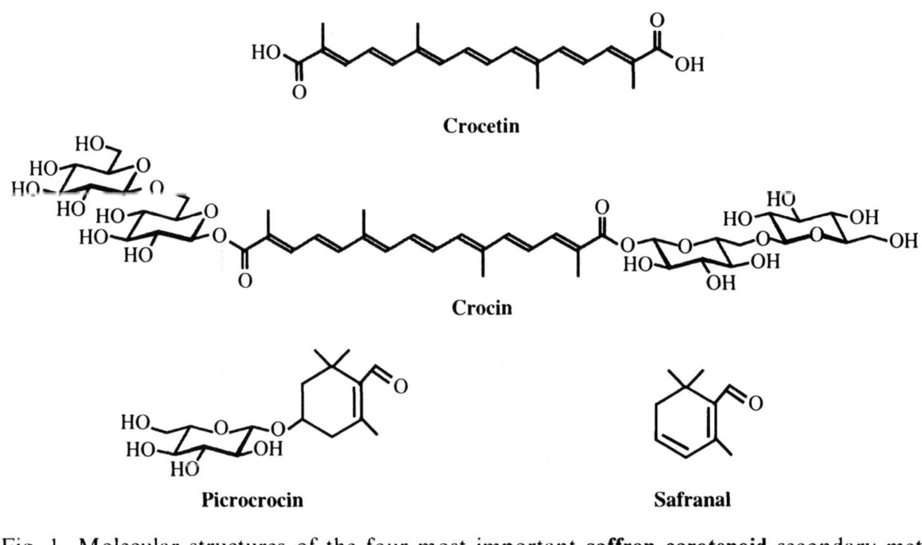

Fig. 1. Molecular structures of the four most important **saffron carotenoid** secondary metabolites.

of these natural agents in cancer therapy and chemopreventive trials (Abdullaev and Espinosa-Aguirre, 2004). Giaccio reviewed on the know properties and components of saffron (Giaccio, 2004). He gave main emphasis on the **crocetin**, a carotenoids (8,8′-diapo-8,8′-carotenoic acid) present in saffron and characterized by a diterpenic and symmetrical structure with seven double bonds and four methyl groups (Giaccio, 2004). Giaccio (2004) reviewed that crocetin increases alveolar oxygen transport and enhances pulmonary oxygenation. It improves cerebral oxygenation in haemorrhaged rats and positively acts in the atherosclerosis and arthritis treatment. It inhibits skin tumour promotion in mice (i.e., with benzo(a)pyrene); it has an inhibitory effect on intracellular nucleic acid and protein synthesis in malignant cells, as well as on protein kinase C (PKC) and proroonсogene in INNIH/3T3 cells, which is most likely due to its anti-oxidant activity (Giaccio, 2004). Furthermore, he also reviewed that crocetin protects against oxidative damage in rat primary hepatocytes. It also suppresses aflatoxin B1-induced hepatotoxic lesions and has a modulatory effect on aflatoxin B1 cytotoxicity, and DNA adduct formation on C3H10/T1/2 fibroblast cells. It also has a protective effect on the bladder toxicity, induced by cyclophosphamide (Giaccio, 2004). Lai and Roy (2004) reviewed the antimicrobial and chemopreventive properties of several herbs and species, including the saffron. The authors reviewed that the saffron, which is used as a food colourant, contains potent phytochemicals, including carotenoids. This compound provides significant protection against cancer (Lai and Roy, 2004).

The interest on **carotenoids** as potential biomedical drugs is significantly growing. **Carotenoids** are one of the most diverse and widely distributed groups of natural terpenoid pigments in plants and microorganisms, and are commonly present in our diet. **Carotenoids** exhibit biological activities as **antioxidants**, affect cell growth regulation, and modulate gene expression and immune response (Rock, 1997). Several studies have pointed out the use of some of them, such as β-carotene, α-carotene, lycopene, zeaxanthin or canthaxantine, in cancer prevention and therapy (Gerster, 1993; Smith, 1998; Lippman and Lotan, 2000; Heber, 2000; Cohen, 2002b; Johnson, 2002).

Saffron's biomedical properties have attracted the interest of researchers during the last decades (see reviews of Basker and Negbi, 1983; Abdullaev, 1993, 2002; Ríos et al., 1996; Souret and Weathers, 1999). Hartwell (1982) reported that in ancient times **saffron** was used as an anticancer agent, and he described the use of preparations containing **saffron** extracts against different kinds of tumours and cancers. Thus, liver, spleen, kidney, stomach and uterus tumours have been treated with pharmaceutical preparations of **saffron**. In the early 1990s, some authors demonstrated that crude **saffron** extracts presented antitumour and anticarcinogenic activities, as well as cytotoxic and antimutagenic effects. The aim of the present review is to summarize the recent research on the active anticancer constituents present in the **saffron** plant, their potential therapeutic applications, and the biotechnological production of these substances.

II. Tumouricidal properties

Gainer and co-workers (1976) noticed that gardenia **crocetin** delayed the onset and decreased the number of skin papillomas and Rous sarcoma tumours. Nair and co-workers (1991a) showed that oral administration of **saffron** ethanolic extracts increased the life span of Swiss albino mice intraperitoneally transplanted with sarcoma-180 (S-180) cells, Ehrlich ascites carcinoma (EAC) or Dalton's lymphoma ascites (DLA) tumours. By this time, the authors did not identify the exact nature of the active compound from **saffron** stigmas, but suggested that this compound showed the presence of glycosidic linkage. Liposome encapsulation of **saffron** effectively enhanced its **antitumour activity** against S-180 and EAC solid tumours in mice, promoting significant inhibition in the growth of these tumours. On the other hand, in the presence of the T-cell mitogen phytohaemagglutinin, **saffron** stimulated non-specific proliferation of lymphocytes *in vitro* (Nair et al., 1992), suggesting that the **antitumour activity** might be immunologically mediated.

García-Olmo and co-workers (1999) examined the effects of long-term treatment with **saffron crocin** (the glycosidic form of **crocetin**) on tumour growth and lifespan of rats bearing syngeneic colourectal tumours, induced by rat adenocarcinoma DHD/K12-PROb cells injected subcutaneously. **Crocin** treatment of those animals increased significantly their survival time and decreased tumour growth rate, more intensely in females. The selective action of **crocin** in female rats as compared with male rats suggests that the effects of **crocin** in animals might be partially dependent on hormonal factors.

Carotenoids are well tolerated, even at high doses, but the insolubility in water of most of them makes their administration difficult, and has impaired their therapeutic use. However, **crocin** is an unusually water-soluble **carotenoid** due to its high glycosylation. Long-term treatment with **crocin** does not result in deleterious metabolic changes in rats, as it has been concluded from several studies that investigated the potential toxicity of **saffron** extracts (see the review by Ríos et al., 1996). The only potentially deleterious effect observed was a slight decrease in glucose serum levels (García-Olmo et al., 1999), in accordance with the results of el Daly, who observed a greater decrease in blood glucose in rats treated with **saffron** extracts and cisplatin, than in rats treated with cisplatin alone (el Daly, 1998). The mechanism of this alteration remains unknown, but it could be related to an increase of insulin levels mediated by pancreatic dysfunction. In a study of genotoxicity of gardenia **carotenoids**, Ozaki and co-workers (2002) did not find any mutagenic activity due to **crocetin**, whereas genipin, formed by geniposide hydrolysis, caused DNA damage and induced tetraploidy.

Carotenoids play essential functions in plant tissues protecting against oxidative damage. Consistently with this function in plants, gardenia **crocetin** decreased lipid peroxidation induced by reactive oxygen species in rat primary hepatocytes (Tseng et al., 1995). Gardenia **crocin**, the water-soluble form, has also shown **antioxidant** properties at concentrations up to 40 ppm. At 20 ppm, the **antioxidant** activity of **crocin** is comparable to that of butylated hydroxyanisole (BHA) (Pham et al., 2000). Extracts from **saffron** and other **carotenoid**-containing spices showed significant hydrogen peroxide scavenging activity as measured by using peroxidase-based assay systems (Martínez-Tomé et al., 2001). Because most **carotenoids** are lipid-soluble and might act as membrane-associated free-radical scavengers, the **antioxidant** properties of these compounds could prevent DNA damage induced by free radicals, and free radical chain reactions. This could explain the **antitumour activity** of **saffron carotenoids** (Abdullaev, 2002).

Vitamin A (retinal) is part of the organism's defense barrier against free radicals. Its **antioxidant** mechanism of action includes scavenging of single oxygen and thiol-free radicals, and it could be related to processes that involve genetic expression and cell differentiation. Nair and co-workers (1994) detected an increase in the levels of β-carotene and vitamin A in the serum of laboratory animals under oral administration of **saffron** extracts. Tarantilis and co-workers (1994) have suggested that **saffron carotenoids** possessed provitamin A activity by according to the hypothesis that the action of **carotenoids** was dependent upon its conversion into retinal (vitamin A), because most of the evidences supporting the anticancer effects of **carotenoids** were referred to β-carotene. However, other molecules with no provitamin A activity, such as lycopene, also show protective effects, and conversion of **carotenoids** into retinoids seems not to be a prerequisite for their anticancer-action (Smith, 1998; Heber, 2000; Rao and Agarwal, 2000; Abdullaev, 2002).

III. Chemopreventive activity

Mathews-Roth (1982) examined the effect of gardenia **crocetin** on experimental skin tumours in nude mice and he found a small inhibitory effect on the development of skin tumours induced by the application of 9,10-dimethyl-1,2-benzanthracene and croton oil. In rats, gardenia **crocins** revealed a great protective effect against hepatocarcinogenic compounds such as aflatoxin B1 and dimethylnitrosamine, partially suppressing chronic hepatic damage (Lin and Wang, 1986). The **cytotoxicity** and DNA-adduct formation of rat microsome-activated aflatoxin B1 in C3H10T1/2 fibroblast cells were significantly inhibited by treatment with gardenia **crocetin**, via a mechanism similar to hepatoprotective action (Wang et al., 1991a, b). Similarly, pre-treatment of C3H10T1/2 cells with **crocetin** (0.1 mM) inhibited the benzo(a)pyrene genotoxic effect, decreasing the covalent binding of B(a)P-diol-epoxide to DNA. Transformation frequencies were also lower than that of cells not treated with **crocetin**. It was suggested that the inhibition of B(a)P-induced genotoxicity and neoplastic transformation was due to a mechanism that increased the activity of glutathion *S*-transferase (GST) and decreased the formation of B(a)P adduct (Chang et al., 1996).

Gardenia **crocetin** was found to be a potent inhibitor of skin tumour promotion induced by 12-*O*-tetradecanoylphorbol-13-acetate (TPA) in mice. When NIH/3T3 fibroblasts were treated with TPA alone, PKC translocated from the cytosolic fraction to the particulate fraction. Pre-treatment with **crocetin** inhibited the TPA-induced PKC activity in the particulate fraction but did not affect the level of PKC protein. **Crocetin** also reduced the level of TPA-stimulated protein phosphorylation and suppressed TPA-induced c-Jun, c-Fos and c-Myc gene expression. Thus, the inhibition of skin carcinogenesis by **carotenoids** may function via their **antioxidant** properties, which, in turn, lead to a reduction of TPA-induced protooncogene expression in the mouse epidermis (Wang et al., 1996; Hsu et al., 1999).

Topical administration of **saffron** extracts inhibited the initiation/promotion of 7,12-dimethylbenz[a]anthracene (DMBA)-induced skin tumours in mice, delaying the onset of papiloma formation and reducing the mean number of papillomas per mouse. Oral administration of the same dose of **saffron** extracts restricted tumour incidence of 20-methylcholanthrene-induced soft-tissue sarcomas in mice (Salomi et al., 1990, 1991a). Extracts from **saffron** stigmas prolonged the lifespan of cisplatin-treated mice and prevented partially the decrease in body weight, leukocyte count and haemoglobin levels (Nair et al., 1991b).

Premkumar and co-workers (2001) assessed the effects of aqueous extracts of **saffron** (composed mainly by **carotenoids**) in Swiss albino mice, and suggested that pre-treatment with **saffron** can significantly inhibit the genotoxicity of cisplatin, cyclophosphamide, mitomycin and urethane. **Crocetin** from **saffron** also ameliorates bladder toxicity of the anticancer agent cyclophosphamide without altering its **antitumour activity** (Nair et al., 1993). In an experiment to evaluate its protective effect on cisplatin-induced toxicity in rats (3 mg/kg body wt), el Daly (1998) showed that treatment of animals with cysteine (20 mg/kg body wt) together with **saffron** extract (50 mg/kg body wt) significantly reduced the toxic effects caused by cisplatin, such as nephrotoxicity and changes in enzyme activity.

One of the most promising strategies for cancer prevention today is chemoprevention using readily available natural substances from vegetables, fruits, herbs and spices. Among the spices, saffron a member of the large family Iridaceae, has drawn attention because apart from its use as a flavouring agent, pharmacological studies have demonstrated many health-promoting properties including radical scavenging, anti- mutagenic and immuno-modulating effects (Das et al., 2004). Das et al. (2004) investigated and reported the effects of an aqueous infusion of saffron on two-stage skin papillogenesis/carcinogenesis in mice initiated by 7-12 dimethyl benz[a] anthracin (DMBA) and promoted with croton oil (Das et al., 2004). They found that, significant reduction in papilloma formation with saffron application in the pre-initiation and post-initiation periods, and particular when the agent was given both pre- and post-initiation (Das et al., 2004). Das et al. (2004) concluded from their studies that, the inhibition appeared to be at least partly due to modulatory effects of saffron on some phase II detoxifying enzymes like GST and glutathione peroxidase (GPx), as well as catalase (CAT) and superoxide dismutase (SOD) (Das et al., 2004).

IV. Cellular effects

Some studies found that **crocetin** from gardenia increased the relative *in vitro* growth of normal rat-muscle-derived cells and tumour cells, hypothesizing that **crocetin** affects cell division enzymatic processes (Wilkins et al., 1977; Wilkins and Gainer, 1979).

Incubation of HeLa cells (derived from a cervical epitheloid carcinoma) with **saffron** extracts resulted in significant inhibition of colony formation and cellular DNA and RNA synthesis, with 50% inhibition obtained at concentrations of 100–150 μg/ml whereas inhibition of protein synthesis was not detected even at high extract concentrations (Abdullaev and Frenkel, 1992a). In a study on the effect of **saffron** extract on macromolecular synthesis in three human cell lines: A549 cells (derived from a lung tumour), WI-38 cells (normal lung fibroblasts) and VA-13 cells (WI-38 cells transformed by SV40 virus), Abdullaev and Frenkel (1992b) found that the malignant cells were more sensitive than the normal cells to the inhibitory effects of **saffron** on both DNA and RNA synthesis. It has been proposed that the inhibitory effect on cellular DNA and RNA synthesis, but not on protein synthesis, is one of the mechanisms of action for the **antitumour** and anticarcinogenic **activities** of **saffron** (Abdullaev, 2002).

Abdullaev (1994) studied the inhibitory effect of **crocetin** on intracellular nucleic acid and protein synthesis in three malignant human cell lines: HeLa, A549 (lung adenocarcinoma) and VA13 (sv-40 transformed foetal lung fibroblasts). Incubation of these cells with **crocetin** for three hours caused a dose-dependent inhibition of nucleic acid and protein synthesis, but he found no effect on colony formation. Other studies described inhibition of growth of human chronic myelogenous leukaemia K562 and promyelocytic leukaemia HL-60 cells by dimethyl-**crocetin**, **crocetin** and **crocin**, with 50% inhibition (ID_{50}) reached at concentrations of 0.8, 2 and 2 μM, respectively (Morjani et al., 1990; Tarantilis et al., 1994). **Cytotoxicity** of dimethyl-**crocetin** and **crocin** to various tumour cell lines (DLA, EAC, S-180, L1210 leukaemia and P388 leukaemia) and to human primary cells from surgical specimens (osteosarcoma 917, fibrosarcoma 1456 and ovarian carcinoma 1998) has been reported (Salomi et al., 1991b; Nair et al., 1995), with concentrations producing 50% **cytotoxicity** ranging from 7 to 30 μg/ml for dimethyl-**crocetin** and from 11 to 39 μg/ml for **crocin**. These authors detected significant inhibition in the synthesis of nucleic acids, and suggested that dimethyl-**crocetin** could disrupt DNA–protein interactions (e.g. toposiomerases II) important for cellular DNA synthesis.

Escribano and co-workers (1996) demonstrated that the inhibitory activity on the in vitro growth of HeLa cells produced by **saffron** extracts ($ID_{50} = 2{,}3$ mg/ml) was mainly due to **crocin** (ID_{50} of 3 μM), whereas picro**crocin** and safranal, with an ID_{50} of 3 and 0.8 mM, respectively, played a minor role in the **cytotoxicity** of **saffron** extracts. These results suggested that sugars might play a role in **saffron**'s cytotoxic effect, since **crocetin** (the deglycosylated **carotenoid**) did not cause cell growth inhibition even at high doses. These findings are in accordance with the results of Abdullaev (1994), who found no effect of **crocetin** on colony formation in HeLa cells and two other solid tumour cell lines, but are, however, in disagreement with other authors who reported **cytotoxicity** for **crocetin** against a cell line derived from a non-solid tumour (Tarantilis et al., 1994) and various tumour cell lines and human primary cells from surgical specimens (Nair et al., 1995). García-Olmo and co-workers (1999) determined an ID_{50} of 0.4 and 1.0 mM for **crocin** on the rat adenocarcinoma DHD/K12-PROb cells and human colon adenocarcinoma HT-29 cells, respectively.

Molnar and co-workers (2000) carried out a study with **saffron**, ginsenoside and cannabinoid derivatives to determine potential membrane-associated antitumour effects of these substances. **Saffron** derivatives were ineffective on the reversal of multidrug resistance of lymphoma cells (the reversal of multidrug resistance is the result of the inhibition of the efflux pump function in the tumour cells). **Crocetin** esters were less potent than **crocin** itself in the inhibition of early antigen expression. However, **crocin** and diglucosyl**crocetin** inhibited early tumour antigen expression of adenovirus-infected cells, being triglucosyl**crocetin** less effective. **Crocin** did not show any antiviral effects on infected vero cells.

Microscopic studies revealed that HeLa cells treated with **crocin** exhibited vacuolated areas, size reduction, cell shrinkage and piknotic nuclei (Escribano et al., 1996; García-Olmo et al., 1999), suggesting that programmed cell death is induced by

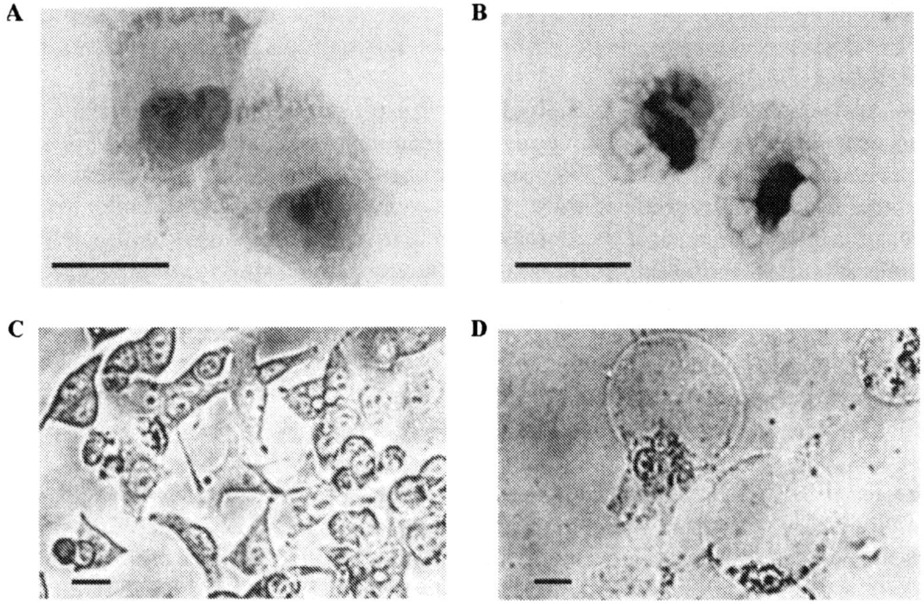

Fig. 2. Effect of **saffron** compounds on tumour cells. HeLa cells were incubated in the absence (A) or in the presence (B) of 10 μM **crocin** for 18 h, and then stained with haematoxylineosin. Apoptotic morphological changes induced in treated cells can be observed. (C) Control MDA-MB-231 cells. (D) Cells incubated with 10 mg/ml of the corm cytotoxic agent for 1.5 h. Note the intense swelling. Bar = 20 μm.

crocin, as was earlier suggested by Morjani and co-workers in 1990 (Figure 2). **Apoptosis** is induced in selected cancerous cell lines by a range of plant agents (Thatte et al., 2000). Several mechanisms have been identified to underlie the modulation of **apoptosis** by plant compounds including endonuclease activation, induction of p53, activation of caspase 3 protease via a Bcl-2-insensitive pathway, potentiation of free-radical formation and accumulation of sphingagine.

Soeda and co-workers (2001) demonstrated antiapoptotic actions of **saffron crocin** in non-cancerous cells. **Crocin** suppresses cell death induced by tumour necrosis factor-α (TNF-α) and expression of Bcl-Xs and cysteine protease (Lice) mRNAs and simultaneously restores the cytokine-induced reduction of TNF-α and Bcl-Xl mRNA expression. The modulating effects of **crocin** on the expression of Bcl-2 family proteins led to a marked reduction of a TNF-α-induced release of cytochrome *c* from the mitochondria. **Crocin** also blocked cytochrome c-induced activation of caspase-3. Likewise, **crocin** inhibited the effect of daunorubicin, an **apoptosis** inducer, suggesting that **crocin** inhibits both internal and external apoptotic stimuli in highly differentiated cells (neurons). This selective behaviour suggests important therapeutic implications; related to the fact that programmed cell death is reduced in cancer and increased in neurodegenerative diseases.

An influence of **carotenoids** has been observed on the expression of genes involved in cell-to-cell communication (Zhang et al., 1991). Diverse natural and synthetic **carotenoids** increase gap junctional intercellular communication and induce connexion 43, a gene that encodes a major gap junction protein, with differences in gene expression depending on the different cellular-type assays (Zhang et al., 1991, 1992, 1995; Pung et al., 1993). These authors proposed that **carotenoid**-enhanced intercellular communication provides a mechanistic basis for the cancer chemopreventive action of **carotenoids**.

V. Bioactive compounds from corm

C. sativus is an autotriploid species (24 chromosomes, $n = 8$), sexually sterile because its pollen, produced by an irregular meiosis, is unviable. The plant is reproduced vegetatively by means of its bulb or **corm**, a modified stem that plays an important role in **saffron** biology (Mathew, 1983). Chemical tests on the **corms** of **saffron** showed the presence of two saponins (one triterpenic and another steroidic), triterpenic acids, sugars (glucose), mucilage, amino acids, fats (fatty acids, sterols) and starch, as the principal storage polysaccharide (Loukis et al., 1983; Chrungoo et al., 1983; Sampathu et al., 1984). **Saffron corms** also contain two high molecular weight proteins, one of them possessing activity as a platelet aggregation inducer and the other functioning as an inhibitor of that activity (Liakopoulou-Kuriakides et al., 1985, 1990), as well as mannan-binding lectins specifically expressed in **corms** (Oda and Tatsumi, 1993; Escribano et al., 2000a). The antitumour effects of plant lectins have been reviewed by Abdullaev and de Mejia (1997).

A remarkable cytotoxic effect of aqueous extracts of **corms** against HeLa cells has been described (Escribano et al., 1999a). Using a three-step chromatographic protocol (size-exclusion, anion-exchange and reserved-HPLC), a bioactive fraction was purified from the extracts. It was composed of 94.5% glycoside and 5.5% polypeptide fractions. The most abundant sugar was rhamnose, which accounted for the 36.4% (mass/mass) of monosaccharides. Fucose, arabinose, xylose, galactose and uronic acids were also present. The amino acid analysis of the polypeptide fraction revealed a high relative molar proportion of aspartic acid/asparagines (15.4%), alanine (13.4%), glutamic acid/glutamine (12.2%), glycine (11%) and serine (8.2%). Relative proportions of carbohydrate and protein moieties, sugar and amino acid compositions, and SDS-PAGE analysis of the deglycosylated moiety led us to suggest a proteoglycan nature for this agent (Escribano et al., 1999a). Nevertheless, more detailed chemical analysis of this highly complex bioactive fraction support that the glycoside fraction is composed by triterpenoid compounds.

When the cytotoxic activity was tested against HeLa cells, the bioactive fraction showed an ID_{50} of 9 μg/ml (Escribano et al., 1999a). The deglycosylated portion did not show any effect, indicating that the presence of the carbohydrate component is essential for **cytotoxicity**. Morphological changes of treated cells showed a sequential pattern. HeLa cells treated for 5 min exhibited evident swelling, local plasma membrane evaginations and loss of star-like morphology (Figure 2). Longer treatments confirmed cell damage, apparently due to swelling and plasma membrane breakage. The swelling pattern of tumour cells exposed to the compound supports the idea of a change in the physical barrier properties of the plasma membrane, resulting in an osmotic like effect that leads to cell lysis. This is in agreement with the observation of an initial calcium influx, further calcium release, and liberation of cytoplasmic proteins. DNA intercalation of propidium iodide, when lysis occurs, is also observed.

Membrane permeabilization of target cells is a widespread mechanism of **cytotoxicity** displayed by a variety of natural compounds, including saponins, plant lectins, amphiphilic polypeptides or cardiotoxins and defensins (Escribano et al., 2000b).

The cytotoxic activity of this agent on human malignant cell lines (HeLa, breast carcinoma MDA-MB-231, and fibrosarcoma HT-1080), a non-malignant cell line (fibroblasts ASJ-4), and blood cells and hair follicles in culture, was also analysed. ID_{50} values ranged from 7 to 22 μg/ml for tumour cells, and 60 μg/ml for fibroblasts. Comparison of ID_{50} values of fibrosarcoma cells and normal fibroblasts, both of mesenchymal origin, showed that this agent is near eight times more toxic on tumour cells than on non-tumour cells. The fact that lysis of 50% of erythrocytes is reached with 100 μg/ml, a concentration about 10 times higher that the ID_{50} calculated for tumour cell lines, also evidenced the specificity of this cytotoxic effect . This selectivity could be due either to interactions with membrane lipids, the composition of which may differ among distinct cell types, or to the existence of cell membrane receptors, which could be more abundant in malignant cells. Other plausible explanations, such as distinctive cell metabolic states or variations in the extracellular matrix composition, cannot be excluded (Escribano et al., 2000b).

VI. Immuno-stimulating activity

Macrophages are cells of the reticuloendothelial system that play an important role in the body's defence against tumours. Selective stimulation of this cell population could be important to the development of therapeutic implications. Escribano and co-workers (1999b) studied the activation of macrophages by the bioactive fraction extracted from **saffron corms** at non-cytotoxic concentrations, measured by the release of nitric oxide (NO). Treatment with 50 μg/ml, doubled the release of nitrate and nitrite by these cells. Higher concentrations (up to 500 μg/ml), resulted in a decreased NO production in parallel with a marked fall in cell viability. In macrophages, NO is synthesized by an inducible nitric oxide synthase (iNOS). Discrete interactions of the agent with the plasma membrane could lead to different modulation of signalling transduction pathways, some involved in iNOS transcription. A concentration of 100 ng/ml did not stimulate cAMP generation in macrophages but rapidly increased cytosolic PKC activity. PKC activation could be the result of the increase of calcium permeability of plasma membrane.

Induction of iNOS expression in macrophages depends upon activation of different transcription factors (Nathan and Xie, 1994). The nuclear factor κB (NF-κB), composed by dimers of various proteins, participates in the regulation of the expression of multiple genes involved in the immune response, including iNOS. Triggering macrophages with concentrations of agents that induce NO production increased the binding of NF-κB proteins to the κB motif of the iNOS gene promoter. The NF-κB activation, consequence of treatment with the **saffron** agent, is similar to that obtained with other macrophage-stimulating factors (Velasco et al., 1997; Escribano et al., 1999b).

It has been suggested that **macrophage activation** may induce the apoptotic death of macrophages (Albina et al., 1993). Incubation of macrophages with non-cytotoxic concentrations of the bioactive fraction of **saffron** extract (50–100 ng/ml) resulted in an increase of DNA laddering, characteristic of apoptotic cells (Escribano et al., 1999b). A dose-dependent increase of apoptotic cells was also observed by flow cytometry, confirming this effect. This behaviour was specific for macrophages, since no **apoptosis** was observed in HeLa cells treated with similar concentrations (Escribano et al., 2000b).

VII. Biotechnological production

Since **chemoprevention** is one of the most promising strategies to control cancer, one of the most important causes of mortality in the world, nutraceuticals are of special interest. There is growing evidence indicating that **saffron carotenoids** possess chemopreventive and tumouricidal properties. However, the scarcity of **saffron** and high cost of obtaining large quantities of these compounds may impair prevention and treatment of cancer using **saffron. Saffron** is one of the world's most labour-intensive crops; it requires the painstaking task of hand-picking stamens from 70,000 flowers to obtain one pound, which explains its high cost. Therefore, to circumvent this handicap, new strategies for larger-scale production of **saffron carotenoids** must be explored.

Applying the methods of plant cell, tissue and organ culture to the production of biomedical interesting compounds has been a goal in research since the late 1950s. Since **crocin** represent only 10% of the dry weight of **saffron** stigmata, and this raw material is highly prized, biotechnology must be applied to obtain **saffron carotenoids** at a reasonable cost for their potential use in medicine. The first report by Koyama and co-workers (1988) on **crocin** and picro**crocin** synthesis in vitro from the formation of stigma-like structures from stigmata explants indicated that both compounds could be isolated, but not quantified, from cell cultures grown in media containing combinations of auxin and cytokinin. Growth of stigma-like structures containing **crocin**, picro**crocin** and safranal has been successfully accomplished using ovary (Himeno and Sano, 1987; Fakhrai and Evans, 1990), stigmata (Sano and Himeno, 1987; Sarma et al., 1990; González-Rumayor, 1998), or anthers and petals (Fakhrai and Evans, 1990) explants as starting material. However, the concentration of saffron molecules extracted from these in vitro structures was lower than that measured in natural stigmata. In vitro regeneration of "red filamentous" structures and red globular callus obtained from buds has also produced crocin, picrocrocin and safranal, in this case being the concentration of crocin slightly lower and the levels of safranal and picrocrocin equal and higher, respectively, than that measured in dry stigmata (Visvanath et al., 1990). More recently, Loskutov and co-workers (1998) reported the production, by stigma-like structures obtained in vitro, of the four most important secondary metabolites (**crocin**, **crocetin**, picro**crocin** and safranal) in a concentration identical or higher than the natural **saffron**.

Despite numerous studies carried out on the chemical characterization of **saffron carotenoid** derivatives, little is known about their biosynthesis pathways. Pfander and Schurtenberger (1982) proposed that the biogenesis of the colour principles and odour-active compounds is derived by bio-oxidative cleavage of zeaxanthin. Using an extract of callus initiated from buds, Dufresne and co-workers (1997) have studied *in vitro* the enzymatic formation of **crocetin** glucosyl esters from all-*trans*-**crocetin**. The reactions involved the step-wise addition of a glucose moiety to a free carboxyl functional group and the 1→6 addition of a glucose moiety to a glucosyl ester functional group. The kinetics of reaction for each glycoside seemed to indicate that two distinct glucosyl transferases were implicated in the synthesis of **crocin**.

Because of the importance of **carotenoids** as a group of plant metabolites, the enzymes responsible for the biosynthetic pathway from phytoene to zeaxanthin have been characterized and their structural genes cloned from a wide array of higher plants (Hirschberg, 2001). However, the identification of genes encoding for the enzymes involved in the biosynthesis of other **carotenoids** derivatives, the apo**carotenoids**, which have important metabolic and hormonal functions in diverse organisms, remained mostly unknown. Some partial cDNA clones have been isolated and characterized by our group which correspond to genes coding for enzymes involved in the biosynthesis of **saffron carotenoids**, such as 3-hydroxy-3-methylglutaryl-CoA reductase, *β*-carotene hydroxylase, and some oxygenases (GenBank accesion numbers: AJA416715, AJ416711, AJ416712, AJ416713, AJ416714). Recently, Bouvier and co-workers (2003) identified and functionally characterized the zeaxanthin 7,8(7′,8′)-cleveage dioxygenase gene (*CsZCD*) and the **carotenoid** 9,10(9′,10′)-cleavage dioxygenase gene (*CsCCD*) from **saffron** plant, and discussed the expression of these two genes in **saffron** tissues. The identification of genes encoding specialized steps in **carotenoid** metabolism is of major interest. This would allow the engineering of **saffron**, and other crops or micro organisms aimed at the biotechnological production of these high-value compounds, a source of research materials and potential useful drugs.

Likewise, a handicap for the study of the properties of the antitumoural agent extracted from **saffron corms** is the limited amount of material present in **corms** (0.5% of the dry weight from **corm** soluble extracts) (Escribano et al., 1999a). The use of **corm tissue culture** methods to optimize the production and purification of this agent to complete functional and therapeutic studies has been considered worthwhile. *In vitro* culture of **saffron corm** has been used as a way to develop an efficient regeneration system for **saffron** breeding (Isa and Ogasawara, 1988; Dhar and Sapru, 1993; Piqueras et al., 1999). A fraction with cytolytic properties can be isolated by reverse HLPC form callus cultures of **saffron corms**. Very likely, this fraction is identical to that present in **corms**, possessing equal cytotoxic activity on HeLa cells (Escribano et al., 1999c). The recovery of this fraction from calli represents, at least, 8.5% of the dry weight of the callus soluble fraction, 17 times greater than the amount recovered from **corm**. The Biotechnology Division (IDR) of the University of Castilla-La Mancha, recently attempted the culture of **corm** cells in liquid media, establishing the first step to scale-up the preparation and purification of this compound with potential application in antitumoural research.

VIII. Conclusions

Saffron plant has shown to be a source of bioactive compounds with cytotoxic, antitumoural, chemopreventive, antimutagenic and immuno-stimulating properties. **Crocins**, the major **carotenoid** components of **saffron** stigma, demonstrated **antitumour**

activity, promoting tumour growth inhibition and increasing the life-span of treated tumour-bearing animals. **Crocins** are well tolerated and present no or minor side-effects. These, together with their water-solubility, make them suitable for chemotherapeutic use.

Crocins and **crocetins** (the deglycosylated forms) were also found to be potent inhibitors of carcinogenesis as well as attenuators of the toxicity of some anticancer agents. Studies about the **cytotoxicity** of **carotenoids** present in **saffron** produced controversial results concerning the comparative effects of glycosydic-and sugar-free **carotenoids**, but revealed that malignant cells are more sensitive than normal cells to the toxic effect of these compounds (Abdullaev and Frenkel, 1992b; Abdullaev, 1994; Tarantilis et al., 1994; Nair et al., 1995; Escribano et al., 1996).

Several mechanisms attempting to explain the antitumour action at the cellular and molecular levels of the **carotenoids** present in **saffron** have been suggested:

(i). Modulation of programmed cell death, selectively promoting apoptosis in tumoural cells and inhibiting both internal and external apoptosis stimuli in non-tumoural cells (Morjani et al., 1990; Escribano et al., 1996; García-Olmo et al., 1999; Thatte et al., 2000).
(ii). Inhibition of cellular DNA and RNA synthesis, but not protein synthesis. Disruption of DNA–protein interactions has been proposed to explain this inhibition of nucleic acid synthesis (Abdullaev and Frenkel, 1992a; Nair et al., 1995).
(iii). Antioxidant activity; inhibition of free-radical chain reactions that could lead to oxidative damage and DNA alterations (Tarantilis et al., 1994; Tseng et al., 1995; Pham et al., 2000; Martínez-Tomé et al., 2001).
(iv). Enzymatic changes (GST, PKC), decreases in the formation of B(a)P adduct, and reduction in expression of protooncogenes (Chang et al., 1996; Hsu et al., 1999).

The **corms** of ***C. sativus*** contain glycosylated compounds, currently under study to determine their chemical structures, possessing remarkable **cytotoxicity** and immuno-stimulating activity (Escribano et al., 1999a, b, 2000b), which make them valuable in anticancer research. The development of efficient protocols for the production and purification of **saffron** bioactive molecules is an essential requirement for their use in **chemoprevention** and cancer treatment.

Acknowledgments

The author thanks Dr. J. Escribano, Dr. H. H. Riese, Dr. M. J.M. Díaz-Guerra, Dr. A. Piqueras, Dr. J. Medina, Dr. Lourdes Gómez-Gómez, Dr. A. Rubio and Dr. M. Álvarez-Ortí for their work in saffron research. Dr. Jorge Laborda is acknowledged by his suggestions and critical reviewing of the manuscript. The research of the author's team in this field has been funded by the Spanish Ministry of Science and Technology (grants SAF-0149-C02, 1FD97-1417-C02-01 and PB98-0317) and by the Regional Government of Castilla-La Mancha (172/IA-35, PAI-02-026 and AGR020023).

References

Abdullaev FI. (1993) Biological effects of **saffron**. Biofactors 4:83–6.

Abdullaev FI. (2002) Cancer chemopreventive and tumouricidal properties of **saffron** (***Crocus sativus*** L.). Exp Biol Med (Maywood) 227:20–5.

Abdullaev FI. (1994) Inhibitory effect of **crocetin** on intracellular nucleic acid and protein synthesis in malignant cells. Toxicol Lett 70:243–51.

Abdullaev FI, de Mejia EG. (1997b) Antitumour effect of plant lectins. Nat Toxins 5:157–63.

Abdullaev FI, Espinosa-Aguirre JJ. (2004) Biomedical properties of saffron and its potential use in cancer therapy and chemoprevention trials. Cancer Detect Prev 28(6):426–32.

Abdullaev FI, Frenkel GD. (1992a) Effect of **saffron** on cell colony formation and cellular nucleic acid and protein synthesis. Biofactors 3:201–4.

Abdullaev FI, Frenkel GD. (1992b) The effect of **saffron** on intracellular DNA, RNA and protein synthesis in malignant and non-malignant human cells. Biofactors 4:43–5.

Albina EA, Cui S, Mateo RB, Reichner JS. (1993) Nitric oxide-mediated **apoptosis** in murine peritoneal macrophages. J Immunol 150:5080–5.

Aruna K, Sivaramakrishnan VM. (1990) Plant products as protective agents against cancer. Indian J Exp Biol 28:1008–11.

Basker D, Negbi M. (1983) Uses of **saffron**. Econ Bot 3:228–35.

Bouvier F, Suire C, Mutterer J, Camara B. (2003) Oxidative remodeling of chromoplast **carotenoids**: identification of the **carotenoid** dioxygenase *CsCCD* and *CsZCD* genes involved in *Crocus* secondary metabolite biogenesis. Plant Cell 15:47–62.

Chang WC, Lin YL, Lee MJ, Shiow SJ, Wang CJ. (1996) Inhibitory effect of **crocetin** on benzo(a)pyrene genotoxicity and neoplastic transformation in C3H10T1/2 cells. Anticancer Res 16:3603–8.

Chrungoo NK, Koul KK, Farooq S. (1983) Carbohydrate changes in **corms** of **saffron** crocus (***Crocus sativus*** L.) during dormancy and sprouting. Trop Plant Sci Res, 1:295–8.

Cohen LA. (2002a) Nutrition and prostate cancer: a review. Ann.NY Acad Sci, 963:148–55.

Cohen LA. (2002b) A review of animal model studies of tomato **carotenoids**, lycopene, and cancer **chemoprevention**. Exp Biol Med (Maywood) 227:864–8.

Croce CM. (2001) How can we prevent cancer? Proc Natl Acad Sci USA 98:10986–8.

da Rocha AB, Lopes RM, Schwartsmann G. (2001) Natural products in anticancer therapy. Curr Opin Pharmacol 1:364–9.

Das I, Chakrabarty RN, Das S. (2004) Saffron can prevent chemically induced skin carcinogenesis in Swiss albino mice. Asian Pac J Cancer Prev 5(1):70–6.

Dhar AK, Sapru R. (1993) Studies on **saffron** in Kashmir. III. In vitro production of **corm** and shoot like structures. Indian J Genet 53:193–6.

Dragsted LO, Strube M, Larsen JC. (1993) Cancer-protective factors in fruits and vegetables: biochemical and biological background. Pharmacol Toxicol 72(Suppl 1):116–35.

Dufresne C, Cormier F, Dorion S. (1997) In vitro formation of **crocetin** glucosyl esters by ***Crocus sativus*** callus extract. Planta Med 63:150–3.

el Daly ES. (1998) Protective effect of cysteine and vitamine E, ***Crocus sativus*** and *Nigella sativa* extracts on cisplatin-induced toxicity in rats. J Pharm Belg 53:87–93.

Escribano J, Alonso GL, Coca-Prados M, Fernández JA. (1996) **Crocin**, safranal and picro**crocin** from **saffron** (***Crocus sativus*** L.) inhibit the growth of human cancer cells *in vitro*. Cancer Lett 100:23–30.

Escribano J, Díaz-Guerra MJM, Riese HH, Alvarez A, Proenza R, Fernández JA. (2000b) The selective cytolytic effect of a proteoglycan isolated from **corms** of **saffron** plant (***Crocus sativus*** L.) on human cell lines in culture. Planta Med 66:157–62.

Escribano J, Díaz-Guerra MJM, Riese HH, Ontañón J, García-Olmo D, García-Olmo DC, Rubio A, Fernández JA. (1999b) In vitro activation of macrophages by a novel proteoglycan isolated from **corms** of ***Crocus sativus*** L. Cancer Lett 144:107–14.

Escribano J, Piqueras A, Medina J, Rubio A, Alvarez-Ortí M, Fernández JA. (1999c) Production of a cytotoxic proteoglycan using callus culture of **saffron corms** (***Crocus sativus*** L.). J Biotechnol 73:53–9.

Escribano J, Ríos I, Fernández JA. (1999a) Isolation and cytotoxic properties of a novel proteoglycan from **corms** of **saffron** plant (***Crocus sativus*** L.). Biochim Biophys Acta 1426:217–22.
Escribano J, Rubio A, Alvarez-Ortí M, Molina A, Fernández JA. (2000a) Purification and characterization of a mannan-binding lectin specifically expressed in **corms** of **saffron** plant (***Crocus sativus*** L.). J Agric Food Chem 48:457–63.
Fakhrai F, Evans P. (1990) Morphogenic potential of cultures floral explants of ***Crocus sativus*** L. for the in vitro production of **saffron**. J Exp Bot 41:47–52.
Furst A. (2002) Can nutrition affect chemical toxicity? Int J Toxicol 21:419–24.
Gainer JL, Wallis DA, Jones JR. (1976) The effect of **crocetin** on skin papillomas and Rous sarcoma. Oncology 33:222–4.
García-Olmo DC, Riese HH, Escribano J, Ontañón J, Fernández JA, Atiénzar M, García-Olmo D. (1999) Effects of long-term treatment of colon adenocarcinoma with **crocin**, a **carotenoid** from **saffron** (***Crocus sativus*** L.): an experimental study in the rat. Nutr Cancer 35:120–6.
Gerster H. (1993) Anticarcinogenic effect of common **carotenoids**. Int Nutr Res 63:92–121.
Giaccio M. (2004) Crocetin from saffron: an active component of an ancient spice. Crit Rev Food Sci Nutr 44(3):155–72.
González-Rumayor V. (1998) Cultivo de tejidos de ***Crocus sativus*** L.: producción de azafrán in vitro. Madrid: Ph.D. thesis, Universidad Complutense.
Hartwell JL. (1982) Plants used against cancer. A survey. Lawrence: Quaterman Publications, pp. 284–289.
Havas S. (1997) Diet and cancer. Md Med J 46:477–80.
Heber D. (2000) Colorful cancer prevention: alpha-carotene, lycopene, and lung cancer. Am J Clin Nutr 72:901–2.
Himeno H, Sano K. (1987) Synthesis of **crocin**, picro**crocin** and safranal by **saffron** stigma-like structures proliferated in vitro. Agric Biol Chem 51:2395–400.
Hirschberg J. (2001) **Carotenoid** biosynthesis in flowering plants. Curr Opin Plant Biol 4:210–8.
Hsu JD, Chou FP, Lee MJ, Chiang HC, Lin YL, Shiow SJ, Wang CJ. (1999) Suppression of the TPA-induced expression of nuclear-protooncogenes in mouse epidermis by **crocetin** via **antioxidant** activity. Anticancer Res 19:4221–7.
Isa T, Ogasawara T. (1988) Efficient regeneration from the callus of **saffron** (***Crocus sativus***). Jp J Breeding 38:371–4.
Johnson EJ. (2002) The role of **carotenoids** in human health. Nutr Clin Care 5:56–65.
Kelloff GJ, Crowell JA, Steele VE, Lubert RA, Malone WA, Boone CW, Kopelovich L, Hawk ET, Lieberman R, Lawrence JA, Ali I, Viner JL, Sigman CC. (2000) Progress in cancer **chemoprevention**: development of diet-derived chemopreventive agents. J Nutr 130(2S):467–71.
Koyama A, Ohmoir Y, Fujioka N, Miyagawa K, Yamasaki K, Kohda H. (1988) Formation of stigma-like structures and pigments in cultures tissues of ***Crocus sativus*** L. Planta Med 54:375–6.
Lai PK, Roy J. (2004) Antimicrobial and chemopreventive properties of herbs and spices. Curr Med Chem 11(11):1451–60.
Liakopoulou-Kuriakides M, Sinakos Z, Kyriakidis DA. (1985) A high molecular weight plaquelet aggregating factor in ***Crocus sativus***. Plant Science 40:117–20.
Liakopoulou-Kuriakides M, Skubas AI. (1990) Characterization of the plaquelet aggregation inducer and inhibitor isolated from ***Crocus sativus***. Biochem Int 22:103–10.
Lin JK, Wang CJ. (1986) Protection of **crocin** dyes on the acute hepatic damage induced by aflatoxin B1 and dimethylnitrosamine in rats. Carcinogenesis 7:595–9.
Lindholm P, Gullbo J, Claeson P, Goransson U, Johansson S, Backlund A, Larsson R, Bohlin L. (2002) Selective **cytotoxicity** evaluation in anticancer drug screening of fractionated plant extracts. J Biomol Screen 7:333–40.
Lippman SM, Lotan R. (2000) Advances in the development of retinoids as chemopreventive agents. J Nutr 130(2S):479–82.

Locock RA. (1995) Alternatives **saffron**. Canadian Pharmaceutical J 127:45–6.
Loskutov A, Beninger C, Hosfield G, Sink K. (1998) Production of **saffron** in **tissue cultures** of ***Crocus sativus*** L, In vitro 34(3-part II): 64–A.
Loukis A, Al-Kofahi A, Philianos S. (1983) Étude des constituants des bulbes de ***Crocus sativus*** L. Plantes Médicinales et Phytothérapie 17:89–91.
Martínez-Tomé M, Jiménez AM, Ruggieri S, Frega N, Strabbioli R, Murcia MA. (2001) **Antioxidant** properties of Mediterranean spices compared with common food additives. J Food Prot 64:1412–9.
Mathew B. (1983) The Crocus. A revision of the genus *Crocus* (Iridaceae). Oregon: Timber Press.
Mathews-Roth MM. (1982) Effects of **crocetin** on experimental skin tumours in hairless mice. Oncology 39:362–4.
Milner JA. (2002) Strategies for cancer prevention: the role of diet. Br J Nutr 87(Suppl 2):S265–72.
Molnar J, Szabo D, Pusztai R, Mucsi I, Berek L, Ocsovszki I, Kawata E, Shoyama Y. (2000) Membrane associated antitumour effects of **crocin**-, ginsenoide- and cannabinoid derivatives. Anticancer Res 20:861–7.
Morjani H, Tarantilis P, Polissiou M, Manfait M. (1990) Growth inhibition and induction of erythroid differentiation activity by **crocin**, dimethyl**crocetine** and β–carotene on K562 tumour cells. Anticancer Res 10:1398–406.
Nair SC, Kurumboor SK, Hasegawa JH. (1995) **Saffron chemoprevention** in biology and medicine: a review. Cancer Biother 10:257–64.
Nair SC, Panikkar B, Panikkar KR. (1991a) **Antitumour activity** of **saffron** (***Crocus sativus***). Cancer Lett 57:109–14.
Nair SC, Panikkar KR, Parathod RK. (1993) Protective effects of **crocetin** on bladder **cytotoxicity** induced by cyclophosphamide. Cancer Biother 8:339–43.
Nair SC, Salomi MJ, Panikkar B, Panikkar KR. (1991b) Modulatory effects of ***Crocus sativus*** and *Nigella sativa* extracts on cisplatin-induced toxicity in mice. J Ethnopharmacol 31:75–83.
Nair SC, Salomi MJ, Varghese CD, Panikkar B, Panikkar KR. (1992) Effect of **saffron** on thymocyte proliferation, intracellular glutathione levels and its **antitumour activity**. Biofactors 4:51–4.
Nair SC, Varghese CD, Pannikar KR, Kurumboor SK, Parathod RK. (1994) Effects of **saffron** in vitamin A levels and its **antitumour activity** on the growth of solid timors in mice. Int J Pharmacog 32:105–14.
Nathan C, Xie QW. (1994) Nitric oxide synthases: roles, tolls and controls. Cell 78:915–8.
Oda Y, Tatsumi Y. (1993) New lectins from bulbs of *Croccus sativum*. Biol Pharm Bull 16:978–81.
Ozaki A, Kitano M, Furusawa N, Yamaguchi H, Kuroda K, Endo G. (2002) Genotoxicity of gardenia yellow and its components. Food Chem Toxicol 40:1603–10.
Pfander H, Schurtenberger H. (1982) Biosynthesis of C20-**carotenoids** in ***Crocus sativus***. Phytochemistry 21:1039–42.
Pham TQ, Cormier F, Farnworth E, Tong VH, Van Calsteren MR. (2000) **Antioxidant** properties of **crocin** from *Gardenia jasminoides* Ellis and study of the reactions of **crocin** with linoleic acid and **crocin** with oxygen. J Agric Food Chem 48:1455–61.
Piqueras A, Han BH, Escribano J, Rubio C, Hellín E, Fernández JA. (1999) Development of **corm**ogenic nodules and micro**corms** by **tissue culture**, a new tool for the multiplication and genetic improvement of **saffron**. Agronomie 19:603–10.
Premkumar K, Abraham SK, Santhiya ST, Gopinath PM, Ramesh A. (2001) Inhibition of genotoxicity by **saffron** (***Crocus sativus*** L.) in mice. Drug Chem Toxicol 24:421–8.
Pung A, Franke A, Zhang LX, Ippendorf H, Martin HD, Sies H, Bertram JS. (1993) A synthetic C22 **carotenoid** inhibits carcinogen-induced neoplastic transformation and enhances gap junctional communication. Carcinogenesis 14:1001–5.
Rao AV, Agarwal S. (2000) Role of **antioxidant** lycopene in cancer and heart disease. J Am Coll Nutr 19:563–9.
Raskin I, Ribnicky DM, Komarnytsky S, Ilic N, Poulev A, Borisjuk N, Brinker A, Moreno DA, Ripoll C, Yakoby N, O'Neal JM, Cornwell T, Pastor I, Fridlender B. (2002) Plants and human health in the twenty-first century. Trends Biotechnol 20:522–31.

Ríos JL, Recio MC, Giner RM, Máñez S. (1996) An update review of **saffron** and its active constituents. Phytotherapy Res 10:189–93.
Robinson A. (1995) Notes on the **saffron** plant (***Crocus sativus*** L.). Pharm Hist (Lond) 25:2–3.
Rock CL. (1997) **Carotenoids**: biology and treatment. Pharmacol Ther 75:185–97.
Rogers AE, Zeisel SH, Groopman J. (1993) Diet and carcinogenesis. Carcinogenesis 14:2205–17.
Salomi MJ, Nair SC, Panikkar KR. (1990) Inhibitory effects of *Nigella sativa* and **saffron** (***Crocus sativus***) on chemical carcinogenesis in mice and its non-mutagenic activity. Proc Ker Sci Congr 3:125–6.
Salomi MJ, Nair SC, Panikkar KR. (1991a) Inhibitory effects of *Nigella sativa* and **saffron** (***Crocus sativus***) on chemical carcinogenesis in mice. Nutr Cancer 16:67–72.
Salomi MJ, Nair SC, Panikkar KR. (1991b) **Cytotoxicity** and non-mutagenicity of *Nigela sativa* and **saffron** (***Crocus sativus***) in vitro. Proc Ker Sci Congr 5:244.
Sampathu SR, Shivashankar S, Lewis YS. (1984) **Saffron** (***Crocus sativus*** L.). Cultivation, processing, chemistry and standardization. CRC Crit Rev Food Sci Nutr 20:123–57.
Sano K, Himeno H. (1987) In vitro proliferation of **saffron** (***Crocus sativus*** L.) stigma. Plant Cell Tiss Org 11:159–66.
Sarma S, Maesato K, Hara T, Sonoda Y. (1990) In vitro production of stigma-like structures from stigma explants of ***Crocus sativus*** L. J Exp Bot 41:745–8.
Smith TAD. (1998) **Carotenoids** and cancer: prevention and potential therapy. Brit J Biomed Sci 55:268–75.
Soeda S, Ochiai T, Paopong L, Tanaka H, Shoyama Y, Shimeno H. (2001) **Crocin** suppresses tumour necrosis factor-alpha-induced cell death of neuronally differentiated PC-12 cells. Life Sci 69:2887–98.
Souret FF, Weathers PJ. (1999) Cultivation, in *vitro* culture, secondary metabolite production, and phytopharmacognosy of **saffron** (***Crocus sativus*** L.). J Herbs, Spices & Medicinal Plants 6:99–116.
Tarantilis PA, Morjani HM, Polissiou M, Manfait M. (1994) Inhibition of growth and induction of differentiation of promyelocytic leukemia (HL-60) by **carotenoids** from ***Crocus sativus*** L. Anticancer Res 14:1913–8.
Thatte U, Bagadey S, Dahanukar S. (2000) Modulation of programmed cell death by medicinal plants. Cell Mol Biol (Noisy-le-grand) 46:199–214.
Tseng TH, Chu CY, Huang JM, Shiow SJ, Wang CJ. (1995) **Crocetin** protects against oxidative damage in rat primary hepatocytes. Cancer Lett 97:61–7.
Unnikrishnan MC, Kuttan R. (1990) Tumour reducing and anticarcinogenic activity of selected spices. Cancer Lett 51:85–9.
Velasco M, Díaz-Guerra MJM, Díaz-Achirica P, Andreu D, Rivas L, Bosca L. (1997) Macrophage triggering with cecropin A and melittin-derived peptides induces type II nitric oxide synthase expression. J Immunol 158:4437–43.
Visvanath S, Ravishankar GA, Venkataraman LV. (1990) Induction of **crocin**, **crocetin**, picro**crocin** synthesis in callus culture of **saffron**. Biotechnol Appl Biochem 12:336–40.
Wang CJ, Cheng TC, Liu JY, Chou FP, Kuo ML, Lin JK. (1996) Inhibition of protein kinase C and proto-oncogene expression by **crocetin** in NIH/3T3 cells. Mol Carcinog 17:235–40.
Wang CJ, Shiah HS, Lin JK. (1991a) Modulatory effect of **crocetin** on aflatoxin B_1 cytotoxocity and DNA adduct formation in C3H10T1/2 fibroblast cell. Cancer Lett 56:1–10.
Wang CJ, Shiow SJ, Lin JK. (1991b) Effects of **crocetin** on the hepatotoxicity and hepatic DNA binding of aflatoxin B_1 in rats. Carcinogenesis 12:459–62.
Watanabe T, Terabe S. (2000) Analysis of natural food pigments by capillary electrophoresis. J Chromatogr A 880:311–22.
Wilkins ES, Gainer JL. (1979) The effect of **crocetin** on the irradiation of Walker-256: in vitro and in vivo studies. Cancer Biochem Biophys 3:71–4.
Wilkins ES, Gainer JL, Wilkins MG. (1977) Increased relative growth rate of normal rat cells in vitro with **crocetin**. Experientia 33:1028.
Willett WC. (1994) Diet and health: what should we eat? Science 264:532–7.
Willett WC. (1999) Goals for nutrition in the year 2000. CA Cancer J Clin 49:331–52.

Winterhalter P, Straubinger M. (2000) **Saffron**-renewed interest in an ancient spice. Food Rev Int 16:39–59.

Zhang LX, Acevedo P, Guo H, Bertram JS. (1995) Upregulation of gap junctional communication and connexin43 gene expression by **carotenoids** in human dermal fibroblasts but not in human keratinocytes. Mol Carcinog 12:50–8.

Zhang LX, Cooney RV, Bertram JS. (1991) **Carotenoids** enhance gap junctional communication and inhibit lipid peroxidation in C3H/10T1/2 cells: relationship to their cancer chemopreventive action. Carcinogenesis 12:2109–14.

Zhang LX, Cooney RV, Bertram JS. (1992) **Carotenoids** up-regulate connexin43 gene expression independent of their provitamin A or **antioxidant** properties. Cancer Res 52:5707–12.

M.T.H. Khan and A. Ather (eds.)
Lead Molecules from Natural Products

Lead compounds and drug candidates from some Turkish plants for human health

İlkay Orhan, Bilge Şener

Abstract

Biodiversity has tremendous potential in having excellent chemical diversity to drug discovery programs and in serving as templates for synthetic drugs. Research activities on plants growing in Turkey, showed that diverse classes of bioactive compounds ranging from simple aromatics, e.g. paeonol, to complex molecules of alkaloids, terpenoids and steroids possess significant biological activities, including anti-cholinesterase, anticholinergic, antihypertensive, antithrombocyter, anti-inflammatory, antimalarial and antibacterial. The results of these investigations and their potential therapeutic application have been reviewed in this chapter.

Keywords: *Biodiversity, Medicinal plants, Biological activities, Natural products*

Abbreviations: AA, arachidonic acid; AChE, acetylcholinesterase; AD, Alzheimer's Disease; HTS, high-throughput screening; NSAID, non-steroidal anti-inflammatory drugs; PAF, platelet-activated factor; ROS, reactive oxygen species.

I. Introduction

Discoveries for the advancement of medicine and understanding of life sciences constitute one of the most powerful ways in which biodiversity can contribute to human society. Bioresources have tremendous potential in providing bioactive compounds for the development of new lead candidates for pharmaceuticals, nutraceuticals and agrochemicals. Therefore, natural products continue to be an important source of modern drugs in clinical use as an active ingredient, starting material to produce semi-synthetic drugs as well as lead compounds for totally synthetic drugs.

Plants produce a diverse range of structurally novel bioactive molecules, making them a rich source of different types of medicines. Traditional medicines are still the mainstay of about 75–80% of the world population, mainly in the developing world.

Living organisms are unknown complex mixtures having potentially large number of secondary metabolites. The most of valuable drugs in use have been obtained from bioresources. Since the discovery of penicillin, a large number of antibiotics have been isolated from scores of microorganisms each year as life-saving drugs. The new important anti-infectives are not only limited to investigating the microorganisms but also include plants. For example, artemisinin was isolated from the Chinese medicinal plant known as *Artemisia annua* to be effective for the deadly malaria.

Natural products have also provided various important drugs including bleomycin, doxorubicin, mitomycin, vincristine, vinbastine, streptozocin and the most recent one, paclitaxel in the chemotherapy of cancer. Irinotecan (Camptothecin derivative), etoposide and tenoposide (Podophyllotoxin derivatives) and Vinorelbin (*Vinca* alkaloids derivative) are the examples of semi-synthetic derivatives of anti-cancer drugs.

It is known that *Digitalis* glycosides (digoxin, digitoxin and the lanatosides) were effective in controlling heart diseases for many years. Besides the cardiac glycosides, a number of alkaloids such as quinidine and theophylline are important drugs in the treatment of various cardiovascular diseases. In addition, some nitrogen-containing compounds including morphine, codeine, papaverine, d-tubocurarine, reserpine, physostigmine, galanthamine, ephedrine and ergotamine are known as important drugs for central nervous system disorders.

The useful cholesterol-lowering agents described as statins were derived from the metabolites of fungi. These drugs inhibit 3-hydroxy-3-methylglutaryl coenzyme A reductase, which is a critical enzyme in the biosynthesis of cholesterol (Şener, 2003).

Given the current research and development performance associated with the understanding of disease process, there is still a great need for novel compounds with unique mechanisms of action to treat diseases such as cancer, Alzheimer's, arthritis, diabetes, etc.

Our researches have been focused to develop bioactive compounds from Turkish plants as leads for drug candidates.

Turkey is one of the rich countries in the world for biological sources depending on different geographical, ecological and aquatic environments as well as passageway between Europe, Asia and Africa. The floristic diversity provides a wide choice of species represented by 11,000 taxa of which 3700 are endemic. Turkish flora is the richest among any country in Europe, North Africa and the Middle East (Güner et al., 2001).

The results indicated that until the end of 1999, over 930 papers were published reporting the chemical compositions of 830 taxa belonging to 237 genera included 58 families, meaning that only 7.7% of the flora of Turkey have been investigated chemically. This is almost half of the number determined for the world. It has been estimated that out of ca. 250,000 species of flowering plants only about 15% have been studied phytochemically (Baytop, 1999; Başer, 2002).

Many new natural product-originated bioactive compounds effective in treating several diseases have been isolated from different plants, fungi and microorganisms. They are unknown complex mixtures having potentially large number of secondary metabolites. Sensitive bioassays for the high-throughput screening (HTS) methods have been developed to screen these plant extracts. The simplest assays are the ones based on the mechanisms of action of a known drug.

The assays have also been incorporated into efficient testing schemes that are useful to HTS. For example, one assay used for Alzheimer's Disease (AD) is based on the inhibition of acetylcholinesterase (AChE). The development of new leads of AChE inhibitors has been realized by Ellman method for screening of biological sources (Ellman, 1958; Ellman et al., 1961).

Recent results obtained from some Turkish plants as lead candidates for the development of drugs have been described.

II. Antiacetylcholinesterase compounds

In our century, AChE inhibitors have become the most popular strategy for increasing cholinergic activity in the brain and have shown the most encouraging results as palliative therapy for AD.

AD is one of the most common mental problems in the aged population (Adams et al., 1984; Arnold and Kumar, 1993; Aisen and Davis, 1997). The basal forebrain and brainstem cholinergic systems play also an important role in the regulation of cortical and thalamic electrical activity (Bachman et al., 1992). The findings from experimental animals, aging and AD research have provided an experimental foundation for the cholinergic hypothesis of learning and memory (Gauthier, 2001; Guardado-Santervas, 2001; Schneider, 2001). On the basis of cholinergic hypothesis, AD results from a defect in the cholinergic system. One goal to the treatment for AD is to increase the acetylcholine level in the brain. Therefore, AChE inhibitors are developed for the treatment of this disease.

AD, a slowly progressive neuropsychiatric illness is principally characterized by memory deficits and it has become the fourth leading cause of death after heart diseases, cancer and stroke in industrialized nations of the US and Europe. Among the AChE inhibitors used in present, physostigmine and tacrine are not ideal drugs for clinical use because of their low bioavailability, narrow therapeutic window and some side effects such as hepatotoxicity (Duvoisin, 1968). More recent drugs like donepezil and rivastigmine have effect only against mild type of AD. Therefore, the search for novel AChE inhibitors with better properties is necessary.

The AChE catalyzes the hydrolysis of the neurotransmitter acetylcholine and it has long been an attractive target for rational drug design and development of mechanism-based inhibitors for the treatment of ADs. Therefore, AChE inhibitors are the only class of drugs that produce improvements in cognitive function.

II.A. Lycopodium species

Huperzine A (**1**) has a special significance among the compounds with AChE inhibitory activity, isolated from natural resources. Huperzine A, [(5R, 9R, 11E)-5-amino-11-etilidin-5,6,9,10-tetrahydro-7-metil-5,9-metanosikloocta-[b]-piridin-2 (1 H)-on], is an alkaloid isolated by the researchers of the Shanghai Institute of Materia Medica in 1986 from the clubmoss *Lycopodium serratum* Thunb. (syn. *Huperzia serrata* (Thunb.) Trev) (Lycopodiaceae). This plant, called as "Qing Ceng Ta", has been used in traditional Chinese medicine for its memory-enhancing property since centuries (Liu et al., 1986).

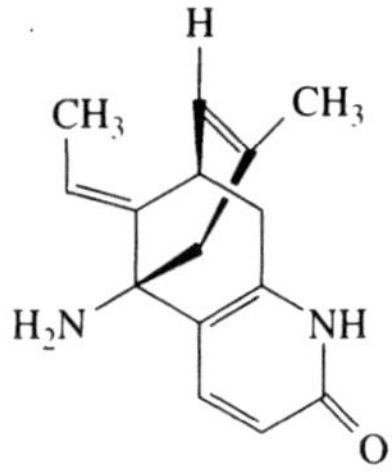

1

Over 100 alkaloids, a number of which comprising a series of Huperzine A–R, have been isolated from the genus *Lycopodium* that is very rich in alkaloid content (Wang and Tang, 1998). Among them, only Huperzine A has possessed remarkable AChE inhibitory activity (Liu et al., 1986).

The activity of Huperzine A has been found to be as high as, or even greater than, physostigmine, galanthamine, donepezil and tacrine, the commercial drugs already used against AD.. In various in vivo and *ex vivo* experiments, it has been shown to inhibit AChE reversibly and also to prevent oxidative cell damage induced by β-amyloid plaques (Wang et al., 1986; Tang, 1996; Jing et al., 1999; Tang and Han, 1999; Ye et al., 1999).

II.B. Amaryllidaceae alkaloids

Amaryllidaceae alkaloids are typically found in all species of Amaryllidaceae that is one of the largest families of very ornamental bulbous plants (Martin, 1987). Amaryllidaceae alkaloids have long held a prominent position in the history of natural products chemistry because of the structural similarity to the essential amino acids phenylalanine and related metabolites of tyrosine (Könükol, 1992).

New alkaloids have been isolated from a variety of sources with increasing frequency and have been characterized using the latest spectroscopic techniques. The medicinal properties of these alkaloids remain of great interest as well as the nature of their structure and stereochemistry. To date, over 190 Amaryllidaceae alkaloids have been isolated. These alkaloids are known to exhibit a variety of biological activities such as antitumor (Zee-Chang et al., 1978; Pettit et al., 1986; Könükol, 1992), antiviral (Ieven et al., 1983) and anticholinergic activities (Ieven et al., 1981) . Some of them have been used in the treatment of myasthenia gravis, myopathy and diseases of the nervous system. The bulbs of *Galanthus*, *Narcissus* and *Leucojum* are of interest because they contain galanthamine used in the treatment of poliomyelitis.

Galanthamine **(2)**, an alkaloid isolated from some *Galanthus* species (Amaryllidaceae), has been used recently in the treatment of AD. It has a reversible action of AChE inhibition and also modulates the nicotinic acetylcholine receptors (Boissier et al., 1960; Thomsen and Kewitz, 1990; Schrattenholz et al., 1996;

Albuquerrque et al., 1997; Levin and Simon, 1998; Tariot et al., 2000). Although the most common side effect of galanthamine is nausea, it is possible to eliminate nausea by increasing the galanthamine dose slowly (Raskind et al., 2000). In addition, galanthamine was shown to have no hepatotoxicity (Guillou et al., 2000). Therefore, Galanthamine (Nivalin®) has been approved as its HBr salt for the first time in Austria, later licensed as Reminyl® in the USA, some European countries as well as in Turkey.

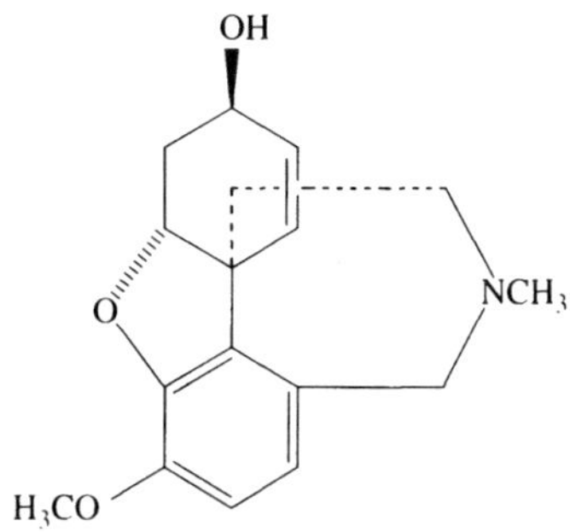

2

Besides, positive effect of galanthamine was found in several learning and memory tests in animals, based on the cholinergic hypothesis that memory impairments in patients suffering from AD result from a defect in the cholinergic system and one approach to the treatment for this disease is to enhance the acetylcholine level in the brain (Könükol, 1992).

This development has prompted us to investigate the anticholinesterase activity of some plant species of Amaryllidaceae growing in Turkey, namely *Galanthus elwesii* Hooker fil., *G. ikariae* L., *Narcissus tazetta* subsp. *tazetta* L., *Leucojum aestivum* L. and *Pancratium maritimum* L. by Ellman method in comparison with galanthamine as the standard drug.

The bulbs of some species of Amaryllidaceae cultivated in Turkey have been exported as ornamental plants. In the aforementioned context, we systematically studied the uninvestigated five Amaryllidaceae plants growing in Turkey for their potential anticholinesterase activities (Orhan, 2002). In total, six Amaryllidaceae-type known alkaloids called lycorine (**3**), tazettine (**4**), crinine (**5**), galanthamine (**2**), 3-epi-hydroxybulbispermine (**6**) and 2-demethoxymontanine (**7**) from the active fractions of *G. ikariae* were obtained by bioactivity-directed fractionation. On the other hand, lycorine, tazettine, 3-epi-hydroxybulbispermine, *N*-nor-galanthamine (**8**) and haemantamine (**9**) were isolated from the active fractions of *N. tazetta* subsp. *tazetta* as the common Amaryllidaceae alkaloids. Although *G. ikariae* and *N. tazetta* subsp. *tazetta* extracts showed 75.56% and 46.62% inhibition, respectively; it made us to think that the activity of the extracts being lower than 50% was due to the synergistic interaction between the alkaloids isolated. Among these active alkaloids, lycorine ($IC_{50} = 3.16\,\mu M$) and galanthamine ($IC_{50} = 3.2\,\mu M$) have been determined to exhibit potent inhibitory activity (Orhan and Şener, 2003).

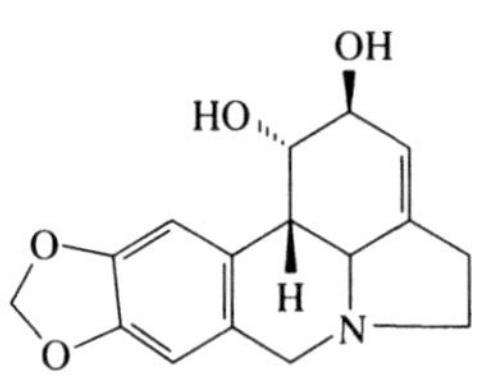

3

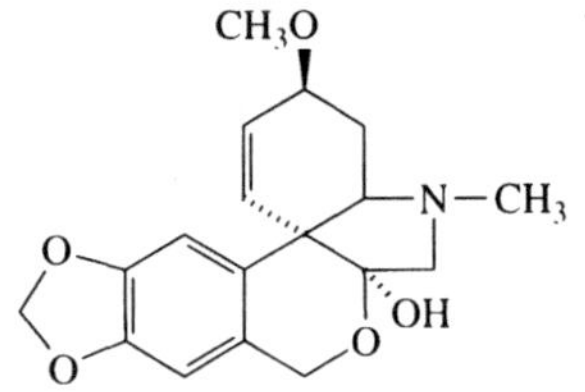

4

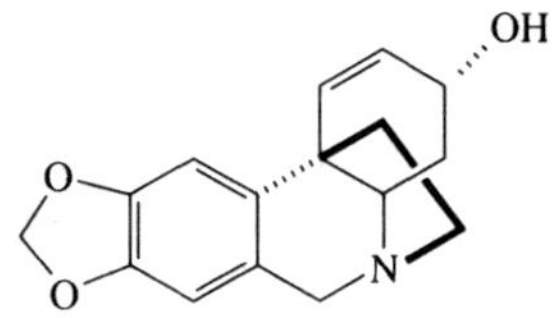

5

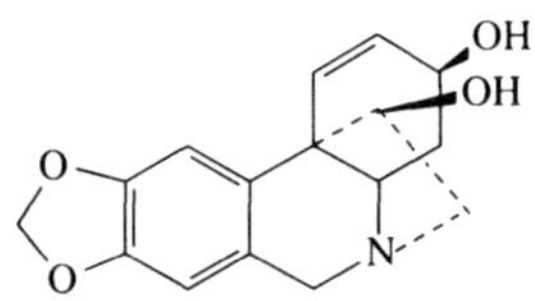

6

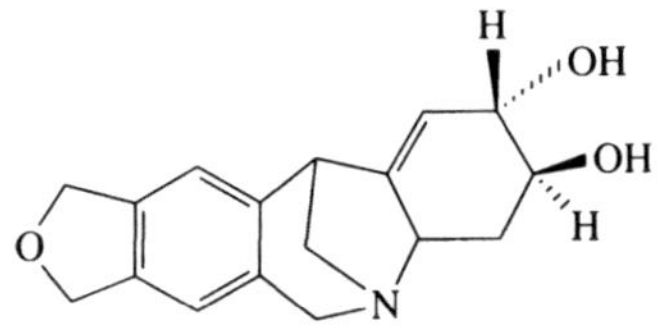

7

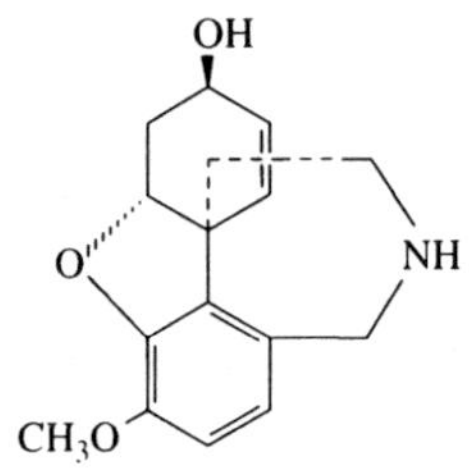

8

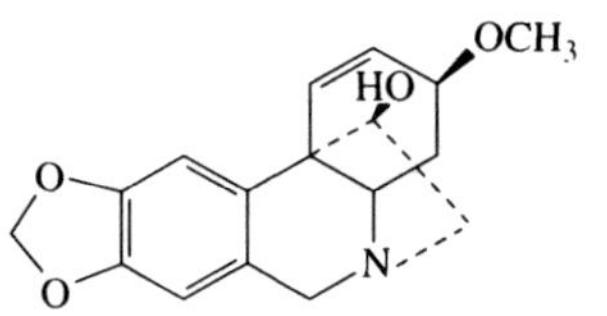

9

These findings showed that a single alkaloid is not responsible for anticholinesterase activity of *G. ikariae* and *N. tazetta* subsp. *tazetta* extracts. However, the activity may depend on the synergistic interaction between the alkaloids isolated. According to these results, the bulbs of Amaryllidaceae plants can also be evaluated as a source of anticholinesterase alkaloids in addition to their ornamental properties.

In a similar study by Lopez et al., they screened 26 extracts prepared from various *Narcissus* species were screened along with 23 pure Amaryllidaceae-type alkaloids against AChE and suggested that the alkaloids having galanthamine and lycorine skeletons possess inhibitory activity (Lopez et al., 2002).

II.C. Isoquinoline alkaloids

The genus *Fumaria* L. (Fumarioideae, Papaveraceae) is represented by 19 species in Turkey (Davis, 1965; Davis et al., 1988). These are small annual herbs and known as "Şahtereotu" in Turkey (Baytop, 1999). Some of them are used in folk medicine against eczema, rheumatism, stomachache and dysentery. In Anatolia, one of them, *Fumaria vaillantii* Lois. is widespread where its extracts are used in folk medicine as a blood purifier in the treatment of skin diseases. Fourteen of them have been investigated in terms of molecular diversity in the alkaloidal content and their biological activities (Şener, 2002).

Fumaria species are the invaluable source of isoquinoline alkaloids, many of which possess a wide diversity of molecular structures and biological activities. According to the standard extraction, isolation and purification procedures described in our previous study (Şener, 1981); 49 isoquinoline alkaloids have been isolated from 14 *Fumaria* species (Blasko et al., 1981, 1982; Şener, 1983, 1984a–d, 1985a–c, 1986, 1988, 1994; Şener et al., 1983; Şener and Temizer, 1991; Küçükboyacı et al., 1998). All alkaloids were identified on the basis of their extensive spectral data reported in the relevant literatures. Morphological characters used for the description of 13 *Fumaria* species were also determined (Şener, 1982).

Within our project on AChE inhibitors from some Turkish plants, we have also screened *Fumaria* species from Fumarioideae subfamily (*Fumaria asepala* Boiss., *F. bastardii* Bor., *F. boissieri* Hausskn., *F. bracteosa* Pomel, *F. capreolata* L., *F. cilicica* Hausskn., *F. densiflora* DC. (syn. *F. micrantha* Lag.), *F. flabellata* Gasp., *F. gaillardotii* Boiss., *F. judaica* Boiss., *F. kralikii* Jordan (syn. *F. anatolica* Boiss.), *F. macrocarpa* Parlatore, *F. microcarpa* Boiss. ex Hausskn., *F. officinalis* L., *F. parviflora* Lam., *F. petteri* Reichb. *subsp. thuretii* (Boiss.) Pugsley (syn. *F. thuretii* Boiss., *F. pikermiaana* Boiss., *F. rostellata* Knaf, *F. schleicheri* Soyer-Willement, *F. vaillantii* Lois. (syn. *F. schrammii* Valenovsky) for their AChE inhibitory activity by Ellman method (Şener and Orhan, 2003).

In the course of our studies on AChE inhibitors (Orhan, 2002), the alkaloidal extracts of *Fumaria* species displayed high inhibitory activity, while galanthamine, the standard drug used in this study, showed 48.80±0.36% inhibitory activity. All of the extracts had much higher activity compared to galanthamine, ranging between 84.98±1.07% and 96.89±0.17% by in vitro using spectrophotometric Ellman method. Out of these species, *F. vaillantii*, having 94.23±0.47% inhibitory activity, was selected for bioactivity-directed fractionation and isolation studies and the common isoquinoline alkaloids named as canadine (**10**), hydrastine (**11**), bulbocapnine (**12**), fumarophycine (**13**), corydaldine (**14**), protopine (**15**), ophiocarpine-*N*-oxide (**16**), β-allocryptopine (**17**), ophiocarpine (**18**) and berberine (**19**) were obtained from the active fractions of *F. vaillantii*. Their AChE inhibitory activities are as follows: Canadine (56.75±0.58%), hydrastine (10.08±0.78%), ophiocarpine (92.97±0.45%), bulbocapnine (65.23±0.42%), fumarophycine (37.90±0.99%), corydaldine (17.11±0.89%), ophiocarpine-*N*-oxide (81.40±0.83%), protopine (80.53±0.59%), β-allocryptopine (89.31±0.41%) and berberine (84.97±0.71%).

Among the alkaloids, ophiocarpine ($IC_{50} = 1.1\,\mu M$) had the most potent inhibitory activity followed by β-allocryptopine ($IC_{50} = 1.3\,\mu M$), berberine ($IC_{50} = 1.6\,\mu M$), ophiocarpine-*N*-oxide ($IC_{50} = 1.79\,\mu M$) and protopine ($IC_{50} = 1.8\,\mu M$). Hydrastine, an alkaloid having phthalideisoquinoline skeleton, and corydaldine, an alkaloid with dihydroisoquinolone structure, were the least potent alkaloids, leading the suggestion that both type of alkaloids do not have remarkable AChE inhibitory activity. On the other hand, fumarophycine with spirobenzylisoquinoline skeleton and bulbocapnine ($IC_{50} = 2.0\,\mu M$) with aporphine skeleton, which have a better IC_{50} value than galanthamine ($IC_{50} = 5.8\,\mu M$) could be considered to contribute to the activity of the extract. Another structure–activity relationship can be established between canadine ($IC_{50} = 2.6\,\mu M$) and ophiocarpine ($IC_{50} = 1.1\,\mu M$). Both of the alkaloids possess the same tetrahydroprotoberberine structure, except a lack of an hydroxyl group in canadine. The much higher activity of ophiocarpine than canadine may be based on the existence of this hydroxyl group by increasing the solubility of this compound. Consequently, we conclude that the responsible compounds for the activity of *F. vaillantii* extract were determined as the alkaloids, namely ophiocarpine, β-allocryptopine, berberine, ophiocarpine-*N*-oxide and protopine that have tetrahydroprotoberberine and protoberberine skeletons and the activity may be due to the synergistic interaction between these alkaloids, which may be of therapeutic value in the treatment of AD (Şener and Orhan, 2003).

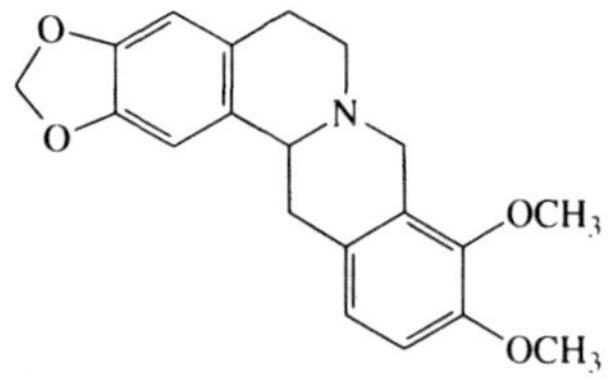

10

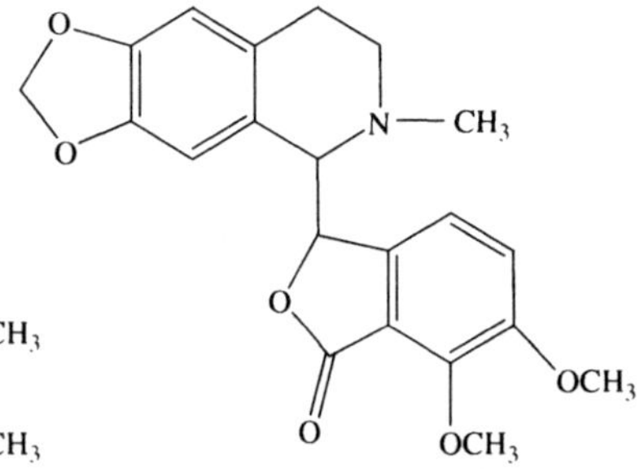

11

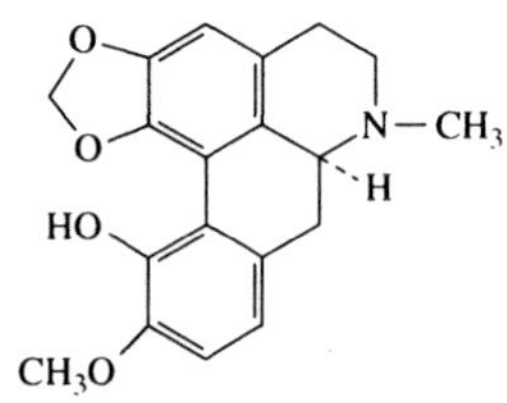

12

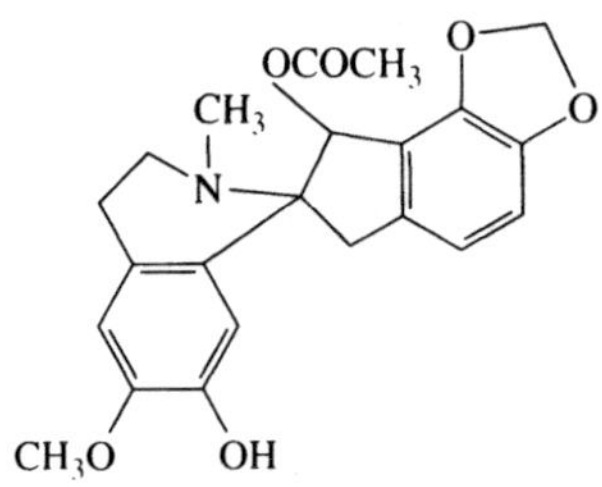

13

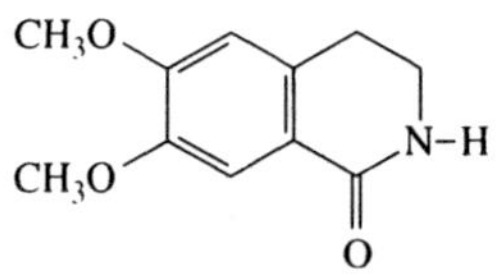

14

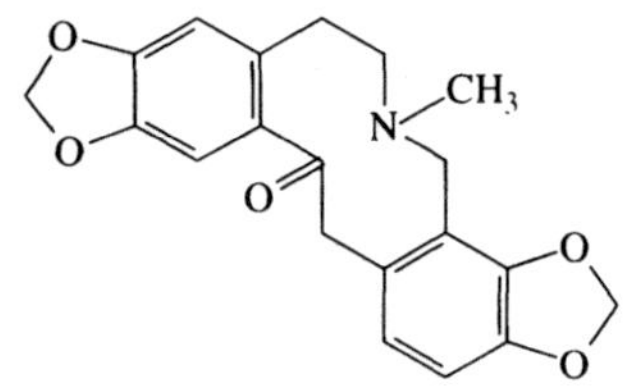

15

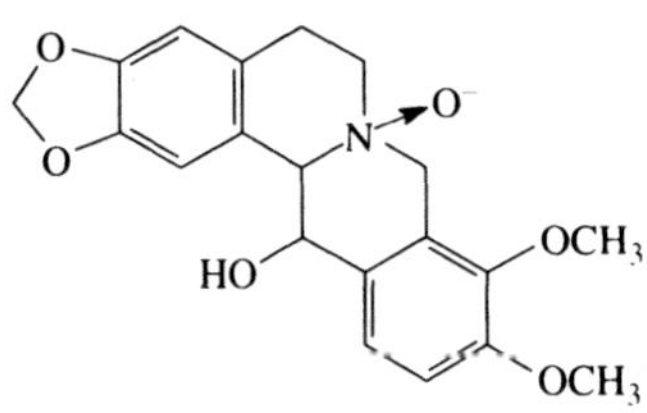

16

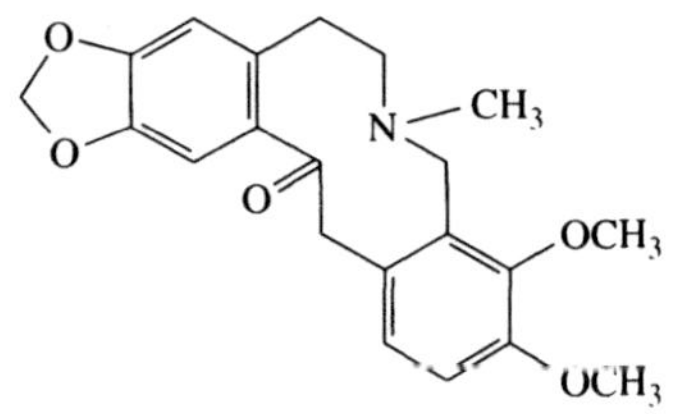

17

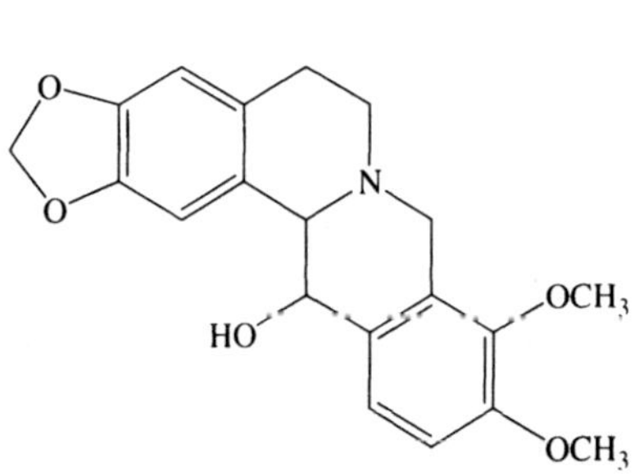

18

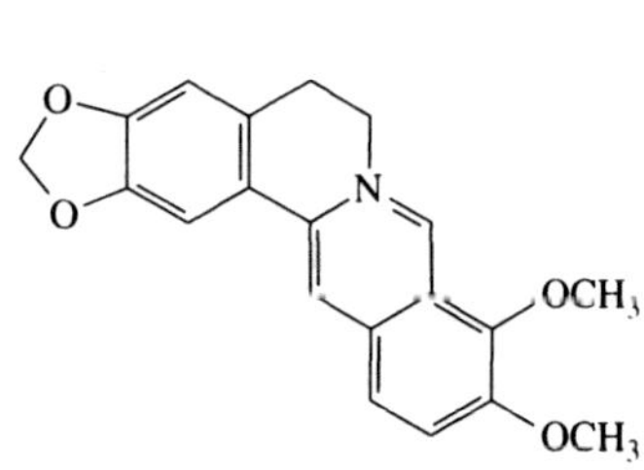

19

II.D. Steroidal alkaloids

The genus *Fritillaria* L. belongs to Liliaceae, which is an important steroidal alkaloid-bearing plant family. Bulbs of the genus *Fritillaria*, commonly known as "Bei-mu" or "Pei-mu" in Chinese and "Bai-mo" in Japanese, have long been known as one of the principal Chinese crude drugs. The dried bulbs or a decoction of *Fritillaria* species are to be prescribed to treat cough, asthma, bronchitis, some tumors and deficiency of milk. *Fritillaria imperialis* L. is cultivated in Turkey as an ornamental plant and the bulbs are exported. Impericine (**20**), forticine (**21**), delavine (**22**), persicanidine A (**23**) and imperialine (**24**) were isolated from the alkaloid extract of the bulbs and their AChE and butyrylcholinesterase inhibitory effects were determined by using eserine as a standard inhibitor. According to their IC_{50} values, these alkaloids were found to be more selective inhibitors of butyrylcholinesterase (Atta-ur-Rahman et al., 2002).

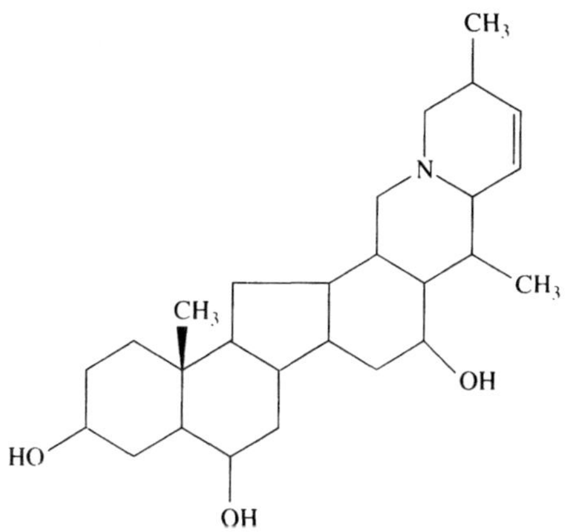

20

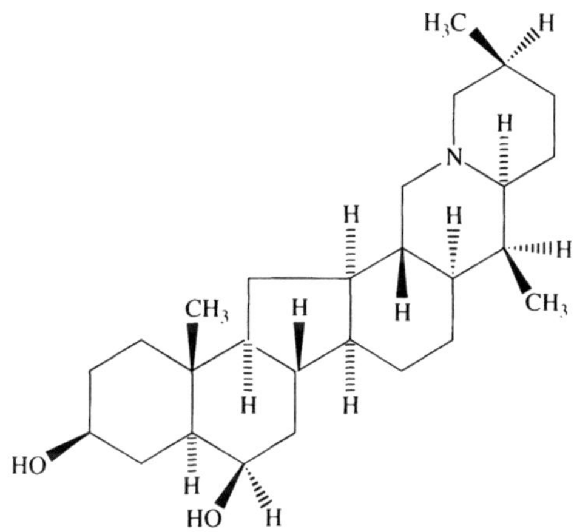

21

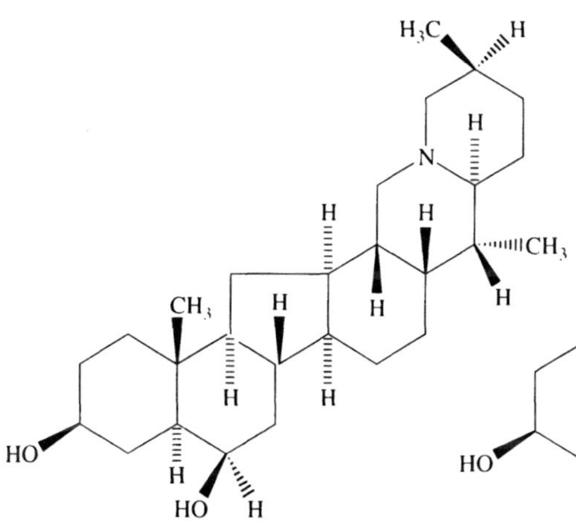

22

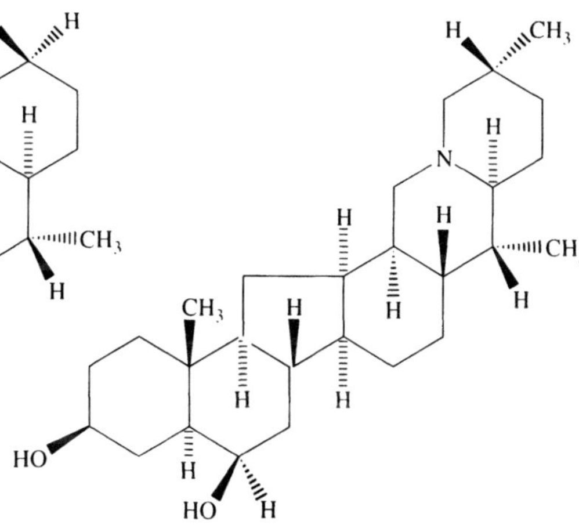

23

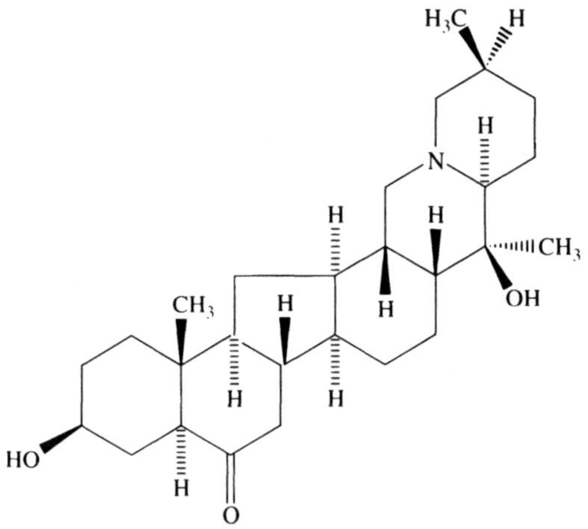

24

III. Anticholinergic compounds

Anticholinergic drugs are considered useful in alleviating asthma. The bulbs of *Fritillaria* species have been used for the treatment of asthma in Chinese medicine. In order to find the active compound/s for the traditional use of the plant in asthma, the anticholinergic activity of the extracts prepared from the bulbs of *F. imperialis* L., cultivated in Turkey, was determined by blockade of acetylcholine responses on isolated guinea-pig ileum and atria. Further isolation studies gave a new bioactive compound determined as ebeinone (**25**). Ebeinone exhibited anticholinergic activity as manifested by blockade of acetylcholine responses on isolated guinea-pig ileum. The antimuscarinic effect of ebeinone was further confirmed when tested in isolated guinea-pig atria, where the same concentration of ebeinone also completely blocked the inhibitory responses of acetylcholine (Atta-ur-Rahman et al., 1994).

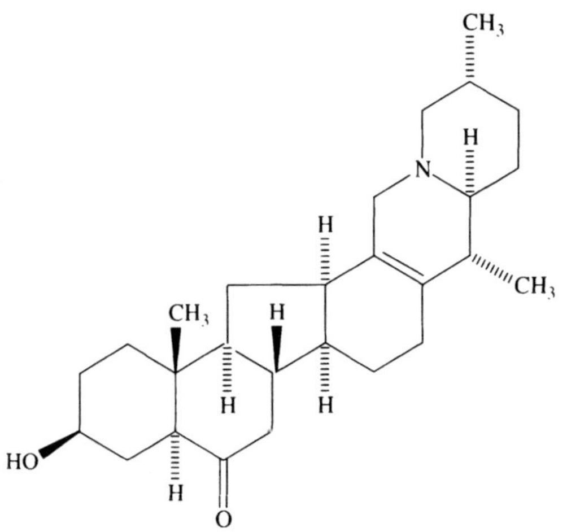

25

Ebeinone may be responsible for the traditional use of *F. imperialis* in the treatment of asthma.

IV. Antihypertensive compounds

Antihypertensive drugs are important for the prevention of coronary diseases, stroke and insufficiency of the kidney. *Veratrum album* L. (Liliaceae) has been used in the treatment of various health disorders such as toothache, herpes and hypertension. During the evaluation of this plant in terms of antihypertensive activity (Şener and Atta-ur-Rahman, 1996), jervinone (**26**), *O*-acetyljervine (**27**) and 1-hydroxy-5,6-dihydrojervine (**28**) were obtained as antihypertensive steroidal compounds. In anesthetized rats, jervinone caused a dose-dependent fall in blood pressure. The systemic blood pressure was recorded from the carotid artery via an arterial cannula connected to a pressure transducer and responses were measured on a Grass model 7D polygraf (Atta-ur-Rahman et al., 1993).

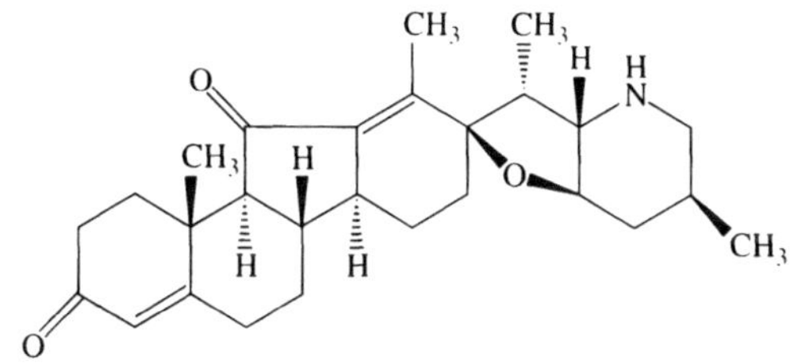

26

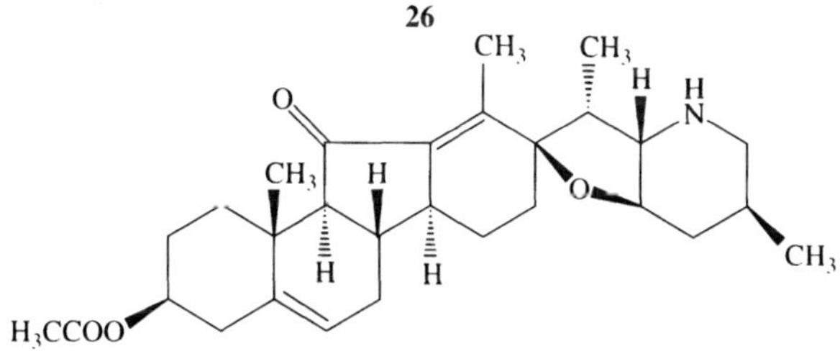

27

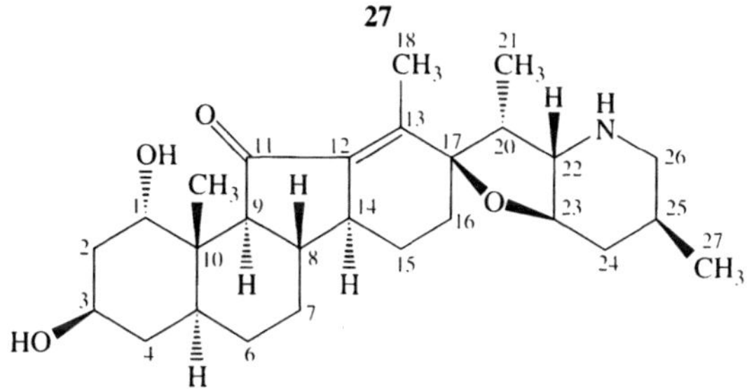

28

Noradrenaline produced a pronounced increase in blood pressure and pre-treatment with jervinone did not alter the vasoconstrictor response to noradrenaline, which rules out the possibility of the involvement of adrenoreceptors. Acetylcholine at a dose of 1 μg/kg produced a decrease in mean arterial blood pressure comparable to that of jervinone at 10 μg/kg. Pre-treatment with atropine (1 mg/kg) completely abolished the antihypertensive response to jervinone (10 µg/kg) as well as to acetylcholine (1 µg/kg). *O*-acetyljervine (**27**) and 1-hydroxy-5,6-dihdrojervine (**28**) also produced fall in blood pressure at 10–300 µg/kg with a small degree of tachycardia. Antihypertensive effects of *O*-acetyljervine and 1-hydroxy-5,6-dihydrojervine were not modified by pre-treatment with atropine, suggesting that the mechanisms of the antihypertensive effect of these compounds are different from that of jervinone or acetylcholine.

V. Antithrombocyter compounds

The role of antithrombocyter drugs in the control of cardiovascular diseases continues to be emphasized. It is well recognized that platelet–vessel wall interactions are important in the development of thrombosis and atherosclerosis. Thus, inhibition of platelet function may be a promising approach for the prevention of thrombosis. Although many agents have been reported to have in vitro anti-platelet effects, only few of them are clinically useful in antithrombotic therapy. Therefore, it is very important in searching for new candidates for this purpose. The effect of some Turkish plants against human platelet aggregation induced by arachidonic acid (AA), collagen and platelet-activated factor has been examined. Among them, *F. vaillantii* showed complete inhibition on platelet aggregation caused by AA and collagen inhibitors of thromboxane formation. Bioassay-guided fractionation of the ethanolic extract of *F. vaillantii* resulted in the isolation of protopine (**15**) as a bioactive compound (Şener, 1994).

Protopine also inhibits human platelet aggregation induced by platelet-activated factor (PAF). Since PAF is an important mediator of inflammation, thrombosis and asthma, it can be deduced that protopine may be a useful compound possessing anti-PAF properties. Therefore, protopine can be used as a lead compound for the development of antithrombocyter drugs.

VI. Antimalarial compounds

Malaria is one of the most important infectious diseases worldwide. Globally, the incidence of malaria, a situation caused by behavioral changes of mosquitoes which result in increased resistance to insecticide, human population movements and changing agricultural practices is on the increase (Şener et al., 1988). Also, while old drugs are losing effectiveness, there has been little emphasis worldwide and funding on development of new, more effective antimalarials. In order to control malaria in the long run, new drugs are certainly required.

Some members of the Amaryllidaceae alkaloids have attracted significant attention as a consequence of their biological activities. Some of them have been used in the treatment of myasthenia gravis, myopathy and the diseases of the nervous system. Galanthamine (**2**), a widespread alkaloid among the Amaryllidaceae plants, exhibits an analgesic activity comparable to that of morphine. It has also been shown to inhibit cholinesterases reversibly. Galanthamine exhibits cytotoxic activity in vitro testing against fibroblastic murine nontumor cells. It has been noticed that some compounds possessing cytotoxic activity will also possess antimalarial activity (Şener et al., 1992).

Four groups of Amaryllidaceae alkaloids, namely lycorine-, crinine-, tazettine- and galanthamine-type, as well as plant extracts of the Amaryllidaceae plants

(*Pancratium maritimum* L., *Leucojum aestivum* L. and *Narcissus tazetta* ssp. tazetta L.) growing in Turkey were evaluated in vitro for their ability to inhibit *Plasmodium falciparum*, chloroquine sensitive (T9.96) and chloroquine resistant (K_1) growth by a HTS method in a 96-well microtiter plate. All four groups of alkaloids exhibited antimalarial activity at different potencies. 6-Hydroxyhaemanthamine, haemanthamine (**9**) and lycorine (**3**) were found to be the most potent alkaloids against *P. falciparum* (T9.96) and galanthamine and tazettine had the least potent activity against *P. falciparum* (K_1).

It is interesting to discover that lycorine- and crinine-type alkaloids have antimalarial potency in the same range. However, tazettine (**4**) and galanthamine (**2**) are about 15–20-fold less potent than lycorine and crinine groups against chloroquine-sensitive strain of *P. falciparum* (T9.96). The antimalarial potency of galanthamine against chloroquine-sensitive strain (T9.96) and chloroquine-resistant strain (K_1) of *P. falciparum* were different. The former strain was about four times more sensitive to galanthamine. The reasons for these different potencies are not known at present (Şener et al., 2003).

Differences in the potency of antimalarial activity of these alkaloids could be associated with the difference in their chemical structures. Except galanthamine (**2**) and tazettine (**4**), the other alkaloids have no *N*-methylated side chain. Lycorine- and crinine-type alkaloids contain methylenedioxy group attached to the benzene ring of the molecule (Şener et al., 1993a–c, 1994, 1998; Ide et al., 1996).

In conclusion, the results of this study suggested that the methylenedioxybenzene part of the molecule and tertiary nitrogen without methyl contributed to higher activity of both lycorine and crinine groups of alkaloids than galanthamine and tazettine-type compounds (Şener et al., 2003).

VII. Anti-inflammatory compounds

Both steroidal and non-steroidal anti-inflammatory drugs (NSAID) currently used in the treatment of inflammatory diseases are known to have various side effects. Thus, investigations for the new anti-inflammatory agents with minimum side effects are still a challenge and studies on both synthetic drugs and plants are conducted to realize this purpose. In folk medicine of Turkey, many plants have been used in the treatment of rheumatoid arthritis and some anti-inflammatory diseases. The aqueous extracts of 29 plant species from 18 families were tested for their anti-inflammatory activities by using carrageenan-induced hind paw edema method in mice (Yeşilada et al., 1993).

VII.A. Paeonia daurica

The roots of some *Paeonia* species (Paeoniaceae) have been used as analgesics, anti-inflammatory agent, antispasmodic, hypotensive and antiepileptic especially in East Asian traditional medicinal systems for centuries. In Turkey, there are seven species of *Paeonia*, however they are not widely used in folk medicine. In our previous work, the anti-inflammatory activity of two Turkish *Paeonia* species, *P. peregrina* Miller and *P. daurica* Andrews were investigated with carrageenan-induced hind paw edema method in mice and a significant activity was observed in the roots of *P. daurica*.

To isolate the active principle (s), after filtration of the precipitate occurred during the concentration of the ethanolic extract of the roots of *P. daurica*, the remaining extract was subjected to a series of successive fractionations and paeonol (**29**) was obtained as an active compound (Yeşilada et al., 1992).

VII.B. Iris germanica

Iris germanica L. (Iridaceae) is widely distributed in most parts of the world, and is also cultivated as an ornamental plant. In the preparation of traditional medicines, mainly the rhizomes of this plant are used. Rhizomes of the plant have been used in dropsy and gall bladder diseases as well as antispasmodic, emmenagogue, stimulant, diuretic and aperient. The juice of the roots of *I. germanica* is applied to sores and for removal of freckles from the skin. It is employed as an ingredient of composition for purifying blood and for venereal diseases (Hanawa et al., 1991).

Human neutrophils are known to be the first line of defense against invading microorganisms. This protection is based on the production of oxidative bursts at the site of microbial invasion. The mechanisms of the site-directed killing of microorganisms and extracellular tumor- and virus-infected cells have been extensively studied. On the other hand, the uncontrolled release of reactive oxygen species (ROS) is suspected to be responsible for certain pathological conditions such as heart attacks, septic shocks, rheumatoid arthritis and ischemia reperfusion injury (Bagchi et al., 1997). In these cases, the administration of agents that can decrease the neutrophils accumulation in the inflamed area might be a remedy for these conditions.

A cell-based in vitro bioassay was used to examine the anti-inflammatory activity of the rhizomes of *I. germanica* L. and to explore their potential as NSAID.

The methanolic extract of the rhizomes of *I. germanica* of Turkish origin has resulted in the isolation of seven bioactive compounds by a spectrophotometric assay using the activated human neutrophils (Atta-ur-Rahman et al., 2003). The structures of these compounds were identified on the basis of spectroscopic methods and were found to be: 5,7-dihydroxy-3-(3′-hydroxy-4′,5′dimethoxy)-8-methoxy-4*H*-1-benzopyran-4-one (**30**), 5,7-dihydroxy-3-(3′-hydroxy-4′,5′-dimethoxy)-6-methoxy-4*H*-1-benzopyran-4-one (**31**), 5-hydroxy-3-(4′-hydroxy)-6,7-methylene-dioxy-4*H*-1-benzopyran-4-one (**32**), 5-methoxy-3-(4′-hydroxy)-6,7-methyenedioxy-4*H*-1-benzopyran-4-one (**33**), 5,7-dihydroxy-3-(3′-hydroxy-4′-methoxy)-6-methoxy-4*H*-1-benzopyran-4-one (**34**), 5,7-dihydroxy-3-(3′-methoxy-4′-hydroxy)-6-methoxy-4*H*-1-benzopyran-4-one (**35**) and isopaeonol (**36**).

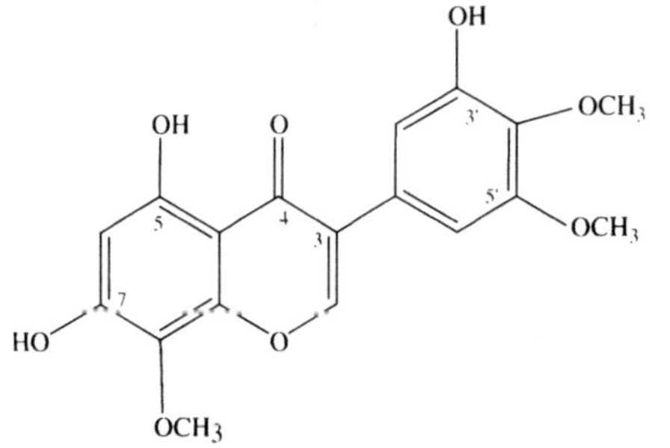

30

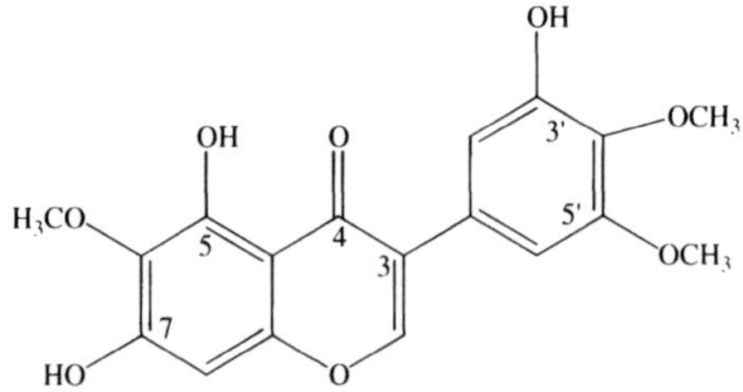

31

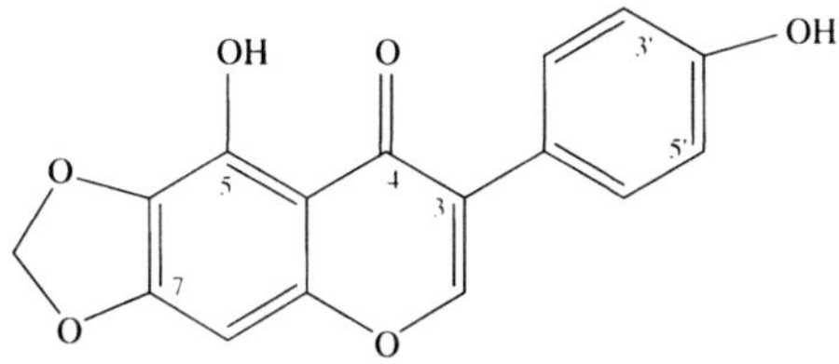

32

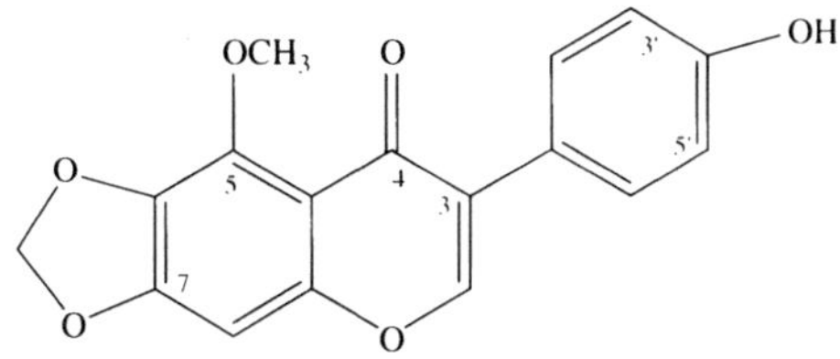

33

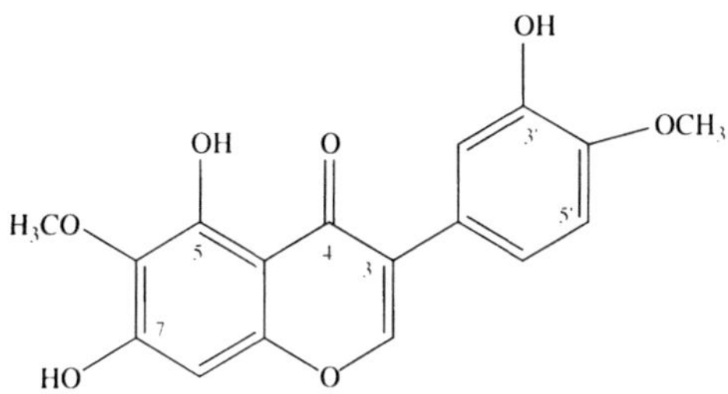

34

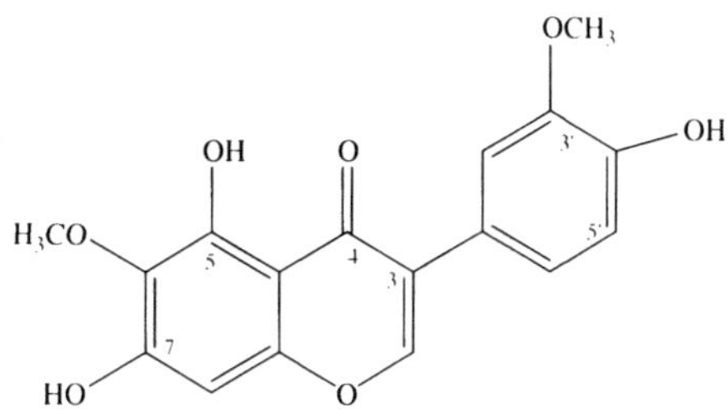

35

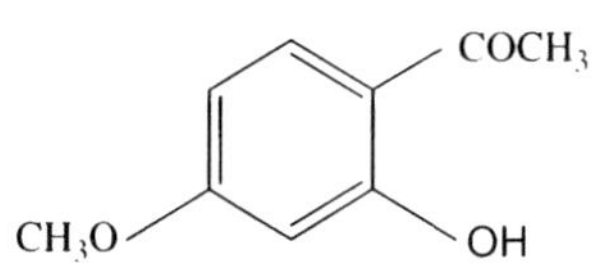

29

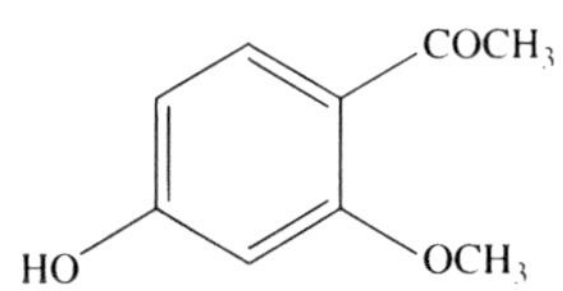

36

Compounds **31**, **34** and **35** were found to be the most active among them. Compound **30** also showed a significant activity, while compounds **32**, **33** and **34** exhibited only low activities when compared with aspirin and indomethacine clinically used NSAID (Atta-ur-Rahman et al., 2003).

VIII. Antibacterial compounds

Some plants are traditionally used to prevent sepsis, which means that they have antibacterial activity. In order to determine their scientific basis for their use, the antibacterial effects of 102 plant species were screened against eight bacteria by using the disc diffusion method. Among them 29 plant species have shown significant antibacterial activity (Akın et al., 1986). During our works in the field of bioactivity-guided isolation studies, some bioactive steroidal alkaloids and isoflavonoids have been obtained from *Buxus longifolia* L. and *I. germanica* L., respectively.

VIII.A. Steroidal alkaloids

Extracts of *Buxus* species (Buxaceae) or "boxwood" have been used in the indigenous system of medicine for the treatment of a number of diseases including skin diseases, rheumatism and malaria. Recently, the ethanolic extract of boxwood-containing cyclobuxine-D and buxamine has been reported to be active against human immunodeficiency virus (HIV) and diseases in which the tumor necrosis factor is involved. Furthermore, buxaminol E, a steroidal alkaloid isolated from *Buxus sempervirens*, has also been reported as a remarkable inhibitor of the enzyme (Choudhary et al., 2003).

N-Benzoylbuxahyrcanine (**37**), *N*-tigloylbuxahyrcanine (**38**) and *N*-isobutyroylbuxahyrcanine (**39**) were found to be active against AChE and butyrylcholinesterase. Compound **38** was determined to be the most potent against both enzymes by using eserine as reference (Atta-ur-Rahman et al., 1998).

N-Benzoyl-*O*-acetyl-buxalongifoline (**40**), buxasamarine (**41**) and (+)-cyclovirobuxeine-F (**42**) isolated from the leaves of *Buxus longifolia* L. showed antibacterial activity against *Salmonella typhi*, *Shigella flexneri* and *Pseudomonas aeruginosa* while (+)-cyclobuxamidine (**43**) was active against *S. typhi* and *Escherichia coli* using the agar-well diffusion method and ampicillin as the reference drug (Atta-ur-Rahman et al., 1997).

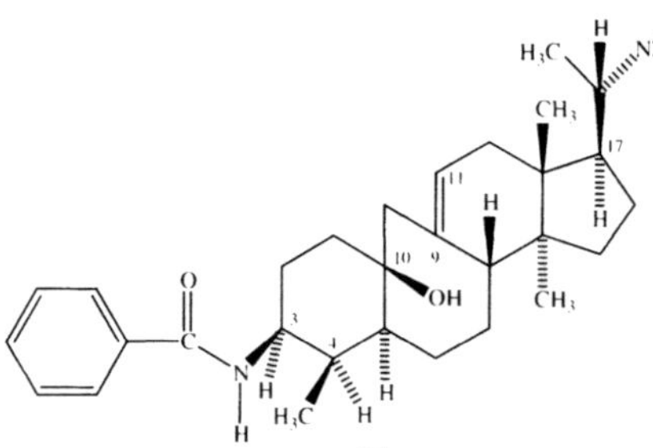

37

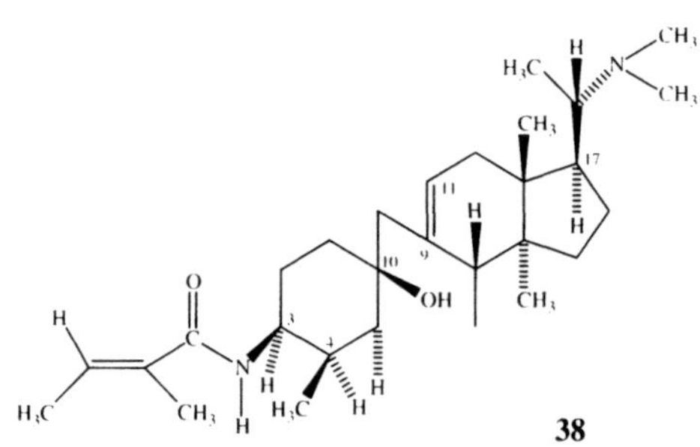

38

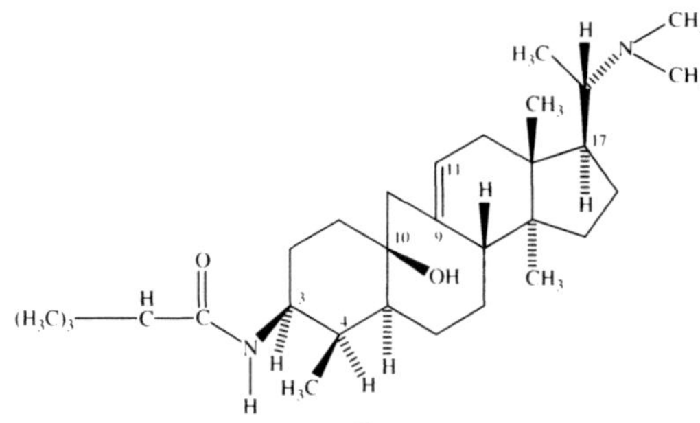

39

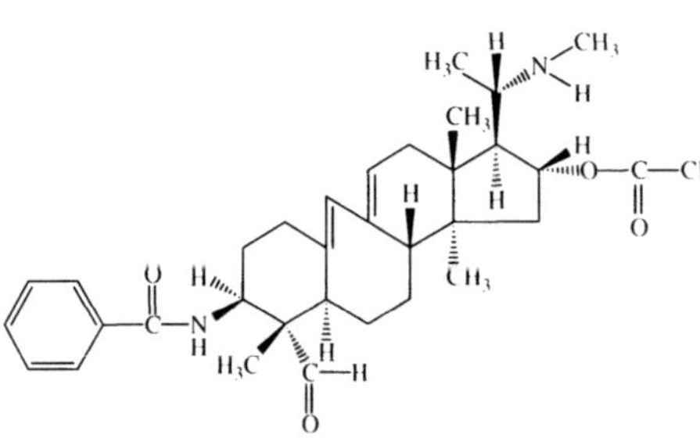

40

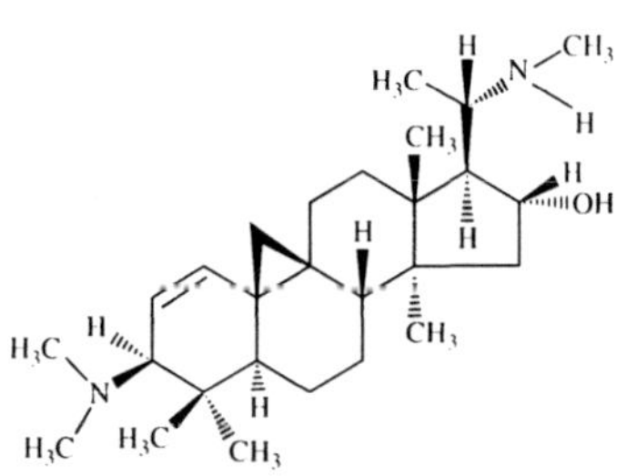

41

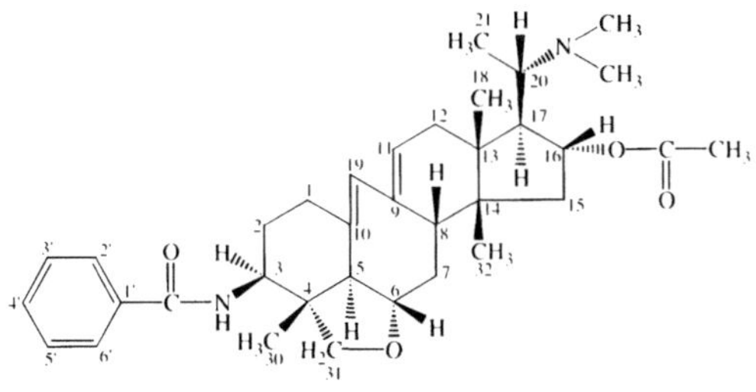

42

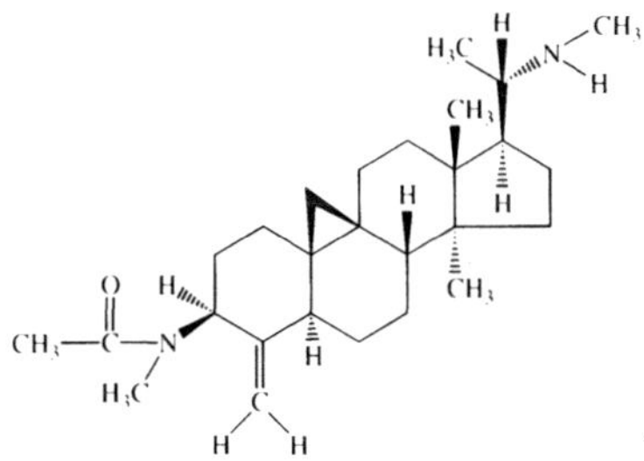

43

VIII.B. Isoflavonoids

Phytochemical investigations on *Iris* species have resulted in the isolation of a variety of compounds including flavonoids, isoflavonoids and their glycosides, benzoquinones, triterpenoids and stilbene glycosides (Atta-ur-Rahmen et al., 2003). The compounds isolated from *Iris* species were reported to have piscicidal, antineoplastic, antioxidant, antitumor, antiplasmodial and antituberculosis properties.

The chloroform extract of the rhizomes of *I. germanica* L. exhibited bactericidal activity against *Staphylococcus aureus* and *P. aeruginosa*, while ethyl acetate extract was found to be active against *Streptococcus pyogenes* and *S. aureus*. Bioassay-guided fractionation, according to the bactericidal activity results, led to the isolation of two isoflavonoids characterized as 5,7,3′-trihydroxy-6,4′,5′-trimethoxyisoflavone (**44**) and 5-hydroxy-4′-methoxy,6,7-methylenedioxyisoflavone (**45**) (Orhan et al., 2003). Compound **44** exhibited activity against *S. aureus* ($IC_{50} =$ 3.33 μg/ml), while compound **45** displayed activity against *S. aureus* ($IC_{50} =$ 3.33 μg/ml) and *P. aeruginosa* ($IC_{50} =$ 3.33 μg/ml) using tetracycline as the reference drug (Orhan et al., 2003).

Therefore, the rhizomes of *I. germanica* L. can be concluded to contain bactericidal, isoflavonoids which might be developed as new bactericidal lead agents for clinical studies.

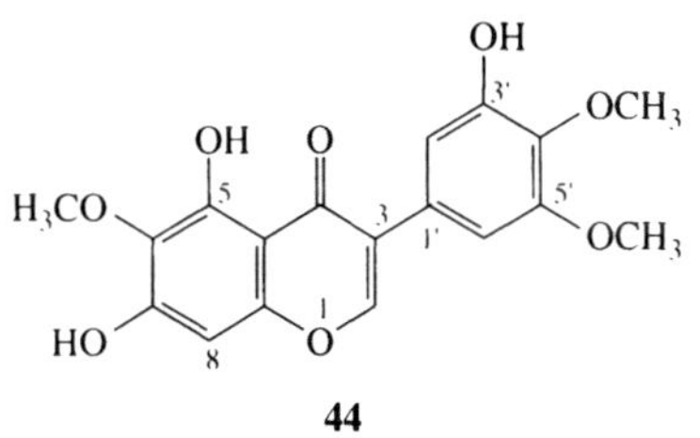

44

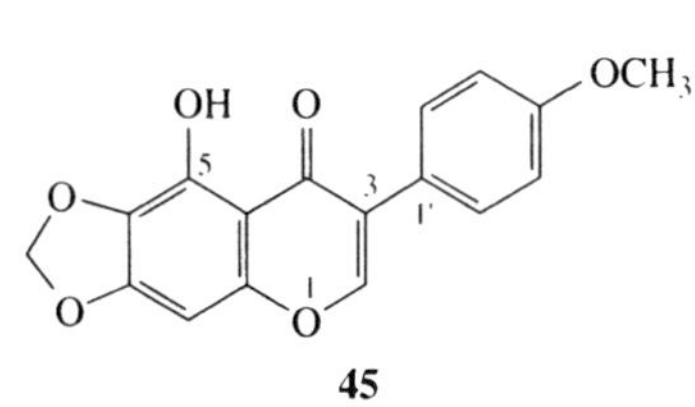

45

IX. Conclusion

Discoveries of lead compounds for the development of new drug candidates from bioresources can help to promote incentives for conservation by providing an economic return to innovative use of those sources.

Screening of natural sources has had an impressive tool of determining active agents. The key to success of discovering therapeutic agents from bioresources is based on bioassay-directed isolation techniques. HTS tests and mechanism-based screening protocols as well as information of folkloric utilization of plants have led to the discovery of lead compounds as drug candidates.

These results show that the available biodiversity of natural sources and the isolated bioactive compounds may act as potential leads for the development of clinically useful pharmaceuticals. All the known AChE inhibiting drugs used in the therapy of AD suffer from several side effects such as high toxicity, short duration of biological action, low bioavailability and narrow therapeutic effects. Consequently, development of new AChE inhibitors with less toxicity and more potent activity are necessary. The research in finding new drugs with AChE inhibitory activity to be used in the treatment of AD from natural resources such as Huperzine A, also yielded some herbal-originated extracts and/or compounds which act by different mechanisms such as *Ginkgo biloba*, *Panax ginseng*, *Davilla rugosa*, (–)-epigallocatechin, ferulic acid, etc. (Jing et al., 1999). However, AChE inhibitors have been accepted to be the most effective in the treatment of AD, so far.

Turkish Amaryllidaceae plants were also tested for their antimalarial activity against *P. falciparum* in vitro culture. Lycorine-, crinine-, tazettine- and galanthamine-type alkaloids exhibit antimalarial activity, however with different potencies. It is found that lycorine has the most potent and galanthamine has the least potent activity. Lycorine, galanthamine and tetrahydroprotoberberine-type alkaloids may be of therapeutic value in the treatment of AD.

Ebeinone, a steroidal-type alkaloid isolated from the bulbs of *F. imperialis* L. may be responsible for the treatment of asthma.

Protopine, an isoquinoline-type alkaloid isolated from the aerial parts of *Fumaria* L. species is a strong inhibitor for thromboxane formation.

Paeonol, a simple aromatic compound isolated from the roots of *Paeonia daurica* Andrews, as well as some isoflavonoids obtained from the rhizomes of *Iris* L. species have a noticeable anti-inflammatory activity when compared with aspirin and indomethacine clinically used as NSAID. In addition, some isoflavonoids may also serve for the development of bactericidal drug candidates for clinical studies.These bioactive natural compounds can also serve as models for synthetic compounds in improving human health. They play valuable leads for nutraceuticals in connection with human nutrition by the general public as well.

References

Adams RL, Craig PL, Parsons OA. (1984) Neuropsychology of dementia. Neurolog Clin 4:387–405.

Aisen PS, Davis KL. (1997) The search for disease-modifying treatment for Alzheimer's disease. Neurology 48:35–41.

Akın S, Bingöl F, Şener B, Şahinkaya H. (1986) Antibacterial effects of some higher plants. GUEDE-J Fac Pharm Gazi 3:65–80.
Albuquerrque EX, Alkondon M, Pereira EFR. (1997) Properties of neuronal nicotinic acetylcholine receptors: pharmacological characterization and modulation of synaptic function. J Pharmacol Exp Ther 280:1117–36.
Arnold SE, Kumar A. (1993) Reversible dementias. Med Clin Nort Am 77:215–25.
Atta-ur-Rahman, Akhtar MN, Choudhary MI, Tsuda Y, Şener B, Khalid A, Parvez M. (2002) New steroidal alkaloids from *Fritillaria imperialis* and their cholinesterase inhibiting activities. Chem Pharm Bull 50:1013–6.
Atta-ur-Rahman, Ali RA, Gilani AH, Choudhary MI, Aftab K, Şener B, Türköz S. (1993) Isolation of antihypertensive alkaloids from the rhizomes of *Veratrum album*. Planta Med 59:569–73.
Atta-ur-Rahman, Farooq A, Choudhary MI, Gilani AH, Shaheen F, Ali RA, Noor-e-ain F, Şener, B. (1994) A new anticholinergic steroidal alkaloid from *Fritillaria imperialis* of Turkish origin. Planta Med 60:377–9.
Atta-ur-Rahman, Nasim S, Baig I, Jalil S, Orhan I, Şener B, Choudhary MI. (2003). Anti-inflammatory isoflavonoids from the rhizomes of *Iris germanica*. J Ethnopharmacol 86:177–180.
Atta-ur-Rahman, Noor-e-ain F, Choudhary MI, Parveen Z, Türköz S, Şener B. (1997) New steroidal alkaloids from *Buxus longifolia*. J Nat Prod 60:976–81.
Atta-ur-Rahman, Parveen S, Khalid A, Farooq A, Ayatollahi SAM, Choudhary MI. (1998) Acetylcholinesterase inhibiting triterpenoidal alkaloids from *Buxus hyrcana*. Heterocycles 49:481–8.
Bachman DL, Wolf PA, Linn RT. (1992) Prevalence of dementia and probable senile dementia of the Alzheimer type in the Framingham study. Neurology 42:115–9.
Bagchi D, Bagchi M, Stohs SJ. (1997) Comparative in vitro oxygen radical scavenging ability of zinc methionine and selected zinc salts and antioxidants. General Pharmacol 28:85–91.
Başer KHC. (2002) Aromatic biodiversity among the flowering plant taxa of Turkey. Pure Appl Chem 74:527–45.
Baytop T. (1999) Therapy with medicinal plants in Turkey (past and present), 2nd edition. Nobel Medical Presshouse: Istanbul.
Blasko G, Elango V, Şener B, Freyer AJ, Shamma M. (1982) Secophthalideisoquinolines. J Org Chem 47:880–5.
Blasko G, Murugesan N, Hussain F, Minard R, Shamma M, Şener B, Tanker M. (1981) Revised structure for fumarofine, an indenobenzazepine type alkaloid. Tetrahedron Lett 22:3135–8.
Boissier JR, Combes G, Pagny J. (1960) La Galanthamine, Puissant Cholinergique Naturel. Ann Pharm Franç 18:888–900.
Choudhary MI, Shahnaz S, Parveen S, Khalid A, Ayattollahi SAM, Atta-ur-Rahman, Parvez M. (2003) New terpenoid alkaloid cholinesterase inhibitors from *Buxus hyrcana*. J Nat Prod 66:739–42.
Davis PH. (1965) Flora of Turkey and the east Aegean islands, Vol. 1, pp. 242–248. Edinburgh: Edinburgh University Press.
Davis PH, Mill RR, Tan K. (1988) Flora of Turkey and the east Aegean islands. Vol. 10, pp. 28–29. Edinburgh: Edinburgh University Press.
Duvoisin RC. (1968) Reversal of central anticholinergic syndrome in man by physostigmine. JAMA 206:1963–7.
Ellman GL. (1958) A colorimetric method for determining low concentrations of mercaptans. Arch Biochem Biophys 74:443–50.
Ellman GL, Courtney D, Andres V, Featherstone RM. (1961) A new and rapid colorimetric determination of acetylcholinesterase activity. Biochem Pharmacol 7:88–95.
Gauthier S. (2001) Alzheimer's disease: current and future therapeutic perspectives. Prog Neuro-Psycoph 25:73–89.
Guardado-Santervas P. (2001) Controversy surrounding new drugs for the treatment of dementia. An unfavorable view. Rev Neurol 32:172–5.
Güner A, Özhatay N, Ekim T, Başer KHC. (2001) Flora of Turkey and the east Aegean islands. Edinburgh: Edinburgh University Press, Supplement 2, Vol. 11, pp. 28–9.

Hanawa F, Tahara S, Mizutani J. (1991) Isoflavonoids produced by Iris pseudacorus leaves treated with cupric chloride. Phytochemistry 30:157–63.

Ide S, Şener B, Temizer H, Könükol S. (1996) The crystal and molecular structure of two drug compounds: tazettine and corydaline. Cryst Res Technol 31:617–24.

Ieven M, Berghe DA, Vlietinck AJ. (1983) Plant antiviral agents, IV. Influence of lycorine on growth pattern of three animal viruses. Planta Medica 49:109–14.

Ieven M, Vlietinck AJ. (1981) Toxic and therapeutically active substances of the Amaryllidaceae. Pharm Weekbl 116:169–78.

Jing L, Zhang HY, Wang LM, Tang XC. (1999) Inhibitory effects of huperzine B on cholinesterase activity in mice. Acta Pharmacol Sin 20:141–5.

Könükol S. (1992) Pharmacognosic researches on the alkaloids of Pancratium maritimum L. Ph.D. thesis, Institute of Health Sciences of Gazi University, Ankara..

Küçükboyacı N, Bingöl F, Şener B, Kutney JP, Stoynov N. (1998) Isoquinoline alkaloids from *Fumaria bastardii*. Nat Prod Sci 4:257–62.

Levin ED, Simon BB. (1998) Nicotinic acetylcholine involvement in cognitive function of animals. Psychopharmacology (Berlin) 138:217–30.

Liu JS, Zhu YL, Yu CM, Zhou YZ, Han YY, Wu FW, Qi BF. (1986) The structures of huperzine A and B[1], two new alkaloids exhibiting marked anticholinesterase activity. Can J Chem 64:837–9.

Lopez S, Bastida J, Viladomat F, Codina C. (2002) Acetylcholinesterase inhibitory activity of some amaryllidaceae alkaloids and *Narcissus* extracts. Life sci 71:2521–9.

Martin SF. (1987) The amaryllidaceae. In: Brossi A editor. The alkaloids, Vol. 30. Academic Press: NewYork, pp. 251–357.

Orhan I. (2002) Investigations on the acetylcholinesterase inhibitory activity of some Turkish plants. Ph.D. thesis, Institute of Health Sciences of Gazi University, Ankara.

Orhan I, Nasim S, Şener B, Ayanoğlu F, Özgüven M, Choudhary MI, Atta-ur-Rahman. (2003) Two isoflavones and bioactivity spectrum of the crude extracts of *Iris germanica* rhizomes. Phytother Res 17:575–7.

Orhan I, Şener B. (2003) Bioactivity-directed fractionation of alkaloids from some amaryllidaceae plants and their anticholinesterase inhibitory activity. Chem Nat Compds (Khim Prir Soedin) 39:383–6.

Pettit GR, Gaddamidi V, Herald DL, Singh SB, Cragg GM, Schmidt JM. (1986) Antineoplastic agents, 120. *Pancratium littorale*. J Nat Prod 49:995–1002.

Raskind MA, Peskind ER, Wessel T, Yuan W and the Galantamine USA-1 Study Group. (2000) A randomized, placebo-controlled trial with a 6-month extension. Neurology 54:2261–8.

Schneider LS. (2001) Treatment of Alzheimer's disease with cholinesterase inhibitors. Clin Geriatr Med 17:337–9.

Schrattenholz A, Pereira EFR, Roth U, Weber KH, Albuquerrque EX, Maelicke A. (1996) Agonist responses of neuronal nicotinic acetylcholine receptors are potentiated by a novel class of allosterically acting ligands. Mol Pharmacol 49:1–6.

Şener B. (1981) Turkish *Fumaria* L. species and their alkaloids. Associate Professorship Thesis, Ankara University, Ankara.

Şener B. (1982) Turkish species of *Fumaria* L. and their alkaloids. I. Morphological characteristics of Turkish *Fumaria* L. species. Ankara Ecz Fak Der 12:83–104.

Şener B. (1983) Turkish species of *Fumaria* L. and their alkaloids. II. Alkaloids of *Fumaria gaillardotii* Boiss. Int J Crude Drug Res 21:135–9.

Şener B. (1984a) Secodensiflorine: a new alkaloid from *Fumaria densiflora DC.* Int J Crude Drug Res 22:79–80.

Şener B. (1984b) Microcarpine: a new alkaloid from *Fumaria microcarpa Boiss.* Int J Crude Drug Res 22:45–6.

Şener B. (1984c) Turkish species of *Fumaria* L. and their alkaloids. III. Alkaloids of *Fumaria judaica* Boiss. Int J Crude Drug Res 22:181–3

Şener B. (1984d) Turkish species of *Fumaria* L. and their alkaloids. IV. Alkaloids of *Fumaria macrocarpa* Parlatore. Int J Crude Drug Res 22:185–7.

Şener B. (1985a) Turkish species of *Fumaria* L. and their alkaloids. V. Alkaloids of *Fumaria capreolata* and *Fumaria asepala*. J Nat Prod 484:670.

Şener B. (1985b) Turkish species of *Fumaria* L. and their alkaloids. VI. Alkaloids of *Fumaria capreolata* L. Int J Crude Drug Res 23:161–3.
Şener B. (1985c) Turkish species of *Fumaria* L. and their alkaloids. VII. Alkaloids of *Fumaria officinalis* L. and *Fumaria cilicica* Hausskn. J Fac Pharm Gazi 2:45–9.
Şener B. (1986) Turkish species of Fumaria L. and their alkaloids. VIII. Alkaloids of *Fumaria asepala* Boiss. Int J Crude Drug Res 24:105–6.
Şener B. (1988) Turkish species of *Fumaria* L. and their alkaloids. IX. Alkaloids of *Fumaria parviflora* Lam., *F. petteri* Reichb. subsp. *thuretii* (Boiss.) Pugsley and *F. kralikii* Jordan. Int J Crude Drug Res 26:61–2.
Şener B. (1994) Recent results in the search for bioactive compounds from Turkish medicinal plants. Pure and Appl Chem 66:2295–8.
Şener B. (2002) Molecular diversity in the alkaloids of Turkish *Fumaria* L. species. Acta Pharmaceut Turcica 44:205–12.
Şener B. (2003) Türkiye'de Yetişen Bazı Bitkilerden Elde Edilen Biyoaktif Bileşikler. In: Meriç li AH, Sariyar G, Mat A, Sutlupinar N, editors. Prof. Dr. Turhan Baytop Anısına. Istanbul: Istanbul University Press.
Şener B., Atta-ur-Rahman. (1996) New bioactive organic compounds from Turkish medicinal plants. Recl Trav Chim Pays-Bas 115:103–7.
Şener B, Gözler B, Minard RD, Shamma M. (1983) Alkaloids of Fumaria vaillantii. Phytochemistry 22:2073–5.
Şener B, Könükol S, Kruk C, Pandit UK. (1993a) New crinine-type alkaloids from *Pancratium maritimum* L. growing in Turkey. Nat Prod Lett 1:287–91.
Şener B, Könükol S, Kruk C, Pandit UK. (1993b) Alkaloids from Amaryllidaceae I. Alkaloids of lycorine and lycorenine class from *Pancratium maritimum* L. Arch Pharm (Weinheim) 326:61–2.
Şener B, Könükol S, Kruk C, Pandit UK. (1993c) Alkaloids from amaryllidaceae IV. Alkaloids from the aerial parts of *Pancratium maritimum*. GUEDE-J Fac Pharm Gazi 10:83–6.
Şener B, Könükol S, Kruk C, Pandit UK. (1994) Alkaloids from amaryllidaceae II. Alkaloids of crinine class from *Pancratium maritimum*. J Chem Soc Pak 16:275–9.
Şener B, Könükol S, Kruk C, Pandit UK. (1998) Alkaloids from amaryllidaceae III. Alkaloids from the bulbs of *Pancratium maritimum*. Nat Prod Sci 4:148–52.
Şener B, Könükol S, Muhtar F. (1988) New alkaloids and biological activities of Turkish amaryllidaceae plants. In: Atta-ur-Rahman, Choudhary MI, editors. New trends in natural product chemistry. Amsterdam: Harwood Academic Publishers, pp. 187 95.
Şener B, Orhan I. (2003) Molecular diversity in the bioactive compounds from Turkish plants evaluation of acetylcholinesterase inhibitory activity of *Fumaria* species. J Chem Soc Pak (March) 26(3):313–5.
Şener B, Orhan I, Satayavivad J. (2003) Antimalarial activity screening of some alkaloids and plant extracts from Amaryllidaceae. Phytother Res. 17: 1220–3.
Şener B, Temizer H. (1991) New alkaloids of some Turkish medicinal plants. In: Pandit UK, Alderweireldt FC, editors. Bioorganic chemistry in healthcare and technology. New York: Plenum Press, pp. 281–5.
Şener B, Temizer H, Könükol S, Koyuncu M. (1992) Alkaloids from some Turkish amaryllidaceae plants. In: Atta-ur-Rahman, editor. Advances in Natural Product Chemistry – Proceedings of the 5th International Symposium and Pakistan–U.S. Binational Workshop on Natural Product Chemistry, Karachi-Pakistan, January 5–9. GmbH, Singapore: Harwood Academic Publishers, pp. 401–12.
Tang XC. (1996) Huperzine A (Shuangyiping): a promising drug for Alzheimer's disease. Acta Pharmacol Sin 17:481–4.
Tang XC, Han YF. (1999) Pharmacological profile of huperzine A, a novel acetylcholinesterase inhibitor from Chinese herb. CNS Drug Rev 5:281–300.
Tariot PN, Solomon PR, Morris JC, Kershaw P, Lilienfeld S, Ding C and the Galantamine USA-10 Study Group. (2000) A 5-month, randomized, placebo-controlled trial of galanthamine in AD. Neurology 54:2269–75.
Thomsen T, Kewitz H. (1990) Selective inhibition of human acetylcholinesterase by galanthamine in vitro and in vivo. Life Sci 46:1553–8.

Wang H, Tang XC. (1998) Anticholinesterase effects of huperzine A, E2020 and tacrine in rats. Acta Pharmacol Sin 19:27–30.
Wang YE, Yue DX, Tang XC. (1986) Anticholinesterase activity of huperzine A. Acta Pharmacol Sin 7:110–3.
Ye JW, Cai JX, Wang LM, Tang XC. (1999) Improving effects of huperzine A on spatial working memory in aged monkeys and young adult monkeys with experimental cognitive impairment. J Pharmacol Exp Therapeut 288:814–9.
Yeşilada E, Mutlugil A, Şener B. (1992) The antiinflammatory principle of the roots of *Paeonia daurica*. Int J Pharmacog 30:66–70.
Yeşilada E, Mutlugil A, Şener B, Sezik E. (1993) *Paeonia daurica* Andrews ve *P. peregrina* Miller'in antienflamatuar aktivitesi. In: Çubukçu B, Sarıyar G, Mat A, editors. VIII. Bitkisel İlaç Hammaddeleri Toplantısı Bildiriler, Cilt II. Istanbul: Istanbul University Press-house, pp. 361–4.
Zee-Chang RKY, Berghe DA, Vlietinck AJ. (1978) Antileukemic activity of ungeremine and related compounds. J Med Chem 21:199–203.

M.T.H. Khan and A. Ather (eds.)
Lead Molecules from Natural Products

Molecular design of multifunctional anti-*Salmonella* agents based on natural products

Isao Kubo, Ken-ichi Fujita, Ken-ichi Nihei, Aya Kubo

Abstract

A series of aliphatic (2*E*)-alkenals and alkanals from C_5 to C_{13} were tested for their antibacterial activity against *Salmonella choleraesuis* ssp *choleraesuis* ATCC 35640. The activity against this food-borne bacterium is correlated with the hydrophobic alkyl chain length from the hydrophilic aldehyde group. (2*E*)-Dodecenal (C_{12}) was most effective against *S. choleraesuis* with the minimum bactericidal concentration (MBC) of 6.25 μg/mL (34 μM), followed by (2*E*)-undecenal (C_{11}) with an MBC of 12.5 μg/mL (77 μM). The time kill curve study showed that (2*E*)-dodecenal was bactericidal against *S. choleraesuis* at any of its growing stages. The antibacterial action of medium-chain aldehydes comes in part from their ability to function as nonionic surfactants. Based on the above head and tail concept, a series of antioxidative alkyl gallates (3,4,5-trihydroxybenzoates), was synthesized and tested for their antibacterial activity against *S. choleraesuis*. Nonyl (C_9) and octyl (C_8) gallates were noted to be the most effective against this food borne bacterium with MBCs of 12.5 μg/mL, followed by decyl (C_{10}) gallate with an MBC of 25 μg/mL. Dodecyl (C_{12}) gallate still exhibited the activity against *S. choleraesuis* with an MBC of 50 μg/mL. Propyl (C_3) gallate showed no activity against *S. choleraesuis* up to 3200 μg/mL. The length of the alkyl group plays a role in eliciting the activity to a large extent.

Keywords: *anti-Salmonella agents, (2E)-alkenals, surfactant, combination effect, alkyl gallates, antioxidant, multifunction*

Abbreviations: CF, carboxyfluorescein; CFU, colony forming units; DPPH, 1, 1-diphenyl-2-picrylhydrazyl; ETC, electron transport chain; MBC, minimum bactericidal concentration; MFC, minimum fungicidal concentration; MIC, minimum inhibitory concentration; PC, liposomes of phosphatidylcholine; SAR, structure–activity relationship.

I. Introduction

Salmonellosis is one of the most frequently occurring bacterial food-borne illnesses. The salmonellae are Gram-negative non-spore-forming rods. There are over 2500 serovars of *Salmonella*, all of which are pathogenic for humans. The salmonellae are Gram-negative non-spore-forming rods that ferment glucose but not lactose or sucrose. The study to search for anti-*Salmonella* agents was initiated by the request to solve the problem of contamination in the fruit of *Piper nigrum* (Piperaceae), commonly known as pepper, by *Salmonella* in the Amazon basin. On the basis of the preliminary survey, it can be suggested that this contamination was likely been caused by increased large-scale poultry farms around the area, since poultry and rodents are important sources of *Salmonella* contamination for food. For example, chickens may be infected with any number of types of *Salmonella*, which are then found in their fecal matter. In addition, infected rodents, rats, mice, and bats may contaminate unprotected pepper with their feces and thus spread *Salmonella* bacteria (Frazier and Westhoff, 1988; Tyrrel and Quinton, 2003). There is a great need for effective antibacterial agents and phytochemicals have the potential of fulfilling this need. The selected phytochemicals from our previous studies were tested for their anti-*Salmonella* activity. In order to search for anti-*Salmonella* agents, *Salmonella choleraesuis* ssp *choleraesuis* ATCC 35640 was selected as an example since this bacterium most frequently causes septicemia, although septicemia can be caused by any *Salmonella* (Frazier and Westhoff, 1988).

In our preliminary screening, several phytochemicals previously characterized as antibacterial agents against Gram-negative bacteria were found to exhibit antibacterial activity against *S. choleraesuis*. Among them, polygodial showed the most potent bactericidal activity, followed by (2*E*)-hexenal. Therefore, further evaluation of these compounds may provide new insights into their antibacterial action on a molecular basis. A series of (2*E*)-alkenals was tested for their antibacterial activity against *S. choleraesuis* using a broth dilution method for comparison. The maximum anti-microbial activity of (2*E*)-alkenals is dependent on the balance of the hydrophobic alkyl (tail) chain length from the hydrophilic aldehyde group (head) (Kubo et al., 1995, 2003b). It is well known that the hydrophobicity of molecules is often associated with biological action (Hansch and Dunn, 1972). However, the rationale for this observation, especially the role of the hydrophobic portion, is still poorly understood and widely debated. To make it more clearer – (2*E*)-alkenals are a superior model for structure and anti-*Salmonella* activity relationship (SAR) study because these molecules possess the same hydrophilic portion, the enal group, which explain the role of the hydrophobic alkyl portion. In addition, a series of (2*E*)-alkenals and their related analogues are common in many plants (Kim et al., 1995; Kubo and Kubo, 1995; Kubo et al., 1996, 1999; Kubo and Fujita, 2001) and readily available. Therefore, a series of aliphatic (2*E*)-alkenals, as well as the corresponding alkanals, from C_5 to C_{13} were tested for their antibacterial activity against *S. choleraesuis* (Kubo et al., 2004).

II. Antibacterial activity

II.A. Test strains

The test strains, *Salmonella choleraesuis* ssp *choleraesuis* ATCC 35640, *Escherichia coli* ATCC 9637, *Pseudomonas aeruginosa* ATCC 10145, *Enterobacter aerogenes* ATCC 13048, and *Proteus vulgaris* ATCC 13315, used for this study were purchased from American Type Culture Collection (Manassas, VA).

II.B. Medium

NYG broth (0.8% nutrient broth, 0.5% yeast extract, 0.1% glucose) was used for the antibacterial assay. Nutrient broth and yeast extract were purchased from Becton, Dickinson and Co. (Franklin Lake, NJ).

II.C. Precultivation

The cells of *S. choleraesuis* were precultured in 3 mL of NYG broth without shaking at 37°C for 16 h. The preculture was used for the following antibacterial assay and time kill study.

II.D. Antibacterial assay

Broth macrodilution methods were used as previously described (Kubo et al., 1996, 2001) with slight modifications. Serial twofold dilutions of the test compounds were prepared in DMF, and 30 µL of each dilution was added to 3 mL of NYG broth. These were inoculated with 30 µL of an overnight culture of *S. choleraesuis*. After incubation of the cultures at 37°C for 48 h, the minimum inhibitory concentration (MIC) was determined as the lowest concentration of the test compound that demonstrated no visible growth. The MBC was determined as follows. After the determination of the MIC, 100-fold dilutions with drug-free NYG broth from each tube showing no turbidity were incubated at 37°C for 48 h. The MBC was the lowest concentration of the test compound that was not visible in the drug-free cultivation.

II.E. Combination study

The combination data were obtained by the broth checkerboard method (Norden et al., 1979; Eliopoulos and Moellering, 1991). A series of twofold dilutions of one compound were tested in combination with twofold dilutions of the other compounds. The assays were performed at least in triplicate on separate occasions.

II.F. Time kill study

The cultivation of bacteria with each compound was done in the same manner as in the above MIC. Samples were withdrawn at selected time points, and serial dilutions were performed in sterile saline before the samples were plated onto NYG agar plates. After the plates were incubated at 37°C for 16 h, colony forming units (CFU) were determined.

III. (2*E*)-Alkenals

In a preliminary screening, the 10 phytochemicals selected from our previous studies and their structurally related compounds were tested for their antibacterial activity against *S. choleraesuis* ssp *choleraesuis* ATCC 35640. Among them, polygodial (**1**), a bicyclic sesquiterpene dialdehyde (see Figure 1 for structures), was found to possess

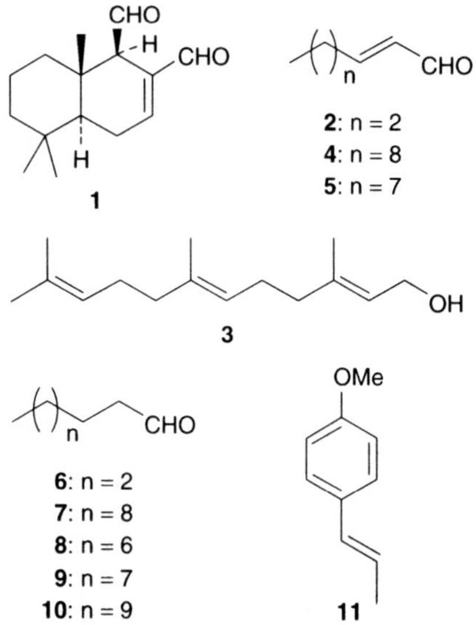

Fig. 1. Chemical structures of polygodial (**1**) and (2*E*)-hexenal (**2**), and related compounds (**3**–**10**).

Table 1
Antibacterial activity (μg/mL) of the selected compounds against *S. choleraesuis* ssp *choleraesuis* ATCC 35640

Compounds tested	MIC	MBC
(2*E*)-Hexenal	100	100
Hexanal	400	800
Hexanol	> 1600	> 1600
Hexanoic acid	400	400
Sorbic acid	400	400
Benzoic acid	800	800
Anisic acid	800	800
Polygodial	50	50
Geraniol	400	800
Farnesol	12.5	400
Anethole	200	200
Eugenol	400	400
Indole	800	800
Gentamycin	12.5	12.5

the most potent activity against this food-borne bacterium, followed by (2*E*)-hexenal (**2**), an aliphatic α,β,-unsaturated aldehyde (Table 1). These two aldehydes were further studied in detail. Polygodial exhibited the activity with both MIC and MBC of 50 μg/mL (213 μM), suggesting that no residual bacteriostatic activity is involved. This sesquiterpene dialdehyde was first isolated as a pungent principle from sprouts of *Polygonum hydropiper* (Polygonaceae) (Ohsuka, 1963), known as "tade" and used as a food spice in Japan. Subsequently it was isolated from two East African *Warburgia* trees, *W. ugandensis* and *W. stuhlmannii* (Kubo et al., 1976), and these two plants are also locally used as food spice (Watt and Breyer-Brandwijk, 1962). The potent fungicidal activity of polygodial, especially against yeasts such as *Candida albicans* and *Saccharomyces cerevisiae*, was subsequently reported, although it possessed little or no activity against several bacteria (McCallion et al., 1982; Taniguchi et al., 1984). In the current experiment, polygodial did not exhibit any antibacterial activity against the four common Gram-negative bacteria tested, *E. coli*, *Pseudomonas aeruginosa*, *Enterobacter aerogenes*, and *Proteus vulgaris* up to 800 μg/mL. Hence, the result obtained against *S. choleraesuis* was unexpected. It should be added that farnesol (**3**), a common sesquiterpene alcohol, was noted to exhibit the most potent activity against *S. choleraesuis* with an MIC of 12.5 μg/mL. However, the lethality was needed at 400 μg/mL, suggesting that residual bacteriostatic activity is involved similar to those found against *S. cerevisiae* (Machida et al., 1998).

The potency of the antibacterial activity against *S. choleraesuis* was followed by (2*E*)-hexenal with both MIC and MBC of 100 μg/mL (1020 μM), indicating that no residual bacteriostatic activity was involved. In contrast to polygodial, (2*E*)-hexenal exhibits broad antimicrobial activity (Corbo et al., 2000; Kubo and Fujita, 2001; Nakamura and Hatanaka, 2002). For example, its antibacterial activity against *E. coli*, *P. aeruginosa*, *E. aerogenes*, and *P. vulgaris* (Kubo et al., 1996) as well as *H. pylori* (Kubo et al., 1999) was previously reported. This aliphatic α,β-unsaturated aldehyde is known as "leaf aldehyde" (Hatanaka, 1993) and is widely distributed in many edible plants (Schauenstein et al., 1977).Because of its availability and broad antimicrobial activity, it was studied in more detail. In our continuing search for antimicrobial agents from edible plants, (2*E*)-hexenal was previously characterized as an antimicrobial agent from the volatile fraction of the cashew apple (Muroi et al., 1993) and olive oil (Kubo et al., 1995; Bisignano et al., 2001). In contrast to (2*E*)-hexenal, hexanol did not show any activity against *S. choleraesuis* up to 1600 μg/mL, but hexanal and hexanoic acid still exhibited some activity, though to a lesser extent than (2*E*)-hexenal. Thus, the conjugated double bond is not essential to elicit the antibacterial activity but is associated with increasing the activity.

In addition to (2*E*)-hexenal, a homologous series of (2*E*)-alkenals and alkanals were assayed for their antibacterial activity against *S. choleraesuis* for comparison. The results are listed in Table 2. As expected, their antibacterial activity against this food-borne bacterium is correlated with the hydrophobic alkyl (tail) chain length from the hydrophilic aldehyde group (head). The activity increased with increasing alkyl chain length up to (2*E*)-dodecenal (C_{12}) (**4**). Thus, (2*E*)-dodecenal is the most effective bactericide against *S. choleraesuis*, followed by (2*E*)-undecenal (C_{11}) (**5**). It appears that *S. choleraesuis* showed different susceptibilities to aldehydes possessing different chain lengths. This result is broadly similar to those of the corresponding alkanols against many microorganisms (Kubo et al., 1995, 2003a), indicating at least in part the similarity of their mode of action. The range of the antibacterial activity of the (2*E*)-alkenals tested against *S. choleraesuis* is between 6.25 and 200 μg/mL, and the MICs and MBCs are markedly the same. Both the MIC and MBC of the most potent (2*E*)-dodecenal are 6.25 μg/mL (34 μM). Notably, this MBC value is slightly more potent than that of gentamicin.

Table 2
Antibacterial activity (μg/mL) of (2*E*)-Alkenal against *S. choleraesuis* ssp *choleraesuis* ATCC 35640

(2*E*)-Alkenal tested	MIC	MBC
C_5	200	200
C_6	100	100
C_7	100	100
C_8	100	100
C_9	25	25
C_{10}	50	50
C_{11}	12.5	12.5
C_{12}	6.25	6.25
C_{13}	25	200

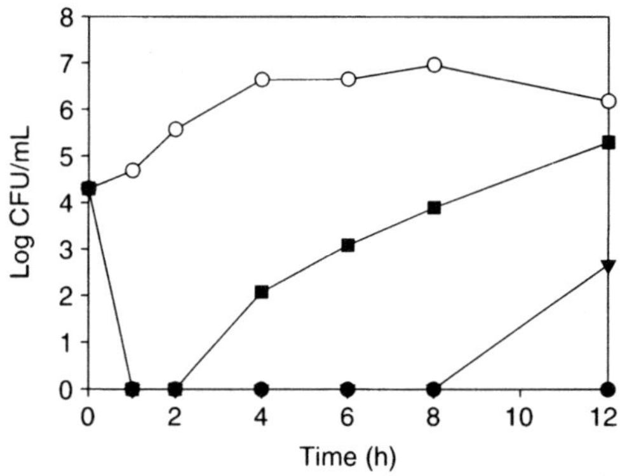

Fig. 2. Bactericidal effect of polygodial against *S. choleraesuis* ssp *choleraesuis* ATCC 35640. Exponentially growing cells of *S. choleraesuis* were inoculated at 37°C in NYG broth with (○) 0, (■) 25, (▼) 50, or (●)100 μg/mL of polygodial. Viability was established by the number of colonies formed on NYG plate after incubation at 30°C in for 24 h.

The bactericidal activity of polygodial against *S. choleraesuis* was confirmed by the time kill curve experiment as shown in Figure 2. Cultures of *S. choleraesuis*, with a cell density of 4.4×10^4 CFU/mL, were exposed to three different concentrations of polygodial. The number of viable cells was determined following different periods of incubation with polygodial. It shows that MIC significantly reduced the growth rate, but that the final cells recover count was not different from the control. It should be noted that lethality occurred quickly, within the first 1 h after the addition of polygodial. This rapid lethality very likely indicates that antibacterial activity of polygodial against *S. choleraesuis* is associated with the disruption of the membrane, similar to its effect found against *S. cerevisiae* (Kubo et al., 2001).

The bactericidal effect of (2*E*)-hexenal was also confirmed by the time kill curve experiment as shown in Figure 3. Cultures of *S. choleraesuis*, with a cell density of 1×10^5 CFU/mL, were exposed to two different concentrations of (2*E*)-hexenal. The number of viable cells was determined following different periods of incubation with

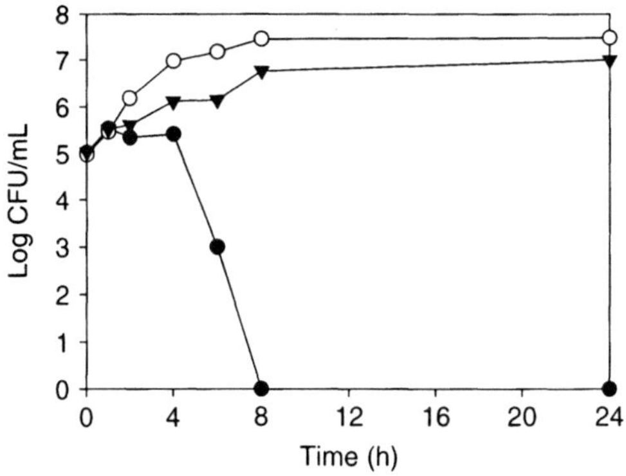

Fig. 3. Bactericidal effect of (2*E*)-hexenal against *S. choleraesuis* ssp *choleraesuis* ATCC 35640. Exponentially growing cells of *S. choleraesuis* were inoculated at 37°C in NYG broth with (○) 0, (▼) 50, or 100 (●) μg/mL of (2*E*)-hexenal. Viability was established by the number of colonies formed on NYG plate after incubation at 30°C in for 24 h.

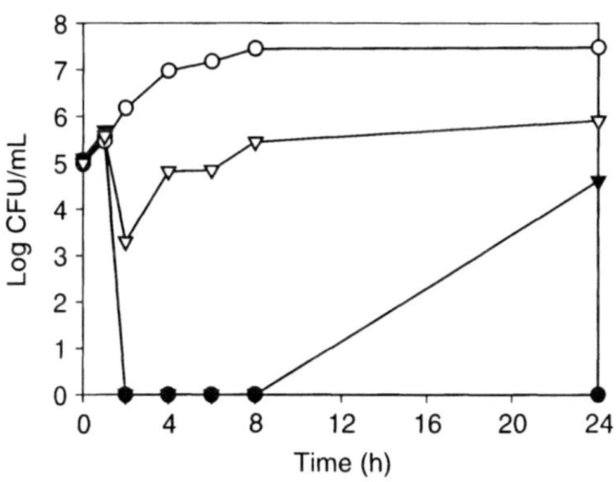

Fig. 4. Effect of (2*E*)-dodecenal on the growth of *S. choleraesuis* ssp *choleraesuis* ATCC 35640. Exponentially growing cells were inoculated into NYG broth and then cultured at 37°C. The arrow indicates the time when the drug was added. (2*E*)-Dodecenal (○) 0, (▽) 1.56, (▼) 3.13, and (●)6.25 μg/mL.

(2*E*)-hexenal. The result verifies that MIC and MBC are the same. Lethality occurred slower than with polygodial. The result obtained indicates that the mode of antibacterial action of polygodial and (2*E*)-hexenal against *S. choleraesuis* differ to some extent.

Subsequently, the bactericidal effect of (2*E*)-dodecenal was confirmed by the time kill curve method as shown in Figure 4. Cultures of *S. choleraesuis*, with a cell density of 1×10^5 CFU/mL, were exposed to three different concentrations of (2*E*)-dodecenal. The number of viable cells was determined following different periods of incubation with (2*E*)-dodecenal. The result verifies that the MIC and MBC are the same. It shows that 1/2MIC slowed growth, but the final cell count was not significantly different from the control. Notably, lethality occurred quickly, within the first 1 h after the addition of (2*E*)-dodecenal. This rapid lethality very likely indicates that antibacterial activity of (2*E*)-dodecenal against *S. choleraesuis* is associated with the disruption of the membrane (Trombetta et al., 2002), similar to its effects found against *S. cerevisiae*.

The bactericidal effect of (2*E*)-hexenal (C_6) occurred slower than that of (2*E*)-dodecenal, which needed 7 h. Such a slow cell death is thought to proceed independent of the membrane disruptive action. The result obtained indicates that the mode of antibacterial action of (2*E*)-hexenal and (2*E*)-dodecenal against *S. choleraesuis* differs to some extent. The effects of (2*E*)-dodecenal and (2*E*)-hexenal against *S. choleraesuis* were further tested by holding the viable cell number in the presence of chloramphenicol. This antibiotic is known to restrict cell division by inhibiting protein synthesis. Figure 5 shows that the effect of chloramphenicol against *S. choleraesuis* cells is bacteriostatic for the first 3 h after the addition of the drug. It should be noted that chloramphenicol is known to be bacteriostatic for a wide range of Gram-positive and Gram-negative bacteria, but this antibiotic expressed a bactericidal effect against *S. choleraesuis* after 8 h incubation. In the presence of chloramphenicol, (2*E*)-hexenal decreased viable cell numbers slightly more

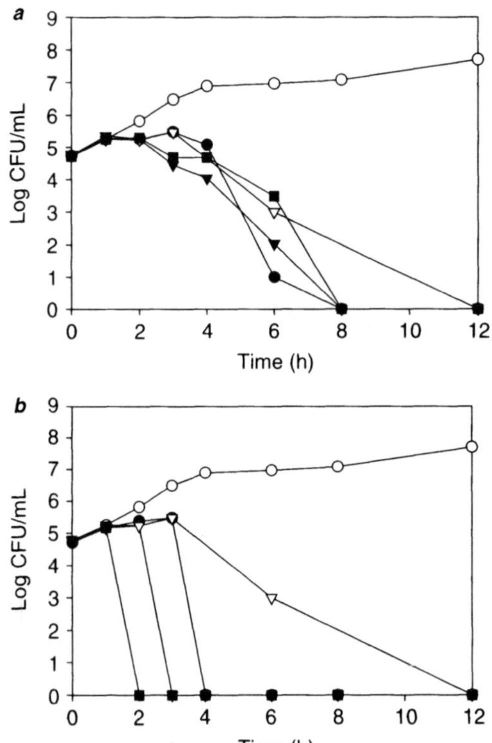

Fig. 5. Effect of (***a***) (2*E*)-hexenal and (***b***) (2*E*)-dodecenal in the presence of chloramphenicol against *S. choleraesuis* ssp *choleraesuis* ATCC 35640. Exponentially growing cells were inoculated into NYG broth and then cultured at 37°C. Chloramphenicol (○) 0 and (▽) 6.25 μg/mL, was added to the culture after 1 h cultivation. (2*E*)-Hexenal (100 μg/mL) or (2*E*)-dodecenal (6.25 μg/mL) and chloramphenicol (6.25 μg/mL) were added at (■) 1, (▼) 2, and (●) 3 h.

quickly than in its absence. (2*E*)-Dodecenal induced rapid decrease in viability regardless of the presence of chloramphenicol. The inhibition of cell division by chloramphenicol did not influence the bactericidal effects of (2*E*)-hexenal and (2*E*)-dodecenal. The reduced viability might not be due to interaction with the biosynthesis of cell wall or plasma membrane components. The synthesis of macromolecules such as DNA, RNA, and proteins was not related to the reduction. The observation that the rapid bactericidal effect of (2*E*)-dodecenal very likely indicates that the primary action of (2*E*)-dodecenal is on the cell membrane.

Subsequently, hexanal (C_6) (**6**) was also found to exhibit the antibacterial activity against *S. choleraesuis* with MIC and MBC of 400 and 800 μg/mL, respectively. It appears that the antibacterial activity against *S. choleraesuis* should not be specific to (2*E*)-alkenals because the conjugated double bond is not essential in eliciting activity, but is involved with increasing the activity. This prompted us to assay the corresponding alkanals for their antibacterial activity against *S. choleraesuis* for comparison. The results are listed in Table 1. The activity of alkanals is weaker than those of the corresponding (2*E*)-alkenals. Similar to (2*E*)-alkenals, their MIC and MBC values are approximately the same and the activity also increased in general with increasing carbon chain length up to dodecanal (C_{12}) (**7**). It should be noted, however, that there is a slight difference between (2*E*)-alkenals and alkanals. For example, decanal (C_{10}) (**8**), undecanal (C_{11}) (**9**), and dodecanal are the most effective but their MIC and MBC values against *S. choleraesuis* are all the same. The increase in the activity as carbon-chain length increases is not distinct in the case of alkanals as compared to those of (2*E*)-alkenals. The bactericidal effect of hexanal and dodecanal were also confirmed by the time kill curve method.

The activity often disappears after the chain length reaches the maximum activity and this phenomenon is known as the cutoff. As expected, dodecanal (C_{12}) was the most effective against *S. choleraesuis* with both MIC and MBC of 100 μg/mL, while tridecanal (C_{13}) (**10**) did not show any activity up to 800 μg/mL. Noticeably, this cutoff was not observed with the (2*E*)-alkenal series against *S. choleraesuis*. That is, (2*E*)-tridecenal exhibits some activity, though to a lesser extent than (2*E*)-dodecenal. This difference in susceptibility of *S. choleraesuis* to (2*E*)-alkenals possessing different chain lengths still remains largely unclear. Since the hydrophobic forces are more favorable than hydrogen bonding forces, this may help to explain the cutoff in that the compound is pulled further into the membrane and loses the orientation required for bilayer disruption. The hydrophilic aldehyde group first binds with an intermolecular hydrogen bond like a "hook" by attaching itself to the hydrophilic portion of the membrane surface, at which point the hydrophobic alkyl portion of the molecule is able to enter into the membrane lipid bilayers (Kubo et al., 1995, 2003b).

IV. Modes of antibacterial action

In our previous papers on structure-antifungal activity relationship (SAR) studies with the same series of acyclic (2*E*)-alkenals, we reported that the antifungal activity of amphipathic medium chain (C_9–C_{12}) (2*E*)-alkenals against *S. cerevisiae* was mediated. This was largely because of their nonionic surface-active properties and hence, the maximum activity can be obtained when balance between hydrophilic and hydrophobic portions becomes the most appropriate, similar to being described for acyclic alkanols (Kubo et al., 1995, 2003a). In other words, the antifungal activity of (2*E*)-alkenals against *S. cerevisiae* is due mainly to biophysical process. This concept can be extended to the antibacterial activity of the same medium chain (2*E*)-alkenals against *S. choleraesuis* because, in the time kill experiment, (i) lethality occurred notably quickly, within the first 1 h after the addition of (2*E*)-dodecenal, (ii) bactericidal activity was found at any growth stage, and (iii) (2*E*)-dodecenal rapidly killed *S. choleraesuis* cells in which cell division was inhibited by chloramphenicol, were observed. Moreover, the antimicrobial activity of (2*E*)-alkenals is nonspecific and the potency of the activity against *S. choleraesuis* was distinctly increased with each additional CH_2 group, up to (2*E*)-dodecenal. The results observed support medium chain (2*E*)-alkenals' ability to function at least in part as nonionic surfactants. It should be noted that the bactericidal activity of (2*E*)-dodecenal against *S. choleraesuis* is eightfold more potent as compared to that of polygodial (Kubo et al., 2001), but the rationale for this still remains largely unclear.

The common nature among these aldehydes should be considered in that the electron negativity on the aldehyde oxygen atom forms an intermolecular hydrogen bond with a nucleophilic group in the membrane, thereby creating disorder in the fluid bilayer of the membrane. The fluidity of the cell membrane can be disturbed maximally by hydrophobic compounds of particular hydrophilic aldehyde group. They could enter the molecular structure of the membrane with the polar aldehyde group oriented into the aqueous phase by hydrogen bonding and nonpolar carbon chain aligned into the lipid phase by dispersion forces. Eventually, when the dispersion force becomes greater than the hydrogen bonding force, the balance is destroyed and the activity disappears. In connection with this, the hydrophobic bonding energy between an average fatty acid ester and a completely hydrophobic peptide is approximately 12 kcal/mol. Addition of a hydrogen bond between a peptide and a fatty ester's carbonyl adds another 3–6 kcal/mol. Furthermore, aldehydes first approach the binding site with the electron negativity of the aldehyde oxygen atom. This hydrogen bond acceptor will affect the hydrogen bonds that regulate the permeability of the lipid bilayer.

Given the surfactant-like properties of medium-chain (2*E*)-alkenals and polygodial, it is possible to suggest that (2*E*)-alkenals and polygodial also act at the lipid–protein interface of integral proteins, such as ion channels and/or transport proteins, denaturing their functional conformation in a similar manner found for alkanols (Kubo et al., 1995, 2003a). The common nature among these (2*E*)-alkenals and polygodial should be considered in that the electron negativity on the aldehyde oxygen atom forms an intermolecular hydrogen bond with a nucleophilic group in the membrane, thereby creating disorder in the fluid bilayer of the membrane. The fluidity of the cell membrane can be disturbed maximally by hydrophobic compounds of particular hydrophilic enal group. Thus, the amphipathic medium-chain (2*E*)-alkenals and polygodial disrupt the hydrogen bonding in the lipid–protein interface in *S. choleraesuis*. The data obtained are consistent with the effect on the bulk membrane rather than a direct interaction of the specific target protein, and (2*E*)-alkenals' nonspecificity of antimicrobial activity supports this assumption. The possibility of the anti-*Salmonella* activity of the medium-chain (2*E*)-alkenals and polygodial is due to their nonionic surfactant property, but this may not be the case for short-chain ($<C_9$) (2*E*)-alkenals. The short-chain (2*E*)-alkenals enter into the cell by passive diffusion across the plasma membrane and/or through porin channels (Schulz, 1996). Once inside the cell, their α,β-unsaturated aldehyde (enal) moiety is chemically highly reactive and hence, they may readily react with biologically important nucleophilic groups such as sulfhydryl, amino, or hydroxyl (Schauenstein et al. 1977). Sulfhydryl groups in proteins and lower molecular weight compounds such as glutathione are known to play an important role in the living cell. Bacteria protect themselves against hydrogen peroxide in various ways (Brul and Coote, 1999), and some of the most ubiquitous systems include glutathione. (2*E*)-Alkenals causes depletion of cytoplasmic and mitochondrial glutathione, which functions in eliminating reactive oxygen species, similar to that found for polygodial (Machida et al., 1999). This (2*E*)-alkenal mediated depletion of intercellular glutathione can be explained by a direct interaction between the enal moiety and the sulfhydryl group of glutathione by a Michael-type addition. This may reveal the reason why (2*E*)-alkenals exhibit in general more potent and broader antimicrobial activity than those of the corresponding alkanals and alkanols. In the case against *S. choleraesuis*, (2*E*)-hexenal exhibited the bactericidal activity against this food borne bacterium with an MBC of 100 μg/mL, whereas hexanol did not show any activity up to 1600 μg/mL.

Moreover, the leakage of carboxyfluorescein (CF) in liposomes of phosphatidylcholine (PC) following exposure to (2*E*)-alkenals was previously reported (Trombetta et al., 2002). Interestingly, (2*E*)-alkenals caused rapid CF leakage from PC liposomes and the effectiveness order correlated well with the alkyl chain length. Thus, (2*E*)-nonenal was more effective in inducing CF leakage from PC liposomes than that of (2*E*)-hexenal. This previous report also supports the surfactant concept.

The process by which antibacterial agents reach the action sites in living bacteria is usually neglected in the cell-free experiment, but this must be taken into account in the current study. The inner and outer surfaces of the membrane are hydrophilic while the interior is hydrophobic, so the increased lipophilicity of (2*E*)-alkenals should affect their movement further into the membrane lipid bilayer portions. It should be logical to assume that most of the lipophilic (2*E*)-alkenal molecules being dissolved in the medium are partially incorporated into the lipid bilayers (Franks and Lieb, 1986) in which they may react with biologically important substances. The amount of (2*E*)-alkenals entering into the cytosol or lipid bilayer is dependent on the length of the alkyl chain. Hence, the length of the alkyl chain is associated with eliciting activity to a large extent (Kubo and Kubo, 1995; Kubo et al., 1995).

V. Combination effects

The combination data were obtained by the broth checkerboard method (Norden et al., 1979; Eliopoulos and Moellering, 1991). In previous reports, the antifungal activity of polygodial against *S. cerevisiae* was described to be significantly enhanced in combination with anethole (**11**), a common phenylpropanoid characterized from the aniseed (Kubo and Himejima, 1992). Hence, polygodial was combined with anethole to see if the same combination effect can also be observed against *S. choleraesuis*. Anethole itself exhibits antibacterial activity against this food-borne

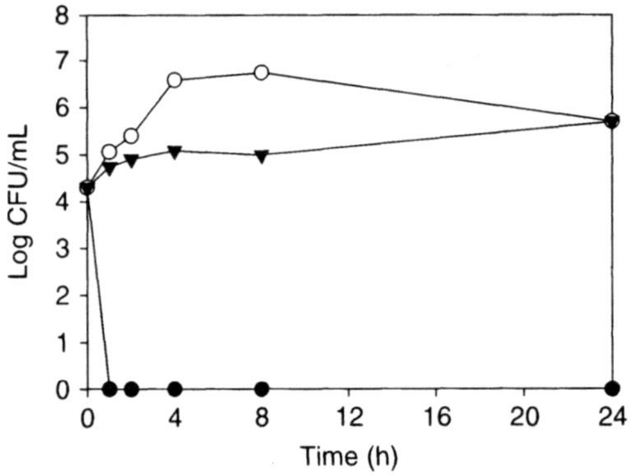

Fig. 6. Bactericidal effect of anethole against *S. choleraesuis* ssp *choleraesuis* ATCC 35640. Exponentially growing cells of *S. choleraesuis* were inoculated at 37°C in NYG broth with (○) 0, (▼) 100, or (●)200 μg/mL of anethole. Viability was established by the number of colonies formed on NYG plate after incubation at 30°C in for 24 h.

bacterium with both MIC and MBC of 200 μg/mL (1.35 mM). Similar to polygodial and (2*E*)-hexenal, no differences in MIC and MBC were noted, suggesting that no residual bacteriostatic activity was involved. The bactericidal effect of anethole was confirmed by the time kill curve method as shown in Figure 6. The lethality occurred slower than that of polygodial, 4 h after the addition of anethole. It shows that 1/2MIC reduced the growth rate, but that the final cell count was not significantly different from the control (Kubo and Fujita, 2001). *S. choleraesuis* is one of the few Gram-negative susceptible bacteria to anethole, which thus resembles polygodial.

The combination of polygodial and anethole synergistically retarded the growth rate of *S. choleraesuis* to a large extent, but this combination showed only marginal synergism on their bactericidal action. Thus, *S. choleraesuis* cells appeared to adapt to this combination stress, eventually recovering and growing normally. These results may indicate possibly different antimicrobial mechanism of the combination between yeasts and bacteria, or more specifically between *S. cerevisiae* and *S. choleraesuis*. Anethole was also combined with (2*E*)-hexenal to see if the combination has any enhancing activity. This combination also exhibited strong synergism on their bacteriostatic action, but only marginal synergism on their bactericidal action. Moreover, anethole was also combined with (2*E*)-dodecenal to see if the combination has any enhancing activity. This combination exhibited strong synergism on their bacteriostatic action, but exhibited only marginal synergism on their bactericidal action. The reason for the residual bacteriostatic activity against *S. choleraesuis* remains unknown.

VI. Molecular design

Safety is a primary consideration for anti-*Salmonella* agents, especially for those present in food products. The anti-*Salmonella* agents isolated from plants for use as food spices and/or characterized, as flavor substances in many edible plants should be superior as compared to nonnatural ones. The knowledge obtained may provide insights into bactericidal action of aldehydes on a molecular basis, and a more rational and scientific approach to use or design efficient and safe anti-*Salmonella* agents. Furthermore, it may be worthwhile to consider the anti-*Salmonella* activity of rather common phytochemicals from an ecological point of view. For example, it should be remembered that chickens used to peck green leaves. Green leaves contain a variety of antibacterial agents against *Salmonella* bacteria, especially those known as green leaf aldehydes and alcohols (Hatanaka, 1993). This indicates that *Salmonella* are very likely controlled in nature when chickens were continuously fed green leaf-based food. In the Amazon basin, *Salmonella* contamination of post-harvest pepper has been increasingly noted with increasing large-scale poultry farms. This may be caused by shifting their food habit from plant-based natural foods to artificial fodders. (2*E*)-Alkenals may have potential as crop preservatives to inhibit or prevent the growth of *Salmonella* bacteria. For example, a minute amount of the medium-chain (2*E*)-alkenals such as (2*E*)-dodecenal and (2*E*)-undecenal can be added to the artificial fodder. On the other hand, the high concentrations needed to cause the loss of viability, but (2*E*)-hexenal may be considered as a genuine anti-*Salmonella* agent because of its high volatility (Wilson and Winiewski, 1989; Corbo et al., 2000) and wide distribution in many edible plants such as apples, pears, grapes, strawberries, kiwi, tomatoes, olives, etc. (Schauenstein et al. 1977). In practical application, interactions with food components may limit the scope of the use of polygodial or (2*E*)-alkenals as preservatives in some foods, since α,β-unsaturated aldehydes are chemically highly reactive substances. Hence, an attempt to search for alternative anti-*Salmonella* agents through synthetic optimization has been made.

VII. Alkyl gallates

On the basis of the data obtained, the hydrophilic aldehyde group can be replaced by any hydrophilic groups as long as the "head and tail" structure is balanced. Hence, a series (C_3–C_{13}) of alkyl gallates (3,4,5-trihydroxybenzoates) was synthesized by one-step esterification utilizing *N,N'*-dicyclohexylcarbodiimide (DCC) as an activating reagent. In brief, to a solution of the gallic acid (1.3 mmol) and alcohol (1.3 mmol) in THF (10 mL) cooled at 0°C was added a solution of DCC (2 mmol) in THF (6 mL). After the solution had been allowed to stir for 20 h, the solvent was removed under reduced pressure. The residue was extracted with ethyl acetate several times and then filtered. The filtrate was washed successively with diluted aqueous citric acid solution, saturated aqueous $NaHCO_3$ solution, and water, dried over $MgSO_4$, and then evaporated. The crude products were purified by chromatography (SiO_2; elution with $CHCl_3$-MeOH, 98:2). In the case of short chain ($>C_8$) alkyl gallates, uric acid needs to be carefully removed. Structures of the synthesized esters were established by spectroscopic methods (UV, IR, MS, and NMR) (Kubo et al., 2002b, 2002c). The head and tail structures are synthetically easily accessible as an ester and, therefore, the construction of a wide range of structurally diverse mimics are available for evaluation.

It should be noted that the "hydrolysable" ester group was selected in order to avoid undesired side effects, particularly endocrine disrupting activity of environmentally persistent estrogen mimics (Soto et al., 1991) such as alkylphenolic

Table 3
Antibacterial activity of gallic acid and its esters against *S. choleraesuis*

Gallates tested	μg/mL	
	MIC	MBC
Gallic acid	1600	1600
C_3	1600	> 3200
C_6	100	100
C_7	50	50
C_8	12.5	12.5
C_9	6.25	12.5
C_{10}	12.5	25
C_{11}	25	50
C_{12}	25	50
C_{13}	> 1600	[a]
Geranyl gallate	50	50

[a]Not tested.

compounds (White et al., 1994). The synthesized alkyl gallates were tested for their antibacterial activity against *S. choleraesuis* using the same broth dilution method (Kubo et al., 2002a). The results are listed in Table 3. This food-borne bacterium showed different susceptibility to alkyl gallates possessing different chain lengths. The range of the antibacterial activity of alkyl gallates against *S. choleraesuis* is between 12.5 and 100 μg/mL, and the MICs and MBCs are nearly the same. The difference in the MIC and MBC of the most potent nonyl (C_9) gallate was twofold, indicating that its activity is bactericidal. Notably, this MBC value is nearly comparable with that of gentamicin and (2*E*)-dodecenal. The potency of the bactericidal activity against this food-borne bacterium was increased with each additional CH_2 group, up to nonyl gallate. However, the activity did not disappear after the chain length reached the maximum activity. Dodecyl (lauryl) gallate shows bactericidal activity, with an MBC of 50 μg/mL, while tridecanyl (C_{13}) gallate showed no activity up to 1600 μg/mL. Thus, the cutoff was made between dodecyl and tridecyl gallates.

Alkyl gallates can be considered as having a head-and-tail structure, similar to (2*E*)-alkenals. Therefore, their mode of antibacterial action was expected to be similar to that of surface-active agents (surfactants). However, the data obtained so far indicate that their antibacterial activity is unlikely to be due mainly to their surfactant property. The activities of the seven alkyl (C_6–C_{12}) gallates against *S. choleraesuis* were comparable. More specifically, the antibacterial activity against *S. choleraesuis* did not distinctly increase with every additional CH_2 group, indicating that the length of the alkyl group is not largely associated with the potency of the activity.

During the study to clarify modes of antibacterial action, alkyl gallates were noted to inhibit bacterial respiratory systems. For example, both nonyl and dodecyl gallates were found to inhibit the oxygen consumption of *P. aeruginosa* cells when the suspensions prepared from the same bacterium cells were incubated with these gallates. The dose-response for respiratory inhibition by dodecyl gallate is shown in

Figure 7. Dodecyl gallate also inhibited *P. aeruginosa* NADH oxidase by a membrane fraction prepared from the same bacterial cells, as shown in Figure 8. The results observed indicate that both alkyl gallates inhibit the bacterial membrane respiratory chain. It seems that the antibacterial activity of alkyl gallates is due primarily to their ability to inhibit respiration. The assay was performed as previously described (Haraguchi et al., 1996, 1999). Dodecyl gallate inhibited the growth of *P. aeruginosa* IFO 3080 strain with an MIC of 12.5 μg/mL but not ATCC 33591 strain up to 800 μg/mL. It should be noted that *P. aeruginosa* IFO 3080 strain used for the experiment is a strict aerobic bacterium. The respiratory inhibition causes bacterial cell death because of lack of anaerobic fermentative ability. The

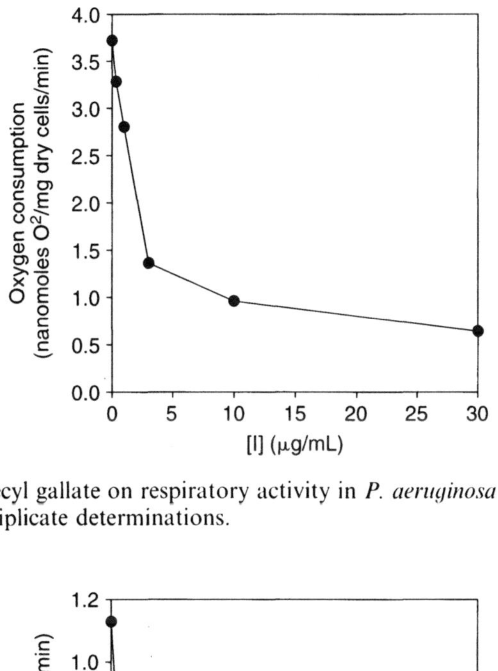

Fig. 7. Effect of dodecyl gallate on respiratory activity in *P. aeruginosa* IFO 3080 cells. Each plot is the mean of triplicate determinations.

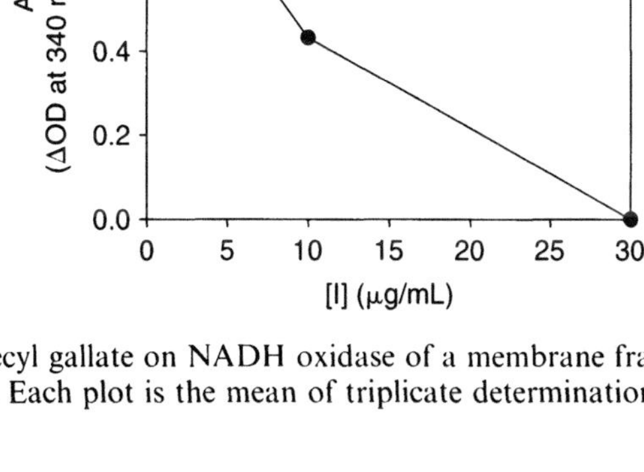

Fig. 8. Effect of dodecyl gallate on NADH oxidase of a membrane fraction isolated from *P. aeruginosa* IFO 3080. Each plot is the mean of triplicate determinations.

concentrations of dodecyl gallate found to inhibit bacterial respiratory systems are comparable to those having bactericidal activity against *S. choleraesuis* as well as *P. aeruginosa* IFO 3080 strains.

The difference in antibacterial actions of alkyl gallates against *S. choleraesuis* needs to be compared with their action against the other Gram-negative bacteria such as *E. coli* and *P. aeruginosa*. In the case of bacteria, various enzymes, especially components of energy-converting systems such as electron transport chains (ETCs) and ATPases, are embedded in the membrane lipid bilayers. In the current study, the process by which alkyl gallates reach the active sites in living microorganisms must be taken into account because this is usually neglected in the cell-free experiment. The inner and outer surfaces of the membrane are hydrophilic while the interior is hydrophobic, so the increased lipophilicity of alkyl gallates should affect their movement into the membrane lipid bilayer portions. It seems reasonable to assume that most of the lipophilic dodecyl gallate molecules dissolved in the medium are incorporated into the lipid bilayers (Franks and Lieb, 1986) without perturbing the lipid (Miller et al., 1989). Once inside the membrane lipid bilayers, alkyl gallates may inhibit the ETC, perhaps by interfering with the redox reactions. The different susceptibilities between *S. choleraesuis* and *P. aeruginosa* may be caused by the different permeability of their outer membrane layer, since this plays a major role in the general resistance of Gram-negative bacteria especially to lipophilic antibiotics. It is known that most Gram-negative bacteria are surrounded by the outer membrane, and this functions as an effective but less specific barrier (Nikaido, 1994). It is logical to assume that most of the lipophilic gallate molecules being dissolved in the medium are incorporated into the outer membrane, and hence cannot reach the ETC in the plasma membrane of *E. coli* and *P. aeruginosa*. This may reveal why alkyl gallates are effective against Gram-positive bacteria but not Gram-negative bacteria such as *E. coli* and *P. aeruginosa*.

Prokaryotic and eukaryotic microorganisms are known to differ in many ways. For example, the ETC involved in the respiratory chain is located in the cytoplasmic membrane in bacteria, while in fungi it is located in the mitochondria. In the case against fungi, dodecyl gallate can rarely enter into the cytoplasm and hence cannot reach the mitochondria. This may explain why dodecyl gallate did not show any effect on eukaryotic microorganisms such as *S. cerevisiae* (Kubo et al., 2002c), in which respiration depends on a mitochondrial ETC. It appears that microorganisms having different membrane structures showed different susceptibilities to alkyl gallates having different chain lengths. The results obtained may provide a more rational and scientific approach to the design of selective and effective antimicrobial agents.

Among the Gram-negative bacteria tested, *S. choleraesuis* and *P. vulgaris* were rare bacteria susceptible to alkyl gallates. In other words, alkyl gallates fall short of the broad spectrum of activity as far as Gram-negative bacteria are concerned. It appears that *S. choleraesuis* and *P. vulgaris* differ from the other Gram-negative bacteria tested in some aspects. If the selective elimination of *S. choleraesuis* and *P. vulgaris* is desirable in food, some alkyl gallates may be considered to be superior. Dodecyl gallate is even more specific against *S. choleraesuis*, compared to octyl gallate (Kubo et al., 2003c). In connection with food, one of the most commonly occurring types of food poisoning is caused by the ingestion of the enterotoxin formed in food during growth of certain strains of *Staphylococcus aureus*. Alkyl gallates were previously reported to be effective against *S. aureus*, including methicillin-resistant *S. aureus* (MRSA) strains (Kubo et al., 2002b).

VIII. Conclusions

The present study shows that polygodial was bactericidal against *S. choleraesuis* and killed the cells quickly. Because of its potent antifungal activity, polygodial's mode of action has been extensively studied using *S. cerevisiae* as a model. In essence, the fungicidal activity of polygodial is likely to be exerted by its multiple functions but primarily comes from its ability to act as a nonionic surface-active agent (surfactant), thereby disrupting lipid–protein interface (Kubo et al., 2001). This surfactant concept can also be applicable in part against *S. choleraesuis* because the lethality against this food-borne bacterium occurred remarkably quickly in the time kill experiment, within the first 1 h after addition of polygodial. This rapid lethality observed supports its ability to function as a nonionic surfactant. If this is the case, polygodial very likely targets the extra cytoplasmic region as a nonionic surfactant and thus does not need to enter the cell, thereby avoiding most cellular pump-based resistance mechanisms. A further study to investigate this idea was not performed since polygodial falls short of the broad spectrum of activity as far as Gram-negative bacteria are concerned. *S. choleraesuis* was the only susceptible Gram-negative bacterium to polygodial, indicating that *S. choleraesuis* differs from other Gram-negative bacteria in some aspects. This difference may be caused by their different permeability of the outer membrane layer since this layer plays a major role in the general resistance of Gram-negative bacteria. Gram-negative bacteria are surrounded by an outer lipopolysaccharidic membrane and this functions as an effective but less specific barrier (Kubo et al., 2003c). In general, antibacterial activity against Gram-negative bacteria decreases by increasing the lipophilicity of molecules. However, the antibacterial agents against *S. choleraesuis* characterized so far are not the case, but they are rather similar to those against Gram-positive bacteria and fungi. If the selective elimination of *Salmonella* bacteria is desirable, polygodial may be considered to be a superior anti-*Salmonella* agent.

Although a number of antimicrobial agents have been characterized from plants, only a few of them showed activity against Gram-negative bacteria, especially the *Pseudomonas* species. Among the compounds we characterized as antibacterial agents, (2*E*)-hexenal is one of the rare examples to possess antibacterial activity against *P. aeruginosa* (Kubo et al., 1999). We first characterized (2*E*)-hexenal as the principal antimicrobial agent from the cashew apple and subsequently olive oil. This common α,β-unsaturated aldehyde is known as "leaf aldehyde" (Hatanaka, 1993) and is widely distributed. It may be a key defense chemical (postinhibiting) against microbial attacks.

Alkyl gallates are known to possess antioxidant activity. Propyl, octyl, and dodecyl gallates are currently permitted for use as antioxidant additives in food (Aruoma et al., 1993). In most foods, the browning process has two components: enzymatic and nonenzymatic oxidation. The nonenzymatic oxidation can be protected by scavengers. As expected, alkyl gallates, regardless of their alkyl chain length, showed potent scavenging activity on the 1,1-diphenyl-2-picrylhydrazyl (DPPH) radical. The alkyl chain length is not directly related to this scavenging activity, so any alkyl gallate can be used as far as antioxidant activity is concerned. In addition to the antioxidant activity, their antimicrobial activity (Kubo, 1999; Kubo et al., 2002c) would appear to be of great overall value to protect foods. Thus, alkyl gallates first protect foods as multifunctional additives, at least as antioxidants and antimicrobials. In connection with this, alkanols and alkyl gallates are effective against food-related microorganisms such as *Zygosaccharomyces bailii* and *S. choleraesuis* (Kubo et al., 2002a). Also, octyl gallate is a good example because it possesses a broad antimicrobial spectrum (Kubo et al., 2002b) with potent antioxidant activity. (2*E*)-Alkenals also possess the similar broad antimicrobial spectrum, but without antioxidant activity. On the other hand, after alkyl gallates are consumed together with the foods to which they are added as additives, alkyl gallates are hydrolyzed to the original gallic acid and the corresponding alcohols, both of which are common in many edible plants. More importantly, the freed gallic acid still acts as a potent antioxidant (Stupans et al., 2002). For example, gallic acid inhibits the generation of superoxide anion by xanthine oxidase with an IC_{50} of 5.7 μM, competitively. This is approximately 78-fold more potent than that of α-tocopherol. Moreover, gallic acid was reported to inhibit squalene epoxidase with IC_{50} of 73 μM (Abe et al., 2000). The multifunction concept has been further extended to geranyl gallate since geraniol has previously been reported to increase glutathione *S*-transferase activity, which is believed to be a major mechanism for chemical carcinogen detoxification (Zheng et al., 1993). As expected, geranyl gallate exhibited a broad antimicrobial spectrum and the results are listed in Table 3. For instance, it showed bactericidal activity against *S. choleraesuis*, with an MBC of 50 μg/mL (Kubo et al., 2002a), and fungicidal activity against *S. cerevisiae* with the minimum fungicidal concentration (MFC) of 50 μg/mL (Kubo et al., 2002c). Incidentally, geraniol is known in a large number (> 160) of essential oils – such as lemon grass, coriander, lavender, and carrot – and is used as food flavoring for baked goods, soft and hard candy, gelatin, pudding, and chewing gum. There is evidence that antioxidants are significantly associated with reduced cancer risks. The primary biological role of antioxidants is in preventing the damage that reactive free radicals can cause to cells and cellular components. The antioxidant gallic acid and the glutathion *S*-transferase inducer, geraniol, may contribute to reduce cancer risk as well as oxidative damage-related diseases.

Acknowledgments

The authors are indebted to ABEP, Pará, Brazil for the opportunity to explore this interesting science and Dr. S. H. Lee for performing antibacterial assay at an earlier stage of the work.

References

Abe I, Seki T, Noguchi H. (2000) Potent and selective inhibition of squalene epoxidase by synthetic galloyl esters. Biochem Biophys Res Commun 270:137–40.

Aruoma OI, Murcia A, Butler J, Halliwell B. (1993) Evaluation of the antioxidant and pro-oxidant actions of gallic acid and its derivatives. J Agric Food Chem 41:1880–5.

Bisignano G, Lagana MG, Trombetta D, Arena S, Nostro A, Uccella N, Mazzanti G, Saija A. (2001) In vitro antibacterial activity of some aliphatic aldehydes from *Olea europea* L. FEMS Microbiol Lett 198:9–13.

Brul S, Coote P. (1999) Preservative agents in foods, mode of action and microbial resistance mechanisms. Int J Food Microbiol 50:1–17.

Corbo MR, Lanciotti R, Gardini F, Sinigaglia M, Guerzoni ME. (2000) Effects of hexanal, (*E*)-2-hexenal, and storage temperature on shelf of fresh sliced apples. J Agric Food Chem 48:2401–8.

Eliopoulos GM, Moellering Jr. RC. (1991) Antimicrobial combinations. In: Lorian V editor. Antibiotics in laboratory medicine. 3rd edition. William & Wilkins: Baltimore, pp. 432–92.

Franks NP, Lieb WR. (1986) Partitioning of long-chain alcohols into lipid bilayers: implications for mechanisms of general anesthesia. Proc Natl Acad Sci USA 83:5116–20.

Frazier WC, Westhoff DC. (1988) Food microbiology, 4th edition. McGraw-Hill: New York, pp. 410–439.

Hansch C, Dunn III. WJ. (1972) Linear relationships between lipophilic character and biological activity of drugs. J Pharm Sci 61:1–19.

Haraguchi H, Kataoka S, Okamoto S, Hanafi M, Shibata K. (1999) Antimicrobial triterpenes from *Ilex integra* and the mechanism of antifungal action. Phytother Res 13:151–6.

Haraguchi H, Oike S, Muroi H, Kubo I. (1996) Mode of antibacterial action of totarol, a diterpene from *Podocarpus nagi*. Planta Med 62:122–5.

Hatanaka A. (1993) The biogeneration of green odour by green leaves. Phytochemistry 34:1201–18.

Kim JM, Marshall MR, Cornell JA, Preston III JF, Wei CI. (1995) Antibacterial activity of carvacrol, citral, and geraniol against *Salmenella typhimurium* in culture medium and on fish cubes. J Food Sci 60:1364–8.

Kubo I. (1999) Molecular design of antioxidative and antimicrobial agents. Chemtech 29:37–42.

Kubo I, Fujita K. (2001) Naturally occurring anti-*Salmonella* agents. J Agric Food Chem 49:5750–4.

Kubo I, Fujita T, Kubo A, Fujita K. (2003a) Modes of antifungal action of alkanols against *Saccharomyces cerevisiae*. Bioorg Med Chem 11:1117–22.

Kubo I, Fujita K, Kubo A, Nihei K, Lunde CS. (2003b) Modes of antifungal activity of (2*E*)-alkenals against *Saccharomyces cerevisiae*. J Agric Food Chem 51:3951–7.

Kubo I, Fujita K, Nihei K. (2002a) Anti-*Salmonella* activity of alkyl gallates. J Agric Food Chem 50:6692–6.

Kubo I, Fujita K, Lee SH. (2001) Antifungal mechanism of polygodial. J Agric Food Chem 49:1607–11.

Kubo I, Fujita K, Nihei K, Kubo A. (2004) Anti-*Salmonella* activity of (2*E*)-alkenals. J Appl Microbiol 96:693–9.

Kubo I, Fujita K, Nihei K, Masuoka N. (2003c) Non-antibiotic antibacterial activity of dodecyl gallate. Bioorg Med Chem 11:573–80.

Kubo I, Himejima M. (1992) Potentiation of antifungal activity of sesquiterpene dialdehydes against *Candida albicans* and two other fungi. Experientia 48:1162–4.

Kubo A, Kubo I. (1995) Antimicrobial agents from *Tanacetum balsamita*. J Nat Prod 58:1565–9.

Kubo J, Lee JR, Kubo I. (1999) Anti-*Helicobacter pylori* agents from the cashew apple. J Agric Food Chem 47:533–7.

Kubo I, Lee YW, Pettei M, Pilkiewicz F, Nakanishi K. (1976) Potent army worm antifeedants from the East African *Warburgia* plants. J C S Chem Comm 1013–1014.

Kubo A, Lunde CS, Kubo I. (1995) Antimicrobial activity of the olive oil flavor compounds. J Agric Food Chem 43:1629–33.

Kubo A, Lunde CS, Kubo I. (1996) Indole and (*E*)-2-hexenal, phytochemical potentiators of polymyxins against *Pseudomonas aeruginosa* and *E. coli*. Antimicrob Agents Chemother 40:1438–41.

Kubo I, Muroi H, Himejima M, Kubo A. (1995) Structural functions of antimicrobial long-chain alcohols and phenols. Bioorg Med Chem 3:873–80.

Kubo I, Xiao P, Fujita K. (2002b) Anti-MRSA activity of alkyl gallates. Bioorg Med Chem Lett 12:113–6.

Kubo I, Xiao P, Nihei K, Fujita K, Yamagiwa Y, Kamikawa T. (2002c) Molecular design of antifungal agents. J Agric Food Chem 50:3992–8.

Machida K, Tanaka T, Fujita K, Taniguchi M. (1998) Farnesol-induced generation of reactive oxygen species via indirect inhibition of the mitochondrial electron transport chain in the yeast *Saccharomyces cerevisiae*. J Bacteriol 180:4460–5.

Machida K, Tanaka T, Taniguchi M. (1999) Depletion of glutathione as a cause of the promotive effects of polygodial, a sesquiterpene on the production of reactive oxygen species in *Saccharomyces cerevisiae*. J Biosci Biotechnol 88:526–30.

McCallion RF, Cole AL, Walker JRL, Blunt JW, Munro MHG. (1982) Antibiotic compounds from New Zealand plants II. Polygodial, an anti-*Candida* agent from *Pseudowintera colorata*. Planta Med 44:134–8.

Miller KW, Firestone LL, Alifimoff JK, Streicher P. (1989) Nonanesthetic alcohols dissolve in synaptic membranes without perturbing their lipids. Proc Natl Acad Sci USA 86:1084–7.

Muroi H, Kubo A, Kubo I. (1993) Antimicrobial activity of cashew apple flavor compounds. J Agric Food Chem 41:1106–9.

Nakamura S, Hatanaka A. (2002) Green-leaf-derived C_6-aroma compounds with potent antibacterial action that act on both Gram-negative and Gram-positive bacteria. J Agric Food Chem 50:7639–44.

Nikaido H. (1994) Prevention of drug access to bacterial targets: permeability barriers and active efflux. Science 264:383–8.

Norden CW, Wenzel H, Keleti E. (1979) Comparison of techniques for measurement of in vitro antibiotic synergism. J Infec Dis 140:441–3.

Ohsuka A. (1963) The structure of tadeonal and isotadeonal components of *Polygonum hydropiper* L. Nippon Kagaku Zasshi 84:748–52.

Schauenstein E, Esterbauer H, Zollner H. (1977) Aldehydes in biological systems. London: Pion, pp. 172–200.

Schulz GE. (1996) Porins: general to specific, native to engineered passive pores. Curr Opin Struct Biol 6:485–90.

Soto AM, Juticia H, Wray JW, Sonnenschein C. (1991) *p*-Nonyl-phenol: an estrogenic xenobiotic released from "modified" polystyrene. Environ Health Perspect 92:167–73.

Stupans I, Kirlich A, Tuck KL, Hayball PJ. (2002) Comparison of radical scavenging effect, inhibition of microsomal oxygen free radical generation, and serum lipoprotein oxidation of several natural antioxidants. J Agric Food Chem 50:2464–9.

Taniguchi M, Adachi T, Oi S, Kimura A, Katsumura S, Isoe S, Kubo I. (1984) Structure–activity relationship of the *Warburgia* sesquiterpene dialdehydes. Agric Biol Chem 48:73–8.

Trombetta D, Saija A, Bisignano G, Arena S, Caruso S, Mazzanti G, Uccella N, Castelli F. (2002) Study on the mechanisms of the antibacterial action of some plant α,β-unsaturated aldehydes. Lett Appl Microbiol 35:285–90.

Tyrrel SF, Quinton JN. (2003) Overland flow transport of pathogens from agricultural land receiving faecal wastes. J Appl Microbiol 94:87S–93S.

Watt JM, Breyer-Brandwijk MG. (1962) Medicinal and poisonous plants of Southern and Eastern Africa. Edinburgh and London: E. & S. Livingstone Ltd., pp. 120–121.

White R, Jobling S, Hoare SA, Sumpter JP, Parker MG. (1994) Environmentally persistent alkylphenolic compounds are estrogenic. Endocrinology 135:175–82.

Wilson CL, Winiewski ME. (1989) Biological control of postharvest diseases of fruits and vegetables. An emerging technology. Ann Rev Phytopathol 27:425–41.

Zheng GQ, Kenney PM, Lam LK. (1993) Potential anticarcinogenic natural products isolated from lemongrass oil and galanga root oil. J Agric Food Chem 41:153–6.

M.T.H. Khan and A. Ather (eds.)
Lead Molecules from Natural Products

Plant growth inhibitory activities by secondary metabolites isolated from Latin American flora

Carlos L Céspedes, Juan C Marín, Mariana Domínguez, J Guillermo Avila, Blanca Serrato

Abstract

This chapter studies phytochemicals that are biodirected, to find botanical-origin biopesticides targeted against weeds. It describes the identification of some monoterpenes, diterpenes, sesquiterpene lactones, limonoids, triterpenes, coumarins, and flavonoids, including their chemical derivatives from plants belonging to the families Asteraceae, Celastraceae, Gomortegaceae, Lauraceae, Meliaceae, Monimiaceae, Poaceae, and Winteraceae. The findings show that some natural compounds or their derivatives possess plant growth regulatory (PGR) or herbicidal activities. Compounds with acetylated and α-β-unsaturated carbonyl derivatives showed mainly PGR activity. It seems that an α-methylene-γ-lactone and a group α, β-unsaturated carbonyl are important for structural activity requirements. Allelopathic activities were assayed with *Triticum aestivum*, *T. vulgare*, *Trifolium pratense*, *T. alexandrinum*, *T. angustifolium*, *Lolium multiflorum*, *Lactuca sativa*, *Physalis ixocarpa*, and *Raphanus sativus* as weed seeds models. All of them are weedy pests in wheat, oat, potatoes, corn, bean, and other important crops of Latin America. Very little is known about the inhibitory activity of these natural compounds and their derivatives. Natural compounds that we have isolated represent a valuable resource for the study toward noxious weedy species of the allelopathic activities of these plants and their control. Progress in biochemical and allelopathic characterization of this pathway is outlined.

Keywords: *allelopathy, secondary metabolites, allelochemicals, plant-growth regulators, weed seedlings, phytotoxicity, herbicide, photosynthesis, monoterpenes, sesquiterpenes, flavonoids, phenylpropanoids, limonoids, tremetones, coumarins*

Abbreviations: ATP, adenosine triphosphate; BHA, butyl hydroxy anisole; DPPH, 2, 2-diphenyl-1-pycrilhydrazyl; EDTA, ethylenediaminetetraacetic acid; HPLC, high-pressure liquid chromatography; MCPBA, m-chloroperbenzoic acid; PGR, plant growth regulatory activity; PSII, photosystem II; RNS, reactive nitrogen species; ROS, reactive oxygen species; THQ, tetrahydroxyquinone; TLC, thin-layer chromatography; YSI, Yellow Spring Instrument; UV, ultraviolet; 2, 4-D, 2, 4-dichlorophenoxyacetic acid.

I. Introduction

Plants often produce bioactive molecules, which in turn may be modified either *in vivo* or in the laboratory. These compounds may enter into the environment and produce direct or indirect effects on growth and development of individuals of their own or other species. This process is called allelopathy (Rizvi and Rizvi, 1992; Seigler, 1996).

Some investigations on the mechanisms of allelopathic action report that seedling growth regulators and metabolic inhibitors in chloroplasts and mitochondria are different natural compounds (Einhellig, 1995; Harborne, 1999; Kozubek, 1999; Mucciarelli et al., 2001). In some cases these compounds enhance, improve, or even promote root growth (Klein et al., 2000).

We have previously demonstrated that diverse secondary metabolites have different mechanisms of action and interact with several molecular targets in plant seed germination, seedling growth, mitochondrial activities, and their effects on the photosynthetic electron transport chain (Céspedes et al., 1998, 1999a,b, 2000a, 2001a–c, 2002; Pardo et al., 1998, 2000). Uncoupling properties on the electron transport rate in chloroplasts by some natural compounds have been reported (Duke & Lyndon, 1993; Castañeda et al., 1996; Mata et al., 1998, 2002; Rimando, 1998; Dayan et al., 1999; Harborne, 1999; Macías et al., 1999b, 2000a).

The aim of our work is to correlate the structure–activity relationships (SARs) of plant secondary metabolites on the inhibitory behavior of these compounds on seed germination, seed respiration, photosynthesis in chloroplasts, and ATP synthesis in mitochondria, these being the main macroscopic parameters of development and growth of weedy plants. In addition, our findings indicate that it is possible to correlate some antioxidant activities (e.g., crocin, DPPH) with activity against germination, respiration, and the named physiological processes. These data and parameters are important for allelopathic dissections (Macías et al., 1992, 1993, 1997, 1999a,b, 2000a–d) and are accepted as indirect measures of other physiological processes affected by the chemicals assayed (Macías, 1999b).

Our findings show that several extracts and secondary metabolites obtained from botanical sources have plant-growth inhibitory effects during germination and other physiological plant activities (shoot–root elongation of seedlings, seed respiration during germination processes, O_2-uptake, H^+-uptake, and redox inhibition in chloroplast and mitochondria).

On the other hand, phenolic compounds such as flavonoids have important and diverse activities like those of as nectar guide components in flowers (Harborne, 1988, 1999), larval growth inhibitors (Morimoto et al., 2000; Calderón et al., 2001; Torres et al., 2003), cytotoxic agents (Miranda-González et al., 1997), phytoalexins (Echeverri et al., 1997; Harborne, 1999), amebicidal activity (Ávila et al., 1999), effects on the oxidative properties of intact plant mitochondria (Creuzet et al., 1988; Ravanel et al., 1990; Albertin et al., 1994), and enzymatic regulation (Stenlid, 1970; Kubo, 2000; Kubo et al., 2003a,b).

As model weeds, dicots that were assayed were: lettuce (*L. sativa* cv. Roman) a classical seed used for this kind of bioassays; green tomato (*P. ixocarpa* syn. *P. philadelphica*) an invasive weed throughout North, Central, and South America; cress (*R. sativus*) a very common weed in South-American crops; pink clover (*T. pratense*), red clover (*T. alexandrinum* cv. Andino), and *T. angustifolium*. Ryegrass (*L. multiflorum* cv. *Gulf annual (susceptible)* and cv. *perennial (resistant)*) varieties were employed as susceptible and resistant models of monocot weeds. To probe selectivity, wheat was used (*T. vulgare* cv. *Salamanca*). Glyphosate, 2,4-dichlorophenoxyacetic acid (2,4-D), and *p*-nitrophenol were used as internal standards in plant growth bioassays. Additionally, parthenolide (**1**), santamarin (**2**), 2′,4′,6′-trihidroxy acetophenone (**3**), rutin (**4**), and (±)-naringenin (**5**) were used as pattern substances. In addition, tetrahydroxyquinone (THQ) (**6**), butyl hydroxy anisole (BHA) (**7**), quercetin (**8**), caffeic acid (**9**), gallic acid (**10**), and tocopherol (**11**) were used as internal standards in tests concerned with the reduction of 2,2-diphenyl-1-pycrilhydrazyl (DPPH) and other antioxidant measurements (Figure 1).

Here we present a review of the extracts and bioactive compounds isolated from selected Mexican and Chilean endemic flora and their bioactivity data, whenever available. Special emphasis has been noted of their allelopathic activity. Based on their structures, several classes of compounds are represented:

I. Terpenoids: (1) monoterpenes; (2) diterpenes; (3) sesquiterpenes and sesquiterpene lactones; and (4) nortriterpenes (limonoids).
II. Phenolic compounds, including flavonoids, coumarins, and simple phenolics.

II. Terpenoids

II.A. Monoterpenes

We investigated the effects of essential oils on germination and respiration of seeds obtained from fresh aerial parts of *Drymis winteri* (Winteraceae), *Laureliopsis philippiana* (Monimiaceae), *Laurelia sempervirens* (Monimiaceae), *Persea lingue* (Lauraceae), and *Gomortega keule* (Gomortegaceae) (Osses et al., 2005). These trees grow on mountain slopes of rainforests in south central Chile, and the main chemical components of their essential oils are monoterpenes (see Tables 1–6). These extracts showed a significant inhibitory action on germination and respiration of seeds of *L. sativa*, *P. ixocarpa*, and *T. pratense*, and low activity on the monocot seeds assayed, suggesting a strong selectivity against dicot weeds (Torres et al., 2006) (Table 7). The inhibitory activity on respiration of weeds occurred in a dose-dependent manner, in similar form to that reported by Macías et al. (1999c).

Moreover, these oils showed pronounced inhibitory activity on the length of dicot seedlings; they also showed significant activity in the reduction of DPPH free radical, suggesting clear action as antioxidants (Torres et al., 2006).

II.B. Diterpenes

MeOH extracts of bark and wood of *Podocarpus saligna* showed inhibitory activity on germination and respiration of *T. angustifolium*, *R. sativus*, and *T. aestivum* seeds. This time, *T. aestivum* seed was used to probe selectivity action. From these extracts six diterpenes were isolated: ferruginol (**12**), hinokiol (**13**), totarol (**14**), totarodiol (**15**), totarolone (**16**), and sugiol (**17**) (Figure 2). All of these diterpenes showed germination and respiratory inhibition activities. Ferruginol, hinokiol, totarol, and totarodiol were the most potent compounds producing 100% inhibition at concentrations as low as 10.0 ppm. A somewhat lower effect was observed for the extracts; most were active at 50.0 ppm. In addition, diterpenes assayed showed strong antioxidant activity against the reduction of DPPH radical (Figure 3). Both activities

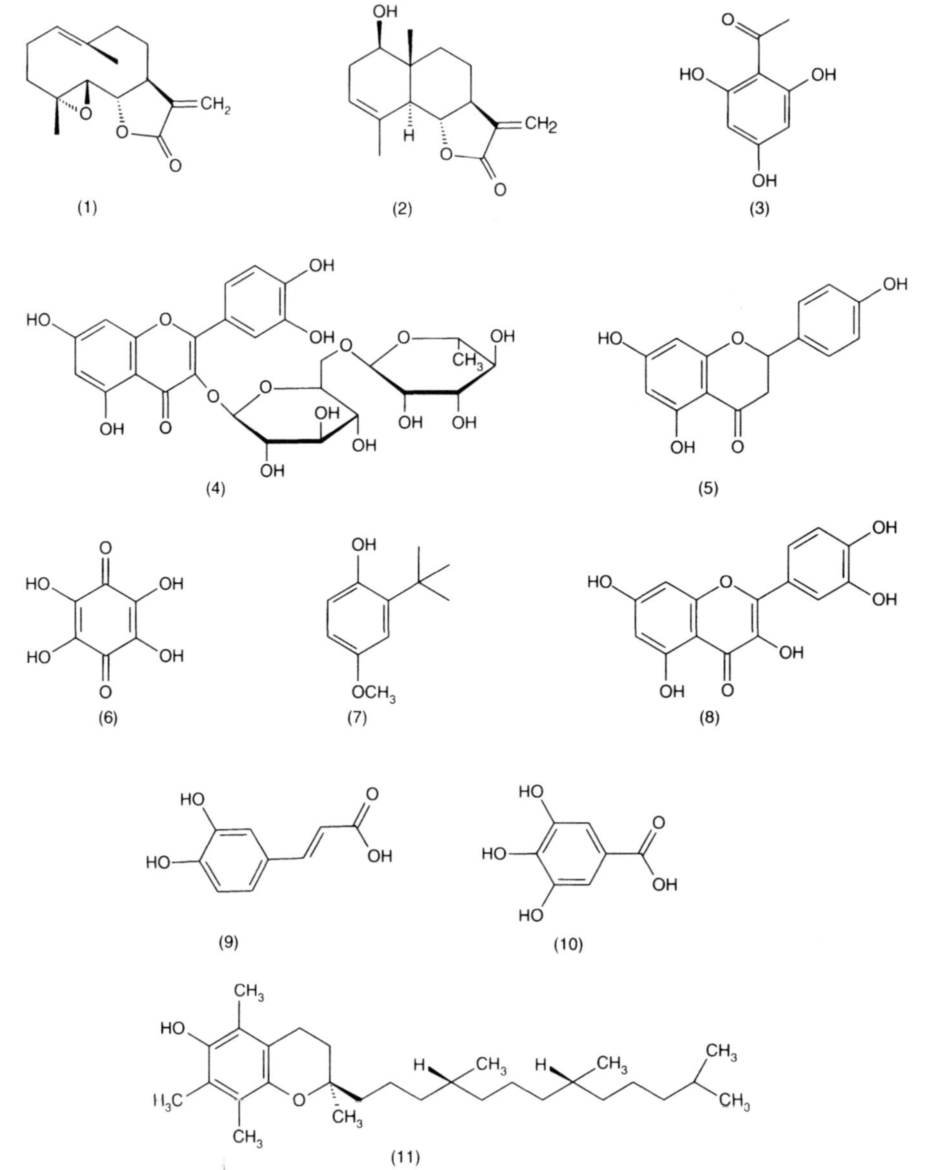

Fig. 1. Chemical structures of pattern and standard compounds used for antioxidant and bioassays.

present a good correlation. The dose-dependent activity shown on weeds is very similar to that for antioxidant activity (Céspedes et al., 2006a) (Table 8).

II.C. Sesquiterpenes and sesquiterpene lactones

From aerial parts of *Cosmos bipinnatus*, *C. scabiosides*, and *C. sulphureum* (Asteraceae), we have isolated dehydrocostuslactone (**18**), costunolide (**19**), and parthenolide

Table 1
Initial fresh weight (g), amount (mg) of essential oils from leaves of the different species studied

Family	Species	Initial fresh weight (g)	Extract (ml)
Winteraceae	Drimys winteri	1114	4.0
Monimiaceae	Laureliopsis philippiana	1000	6.0
Monimiaceae	Laurelia sempervirens	1000	17.0
Lauraceae	Persea lingue	1000	1.4
Gomortegaceae	Gomortega keule	2000	8.8

Table 2
Percentage and retention time of compounds detected by GC present in *Drymis winteri* (Dw)

Peak	Compounds	(%)	*T* (min)
1	Carene	5.908	4.924
2	3-Carene	7.833	5.950
3	3-Methyl-4-methylethyl-cyclohexene	2.403	6.594
4	Terpineol	17.356	8.372
5	3-Methyl-6-methylethyl-cyclohexene	8.048	8.610
6	Farnesene	5.262	12.172
7	α-Neoclovene	3.500	13.015
8	Muurolene	1.590	13.448
9	Citral	10.247	13.570
10	Piperonal	2.466	13.986
11	Episonarene	0.737	14.239
12	1,2,3,4α, 7-Hydrohexane-naphthalene	1.316	14.881
13	Germacrene	8.160	15.050
14	Drimenol	12.103	15.210
15	Cedrene	4.261	15.640

(**1**) (Figure 4). Together with these sesquiterpene lactones, a mixture of phenylpropanoids showed a strong inhibitory activity against electron transport rate in isolated chloroplasts obtained from commercial spinach (Figure 5) (Céspedes et al., 2000a). Mata's group previously reported photophosphorylation and uncoupling behavior of some of the lactones from *Cosmos pringlei* (Robles-García et al., 1997; Mata et al., 2002). Although Macías' group did not report the behavior of sesquiterpene lactones on photosynthetic electron transport rate, they reported the inhibitory action of guaianolide and germacranolide on seedling growth development, mitosis inhibition, and on impairment to reproductive cell, evidencing the ability of these molecules to produce cell death (Macías et al., 2000a–d).

Similarly, we have reported inhibitory activity produced by ovatifolin (**20**) obtained from *Podanthus ovatifolius* (Asteraceae), and several hemisynthetic analogs (**20**)–(**26**) (Figure 6), on seedling growth of *T. pratense*, *L. sativa*, *P. ixocarpa*, *L. multiflorum*, and *T. vulgare* (Céspedes et al., 2001c) (Figure 7). In addition to the effects on growth and development, we have reported electron transport rate inhibition on fresh

Table 3
Percentage and retention time of compounds detected by GC present in *Laureliopsis philippiana* (Lp)

Peak	Compounds	(%)	*T* (min)
1	β-Phelandrene	0.833	4.152
2	β-Pinene	2.152	4.227
3	2-Carene	0.182	5.292
4	Eucalyptol	24.407	5.767
5	1-Methyl-4-(1-methylethyl)-1,3-cyclohexadiene	0.203	7.109
6	3-Carene	55.082	7.757
7	1-Methanol-3-cyclohexadiene	5.675	9.557
8	1,3-Benzodiazole-(α,α)-5,1-propenyl	2.577	11.454
9	Phenol 2, methoxy, 4,1-propenyl	0.200	12.717
10	Benzene, 1,2- dimethoxy-4-2 propenyl	1.864	13.583
11	α-Cubebene	0.187	14.814
12	Benzocycloheptene, 2,4α,5,6,7,8,9α	0.184	15.063
13	1,H-cycloprop(e-azulene)	0.194	15.248

Table 4
Percentage and retention time of compounds detected by GC present in *Laurelia sempervirens* (Ls)

Peak	Compounds	(%)	*T* (min)
1	Bicyclo, (3,1),(O)-hex-2-N-2 methyl	3.324	4.598
2	Cyclohexene, 1-methylene, 4,1 methyl ethyl	19.951	5.234
3	4-Carene	2.128	5.731
4	1,3-Benzodiozole, 5,1 propenyl	74.597	10.290

Table 5
Percentage and retention time of compounds detected by GC present in *Persea lingue* (Pl)

Peak	Compounds	(%)	*T* (min)
1	α-Pinene	13.609	4.234
2	β-Pinene	10.732	4.302
3	β-Myrcene	3.679	4.703
4	α-Phelandrene	17.578	5.057
5	1,3-Cyclohexadiene, 1 methyl, 4 methyl	2.343	5.324
6	Limonene	24.983	5.735
7	Santolin triene	0.340	6.177
8	Ocimene	0.893	7.124
9	3-Carene	0.401	7.460
10	1-ol-4 methyl-1,1-methylene (terpinen-4-ol)-3-cyclohexene	5.869	9.234
11	β-Caryophyllene	6.469	13.800
12	α-Caryophyllene	0.462	14.363
13	δ-Cadinene	1.560	14.839
14	Germacrene	0.780	15.087

Table 6
Percentage and retention time of compounds detected by GC present in *G. keule* (Gk)

Peak	Compounds	(%)	*T* (min)
1	2-Carene	5.266	2.155
2	α-Pinene	6.274	3.786
3	Eucalyptol	55.9195	5.727
4	Bicyclo [1,4,O] hept-2-ene-3,7,7-trimethyl	1.245	6.885
5	Terpinen-4-ol	14.046	8.723
6	α-Terpineol	3.449	8.951
7	4,1 Methyl, ethyl-benzaldehyde	0.300	9.723
8	4-Methylethyl-benzenemethanol,	0.317	10.572
9	α-Farnesene	0.144	12.445

mitochondria isolated from roots or growth radicles of recently germinated beans (Céspedes et al., 2006b). We observed dose-dependent behavior of these compounds (Figure 8), mainly displayed by compounds with two hydroxyl groups such as in Céspedes et al. (2001c). This inhibitory activity was strongly correlated with DPPH-radical reduction in a concentration-dependent manner (Figure 9).

Two other *Parthenium* spp. (Asteraceae) also showed a strong correlation between structure and biological activity (Céspedes et al., 2005). This allelopathic activity is also supported by literature reports on activity of extracts and compounds obtained from this plant by other investigators (Narwal et al., 2003; Venkataiah et al., 2003). Plants of both *Parthenium bipinnatifidium* and *P. hysterophorus* are rich in sesquiterpene lactones of pseudoguaianolide type. In the present investigation, we could observe that ambrosin (**27**) and parthenin (**28**), the two main compounds isolated from these plants, together with other pseudoguaianolides and their derivatives, showed pronounced inhibitory effect on germination and seed respiration of weedy seeds of dicots and, additionally, had pronounced inhibitory activity on oxygen evolution of isolated chloroplasts (Figure 10) (Kalsi et al., 1977; Singh et al., 2002; Céspedes et al., 2005). Some of these compounds and their derivatives (**27**)–(**40**) (Figure 11) also showed good activity for reduction of DPPH free radicals (Figure 12).

We have worked with the sacred tree "Mayten", which belongs to the family Celastraceae, from the rain forests of Southern Chile. "Araucanian" people employ this tree for treatment of diverse ailments and religious-ritual ceremonies. In addition to these properties, plants from the Celastraceae have shown allelopathic effects on other species of the associated flora. We isolated two main components, from MeOH extracts of *Maytenus disticha* and *M. boaria*, both β-agarofuran sesquiterpenes: 9-β-benzoyloxy-1α,2α-6β-8α,15-pentaacetoxydihydro-β-agarofuran (**41**) and 9-β-furanoxy-1α,6β,8α-triacetoxydihydro-β-agarofuran (**42**) (Figure 13) (Alarcón et al., 1995). These compounds showed high potency as inhibitors of electron transport in chloroplasts isolated from spinach (Table 9) (Céspedes et al., 2000b).

II.D. Nortriterpenes (Limonoids)

There are several reports of allelopathic activity of triterpenes and sesquiterpenes from Meliaceae and Rutaceae. Macías' group has interesting papers on this subject,

Table 7
IG_{50} and IR_{50} values of essential oil components on seed germination and respiration of weed seeds, respectively, compared to CH_2Cl_2 extracts from *Baccharis linearis*, *Cosmos bipinnatus*, and *Podanthus ovatifolius*[a,b]

Essential oil[c]	IG_{50}				IR_{50}			
	L. multiflorum	*L. sativa*	*P. ixocarpa*	*T. pratense*	*L. multiflorum*	*L. sativa*	*P. ixocarpa*	*T. pratense*
1	509	588	384	455	589	625	455	563
2	1566	> 2000	> 2000	> 2000	1630	> 2000	> 2000	> 2000
3	1450	> 2000	> 2000	> 2000	1479	> 2000	> 2000	> 2000
4	430	225	230	190	450	200	250	210
5	330	150	250	95	290	120	220	95
6	25.2	12.5	21.0	18.5	25.0	15.0	21.0	19.0
7	6.7	13.7	10.0	11.0	12.0	16.0	12.5	26.3
8	25.2	9.5	21.0	15.0	25.0	15.0	21.0	19.0

Notes: Essential oils of : 1, *D. winteri*; 2, *L. philippiana*; 3, *L. sempervirens*; 4, *P. lingue*; 5, *G. keule*; and IG_{50} and IR_{50} values of: 6, CH_2Cl_2 extracts of *B. linearis*; 7, CH_2Cl_2 extracts of *C. bipinattus*; and 8, CH_2Cl_2 extracts of *P. ovatifolius*.
[a]Means of three experiments, 50 seeds for each assay.
[b]Each value corresponds to the concentration that inhibits 50% of either root or coleoptile/hypocotyl development during seedling stage and was calculated as the dose corresponding to midpoint between complete inhibition (100% of control) and no effect by PROBIT analysis ($p < 0.05$) and ANOVA under the computer program Microcal Origin 6.0.
[c]Values in ppm.

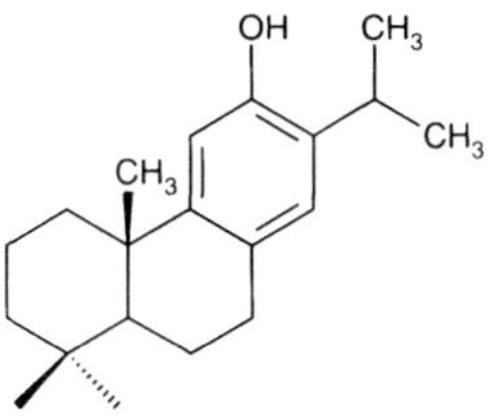
(12)

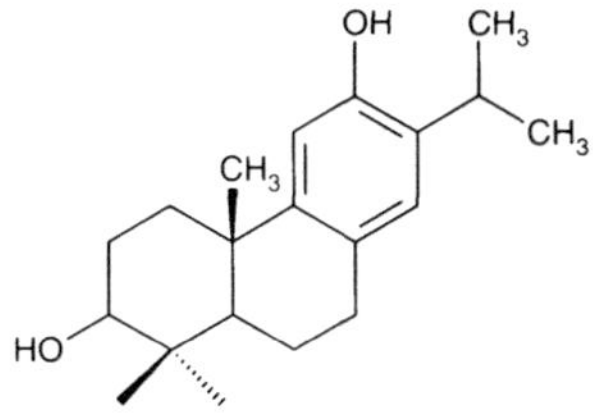
(13)

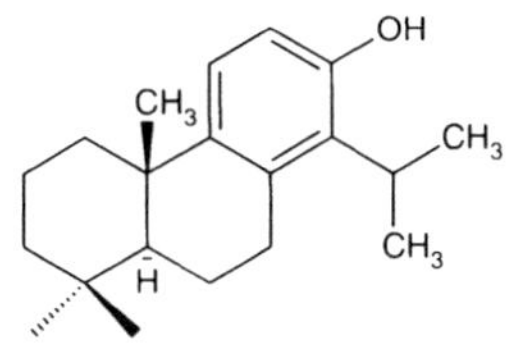
(14)

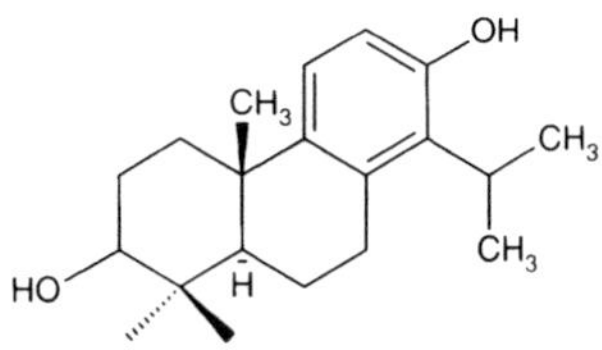
(15)

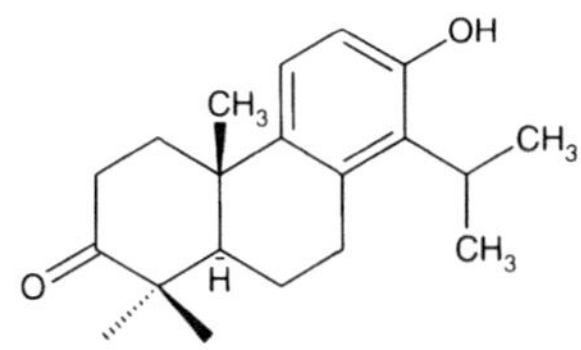
(16)

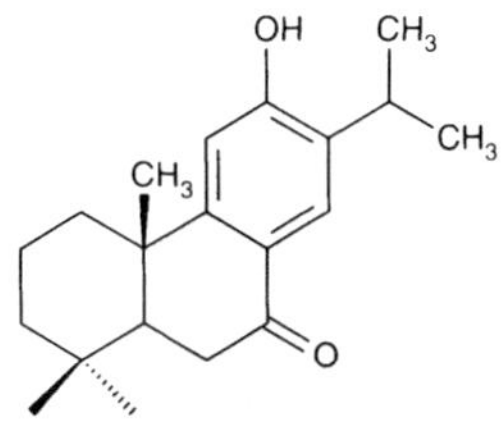
(17)

Fig. 2. Diterpenes present in wood of *Podocarpus saligna*.

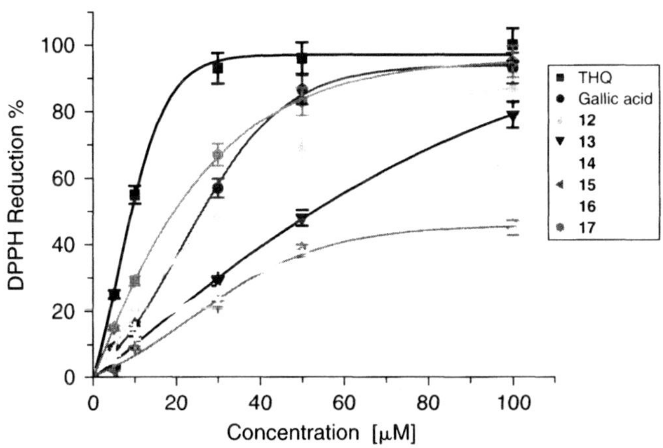

Fig. 3. Scavenging activity of compounds 12–17 on DPPH radical. Measurements at 517 nm, determination after 30 min. Gallic acid (●), THQ (■), ferruginol 12 (▲), hinokiol 13 (▼), totarol 14 (◆), totarodiol 15 (▶), totarolone 16 (◀), and sugiol 17 (✹).

reporting allelopathic triterpenes and sesquiterpenes (Macías et al., 1992, 1993, 1997; 1999a, 1999b, 2000c).

We have isolated several limonoids with inhibitory activity on germination and respiration of weed seeds and partial inhibition of photophosphorylation, H^+-uptake, and noncyclic electron flow, from "cedro" *Cedrela salvadorensis*, *C. ciliolata*, and *C. oxacacensis* and among others (Meliaceae), an interesting group of tree species found in different states of México, such as Veracruz, Chiapas, Tabasco, Oaxaca, and Michoacán. In all cases, inhibition of photosynthetic< autind: Achnine, L.|Mata, R.|Iglesias-Prieto, R.|Lotina-Hennsen, B.",4,0,2?>; Céspedes et al., 1998, 1999a,b).

Table 8
Percentage of inhibitory activity on seed germination of MeOH extracts from wood, and bark from *Podocarpus saligna*[a,b]

Samples	Conc.[c]	*T. angustifolium*			*R. sativus*			*T. aestivum*		
		A	B	C	A	B	C	A	B	C
Control		100	100	100	100	100	100	100	100	100
Wood	10.0	86	11.9	0.0	2	0.2	0.0	72	2.6	2.7
Wood	5.0	92	13.7	0.0	18	11.2	0.0	52	6.9	11.9
Bark	10.0	90	11.9	0.0	20	11.8	0.0	62	5.6	34.8
Bark	5.0	88	12.9	0.0	54	9.1	0.0	52	25.6	61.2
2,4-D	1.0	0.0	0.0	0.0	0.0	0.0	0.0	0.0	0.0	0.0

Note: A: % germination; B: % root length; C: % shoot length.
[a]Means of three experiments. Three times assays for each sample, 50 seeds for each assay.
[b]Each value corresponds to the concentration that inhibits 50% of either root or coleoptile/hypocotyl development during the seedling stage and was calculated as the dose corresponding to midpoint between complete inhibition (100% of control) and no effect by PROBIT analysis ($p < 0.05$) and ANOVA with the computer program Microcal Origin 6.0.
[c]Values in ppm.

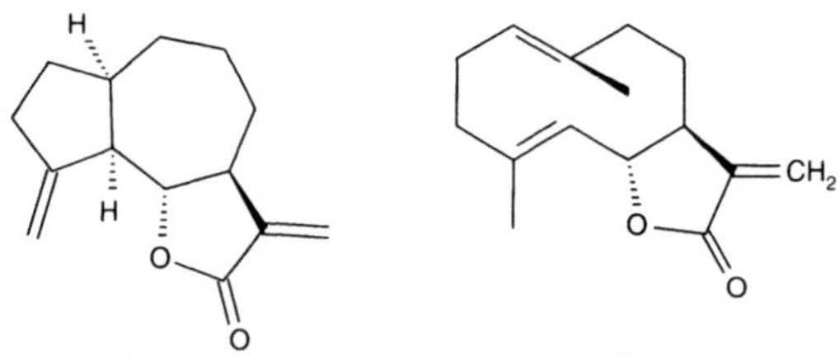

(18) (19)

Fig. 4. Sesquiterpenes isolated from *Cosmos* species.

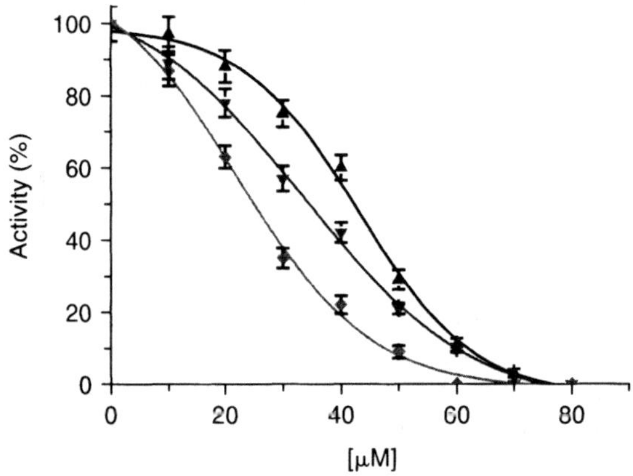

Fig. 5. Effects on basal electron rate of dehydrocostuslactone **18** (◆), costunolide **19** (▼), and parthenolide **1** (▲) in spinach chloroplasts.

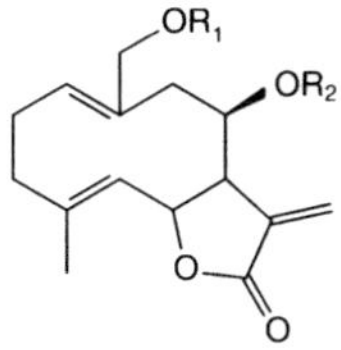

(20) (R_1 = Ac, R_2 = H)
(21) (R_1 and R_2 = Ac)
(22) (R_1 and R_2 = H)

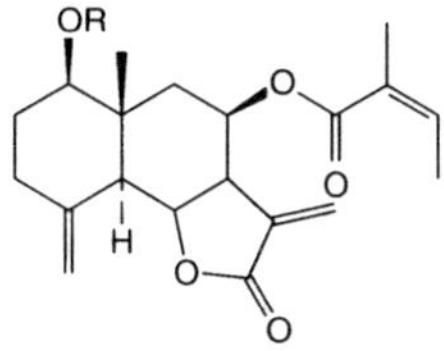

(23) (R = H)
(24) (R = Ac)

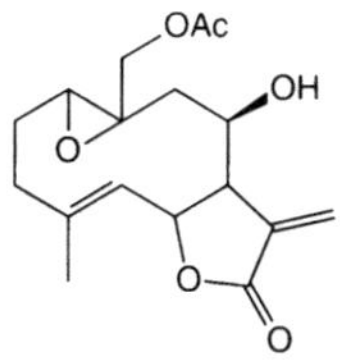

(25)

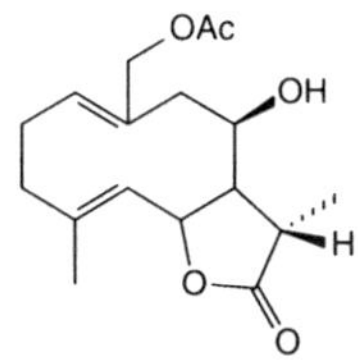

(26)

Fig. 6. Natural sesquiterpene lactones present in aerial parts of *Podanthus ovatifolius* and some derivatives.

Furthermore, these compounds inhibit germination, seed respiration, and seedling growth of weedy plants (Céspedes et al., 1998, 1999a,b, 2001a). The most active compounds were photogedunins A (**43**) and B (**44**), cedrelanolide (**45**), gedunin (**46**), odoratol (**47**), 7-oxogedunin (**48**), 7-deacetoxygedunin (**49**), and photogedunin acetates A (**50**) and B (**51**) (Figure 15). All of these compounds showed interesting activity at concentration ranges from 1 to 50 μM for plant growth regulatory (PGR) activity. They acted as strong herbicides against both monocot and dicot species at higher concentrations (≥75 μM). This effect is shown clearly in oxygen uptake inhibition of seed respiration (Figure 14). For this reason, these nortriterpenes could act as PGRs and as insecticides at the same time (Céspedes et al., 1998, 1999a,b, 2000c).

These nortriterpenes have certain similarities to sendanin (**52**), toosendanin (**53**), and their derivatives (Isman et al., 1995; Kraus, 1995) (Figure 15). We do not know of other reports of the allelopathic effects of nortriterpenes on growth, germination, and on photosynthetic electron transfer rate.

III. Phenolics

As a result of our search for bioactive compounds with agronomic characteristics from several Asteraceae, we isolated a variety of interesting phenolic molecules.

(a) *Tagetes lucida* Cav., known as "pericón", is a medicinal plant used from pre-Columbian times by Aztecs and other Mesoamericans. This species is distributed from Northern México to Northern Nicaragua (Rzedowski, 1972, 1991). Their leaves and florescences are used for stomachache treatment and for treatment of diverse anti-inflammatory ailments. We isolated several coumarins from a dichloromethane extract: scopoletin (6-methoxy-7-hydroxycoumarin) (**54**), 6,7-dihydroxycoumarin (**55**), 7-methoxy-6-hydroxycoumarin (**56**), scoparone (6,7-dimethoxy-coumarin) (**57**), and their derivatives 6,7-diacetoxy coumarin (**58**) and 6-methoxy-7-acetoxycoumarin

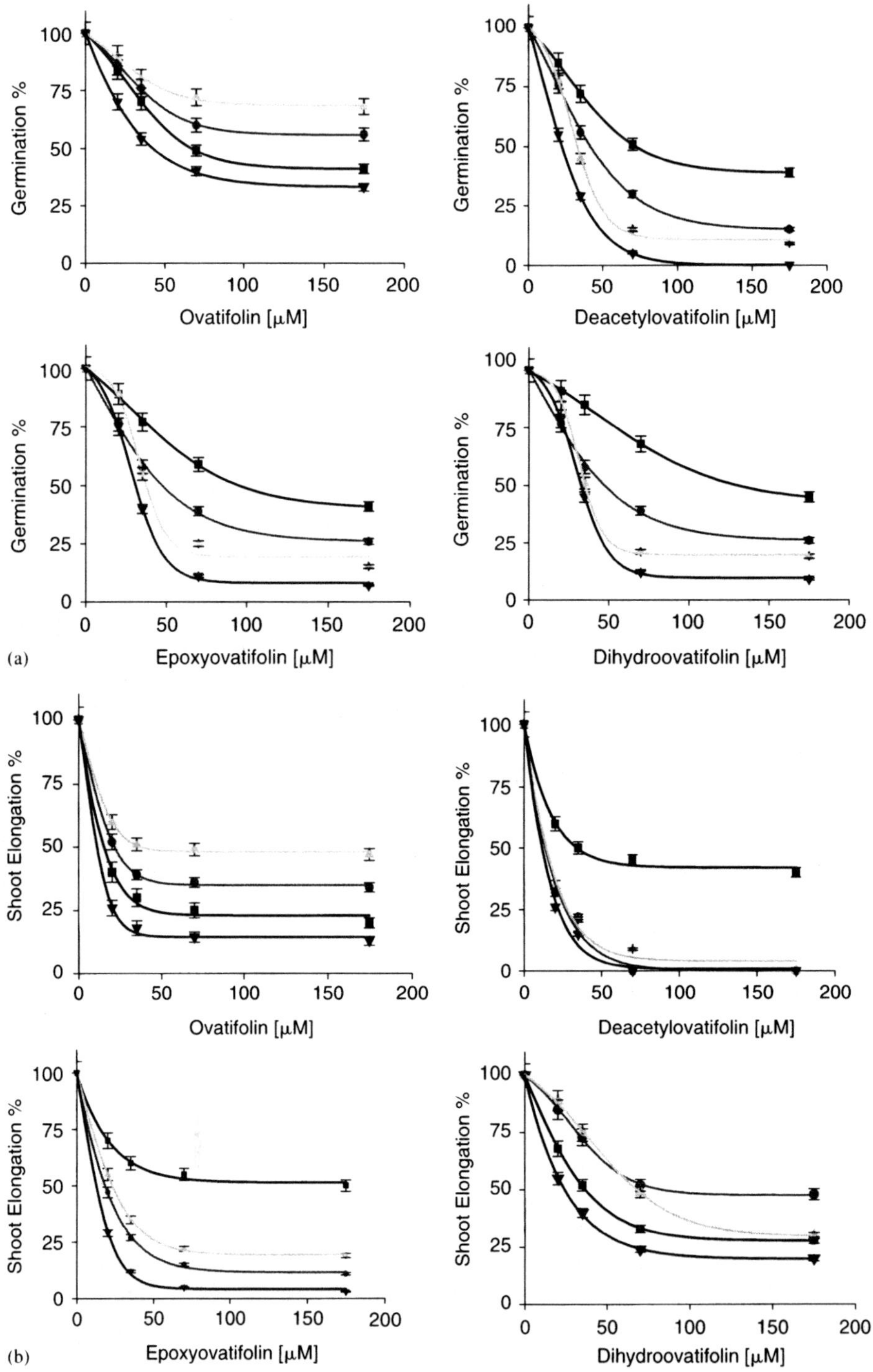

Fig. 7. (a) Effects of solutions of ovatifolin **20**, deacetylovatifolin **22**, epoxyovatifolin **25**, and dihydroovatifolin **26** on germination of *L. multiflorum* (■), *L. sativa* (●), *P. ixocarpa* (▼), and *T. Pratense* (▲) seeds, expressed as percent of control germination. (b) Effects of solutions of compounds **20**, **22**, **25**, and **26** on shoots development of *L. multiflorum* (■), *L. sativa* (●), *P. ixocarpa* (▼), and *T. pratense* (▲) seeds, expressed as percent of control.

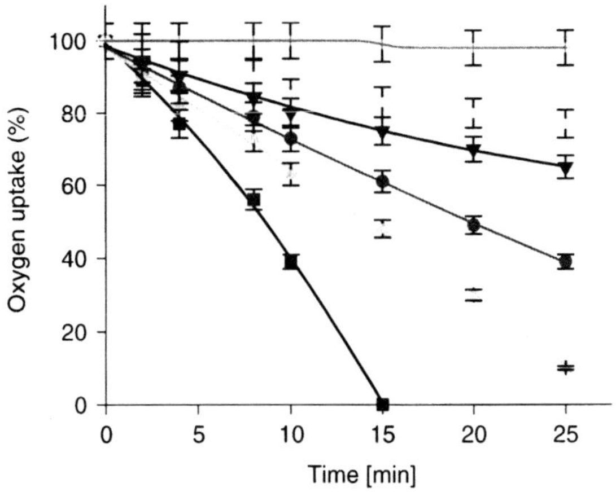

Fig. 8. Oxygen uptake of mitochondrial respiration against the presence of **20** (●), **22** (♦), **23** (▲), **25** (+), **26** (▼), and at 50 μM, control (■) (conditions in Section V).

(**59**) (Figure 16). Extracts and compounds from this plant have inhibitory activity on germination, respiration, and seedling growth, specifically against the dicot seeds assayed (*L. sativa*, *T. pratense*, and *P. ixocarpa*) (Figure 17) (Serrato et al., 2005).

(b) We acquired a number of phenolic compounds from *Baccharis* specimens of México and Chile. We obtained tremetones (benzofurans) from MeOH and CH_2Cl_2 extracts of aerial parts of Chilean *Baccharis* species (*B. linearis*, *B. magellanica*, and *B. umbelliformis*), and additional synthesis of some acetylated and epoxy analogs was carried out. The compounds obtained were tremetone (**60**), 10,11-epoxytremetone (**61**), 10,11-dihydroxytremetone (**62**), 4′-hydroxytremetone (**63**), 4′-acetyltremetone (**64**), and 10,11-epoxy-4′-hydroxytremetone (**65**). In addition, we isolated several acetophenones: *p*-hydroxyacetophenone (**66**), *p*-acetophenone-*O*-glycoside (**67**), *p*-methoxyacetophenone (**68**), 4-acetylacetophenone (**69**), acetophenone (**70**), and *p*-bromoacetophenone (**71**) (Figure 18).

These compounds showed strong pre-emergence activity in a dose-dependent manner with an I_{50} moderately potent (i.e., seed germination and seed respiration inhibition) in the monocot and dicots seeds assayed (Table 10 and Figure 19). Moreover, some of these compounds possessed significant reducing activity against the DPPH free radical, together with partial inhibition of mitochondrial respiration, ATP formation, H^+-uptake, and electron transport rate in isolated chloroplasts; they also inhibited oxygen uptake of fresh mitochondria. These facts indicate that tremetones could be good energetic inhibitors of growth and development of dicot plants (Céspedes et al., 2002). In addition to these tremetones, we isolated simple acetophenones; *p*-hydroxyacetophenone (**66**) appears to be a special case, and this compound had powerful antioxidant and pre-emergent activities (Figure 20).

Additionally, we have studied two Mexican species of *Baccharis* (*B. conferta* and *B. salicifolius*). So far, we have obtained three known flavonoids: quercetin (**8**), naringenin (**5**), and apigenin (**72**) from MeOH extract of aerial parts (Figure 21). These compounds, together with MeOH and CH_2Cl_2 extracts obtained from the plant, showed significant inhibitory activity on germination, respiration, and seedling growth of dicot plants (Figure 22) (Domínguez, 2002; Domínguez et al., 2004).

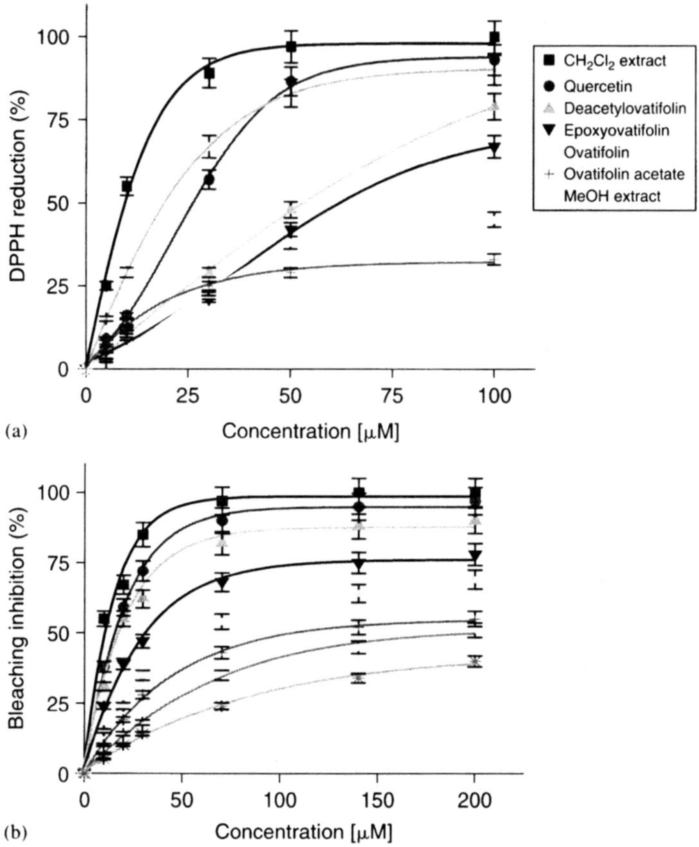

Fig. 9. (a) Scavenging activity of compounds **20–26**, MeOH, and CH_2Cl_2 extracts on DPPH radical. Measurements at 517 nm, determination after 30 min. Quercetin **8** (●), CH_2Cl_2 extract (■), deacetylovatifolin (▲), epoxyovatifolin (▼), ovatifolin (◆), ovatifolinacetate (), and MeOH extract (×;). Values of MeOH and CH_2Cl_2 extracts in ppm. (b) Inhibitory activity of compounds **20–26**, MeOH, and CH_2Cl_2 extracts on the bleaching of crocin measurement at 440 nm of fluorimetric emission, determination after 20 min. CH_2Cl_2 extract (■), deacetylovatifolin (●), epoxyovatifolin (▲), gallic acid **10** (▼), ovatifolin (◆), dihydroovatifolin (), ovatifolinacetate (×;), and arturin (). Values of CH_2Cl_2 extract in ppm.

(c) Interesting effects were found with flavonoids and biflavonoids isolated from *Tephrosia* spp. These plants belong to the family Leguminoseae and grow in both tropical and subtropical lands of México. The main compounds obtained were biflavonoids: tepicanol **(73)** and crassifolin **(74)**, and the simple flavonoids: elongatin **(75)**, tephrobotin **(76)**, tephroleocarpin **(77)**, glabranin **(78)**, and methylglabranin **(79)** (Figure 23). These compounds, isolated from *Tephrosia tepicana*, *T. crassifolia*, *T. viridiflora*, *T. abbottiae*, *T. leiocarpa*, and *T. madrensis*, respectively (Leguminoseae), showed different photosynthetic activities. All compounds, except tepicanol and elongatin, induced concentration-dependent inhibition of photophosphorylation. Compounds **(74)**, **(76)–(79)** interfered with the energetic metabolism and also with respiration of plants at the level of photosynthesis with different mechanisms of action (Figure 24): acting as Hill reaction inhibitors or as energy transfer inhibitors (Céspedes et al., 2001b).

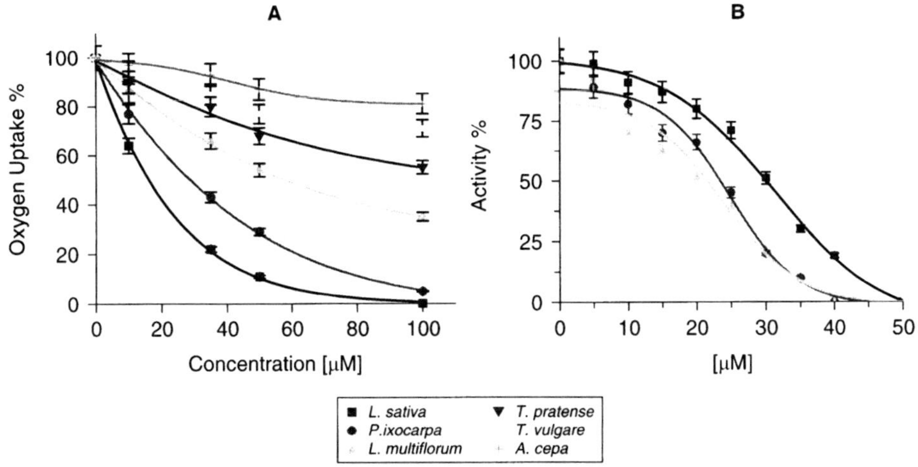

Fig. 10. Graph A: oxygen evolution of the pseudoguaianolide ambrosin **27** on seed respiration of different seeds assayed. Means of three experiments. Values correspond to the concentration that inhibits seed respiration during germination. Values at 72 h. Graph B: electron transport rate (oxygen evolution) in isolated chloroplasts from spinach. Uncoupled (■), phosphorylating (●), and basal (▲). Control rate values in μequiv. e h^{-1} (mg of Chl)$^{-1}$ are 200, 700, and 1100, respectively.

(d) Two phenylpropanoid compounds were isolated from methanol extracts of *Lolium perenne*, *L. multiflorum*, and *Festuca arundinacea* (Poaceae) infected with the endophytic fungus *Acremonium lolii*, growing in cattle pastures of Southern Chile. These compounds were the methyl esters of 4-hydroxy-*trans*-cinnamic acid (**80**) (*p*-coumaric acid) and 3,4-dihydroxy-*trans*-cinnamic acid (**81**) (caffeic acid) (Figure 25). Compound (**81**) inhibited seed respiration of *T. vulgare*, *T. alexandrinum*, *L multiflorum*, and *P. ixocarpa* in the concentration range of 70–150 μM. Compound (**80**) (monohydroxy) is less active than compound (**81**) (Figure 26). These results indicate that cinnamic derivatives may act as allelopathic agents on cattle pastures for both grazing and forage production (Kozubek, 1999; Céspedes et al., 2006c).

IV. Discussion

From the results obtained, some general observations could be made. First, not all compounds tested showed inhibitory activity. Second, the modes of action shown by these compounds depended on whether the target species was a monocotyledonous or dicotyledonous plant, even for compounds isolated from the same family. Whereas monocots are generally inhibited at low concentrations, dicots are stimulated, although they are also inhibited at higher concentrations. For sesquiterpene lactones, low activity was observed with *P. ixocarpa* as the target. Previously observed results for other sesquiterpene lactones indicate similar behavior (Macías et al., 1992, 1999b, 2000a–c). Furthermore, comparison of the values for activity and their profiles, as indicated by compounds that we isolated as well as those from commercial sources, indicates that all these compounds are less effective on dicots,

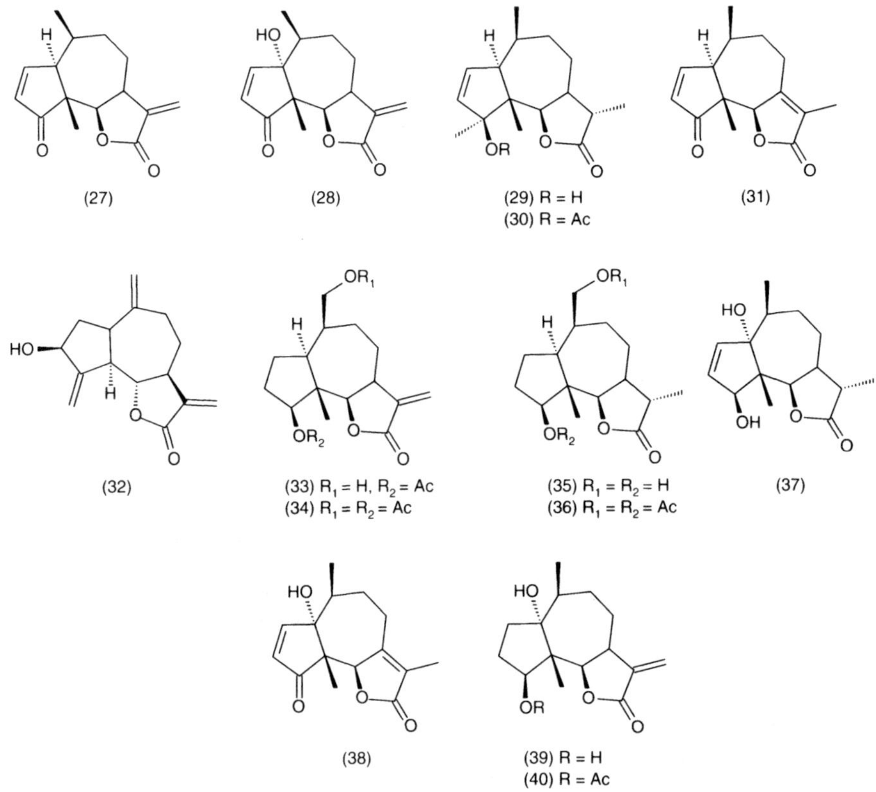

Fig. 11. Pseudoguaianolides isolated from *Parthenium hysterophorus*, *P. bipinnatifidum*, and their derivatives.

but show similar activity values on monocots to those obtained by Macías and his group.

Behaviors vary with the level of accumulation and the types of activity depend on the ability of the compounds to act as inhibitors or as stimulants (i.e., phytohormone-like activity). In some cases, activity reaches a maximum and, if the concentration is increased, the effect is reverted to that of the control.

IV.A. Monocot and Dicot Growth

The I_{50} values of pure compounds and extracts were obtained by determining the concentration that induced 50% of growth inhibition of the root development. This fact is affected to a larger extent, as indicated by low I_{50} values, when compared to coleoptile or hypocotyl development. Growth of monocots was less sensitive to inhibition by pure compounds, as shown by higher I_{50} values (see Table 8, Figures 5, 7, 8, 11, 14, 17, 19, 22, 24, and 26).

The acetate derivatives (**21**, **24**, **30**, **34**, **36**, **40**, **50**, **51**, **58**, **59**, **64**, **69**) promote root and hypocotyle development. Moreover in dicots such as in *T. pratense* and *L. sativa*, it enhances, rather than inhibits, the growth of roots, with the increase in concentration from 50 to 300 μM. In the case of *P. ixocarpa*, root development was only

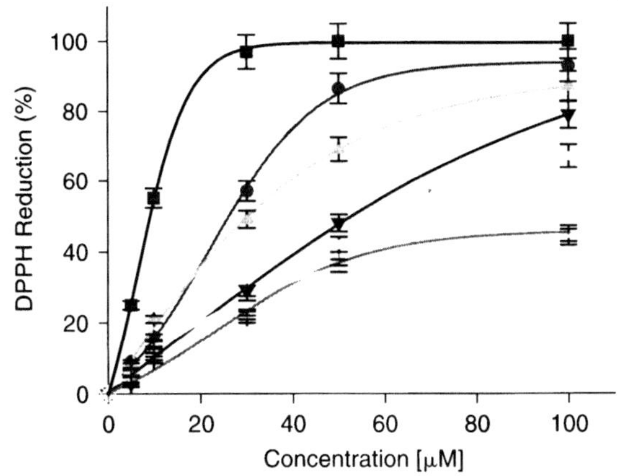

Fig. 12. DPPH reduction of THQ (■), BHA (●), **27** (×;), **35** (), **36** (♦), **37** (▼), and **39** (▲).

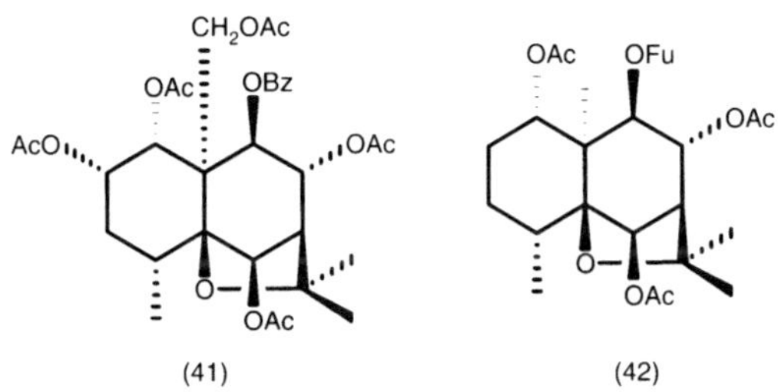

Fig. 13. Main β-agarofurans isolated from *Maytenus boaria* and *M. disticha*.

Table 9
Comparison of the effects of 9-benzoyloxy-1,2,6,8,15-pentaacetoxydihydro-β-agarofuran (41) and 9-furanoxy-1,2,6,8,15-pentaacetoxydihidro-β-agarofuran (42) on different photosynthetic activities of isolated spinach chloroplasts. I_{50} is the concentration producing 50% inhibition of electron transport rates

Activities tested	β-Agarofurans Compound 41	(I_{50}, μM) Compound 42
Photophosphorylation	2.25	78
Basal electron flow	1.4	63
Phosphorylating electron flow	–	26
Uncoupled electron flow	2.5	65
Photosystem II activity	2.6	72
Photosystem I activity	5.0[a]	85[a]

[a]No effects on concentrations of photosystem I activity.

partially inhibited by some acetate derivatives (**50**, **51**, **58**, **59**) in the concentration between 50 and 200 μM whereas shoot development was strongly inhibited. Hypocotyl growth was slightly stimulated at lower concentrations from 0.1 to 50 μM for *P. ixocarpa*, and 200 μM for *T. pratense*, and, thereafter, partially inhibited by these compounds.

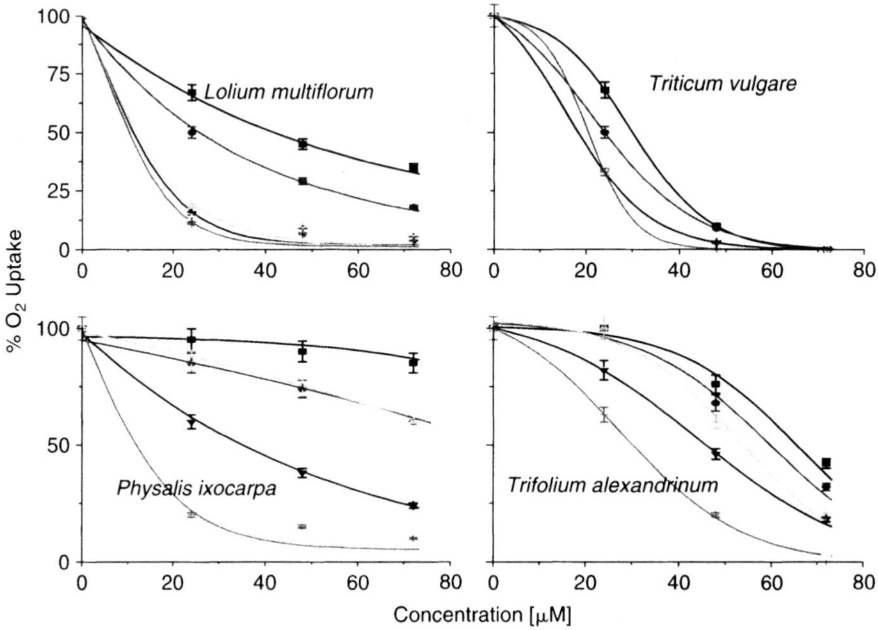

Fig. 14. Inhibition of respiration of *T. vulgare*, *L. multiflorum*, *P. ixocarpa*, and *T. alexandrinum* seeds by limonoids **43** (▼), **44** (♦), **45** (), **48** (●), **49** (■), and **53** (▲). Expressed as percent of O_2-uptake rate as a function of control seed respiration at 75 μM. Each value represents mean ± SE ($N = 5$) in error bars. Statistical analysis by ANOVA, PROBIT, and SNK, under Microsoft Origin 6.0. Values taken after 72 h.

Similar to the effects shown on seed germination, pure compounds also have variable inhibitory effects on growth (80%, 50%, 30% inhibition, or lower) (Table 10, Figures 5, 7, 14, 17, 19, and 22) between 50 and 500 μM for monocots, whereas with some sesquiterpenes, 100% inhibition was achieved above 100 μM for *P. ixocarpa* and above 175 μM for *T. pratense* and *L. sativa*. Growth inhibition followed a dose-dependent pattern in which either stimulation or inhibition of germination was observed.

The results presented regarding differences in behavior of the acetate derivatives and pure hydroxy, methoxy, epoxy, and dihydro derivatives indicate that action mechanisms of phytotoxins may be different for growth and germination (Einhellig, 1986, 1992, 1995; Fischer, 1986; Fischer et al., 1989, 1990, 1991) and confirm the findings obtained by Marles' group (Marles et al., 1995). The mechanism of action established by those authors claims that compounds that involve two functional groups ($O{=}C{-}C{=}CH_2$), are most commonly found as a part of the α-methylene-γ-lactone, are necessary for activity, but may be present in a β-position, e.g., compounds such as α,β-unsubstituted-γ-lactone with a hydroxyl group very near to carbon or other ester or ketone, reacting by a Michael-type addition with a biological nucleophile, in particular, the sulfhydryl group of a reduced glutathione and L-cysteine, or undergo a Wagner–Meerwein rearrangement. These features may also occur with our compounds. On the other hand, greater phytotoxic effects are observed with cyclic alcohols to cyclic ketones (<0.1 mM) (Reynolds, 1987). The

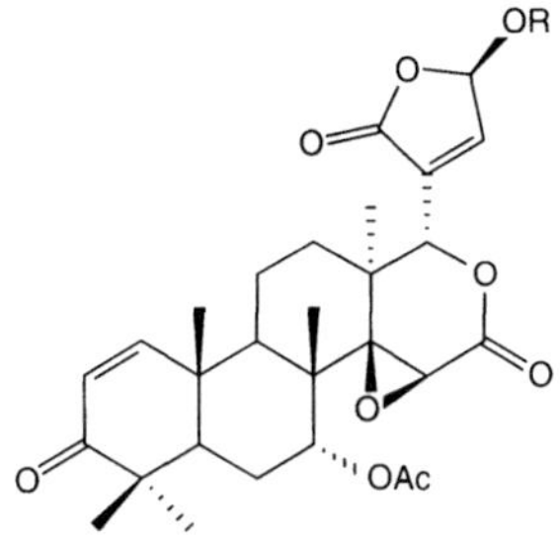

(43) R = H
(50) R = Ac

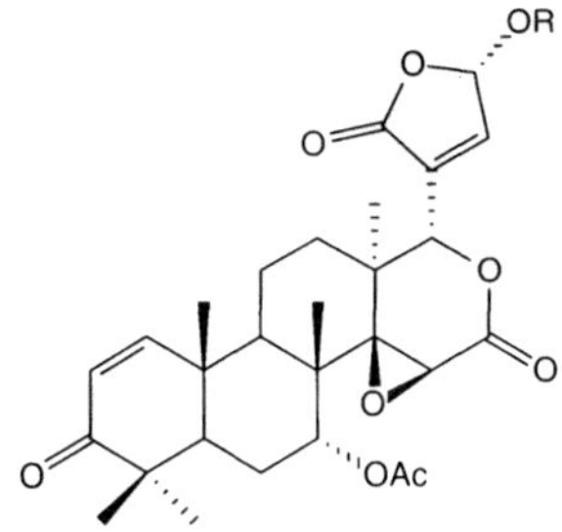

(44) R = H
(51) R = Ac

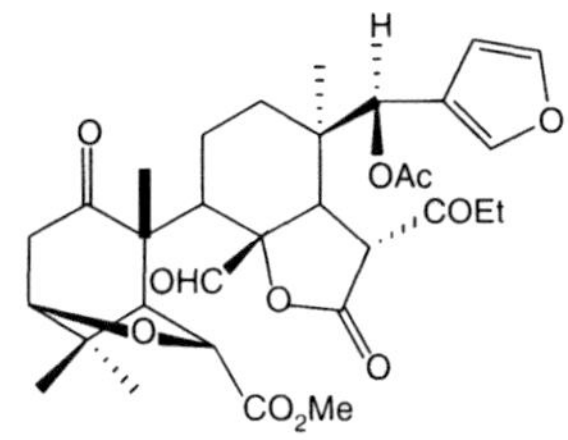

(45)

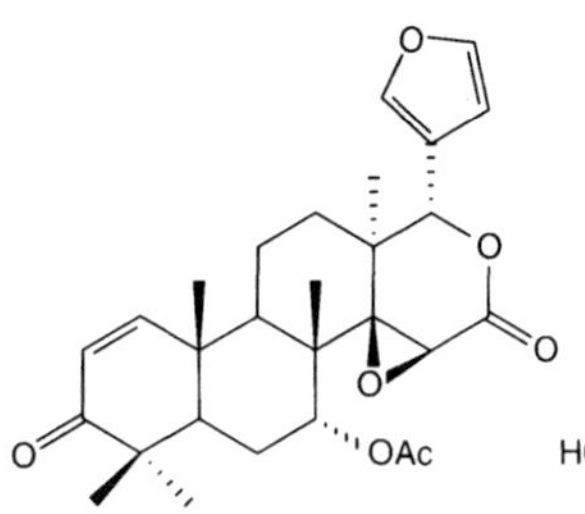

(46)

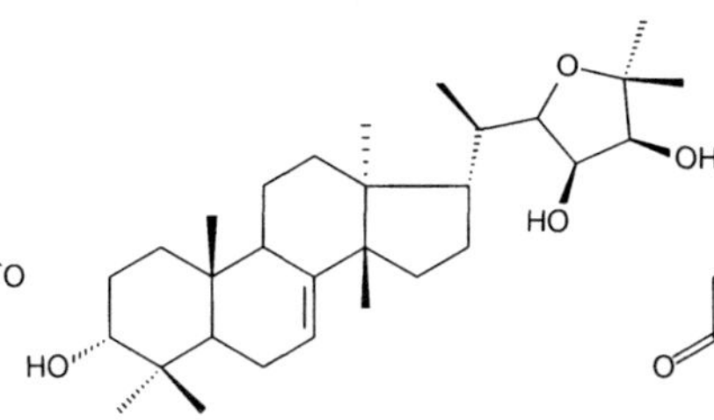

(47)

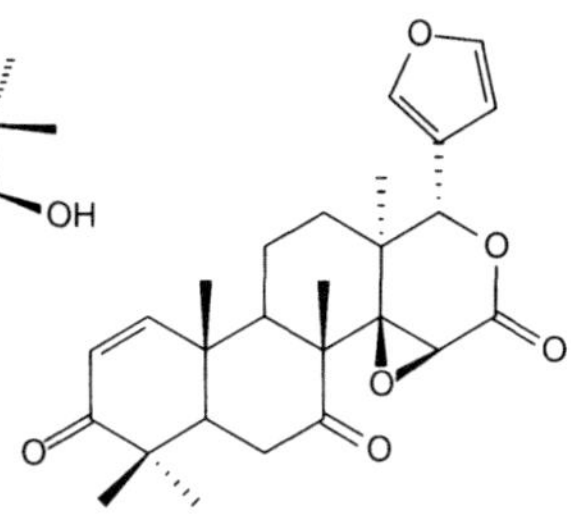

(48)

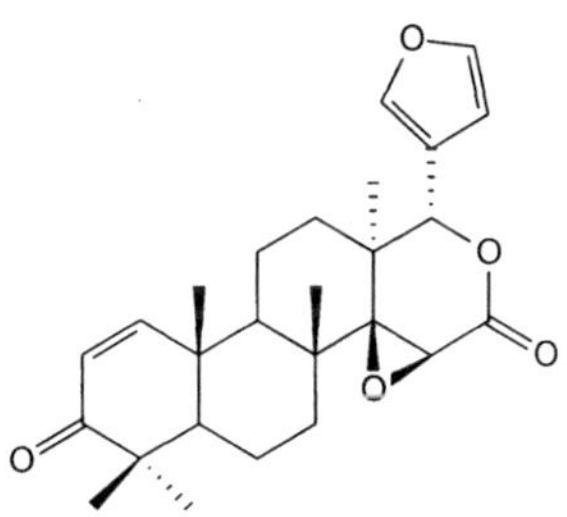

(49)

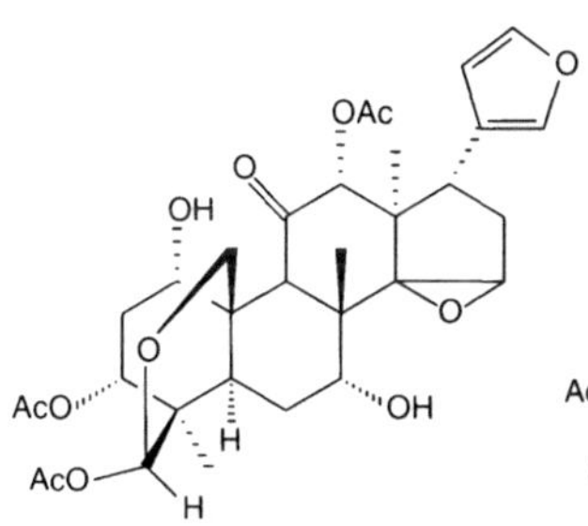

(52)

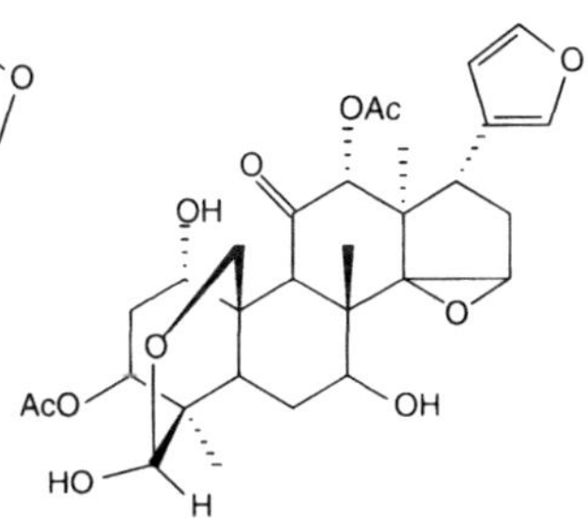

(53)

Fig. 15. Limonoids isolated from Mexican *Cedrela* species and from the exotic tree *Melia toosendan*.

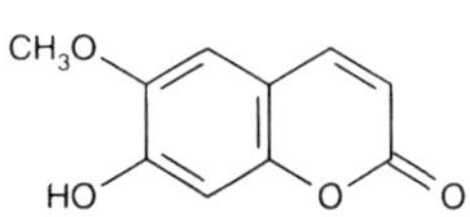

(54)

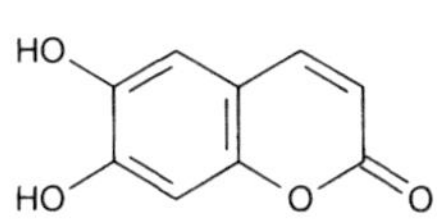

(55)

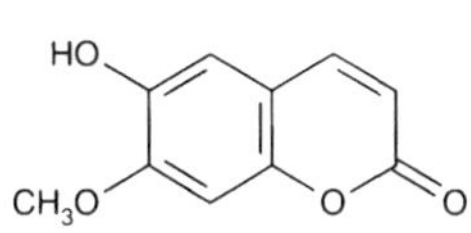

(56)

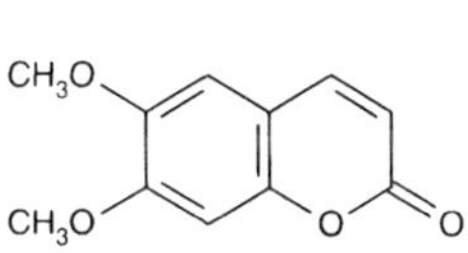

(57)

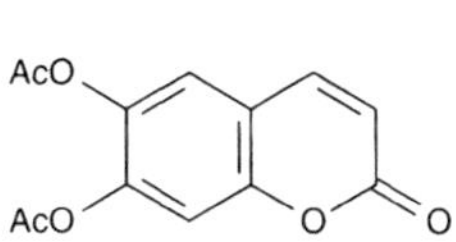

(58)

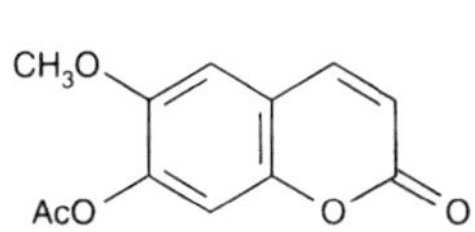

(59)

Fig. 16. Coumarins and derivatives from *Tagetes* species (Asteraceae).

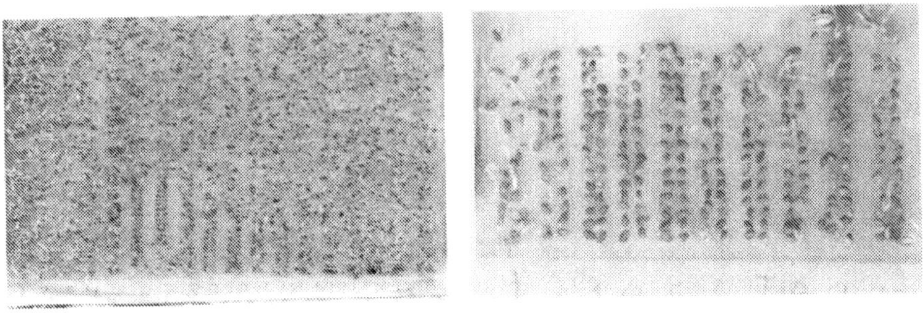

Fig. 17. Bioautographic bioassay on TLC of the effect of coumarins from *Tagetes lucida* on seed germination of *Trifolium pratense*. In the photograph on the left: column 6 fraction with compound **56**, column 7 fraction with compound **55**, column 8 fraction with compound **54**, column 9 fraction with compound **57**, the other columns correspond to other extracts and fractions. In the photograph on the right: column 1 MeOH extract, then compounds **54**, **55–59**, parthenolide **1**, and two fractions from MeOH extract with coumarins as main components. Photographs taken at 3 days of incubation, 28 °C, 80% relative humidity, without photoperiod light.

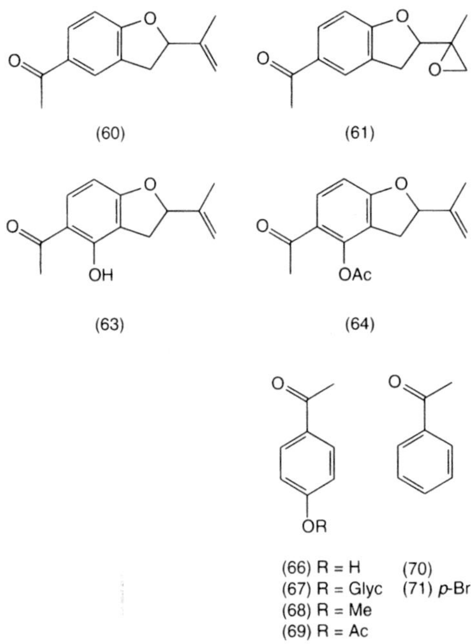

Fig. 18. Natural tremetones, acetophenones, and their derivatives from Chilean *Baccharis* species.

greater toxicity of terpenoid ketones in comparison to alcohols has been demonstrated in several cell systems (Reynolds, 1987; Fischer, 1991). Compounds capable of Michael addition, but lacking a hydroxyl group, or with hydroxyl groups in other positions are inactive. These compounds inhibit metabolism and cellular enzymatic activities, which may be related to hydrogen bonding requirements for appropriate

Table 10
Effect of acetophenones, tremetones, and their derivatives on seed germination[a]

Compound	GI_{50} values[b] (μM)			
	Monocot plants		Dicot plants	
	L. multiflorum	*L. sativa*	*P. ixocarpa*	*T. pratense*
60	37.5	45.0	31.8	35.5
61	65.6	36.1	19.2	22.3
62	31.5	25.9	24.7	27.9
63	25.5	21.0	16.6	17.8
64	79.9	85.5	55.2	40.0
65	10.4	11.5	8.7	7.1
66	14.4	45.3	15.0	28.5
67	c	c	c	c
68	18.1	51.2	23.5	125.9
69	75.3	41.0	c	233.2
70	127.5	47.0	c	137.3
71	c	c	c	c

[a]Means of three experiments.
[b]Concentration that inhibits 50% of seed germination.
[c]GI_{50} was nondetermined owing to lack of response.

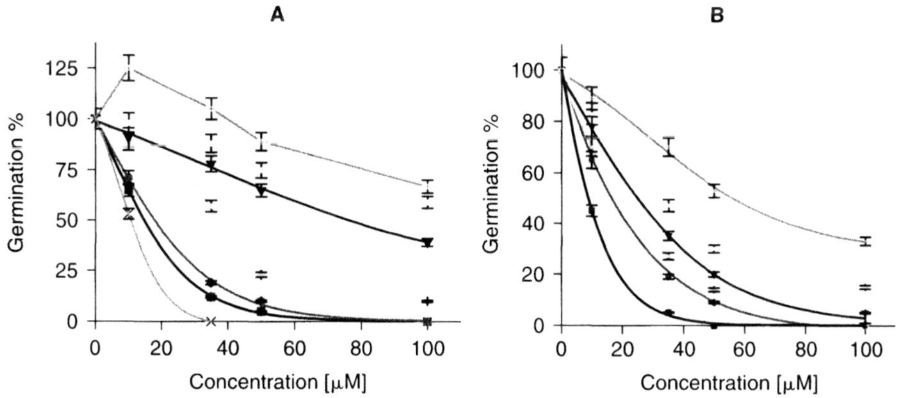

Fig. 19. Graph A: effects of compounds **65** (×;), **66** (■), **67** (), **68** (●), **69** (▼), **70** (▲), and **71** (♦) on *Lolium multiflorum* seed germination. Graph B: effects of **60** (♦), **61** (▲), **62** (▼), **63** (●), **64** (), and **65** (■) on *Physalis ixocarpa* seed germination.

fit into a receptor molecule. When a carbonyl group is adjacent to a hydroxyl group, as in the sesquiterpene lactones isolates by our group for instance.

The 2,4-D shows a pronounced effect on root and shoot length, as well as on seed germination. Total inhibition was obtained at 70 μM with ID_{50} values of 0.4 and 0.5 μM for root and shoot, respectively (data not shown). Anaya et al. (1995) previously reported similar findings.

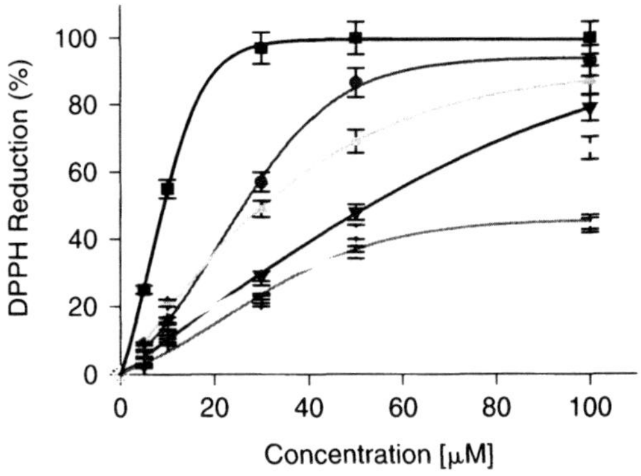

Fig. 20. DPPH reduction of THQ (■), BHA (●), **60** (), **61** (♦), **63** (▼), **65** (▲), and **66** (×;).

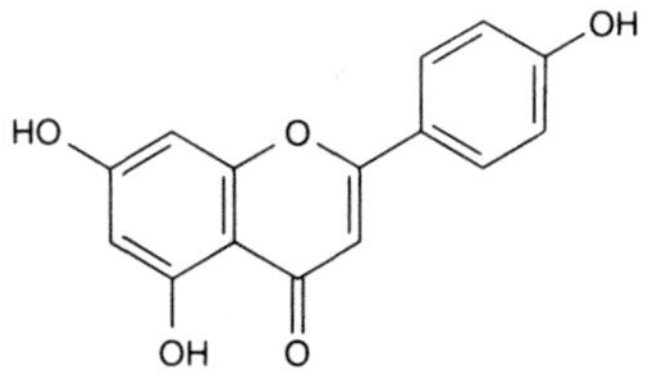

(72)

Fig. 21. Apigenin isolated from Mexican *Baccharis* species.

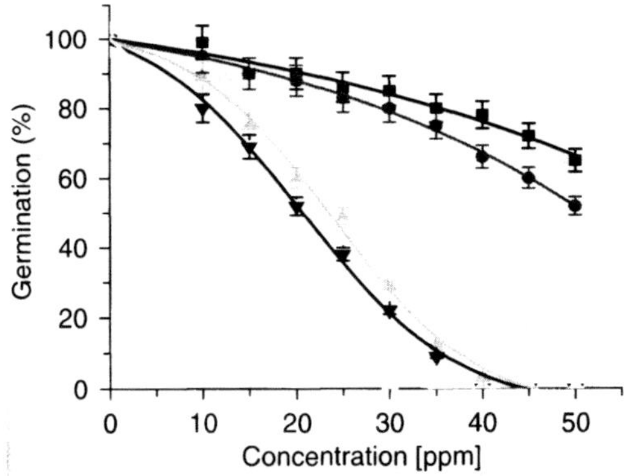

Fig. 22. Inhibitory activity of MeOH (■), CH_2Cl_2 (●) extracts, **5** (▲), **8** (♦), and **72** (▼) on seed germination of *Lactuca sativa*.

IV.B. Dry weight of monocotyledonous and dicotyledonous plants

The results indicate that seedling biomass (dry weight) diminishes with the increase in concentration of the compounds assayed in a similar way to that observed for the inhibition of germination. In general, monocots are more susceptible than dicots to decreases in dry weight. This is reflected in their I_{50} values (Tables 11 and 12). The

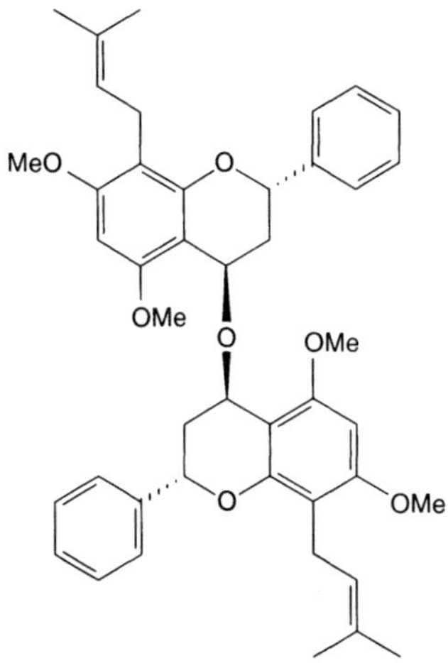

(73)

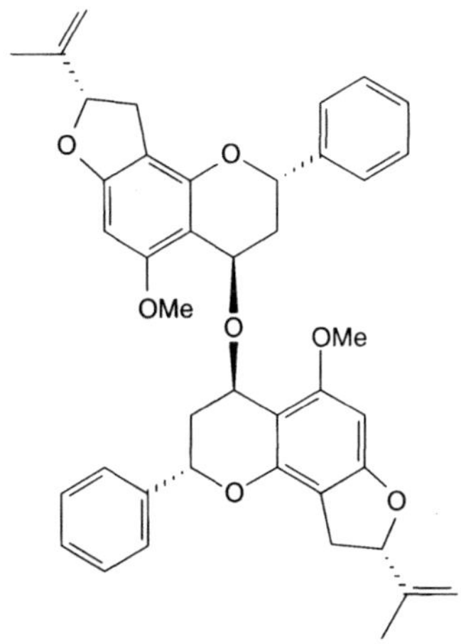

(74)

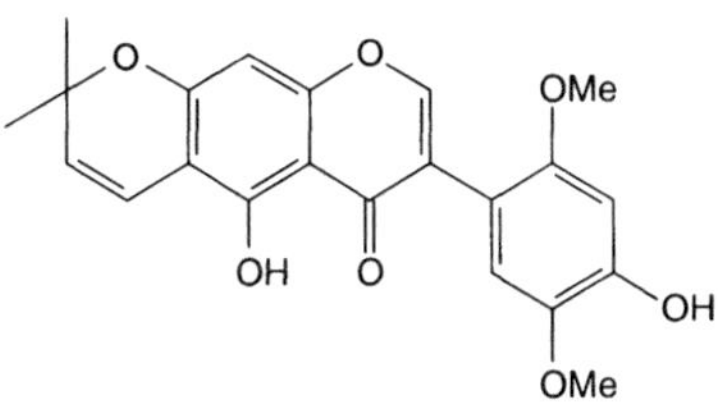

(75)

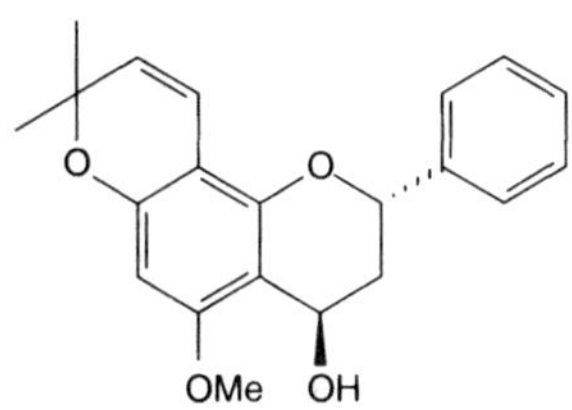

(76)

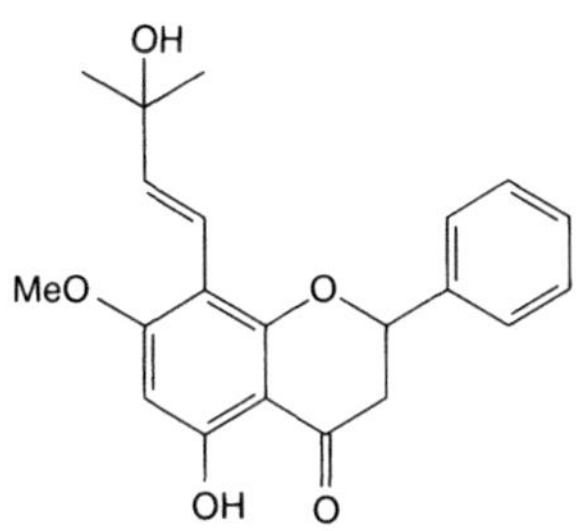

(77)

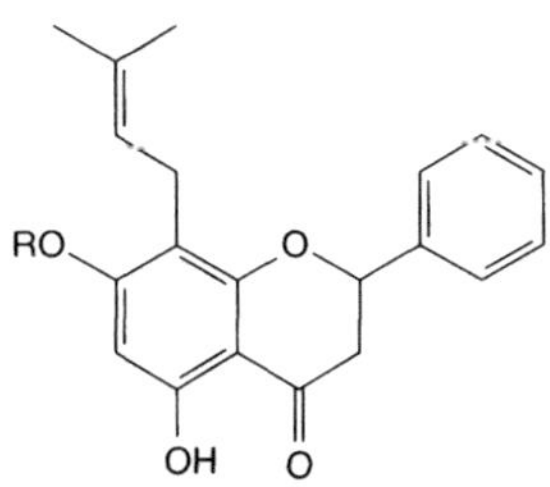

(78) R = H
(79) R = Me

Fig. 23. Flavonoids and biflavonoids from Mexican *Tephrosia* species.

growth enhancement activity of *T. alexandrinum* and *P. ixocarpa* seedling is not correlated with dry weight during seedling growth. The diameter of roots and hypocotyl decreased when tested compounds (Tables 11 and 12) were added.

IV.C. Seed respiration during seed germination

The respiratory rate of some seeds decreases with the concentration of the phytochemicals assayed in a concentration-dependent manner (Tables 8 and 10, and Figures 8, 9, and 16). The only exception is for compound **38** at 50 μM on

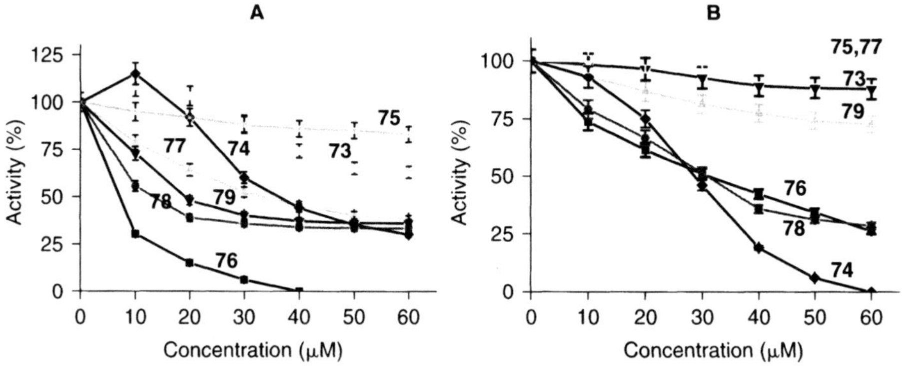

Fig. 24. Graph A: effects of compounds **73–79** on uncoupled electron transport from water to methylviologen. Control rate value was 1797 μequiv. $e^-/(h^{-1}$ mg Chl). Control = 100% of activity. Each point represents the mean of three determinations. Graph B: effects of the compounds **73–79** on phosphorylating electron transport, from water to methylviologen. Control rate value was 1190 μequiv. $e^-/(h^{-1}$ mg Chl). Control = 100% of activity. Each point represents the mean of three determinations.

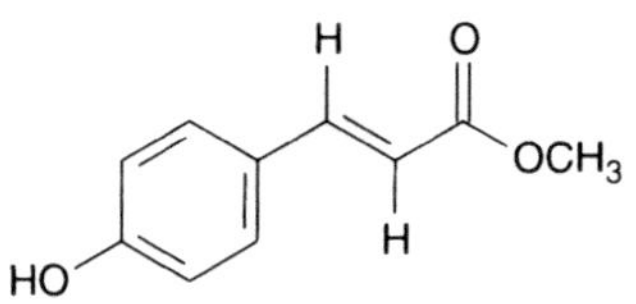

(80)

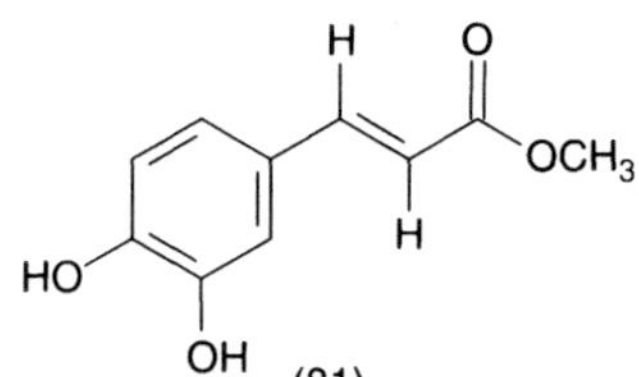

(81)

Fig. 25. Methyl esters of *p*-coumaric and caffeic acid, isolated from both *Lolium multiflorum and L. perenne* (cv. *Embassy*, *Gulf*, and *Nui*), infected with the endophytic fungus *Acremonium lolii*.

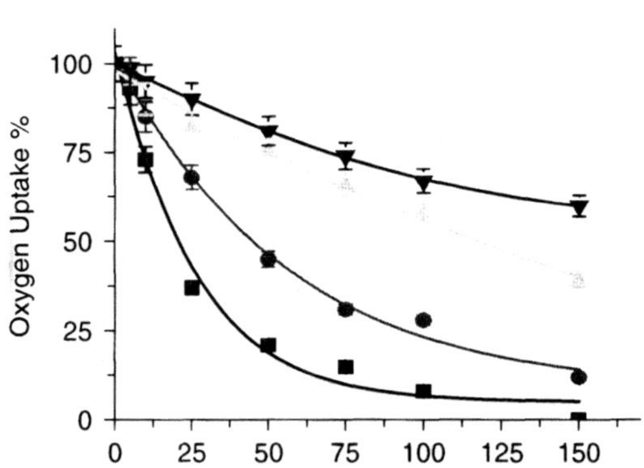

Fig. 26. Effects of solutions of compound **81** on seed respiration during germination of *P. ixocarpa* (■), *L. sativa* (●), *L. multiflorum* (▼), and *T. pratense* (▲) seeds, expressed as percent of control. Values taken at 72 h of imbibition. Means of three replicates.

Table 11
Effect of cedrelanolide on seed growth values (elongation, germination, respiration, and dry weight) (μM)[a]

	Elongation		Germination GI_{50}	Respiration[b] RI_{50}	Dry weight DWI_{50}
Species tested	Shoot	Root			
T. vulgare	297.0	–	157.0	100	155.0
L. multiflorum	142.0	114.0	125.0	100	351.0
T. alexandrinum	452.0	393.0	289.0	400[c]	–
P. ixocarpa	500.0	466.0	370.0	300	–

[a]Means of three experiments. Values are expressed as the dose (μM concentration/replicate) that gives 50% inhibition of seed germination, seedling length or dry weight.
[b]Value that gives 50% inhibition to the 24 h of imbibing.
[c]Value at 30 h.

L. multiflorum, where seed respiration enhancement was observed as the time of inhibition increases. However, at higher concentration (> 500 μM, for sesquiterpene lactones), respiration was also inhibited (Table 12). These results suggest that the pseudoguaianolides and germacranolides may act as phosphorylation uncoupling compounds at low concentration, but at high concentrations they either inhibit energy transduction or respiration redox enzymes. The RI_{50} values (concentration of phytochemicals that cause 50% seed respiration inhibition) for all compounds tested are shown in Table 12. Based on their RI_{50} values, *L. sativa* and *P. ixocarpa* seeds are the most sensitive to inhibition by all the compounds assayed. On the other hand, *L. multiflorum* and *T. vulgare* seeds showed the highest resistance to inhibition of respiration.

IV.D. Chloroplasts and mitochondrial inhibition

The effect of acetophenone, pseudoguaianolides, and other sesquiterpenes on mitochondrial respiration showed that the data are corroborated with the results obtained for seed respiration of *P. ixocarpa*, *T. pratense*, *T. alexandrinum*, and *L. sativa*. We observed an increase in oxygen uptake for concentrations of some compounds **67** and **69** as high as 70 and 9 μM, respectively, and for other compounds **32**, **40**, and, **64**, applications up to 70, 80, and 50 μM, respectively. As for seed respiration, compounds **29**, **37**, **39**, **55**, **65** were the most active. Our results showed that tremetones, sesquiterpene lactones; especially pseudoguaianolides are powerful mitochondrial respiration inhibitors, with concentrations similar to those reported by Moreland and Novitzky (1987).

The effects of tremetones, limonoids, sesquiterpene lactones, and pseudoguaianolides on different photosynthetic reactions were tested. The results showed that these compounds partially inhibit ATP synthesis, H^+-uptake, PSII, and electron transport in a concentration-dependent manner from 1 to 500 μm. These results suggest that the presence of an oxygenated function at certain strategic positions (hydroxyl, methoxy, and epoxy substituents) significantly enhances the inhibitory effects of these compounds as energetic inhibitors.

Table 12
Mean concentrations of different compounds, MeOH and CH_2Cl_2 extracts expressed as concentration that inhibit 50% of O_2-uptake as function of control of seeds respiration[a]

	I_{50} values[b] (μM)				
	Monocot		*Dicot plants*		
Compound	*L. multiflorum*	*T. vulgare*	*L. sativa*	*P. ixocarpa*	*T. pratense*
43	300	340	–[e]	[c]	[c]
44	320	340	–	[c]	[c]
Mixture **43** + **44**	125	90	–	154	[c]
41	5.9	6.7	–	10.9	22.0
42	65	79.0	–	125.0	150.0
20	32	–	[c]	45	[c]
22	51	–	42	25	33
25	93	–	41	30	34
26	120	–	39	35	39
45	100	100	–	300	400
66	41.0	–	43.0	52.9	53.0
68	40.0	–	55.9	54.6	57.6
61	67.9	–	58.0	57.5	49.1
65	39.0	–	27.0	28.7	22.0
MeOH B linearis[d]	119.0	–	[c]	125.0	[c]
CH_2Cl_2 ext. *B. linearis*[d]	25.0	–	15	21	19
MeOH ext. C. bipinnatus[d]	13.0	12.0	–	25.0	2.0
CH_2Cl_2 ext. *C. bipinnatus*[d]	12.0	11.0	–	21.0	20.0
MeOH ext. C. scabiosoides[d]	11.0	11.5	–	30.0	32.5
CH_2Cl_2 ext. C. scabiosoides[d]	10.0	9.5	–	25.0	27.5
MeOH ext. C. sulphureus[d]	11.7	10.9	–	22.0	21.0
CH_2Cl_2 ext. *C. sulphureus*[d]	9.8	8.5	–	10.0	8.9
MeOH ext. *Podanthus*[d]	119	–	[c]	125	[c]
CH_2Cl_2 ext. *Podanthus*[d]	25	–	15	21	19

[a]Means of three experiments.
[b]Each value corresponds to the concentration that inhibits 50% of seed respiration during germination. Values at 72 h.
[c]I_{50} was not determined owing to lack of seed respiration response over a 24 h period.
[d]Values in ppm
[e]Not determined.

In conclusion, our data indicate that pseudoguaianolides are more selective and potent toward dicots than toward monocots. Respiration processes are likely to be involved in the interference action, as this process was inhibited in a parallel way by mixtures and compounds assayed. Because germination is inhibited by lower doses than respiration, it is possible that these mixtures and compounds have more than one interference target. When compounds are acetylated, the potency of inhibition decreases 10 times. Thus, we observed that derivatives of natural products are also active *in vivo*.

The treatment concentrations for all the compounds that reduced seedling growth were low (25–100 μM) compared to the values of allelopathic chemicals that have been previously studied under laboratory conditions (Einhellig, 1986). Reported values for potency of secondary metabolites toward growth reduction (i.e., phenolic acids) is in the range from 100 to 1000 μM, although sorgoleone (10 μM) and juglone (micromolar level), which suppress the growth of several herbaceous species, are much more powerful (Rietveld, 1983; Einhellig and Souza, 1992).

Although the mechanism of action of epimeric photogedunin and its derivatives remains unknown, these nortriterpenes have proved to be good inhibitors of plant growth. They show pre-emergent phytotoxic properties by inhibiting both germination and growth. They also show some degree of selectivity by inhibiting monocotyledonous species more drastically than dicotyledonous species. Nevertheless, these compounds may also induce postemergent herbicidal effects, as they were found to act as Hill-reaction inhibitors (Céspedes et al., 1998).

IV.E. Radical scavenging properties and antioxidant activity

Radical scavenging properties of the compounds were evaluated against the DPPH radical, using DPPH as a thin-layer chromatography (TLC) spray reagent. These compounds (50 μM) appeared as yellow spots against a purple background. In some cases, the same amount of other compounds did not react with the radical. Compounds were also tested against DPPH in a spectrophotometric assay. This method confirms TLC observations regarding which of the active compounds exhibited the strongest radical-scavenging activity in this assay. Quercetin, a flavonol with five hydroxyl groups, was used as a reference compound that possesses strong antioxidant properties; THQ and BHA, commonly useful as antioxidant in foods, were also employed in spectrophotometric assays. On all instances, the same effects were observed (Figures 3, 9, 12, and 20).

The antioxidant activity of these compounds was also evaluated fluorimetrically by measurements of bleaching of the H_2O-soluble crocin (Bors et al., 1992). Alkoxyl radicals were generated from *t*-BuOOH by UV photolysis of aqueous solutions containing 10 μM crocin and 1 mM *t*-BuOOH. *t*-BuOH (0.5 M) was added to scavenge the HO^- radicals produced. Gallic acid was used as a reference compound. Many compounds that were active had activity comparable to that of gallic acid (Figure 9).

IV.F. Concluding remarks

Comparison of the inhibitory activities of terpenes (diterpenes and limonoids) with benzofurans, sesquiterpenes, and sesquiterpene lactones (pseudoguaianolides) shows that the latter compounds were the most effective inhibitors of the respiration pathway (Table 12). The necessary concentration to achieve complete inhibition of these physiological process is ~10-fold greater in terpenes than the concentration of the second group required for the same effect. We think that α-methylene-γ-lactones and an α, β-unsaturated carbonyl function are necessary for activity. Those moieties seem to be important structural requirements for the observed inhibitory activities. The high inhibitory potential of compounds **1**, **3**, **17**, **20**, **26–28**, **35**, **39**, **54–56**, **63**, **65**, **73**, and **79** is probably because of the presence of those moieties.

The presence of these substituents was shown to be responsible for an increase in the ability to inhibit NO production (Matsuda et al., 2000). The interaction between the natural compounds isolated by our group (pseudoguaianolides and germacranolides, for instance) with the generation of ROS (reactive oxygen species) or RNS (reactive nitrogen species) could explain the inhibitory activities observed (the sites and mechanisms of action are being studied).

Considering that those compounds are the major constituents isolated from the aerial parts of plants and the low concentration needed to induce the total inhibition of seed respiration, it is possible to postulate the importance of these allelochemicals in allelopathic studies.

We emphasize that many of these compounds could not be detected in the ethyl acetate extracts. In contrast, lipophylic compounds are abundant in the CH_2Cl_2 extracts and in lower amounts in MeOH extract. More importantly, these compounds seem to be accumulated in cells and are exuded slowly (through trichomes, for instance). For this reason, it may be logical to assume that these secondary metabolites are synthesized for defensive process. It appears that the release mechanism of these compounds and their derivatives may be one of the key processes for understanding plant defenses.

Macías' group investigated inhibitory effects of some guaianolides. They reported that these natural products show enhancement and inhibition effects on different pre-emergence properties (e.g., germination and root length) at low and higher concentrations, respectively. Similar findings were observed by Fischer et al. (1989, 1990). However, the effects of other guaianolides and pseudoguaianolides on seed germination and root length were unclear. The presence of **1** and **9** is unique among the natural products reported to date from *Parthenium* species (Rodríguez et al., 1976; Fraga, 2000).

The effects of our isolated sesquiterpene lactones have certain similarities with zaluzanin, costunolide, parthenolide, and their derivatives (Macías et al., 1999b, 2000a).

In addition to the inhibitory activity of bleaching of crocin induced by alkoxyl radicals, our compounds and extracts also demonstrated scavenging properties toward DPPH in TLC autographic and spectrophotometric assays. It was possible to correlate antioxidant activity with the seedling growth inhibitory activity.

The levels of radicle inhibition obtained with some compounds on monocots (*L. multiflorum*, *T. vulgare*) and dicots (*P. ixocarpa*, *L. sativa*, and *T. pratense*) are totally comparable to those of zaluzanin C, a known natural growth inhibitor.

V. Material and methods

V.A. Plant material

Aerial parts (stem, leaf, and flowers) from all plants were collected in the field in Mexico and Chile. Voucher specimens can be found in the ethnobotanical collections of the Herbarium MEXU and CONC, in Biology Institute, UNAM, México D. F., México and Botany Department, Faculty of Natural Sciences, Universidad de Concepción, Concepción, Chile, respectively.

V.B. Chemicals and solvents

All reagents used were either analytical reagent grade or chromatographic grade. *m*-chloroperbenzoic acid (MCPBA), $NaBH_4$, pyridine, 2,4-D, 1,1′-dimethyl-4,4′-bipyridinium dichloride (methyl viologen), DPPH, THQ, BHA (2[3]-*t*-butyl-4-hydroxyanisole; 2[3]-*t*-butylhydroquinone monomethyl ether), ethylenediaminetetraacetic acid (EDTA), bovine serum albumin, Percoll, 4′-methoxyacetophenone, 4′-bromoacetophenone, acetophenone, gallic acid, quercetin, saffron, sorbitol, tricine, and trizmahydrochloride were purchased from Sigma-Aldrich Química, S.A. de C.V., Toluca, Mexico. Methanol, CH_2Cl_2, $CHCl_3$, NaCl, KCl, NaOH, KOH, *t*-butanol, *t*-butyl hydroperoxide, $CuSO_4$, NH_4Cl, $MgCl_2$, acetic anhydride, silica gel GF_{254} analytical chromatoplates, silica gel grade 60, (70–230, 60 Å) for column chromatography, *n*-hexane, and ethyl acetate were purchased from Merck-Mexico, S.A., Mexico. Pyridine and acetic anhydride were distilled prior to use.

V.C. Apparatus

^{1}H-NMR spectra were recorded at 300 MHz and ^{13}C-NMR spectra at 75 MHz, respectively, on Varian VXR-300S and VXR-500S spectrometers. Chemical shifts (ppm) are related to $(CH_3)_4Si$ as internal reference. $CDCl_3$ and acetone-d_6, from Aldrich Chemical Co., were used as solvents. Coupling constants are quoted in hertz. IR spectra were obtained in KBr or $CHCl_3$ on Perkin Elmer 283-B and FT-IR Nicolet Magna 750 spectrophotometers. A Spectronic model Genesys 5 spectrophotometer was used for biological activities. Optical rotation was measured on a JASCO DIP-360 spectropolarimeter. Melting points were obtained on a Fisher–Johns hot-plate apparatus and remain uncorrected. Oxygen evolution (uptake) was determined with a Clark-type electrode connected to Yellow Spring Instrument (YSI) Oxygraph Model 5300. Fluorimetric measurements were determined with TURNER Barnstead-Thermolyne, model Quantech S5 Fluorometer, with 420, 440, 470, 550, and 650 Turner filters.

V.D. General experimental procedures

HPLC was performed on a WATERS® Model 600E, equipped with μBondapack RP-18 column, 250 × 8 mm, flow rate 1.5 ml/min, UV detector 280 nm, mobile phase $MeOH/H_20$ 7:3 v/v. Analytical TLC was performed on Silica gel 60 F_{254} Merck plates and the spots were visualized by spraying with a 10% solution of H_2SO_4, followed by heating at 110 °C for 3 min.

V.E. Bioactivity-guided isolation and purification of acetophenones, tremetones, and their derivatives

Plant materials of all species were dried, milled, and extracted with MeOH. Furthermore, these extracts were partitioned between CH_2Cl_2 and $MeOH/H_2O$ (1:1 v/v). Evaporation of the solvents in a rotatory evaporator afforded the crude extracts, which were weighed, in order to calculate the yield.

The respective MeOH, CH_2Cl_2, and ethyl acetate extracts of milled aerial parts of all species were assayed for initial phytotoxic growth effects (roots and hypocotyl development) with *L. multiflorum*, *T. aestivum*, *T. vulgare*, *R. sativus*, *T. angustifolium*, *T. pratense*, *L. sativa*, and *P. ixocarpa* seeds and seedlings, using the Petri dish bioassay (Céspedes et al., 2002).

The most active extract of each of the species was tested for phytogrowth activity and then submitted to column chromatography using SiO_2 (G 60, Merck) as solid phase. Elution of each of the active extracts with *n*-hexane:ethyl acetate mixtures and ethyl acetate followed by methanol afforded between 20 and 100 fractions that, after evaporation of the solvent, were analyzed by TLC and bioautographic assay (Céspedes et al., 2000a) using different solvent systems (*n*-hexane:ethyl acetate and CH_2Cl_2:MeOH mixtures).

Repeated TLC of the active fractions led to the isolation of the secondary metabolites, which were further purified by prep-TLC. Compounds were collected and identified by TLC with authentic samples. The compounds were obtained as pure natural products, which were analyzed and characterized by their R_f values, IR, UV, ^{1}H-NMR, and ^{13}C-NMR data.

The identification of all compounds was made by these spectroscopic methods and direct comparison with authentic samples. All compounds were purified in sufficient amount, which were used for both bioassays and derivatization. Both chromatographic and spectral data were available from previous studies

V.F. Derivatization of compounds

Several compounds were characterized as the dihydro derivative or as the acetate and epoxy derivatives, respectively. The acetylation, epoxidation, and reduction of compounds were done as previously reported (Céspedes et al., 1999b, 2001c, 2002).

V.G. Post- and pre-emergence activities

V.G.1. Seeds germination bioassays

L. sativa, *L. multiflorum*, *T. pratense*, and *P. ixocarpa* were purchased from "SEMILLAS-COBO" S. A. de C. V., Mexico D.F., México. For these experiments, 25 seeds of *L. multiflorum*, *T. pratense*, and *P. ixocarpa* were placed in a Petri dish; however, 50 seeds of *L. sativa* were required for the assays. The number of seeds used for each experiment was selected so that an appreciable change in O_2-uptake could be detected by the oxygraph. Seeds were placed on filter paper (Whatman No. 1) in Petri dishes (85 mm diameter). In three replicate experiments, the paper was wet with 8 or 10 ml deionized water or test solution (MeOH < 1%). The dishes were wrapped with parafilm and incubated at 28 °C in the dark at intervals of 48 h. Seeds were considered to have germinated when there was a 1 mm extrusion of the radicle. Each germination assay was replicated thrice. Control seed dishes contained the same number of seeds, volume of water and methanol as the test solutions. Seeds were selected for uniformity of size; damaged seeds were discarded (Céspedes et al., 1999b).

V.G.2. Growth bioassays
Coleoptile, hypocotyl, and root lengths for all germinated seeds were measured after 120 h, in three replicates following the above design, after which the germinated seeds were dried to constant weight at 40 °C (Céspedes et al., 2001a).

V.G.3. Seed respiration
Seed respiration was measured polarographically at 25°C with a Clark-type electrode as oxygen uptake during the germination process using an YSI Oxygraph Model 5300. Oxygen uptake, in the presence of different concentrations of tested compounds, was evaluated over 5 and 10 min in a nonilluminated cell. The requirement for oxygen was plotted as percentage, taking the control as 100% (Céspedes et al., 1999a).

V.G.4. Preparation of thylakoid membranes
Isolated chloroplasts were prepared from spinach leaves (*Spinacia oleracea*) as previously reported (Céspedes et al., 1998) and the pellet was resuspended. These pellets were homogenized in ice-cold extraction buffer (330 mM sorbitol, 50 mM tricine–NaOH, 10 mM $MgCl_2$, 5 mM Na–ascorbate, 2 mM EDTA, and 0.05% bovine serum albumin, pH 7.8) and filtered through a layer of Miracloth and four layers of cheesecloth. The filtrate was centrifuged at 1100 *g* for 5 min at 4°C. The chloroplast pellet was resuspended in ice-cold lysing buffer (25 mM tricine–NaOH, 10 mM $MgCl_2$, pH 7.8) and centrifuged at 3000 *g* for 5 min. The pellet was resuspended, unless indicated, in a buffer (100 mM sorbitol, 25 mM tricine–NaOH, 10 mM $MgCl_2$, 10 mM NaCl, pH 7.8). For the oxygen evolution experiments, the thylakoid membranes were diluted to 4 mg of chlorophyll/ml according to Arnon and as modified by Hiscox and Israelstam (1979) (Rimando et al., 1998).

V.G.5. Measurement of proton uptake and ATP synthesis
Proton uptake was measured as the pH rose from 8.0 to 8.1 (Céspedes et al., 2002) with a combination microelectrode connected to a potentiometer (Model 225 Research pH/ion-meter, Denver Instrument Company, Arvada Colorado, USA) with expanded scale. The reaction medium was 100 mM sorbitol, 5 mM $MgCl_2$, 10 mM KCl, and 1 mM Na^+-tricine pH 8. ATP synthesis was measured titrimetrically according to the procedure of Dilley (1972). 50 μM MV was added as an electron acceptor for the Hill reaction.

V.G.6. Measurement of electron transport
Photosynthetic noncyclic electron transport rates from water to MV were monitored with an YSI Model 5300 oxygen monitor connected to a Clark-type electrode. The reaction medium was the same as that used in the H^+-uptake assay except for the tricine concentration (15 mM) and, in the case of the uncoupled electron transport measurement, 6 mM NH_4Cl was added. All reaction mixtures were illuminated with light from a projector lamp (GAF 2660) filtered through 5 cm of 1% $CuSO_4$ solution at 20 °C (Mills et al., 1980; Céspedes et al., 2002).

V.G.7. Determination of mitochondrial oxygen consumption

The procedure for the isolation of bean root mitochondria is based on previously described protocols (Winning et al., 1995) with minor modifications, as follows. Approximately 700 g of *Phaseolus vulgaris* seeds were sterilized for 30 min in a 10% (v/v) solution of sodium hypochlorite and then washed with sterile water successively. Seeds were growing in a sterile sphagnum Peat-Moss mixed with vermiculite (1:1) (purchased from Hummert de México, S.A. de C.V., Cuernavaca, México) and grown in complete darkness for 72 h at 28°C. Roots were obtained by cutting them from the coleoptile with scissors. All subsequent procedures were carried out at 4 °C, as rapidly as possible. The chilled tissue was ground in a mortar and pestle with silica gel 60 (0.063–0.200 mm) in two volumes of grinding buffer (0.4 M sorbitol, $MgCl_2$ 5 mM, KCl 10 mM, Tricine 30 mM, pH 8.0 with KOH) and filtered through Miracloth into 250 ml centrifuge bottles. The filtrate was then centrifuged at 3000 *g* for 5 min in a Sigma-B. Braun Model 2-15 rotor, and then the supernatant was removed, and the mitochondria was pelleted by centrifugation at 10,000 *g* for 15 min. One to 5 ml of wash buffer (0.4 M sorbitol, $MgCl_2$ 5 mM, KCl 10 mM, tricine 30 mM, pH 7.8 with KOH) was added per tube and the pellet was resuspended using a soft paintbrush. The suspension was placed in 50 ml tubes and centrifuged at 3000 *g* for 5 min in a Sigma-B. Braun Model 2-15 rotor. The supernatant was then carefully transferred to another 50 ml tube and centrifuged at 10,000 *g* for 15 min to repellet the mitochondria, which were then resuspended in 1 to 2 ml of resuspension buffer (0.4 M sorbitol, $MgCl_2$ 5 mM, KCl 10 mM, tricine 30 mM, pH 7.2 with 5 M KOH). To purify the mitochondria from residual contaminating plastids, the crude mitochondrial suspension was loaded onto 26% (v/v) Percoll (SIGMA) in resuspension buffer (ca. 40 ml) in a 50 ml polycarbonate tube and a density gradient was generated by centrifugation at 40,000 *g* for 90 min in a Sigma-B. Braun Model 2–15 rotor. A buff-colored band of mitochondria was visible below a band containing the plastids. The upper layer was removed by aspiration, and the mitochondrial band was recovered (ca. 1–2 ml) and diluted with 20 ml resuspension buffer. Mitochondria were then recovered by centrifugation at 12,000 *g* for 15 min in a Sigma-B. Braun Model 2-15 rotor. The resulting pellet was very loose and so the supernatant had to be removed by aspiration with care.

This wash procedure was repeated and the final pellet was resuspended in 0.4–1 ml of resuspension buffer and stored on ice. Total mitochondrial protein concentration was determined using the Bradford procedure (Bradford, 1976), adjusted to 0.3–0.5 mg for each experiment and the freshly prepared mitochondria were used directly to measure inhibition. The O_2-uptake of mitochondria was monitored with an YSI Model 5300 oxygen monitor connected to a Clark-type electrode.

The integrity of mitochondria was verified using HCN as respiratory chain inhibitor. All compounds were dissolved in 0.5% methanol. Respiration control and ADP/O ratios were calculated for every mitochondrial isolation (Estabrook, 1967). A constant state and three respiratory rates (Chance and Williams, 1955) with 10 μM succinate were obtained before measuring with our compounds. The ADP/O ratio for succinate was 1.5.

V.G.8. Reduction of 2,2′-diphenyl-1-picrylhydrazyl [2,2′-diphenyl-1-(2,4,6-trinitrophenyl) hydrazyl; DPPH] Radical

TLC autographic assay: After developing and drying, TLC plates were sprayed with 0.2% DPPH solution in MeOH. The plates were examined 30 min after spraying. Active compounds appear as yellow spots against a purple background. In similar form, TLC plates were sprayed with 0.05% β-carotene solution in $CHCl_3$. The plates were examined under UV_{254} light until the background became discolored (bleached). Active compounds appear as pale yellow spots against a white background. *Spectrophotometric assay* (Bors, 1992; Cuendet et al., 1997): 50 μl of a solution containing the compound to be tested were added to 5 ml of a 0.004% MeOH solution of DPPH, quercetin was used as internal standard. Absorbance at 517 nm was measured after 30 min, and the percent of activity was calculated.

V.G.9. Bleaching of crocin

Crocin was isolated from commercial saffron (SIGMA) by extraction with MeOH followed by HPLC (RP-18, MeOH/H_2O 1:1) and identified by its ^1H- and ^{13}C-NMR data. The bleaching assay was carried out according to Bors (1992). An aqueous solution containing 10 μM crocin, 1 mM *t*-BuOOH, 0.5 M *t*-BuOH, and various dilutions of the compounds to be tested were prepared. The solutions were placed under UV_{254} light. Following the decrease of absorbance, bleaching of crocin and fluorescence emission at 440 and 470 nm for every 5 min was monitored.

V.G.10. Statistical analysis

Data shown in figures and tables are the mean results obtained. Means were for three replicates and independent seeds, crocin and DPPH preparations, and are presented as the mean ± standard errors of the mean (SEM). Data were subjected to analysis of variance (ANOVA) with significant differences between means identified by GLM procedures. The results are given in the text as probability values, with $p<0.05$ adopted as the criterion of significance. Differences between treatment means were established with a Student–Newman–Keuls (SNK) test. The GI_{50}, RI_{50}, and I_{50} values for each activity were calculated by "Probit analysis" on the basis of the percentage of inhibition obtained at each concentration of the samples. I_{50} is the concentration producing 50% inhibition. Complete statistical analysis was performed by means of the MicroCal Origin 5.1 and 6.1 statistical and graphs PC programs.

Acknowledgments

This chapter was supported in part by Grants: DGAPA-PAPIIT-UNAM IN243802, IN211105-3, and internal grant from Instituto de Química, UNAM, México. The authors acknowledge the great help given by Prof. David S. Seigler, Plant Biology Department, University of Illinois at Urbana-Champaign, in the correction of the Manuscript.

References

Achnine L, Mata R, Iglesias-Prieto R, Lotina-Hennsen B. (1998) Impairment of photosystem II donor side by natural product odoratol. J Agric Food Chem 46:5313–7.

Achnine L, Mata R, Lotina-Hennsen B. (1999) Interference of the natural product 7-oxo-7-deacetoxygedunin with Cfo of H + -ATPase of spinach chloroplasts. Pest Biochem Physiol 63:139–49.

Alarcón J, Becerra J, Silva M, Morgenstren T, Jakupovic J. (1995) β-Agarofurans from seeds of *Maytenus boaria*. Phytochemistry 40:1457–60.

Albertin H, Nurit F, Ravanel P, Tissut M. (1994) Uncoupling activities of monesin in isolated mitochondria, chloroplasts and cells. Phytochemistry 35(5):1105–10.

Anaya AL, Hernández B, Pelayo H, Calera M, Fernández E. (1995) Allelopathy in Mexican plants. In: Inderjit, Dakshini KMM, Einhellig FA, editors. Allelopathy organisms, processes and applications. ACS Symposium Series 582, pp. 224–241.

Ávila JG, Martínez G, Muñoz JL. (1999) In vitro antimicrobial activity of various plant extracts used by *Purepecha* against some enterobacteriaceae. Int J Pharmacol 31(1):61–4.

Bors W, Saran M, Eltsner EF. (1992) Modern methods plant analysis. New Series. NY: Academic Press, Vol. 13, p. 277.

Bradford MM. (1976) A rapid and sensitive method for the quantities of protein utilizing the principle of protein–dye binding. Anal Biochem 72:248–54.

Calderón J, Céspedes CL, Rosas R, Gómez-Garibay F, Salazar JR, Lina L, Aranda E, Kubo I. (2001) Acetylcholinesterase and insect growth inhibitory activities of *Gutierrezia microcephala* on fall armyworm *Spodoptera frugiperda*. Zeit Naturforsch C: Biosci 56c:382–94.

Castañeda P, Gómez L, Mata R, Lotina-Hennsen B, Anaya AL, Bye R. (1996) Phytogrowth-inhibitory and antifungal constituents of *Helianthella quinquenervis*. J Nat Prod 59:323–6.

Céspedes CL, Calderón JS, King-Diaz B, Lotina-Hennsen B. (1998) Phytochemical and biochemical characterization of epimeric photogedunin derivatives. Their different sites of interaction on the redox electron transport carrier of *Spinacea oleracea* L. J Agric Food Chem 46:2810–6.

Céspedes CL, Alvarado G, Ramos E, Salazar JR, Gómez-Garibay F, Calderón JS. (2000a) Allelopathic activity of endemic *Cosmos* spp. (Asteraceae) from Mexico. TIP Rev Especial Ciencias Químico-Biol 3(2):69–74.

Céspedes CL, Achnine L, Alarcón J, Becerra J, Lotina-Hennsen B. (2000b) Photosynthetic inhibitory activity of dihydro-*β*-agarofurans sesquiterpene from *Maytenus disticha* and *Maytenus boaria* (Celastraceae). Zeit Naturforsch C: Biosci 55c:631–7.

Céspedes CL, Calderón JS, Salazar JR, King-Diaz B, Lotina-Hennsen B. (1999a) Allelopathic activity of photogedunins from *Cedrela ciliolata* (Meliaceae). Bol Soc Chilena Química 44:173–83.

Céspedes CL, Calderón JS, Gómez-Garibay F, Segura R, King-Díaz B, Lotina-Hennsen B. (1999b) Phytogrowth properties of limonoids Isolated from *Cedrela ciliolata*. J Chem Ecol 25:2665–76.

Céspedes CL, Calderón JS, Salazar JR, Lotina-Hennsen B, Segura R. (2001a) Plant-growth inhibitory activity of cedrelanolide from *Cedrela salvadorensis*. J Chem Ecol 27:137–49.

Céspedes CL, Achnine L, Lotina-Hennsen B, Salazar JR, Gómez-Garibay F, Calderón JS. (2001b) Inhibition of photophosphorylation and electron transport by flavonoids and biflavonoids from endemic *Tephrosia* spp. of Mexico. Pest Biochem Physiol 69:63–76.

Céspedes CL, Hoeneisen M, Bittner M, Becerra J, Silva M. (2001c) Comparative study of ovatifolin antioxidant and growth inhibition activities. J Agric Food Chem 49:4243–51.

Céspedes CL, Calderon JS, Lina L, Aranda E. (2000c) Growth inhibitory effects on fall army worm *Spodoptera frugiperda* of some limonoids isolated from *Cedrela* spp. (Meliaceae). J Agric Food Chem 48:1903–8.

Céspedes CL, Torres P, Marín JC, Becerra J, Bittner M, Hoeneisen M, Silva M, Lina L, Aranda E. (2006a) Insect growth regulatory activity of phenolic diterpenes from *Podocarpus saligna* D. Don. J Agric Food Chem (submitted).

Céspedes CL, Marin JC, Hoeneisen M, Bittner M, Becerra J, Silva M. (2006b) Mitochondrial, antioxidant and plant growth inhibition activities of ovatifolin. J Agric Food Chem (submitted).

Céspedes CL, Salazar JR, Esquivel B, Perich F, Pardo F. (2006c) Allelopathic activity of ryegrass extracts and compounds in pastures of Southern Chile. J Agric Food Chem (submitted).

Céspedes CL, Uchoa A, Calderón JS. (2005) Plant-growth regulatory properties of natural sesquiterpene lactones from *Parthenium* weedy species and analogous derivatives. A comparative study. Int Biopest (accepted).

Céspedes CL, Uchoa A, Salazar JR, Perich F, Pardo F. (2002) Plant growth inhibitory activity of *p*-hydroxyacetophenones and tremetones from Chilean endemic *Baccharis* species and some analogous: a comparative study. J Agric Food Chem 50:2283–92.

Chance B, Williams GR. (1955) Respiratory enzyme in oxidative phosphorylation III. The steady state. J Biol Chem 217:409–28.

Creuzet S, Ravanel P, Tissut M, Kaouadji M. (1988) Uncoupling properties of three flavonols from plane-tree buds. Phytochemistry 27(10):3093–9.

Cuendet M, Hostettmann K, Potterat O, Dyatmiko W. (1997) Iridoid glucosides with free radical scavenging properties from *Fagraea blumei*. Helv Chim Acta 80:1144–52.

Dayan FE, Hernández A, Allen SN, Moraes RM, Vroman JA, Avery MA, Duke SO. (1999) Comparative phytotoxicity of artemisinin and several sesquiterpene analogues. Phytochemistry 50:607–14.

Dilley RA. (1972) Ion transport (H^+, K^+, Mg_2^-) exchange phenomena. In: San Pietro A, editor. Methods in enzymology. Vol. 24, Section I, "Methodology". New York: Academic Press, pp. 68–74.

Domínguez M. (2002) Comparación de la actividad herbicida de dos especies de la familia Asteraceae de Mexico: *Baccharis salicifolia* y *Baccharis conferta*. Thesis of B.Sc, Chemistry Faculty, UNAM, Mexico D.F., Mexico

Domínguez M, Marín JC, Céspedes CL. (2004) Plant growth regulatory activity of Mexican Asteraceae: *Baccharis salicifolia* and *B. conferta*. J Chem Ecol (submitted).

Duke SO, Lyndon J. (1993) Natural phytotoxins as herbicides. Pest control with enhanced environmental safety, Chapter 9, ACS Symposium Series, pp. 110–124.

Echeverri F, Torres F, Quiñones W, Cardona G, Archbold R, Roldan J, Brito L, Luis J, Lahlou EH. (1997) Danielone, a phytoalexin from papaya fruit. Phytochemistry 44:255–6.

Einhellig FA. (1986) Mechanisms and modes of action of allelochemicals. In: Putnam RA, Tang Ch-Sh editors. The science of allelopathy. New York: Wiley, pp. 171–88.

Einhellig FA. (1995) Mechanism of action of allelochemicals in Allelopathy. In: Inderjit, Dakshini KMM, Einhellig FA, editors. Allelopathy organisms, processes, and applications, ACS Symposium Series 582, pp. 96–116.

Einhellig FA, Souza IF. (1992) Phytotoxicity of sorgoleone found in grain *Sorghum* root exudates. J Chem Ecol 18:1–11.

Estabrook RW. (1967) Mitochondrial respiratory control and the polarographic measurements of ADP:O ratios. In: San Pietro A editor. Methods in enzymology, Vol. 10. New York: Academic Press, pp. 41–7.

Fischer NH. (1986) The function of mono and sesquiterpenes as plant germination and growth regulators. In: Putnam AR, Tang C-S editors. The science of allelopathy. New York: Wiley-Interscience, pp. 203–18.

Fischer NH. (1991) Plant terpenoids as allelopathic agents. In: Harborne JB, Tómas-Barberán FA editors. Ecological chemistry and biochemistry of plant terpenoids. Oxford: Oxford Science Publications, pp. 377–98.

Fischer NH, Weidenhamer JD, Bradow JM. (1989) Inhibition and promotion of germination by several sesquiterpenes. J Chem Ecol 15:1785–93.

Fischer NH, Weidenhamer JD, Riopel JL, Quijano L, Menelaou MA. (1990) Stimulation of witchweed germination by sesquiterpene lactones: a structure–activity study. Phytochemistry 29:1479–83.

Fraga BM. (2000) Natural sesquiterpenoids. Nat Prod Rep 17:483–504.

Harborne JB. (1988) Introduction to ecological biochemistry, 3rd edition. London: Academic Press.
Harborne JB. (1999) Recent advances in chemical ecology. Nat Prod Rep 16:509–23.
Hiscox JD, Israelstam GF. (1979) A method for the extraction of chlorophyll from leaf tissue without maceration. Can J Bot 57(12):1332–4.
Isman MB, Arnason JT, Towers GHN. (1995) Chemistry and biological activity of ingredients of other species of Meliaceae. In: Schmutterer H, Ascher KRS, Isman MB, Jacobson M, Ketkar CM, Kraus W, Rembold H, Saxena R editors. The neem tree *Azadirachta indica* A. Juss. and other Meliaceous plants. Sources of unique natural products for integrated pest management, medicine, industry and other purposes, Vol. XXIII. Eschborn, Germany: GTZ, pp. 652–66.
Kalsi PS, Vij VK, Singh OS, Wadia MS. (1977) Terpenoid lactones as plant growth regulators. Phytochemistry 16:784–6.
Klein JD, Cohen S, Hebbe Y. (2000) Seasonal variation in rooting ability of myrtle (*Myrtus communis* L.) cuttings. Sci Horticult, 83:71–6.
Kozubek A. (1999) Resorsinolic lipids, the natural non-isoprenoid phenolic amphiphiles and their biological activity. Chem Rev 99:1–26.
Kraus W. (1995) Biologically active ingredients. In: Schmutterer H editor. The neem tree *Azadirachta indica* A. Juss. and other Meliaceous plants. Sources of unique natural products for integrated pest management, medicine, industry and other purposes, Vol. XXIII. Weinhein, Germany: VCH Verlagsgesellschaft mbH, pp. 35–88.
Kubo I. (2000) Tyrosinase inhibitors from plants. Rev Latinoamer Química 28:7–20.
Kubo I, Chen Q-X, Nihei K-I, Calderón JS, Céspedes CL. (2003a) Tyrosinase inhibition kinetics of anisic acid. Zeit Naturforsch C: Biosci 58c:713–8.
Kubo I, Kins-Hori I, Nihei K-I, Soria F, Takasaki M, Calderón JS, Céspedes CL. (2003b) Tyrosinase inhibitors from galls of *Rhus javanica* leaves and their effects on insects. Zeit Naturforsch C: Biosci 58c:719–25.
Macías FA, Galindo JC, Massanet GM. (1992) Potential allelopathic activity of several sesquiterpene lactone models. Phytochemistry 31:1969–77.
Macías FA, Galindo JC, Massanet GM, Rodríguez-Luis J, Zubia E. (1993) Allelochemicals from *Pilocarpus goudotianus* leaves. J Chem Ecol 19:1371–9.
Macías FA, Simonet AM, Galindo JC, Castellano D. (1997) Bioactive steroids and triterpenes from *Melilotus messanensis* and their allelopathic potential. J Chem Ecol 23:1781–803.
Macías FA, Simonet AM, Galindo JC, Castellano D. (1999a) Bioactive phenolics and polar compounds from *Melilotus messanensis*. Phytochemistry 50:35–46.
Macías FA, Galindo JC, Castellano D, Velasco R. (1999b) Sesquiterpene lactones with potential use as natural herbicide models (I): *trans,trans*-Germacranolides. J Agric Food Chem 47:4407–14.
Macías FA, Molinillo JMG, Varela RM, Torres A, Galindo JC. (1999c) Bioactive Compounds from the Genus *Helianthus*. In: Macías FA, Galindo JC, Molinillo JMG, Cutler HG editors. Recent advances in allelopathy, Vol. I. Spain: Servicio de Publicaciones Universidad de Cádiz, pp. 121–48.
Macías FA, Galindo JC, Molinillo JM, Castellano D. (2000a) Dehydrozaluzanin C: A potent plant growth regulator with potential use as a natural herbicide template. Phytochemistry 54:165–71.
Macías FA, Galindo JC, Castellano D, Velasco RF. (2000b) Sesquiterpene lactones with potential use as natural herbicide models. 2. Guaianolides. J Agric Food Chem 48:5288–96.
Macías FA, Simonet AM, Pacheco P, Barrero A, Cabrera E, Jiménez-González D. (2000c) Natural and synthetic podolactones with potential use as natural herbicide models. J Agric Food Chem 48:3003–7.
Macías FA, Castellano D, Molinillo JMG. (2000d) Search for a standard phytotoxic bioassay for allelochemicals. Selection of standard target species. J Agric Food Chem 48:2512–21.
Marles RJ, Pazos-Sanou L, Compadre CM, Pezzuto JM, Blosyk E, Arnason JT. (1995) Sesquiterpene lactones revisited: recent developments in the assessment of biological activities and structure relationships. In: Arnason JT, Mata R, Romeo JT editors. Recent advances in phytochemistry, Vol. 29. New York: Plenum Press, pp. 333–56.

Mata R, Macías ML, Rojas IS, Lotina-Hennsen B, Toscano RA, Anaya AL. (1998) Phytotoxic compounds from *Esenbeckia yaxhoob*. Phytochemistry 49:441–9.
Mata R, Rivero-Cruz I, Rivero-Cruz B, Bye R, Timmermann BN. (2002) Sesquiterpene lactones and phenylpropanoids from *Cosmos pringlei*. J Nat Prod 65(7):1030–2.
Matsuda H, Kagerura T, Toguchida I, Ueda H, Morikawa T, Yoshikawa M. (2000) Inhibitory effects of sesquiterpenes from bay leaf on nitric oxide production in lipopolysaccharide – activated macrophages: structure requirement and role of heat shock protein induction. Life Sci 66(22):2151–7.
Mills JD, Mitchell P, Schurmann P. (1980) Modulation of coupling ATPase activity in intact chloroplasts. FEBS Lett 112:173–7.
Miranda-González V, Kazimir J, Chandrashekhar N, Weston L, Cheniae GM. (1997) Inhibition of a photosystem II electron transfer reaction by the natural product sorgoleone. J Agric Food Chem 45:1415–21.
Moreland DE, Novitzky WP. (1987) Effects of phenolics acids, coumarins and flavonoids on isolated chloroplasts and mitochondria. Allelochemicals: role in agriculture and forestry. Washington, DC: American Chemical Society, Symposium Series 330, pp. 247–261.
Morimoto M, Kumeda S, Komai K. (2000) Insect antifeedant flavonoids from *Gnaphalium affine* D. Don. J Agric Food Chem 48:1888–91.
Mucciarelli M, Camusso W, Bertea CM, Bossi S, Maffei M. (2001) Effect of (+)-pulegone and other oil components of *Mentha* × *piperita* on cucumber respiration. Phytochemistry 57:91–8.
Narwal SS, Palaniraj R, Sati SC, Kadian HS, Dahiya DS. (2003) Allelopathic plants: 8. *Parthenium hysterophorus* L. Allelopathy J 11(2):151–70.
Osses R, Torres P, Becerra J, Hoeneisen M, Silva M, Bittner M. (2005) Chemical characterization of essential oils occurring in five native tree species in Chilean temperate forests. J Chilean Chem Soc (submitted).
Pardo F, Perich F, Torres R, Delle-Monache F. (1998) Phytotoxic iridoid glucosides from the roots of *Verbascum thapsus*. J Chem Ecol 24:645–53.
Pardo F, Perich F, Torres R, Delle-Monache F. (2000) Stigmast-4-ene-3,6-dione, an unusual phytotoxic sterone from the roots of *Echium vulgare* L. Biochem Syst Ecol 28:911–3.
Ravanel P, Creuzet S, Tissut M. (1990) Inhibition effect of hydroxyflavones on the exogenous NADH dehydrogenase of plant mitochondrial inner membranes. Phytochemistry 29(2):441–5.
Reynolds T. (1987) Comparative effects of alicyclic compounds and quinones on inhibition of lettuce fruit germination. Ann Bot 60:215–23.
Rietveld WJ. (1983) Allelophatic effects of juglone on germination and growth of several herbaceous and weedy species. J Chem Ecol 9:295–308.
Rimando AM, Dayan F, Czarnota MA, Weston LA, Duke SO. (1998) A new photosystem II electron transfer inhibitor from *Sorghum bicolor*. J Nat Prod 61:927–30.
Rizvi SJH, Rizvi V. (1992) Allelopathy: basic and applied aspects. London: Chapman & Hall.
Robles-García K, Cuevas N, Mata R, Bye R, Lotina-Hennsen B. (1997) Photophosphorylation inhibition and uncoupling behavior of sesquiterpene lactones on photosynthesis. In: Proceedings of X Congreso de Bioenergetica y Biomembranas BEB 97, Soc. Mex. Bioquím, Vol. 16, No. Especial, p. 61.
Rodríguez E, Towers GHN, Mitchell JC. (1976) Biological activities of sesquiterpene lactones. Phytochemistry 15:1573–80.
Rzedowski J. (1972) Contribuciones a la fitogeografia florística e historia de Mexico III. Algunas tendencias en la distribución geográfica y ecología de las Compositae mexicanas. Ciencia (Mexico) 27:123–32.
Rzedowski J. (1991) Diversidad y orígenes de la flora fanerogámica de Mexico. Acta Botánica Mexicana 14:3–21.
Seigler DS. (1996) Chemistry and mechanisms of allelopathy interactions. Agronomy J 88: 876–85.
Serrato B, Vásquez M, Calderón-Mugica JC, Marín JC, Céspedes CL. (2005) Antioxidant and plant growth regulatory activities of extracts and coumarins from Mexican Tarragon (Scent Mexican Marigold), *Tagetes lucida* (Asteraceae). J Agric Food Chem (submitted).

Singh HP, Batish DR, Kohli RK, Saxena DB, Vandana Arora R. (2002) Effect of parthenin – a sesquiterpene lactone from *Parthenium hysterophorus* – on early growth and physiology of *Ageratum conyzoides*. J Chem Ecol 28(11):2169–79.

Stenlid G. (1970) Flavonoids as inhibitors of the formation of adenosine triphosphate in plant mitochondria. Phytochemistry 9:2251–6.

Torres P, Ávila JG, Romo de Vivar A, García AM, Marín JC, Aranda E, Céspedes CL. (2003) Antioxidant and Insect growth regulatory activities of Stilbenes and extracts from *Yucca periculosa*. Phytochemistry 64:463–73.

Torres P, Becerra J, Hoeneisen M, Bittner M, Silva M, Cespedes CL. (2006) Antioxidant and plant growth regulatory activities of essential oils occurring in five native tree species in Chilean temperate forests. J Agric Food Chem (submitted).

Venkataiah B, Ramesh C, Ravindranath N, Biswanath Das. (2003) Charminarone, a seco-pseudoguaianolide from *Parthenium hysterophorus*. Phytochemistry 63(4):383–6.

Winning BM, Sarah CJ, Leaver CJ. (1995) Protein import into plant mitochondria. In: San Pietro A, editor. Methods in enzymology. New York: Academic Press, Vol. 260, Section I, "Methodology", pp. 293–302.

M.T.H. Khan and A. Ather (eds.)
Lead Molecules from Natural Products

Metabolomics – systematic studies of the metabolic profiling

MAHMUD TAREQ HASSAN KHAN, ARJUMAND ATHER

Abstract

Metabolomics is the systematic study of the unique chemical characteristics of an organism. It may be higher or lower plants or may be prokaryotes. Phytometabolomics is the science dealing with the metabolic profiling of plants. Here, we discuss about the rational and analytical technologies used in metabolomics. We will also discuss in general about some of the key applications of metabolomics, for instance in toxicology, functional genomics, nutrigenomics, aging research, and lipid profiling. The metabolic profiling is still under development, but it is likely to become quite an important issue.

Keywords: *metabolomics, phytometabolomics, metabolic profiling, nutrigenomics, omics, GC–MS, LC–MS, functional genomics, nutrigenomics*

Abbreviations used: CE, capillary electrophoresis; EI, electron impact ionization; EST, expressed sequence tag; GC–MS, gas chromatography–mass spectrometry; HPLC, high-performance liquid chromatography; HT, high-throughput; LC–MS, liquid chromatography–mass spectrometry; m/z, mass-to-charge ratio; NMR, nuclear magnetic resonance; TOF, time-of-flight.

I. What is metabolomics?

'Metabolomics' is the 'systematic investigation of the unique chemical fingerprints that specific cellular processes leave behind' – specifically, the study of their small-molecule metabolite profiles (Daviss, 2005). The 'metabolome' represents the collection of all metabolites in a biological organism, which are the end products of its gene expression. Thus, while mRNA gene expression data and proteomic analyses do not tell the whole story of what might be happening in a cell, metabolic profiling can give an instantaneous 'snapshot' of the physiology of that cell. One of the challenges of systems biology is to integrate proteomics, transcriptomics and metabolomics information to give a complete picture of living organisms. The word 'metabonomics' is also used, particularly in the context of drug-toxicity assessment. There is some disagreement over the exact differences between 'metabolomics' and 'metabonomics'; in general, the term 'metabolomics' is more commonly used (Anonymous, 2005). So according to these definitions 'phytometabolomic' is the study of the metabolic profiling of plants.

II. What is the rationale for doing this analysis?

The classical genetic paradigm states that the gene encodes a protein that affects a trait. The protein affects the trait directly or through the chemical conversion of a metabolite. The flow diagram of Figure 1 shows this genetic paradigm.

Therefore, by knowing the metabolites we are close to the trait. Once we know this we can:

1. specifically monitor that metabolite while developing new germplasm optimized for the trait; and
2. define genes that affect that metabolite and use targeted genetic strategies to affect the trait.

III. Analytical technologies used in phytometabolomics

Metabolic profiling, also called metabolomics, is a high-throughput method to characterize many metabolites in (tissues or organs of) an organism.

The analysis is typically performed using gas chromatography (GC) or liquid chromatography (LC) followed by mass spectrometry (MS) for identification of compounds.

By performing detailed chemical profiles of the samples, results in very large datasets. These datasets are then subjected to statistical analyses to identify (sets of) compounds that differ significantly between samples. From these analyses one will be able to distinguish the samples based on their chemical differences.

There are two issues to be addressed for metabolite analysis:

1. Separation of the analytes, usually by. Electrophoresis, particularly capillary electrophoresis, is also used.

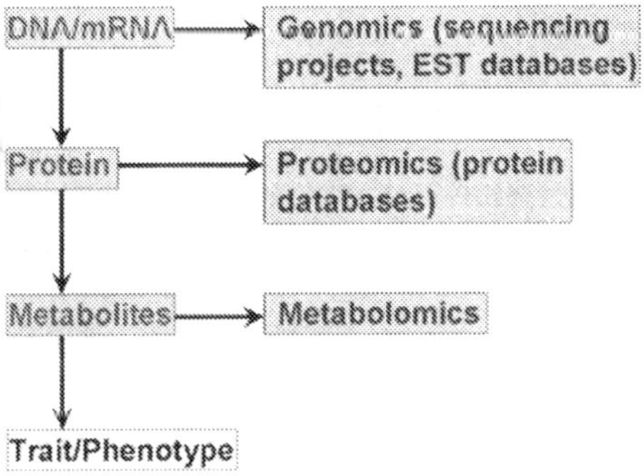

Fig. 1. Flow chart showing the classical genetic paradigm (various 'omics). Here gene encodes a protein that ultimately affects the trait. This trait is affected directly by the protein or by chemical conversion of a specific metabolite.

2. Detection of the analytes, following separation by chromatographic or other methods.

III.A. Separation methods

- *Gas chromatography.* (Figure 2 shows a typical setup for GC), especially when interfaced with MS (GC–MS), is one of the most widely used and powerful methods. It offers very high chromatographic resolution, but requires chemical derivatization for many biomolecules: only volatile chemicals can be analyzed without derivatization. Some large and polar metabolites cannot be analyzed by GC. The GC is basically an oven that contains a 20–60 m long capillary column (wound into a coil). The internal diameter is 0.2–0.5 mm. The column on the inner side contains a thin film of resin. The column is connected to a pressurized tank of an inert gas (usually helium) that creates a flow of gas through the column. The sample (containing a mixture of compounds) is injected at the top of the column when the temperature is low (typically between 40 and 80°C). The temperature is then raised and the different compounds are separated based on their volatility and their affinity for the column. The retention time is the time taken by a compound to elute through the column.
- *High-performance liquid chromatography (HPLC).* Compared to GC, HPLC has lower chromatographic resolution, but it is of an advantage to potentially measure much wider range of analytes.
- *Capillary electrophoresis (CE).* So far, there are few publications on the use of CE for metabolite profiling. This will no doubt change, as there are a number of advantages of CE: it has a higher theoretical separation efficiency than the HPLC, and is suitable for use with a wider range of metabolite classes than is GC. As for all electrophoretic techniques, it is the most appropriate one for charged analytes.

III.B. Detection methods

- *Mass spectrometry (MS)* is used to identify and to quantify metabolites after separation by GC, HPLC or CE. GC–MS is the most 'natural' combination of the

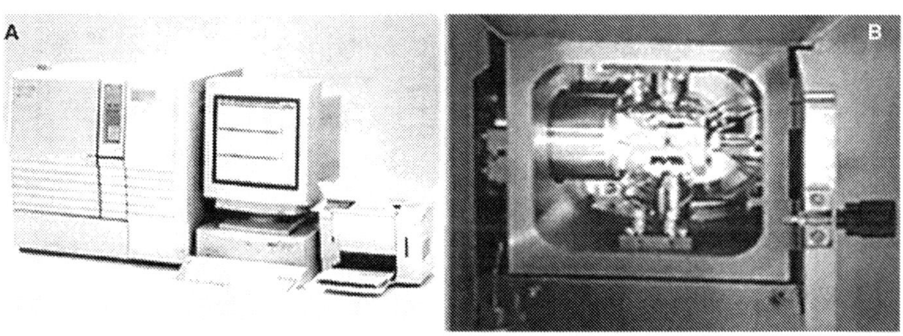

Fig. 2. (A) Shows a typical gas chromatograph and (B) a sample chamber.

three, and was the first to be developed. In addition, mass spectral fingerprint libraries exist or can be developed that allow identification of a metabolite according to its fragmentation pattern. MS is both sensitive (although, particularly for HPLC–MS, sensitivity is more of an issue as it is affected by the charge on the metabolite, and can be subject to ion suppression artifacts) and can be very specific. There are also a number of studies which use MS as a stand-alone technology: the sample is infused directly into the mass spectrometer with no prior separation, and the MS serves to both separate and to detect metabolites. The traditional (not the time-of-flight (TOF)) mass spectrometer ionizes the sample (in this case an individual compound) with the use of electrons (electron impact ionization, EI, also called EIMS). The resulting ions are then accelerated in an electric field and then subjected to a magnetic field that causes the ions to deviate from a straight course. The deviation is a function of the mass-to-charge (m/z) ratio. Many of the ions fragment upon EI. Between the m/z ratio of the molecular ion (unfragmented) and the fragmentation pattern that is observed it is possible to determine the identity of the compound that eluted off the GC column. Figure 3 shows a typical example of GC coupled with MS, two peaks of a GC chromatogram have been converted in to the mass spectra of two individual compounds.

- *Nuclear magnetic resonance (NMR) spectrometry.* NMR is the only detection technique that does not rely on the separation of the analytes, and the sample can

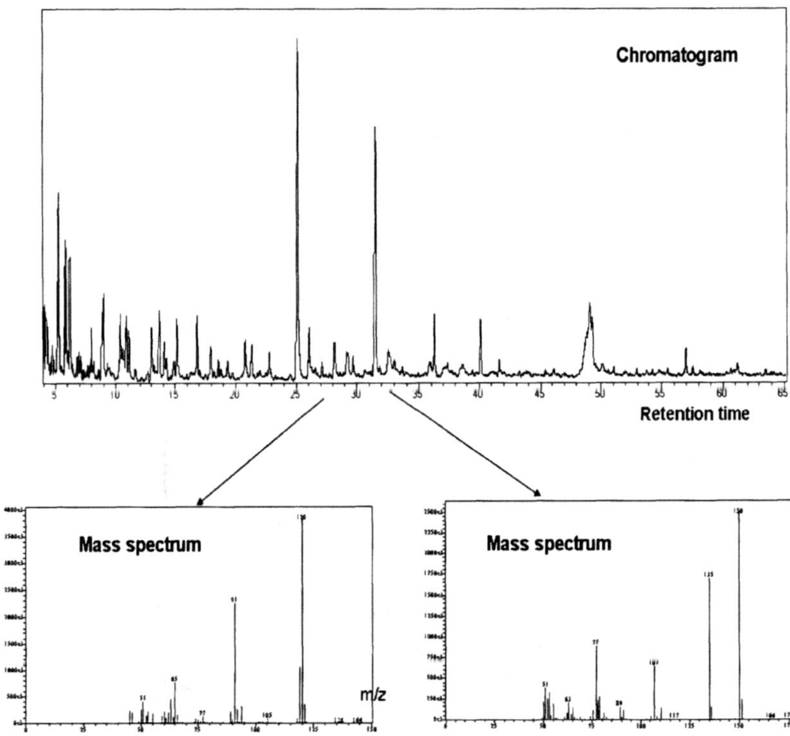

Fig. 3. A typical example of GC–MS. Two peaks from a GC chromatogram have been converted into the mass spectra of two individual compounds.

thus be recovered for further analyses. All kinds of small molecule metabolite can be measured simultaneously – NMR is close to being a universal detector. However, it also possesses one major disadvantage that it is relatively insensitive compared with MS-based techniques.

- *Other techniques.* MS and NMR are by far the two leading technologies for metabolomics. Other methods of detection used include electrochemical detection (coupled with HPLC) and radiolabel (when combined with thin-layer chromatography).

IV. Analysis of the data from different MS techniques

Extensive data are usually found in the LC–MS or GC–MS type of experiments. It is very crucial to analyze these data to understand the whole profile of that particular sample. Different statistical techniques are used to analyze this kind of data. The main type of technique is discriminant analysis. This technique is very useful to analyze a large amount of complex data sets. Figure 4 shows an example of discriminant analysis of a large and complex set of data.

The LC–MS technology has been increasingly used for differential profiling, i.e., broad screening of biomolecular components across multiple samples in order to elucidate the observed phenotypes and discover biomarkers. One of the major challenges in this domain remains the development of better solutions for processing of

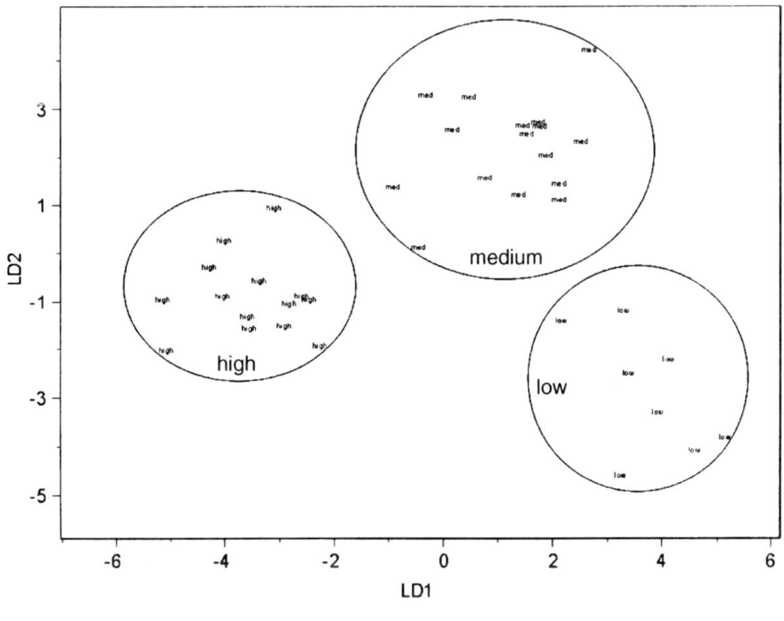

Fig. 4. An example of discriminant analysis of a large and complex set of data, where chemical profiles are indicated by 10 maize lines with different feed qualities (low, medium and high quality as determined by feeding trials). The chemical profiles can be used to classify the lines into the different quality groups. This helps to understand the chemical basis for feed quality.

LC–MS data. More recently, Katajamaa and Oresic (2005) have developed and reported a software package ***MZmine*** that enables differential LC–MS analysis of metabolomics data.

The author reported that this software package enables differential LC–MS analysis of metabolomics data. This software is a toolbox containing methods for all data-processing stages preceding differential analysis: spectral filtering, peak detection, alignment and normalization (Katajamaa and Oresic, 2005). Specifically, they have developed and implemented a new-recursive peak search algorithm and a secondary peak picking method for improving already aligned results, as well as a normalization tool that uses multiple internal standards. Visualization tools enable comparative viewing of data across multiple samples. Peak lists can be exported into other data-analysis programs. They also reported that the toolbox has already been utilized in a wide range of applications, which demonstrated the utilities of the program on an example of metabolic profiling of *Catharanthus roseus* cell cultures (Katajamaa and Oresic, 2005).

The software is freely available under the GNU General Public License and it can be obtained from the project web page at: http://mzmine.sourceforge.net/ (Katajamaa and Oresic, 2005).

V. Major applications of the metabolomics

The key applications of the metabolomics or metabolic profiling could be in toxicology, functional genomics, nutrigenomics, etc.

- *Toxicity assessment/toxicology.* Global metabolic profiling (metabonomics/metabolomics) has shown particular promise in the area of toxicology and drug development. In both pre-clinical screening and mechanistic exploration, metabolic profiling can offer rapid, non-invasive toxicological information that is robust and reproducible, with little or no added technical resources to existing studies in drug metabolism and toxicity (Keun, 2005). In many cases, the observed changes can be related to specific syndromes, e.g. a specific lesion in liver or kidney (Heijne et al., 2005). This is of particular relevance to pharmaceutical companies wanting to test the toxicity of potential candidates: if a compound can be eliminated before it reaches on the grounds of adverse toxicity, it saves the enormous expense of the trials. Extended into the assessment of efficacy and toxicity in the clinic, metabonomics may prove crucial in making personalized therapy and pharmacogenomics a reality (Keun, 2005).
- *Functional genomics.* Metabolomics can be an excellent tool for determining the caused by a genetic manipulation, such as gene deletion or insertion. Sometimes this can be a sufficient goal in itself – for instance, to detect any phenotypic changes in a genetically modified plant intended for human or animal consumption. More exciting is the prospect of predicting the function of unknown by comparison with the metabolic perturbations caused by deletion/insertion of known genes. Such advances are most likely to come from such as *Saccharomyces cerevisiae* and (Daviss, 2005).
- *Nutrigenomics.* It is a generalized term that links genomics, transcriptomics, proteomics and metabolomics to human nutrition. In general, a metabolome in a

given body fluid is influenced by endogenous factors such as age, sex, body composition and genetics as well as underlying pathologies. The large bowel microflora is also a very significant potential confounder of metabolic profiles and could be classified as either an endogenous or exogenous factor. The main exogenous factors are diet and drugs. Diet can then be broken down into nutrients and non-nutrients. Metabolomics is one means to determine a biological endpoint, or metabolic fingerprint, which reflects the balance of all these forces on an individual's metabolism (Matsuzaki et al., 2005; Zeisel et al., 2005).

- *In aging research.* Metabolomics is based on the synchronized investigation of numerous low-molecular weight metabolites from an experimental sample. Metabolomics seeks the most up-to-date information about the state of interaction between an organism and its environment. The ability to use metabolomics approaches for classification and mechanistic studies may influence and augment our ability to study and address the aging process scientifically and clinically (Kristal and Shurubor, 2005).
- *Lipid profiling in the process of drug development.* Recently, Morris and Watkins (2005) reviewed the utilities of advances in lipid profiling through focused metabolomics in the process of drug development. They also stated that, this kind of highly parallel analytical technologies comprising 'omics promised to improve drug development efficiency dramatically by increasing knowledge and improving decision-making capabilities (Morris and Watkins, 2005).

VI. Genomics and metabolomics

Hocquette (2005) in a review stated that, studying genomics enables scientists to analyze genes and their products on a large scale. The first high-throughput (HT) techniques to be developed were sequencing methods. A great number of genomes from different organisms have thus been sequenced. Genomics is now shifting to the study of gene expression and their related functions. In the past 5–10 years genomics, proteomics and HT microarray technologies have fundamentally changed our ability to study the molecular basis of cells and tissues in health and diseases, giving a new comprehensive view. For example, in cancer research we have seen new diagnostic opportunities for tumor classification, and prognostication. A new exciting development is metabolomics and lab-on-a-chip techniques for metabolic studies. These techniques combine miniaturization and automation (Hocquette, 2005).

However, to interpret the large amount of data arising from HT experiments, extensive computational development is required. In the coming years, we will see the study of biological networks dominating the scene in physiology. The great accumulation of genomics information will be used in computer programs to simulate biologic processes. Originally developed for genome analysis, bioinformatics now encompasses a wide range of fields in biology from gene studies to integrated biology (i.e. combination of different data sets from genes to metabolites). This is systems biology, which aims to study biological organisms as a whole. In medicine, scientific results and applied biotechnologies arising from genomics will be used for effective prediction of diseases and risk associated with drugs. Preventive medicine and medical therapy will be personalized. Widespread applications of genomics for personalized medicine will require associations of gene expression pattern with diagnoses, treatment and clinical data. This will help in the discovery and development of news drugs for important clinical problems as well as disease conditions (Hocquette, 2005).

In the field of agriculture and animal sciences, the upshots of genomics will comprise enhancement in food safety, crop yielding, traceability and the quality of animal products (like dairy products as well as meat) through augmented effectiveness in breeding and healthier informations of animal physiology. Genomics and the integrated biology are huge tasks and no single lab can pursue this alone. We are probably at the end of the beginning rather than at the beginning of the end because genomics will probably change biology to a greater extent than previously forecasted. In addition, there is a great need for more information and better understanding of genomics before complete public acceptance (Hocquette, 2005).

The integrated functional genomics and metabolomics nowadays have been approached even to prognose several disease conditions. For instance, Ippolito and coworkers (2005) reported this kind of an integrated approach to define prognosis in human neuroendocrine cancers.

VII. Conclusion

The global metabolic profiling (metabonomics/metabolomics) is a technique that is still under development, but likely to become quite important. Metabolomics will be a major tool in recognizing compounds connected with activity in the traditional medicines, and will also be very useful in recognizing the effect on the test organism, which can be the patient in case of clinical trials with well-established traditional medicines (Verpoorte et al., 2005; Wang et al., 2005).

Frequently, metabolic profiling is done by utilizing either GC coupled with MS or LC coupled with MS, which ultimately produce large amount of data. To handle large and complex data several computational tools are being developed. Research is going on in this context.

More interestingly several scientists are utilizing integrated metabolomics approaches to study the aging processes.

References

Anonymous. (2005) http://en.wikipedia.org/wiki/Metabolomics#endnote_title_definition.

Daviss B. (2005) Growing pains for metabolomics. The Scientist 19(8):25–8.

Heijne WH, Lamers RJ, van Bladeren PJ, Groten JP, van Nesselrooij JH, van Ommen B. (2005) Profiles of metabolites and gene expression in rats with chemically induced hepatic necrosis. Toxicol Pathol 33(4):425–33.

Hocquette JF. (2005) Where are we in genomics? J Physiol Pharmacol 3:37–70.

Ippolito JE, Xu J, Jain S, Moulder K, Mennerick S, Crowley JR, Townsend RR, Gordon JI. (2005) An integrated functional genomics and metabolomics approach for defining poor prognosis in human neuroendocrine cancers. Proc Natl Acad Sci USA 102(28):9901–6.

Katajamaa M, Oresic M. (2005) Processing methods for differential analysis of LC/MS profile data. BMC Bioinform 6:179.

Keun HC. (2005) Metabonomic modeling of drug toxicity. Pharmacol Ther (July 25).

Kristal BS, Shurubor YI. (2005) Metabolomics: opening another window into aging. Sci Aging Knowledge Environ 2005(26):pe19.
Matsuzaki K, Kato H, Sakai R, Toue S, Amao M, Kimura T. (2005) Transcriptomics and metabolomics of dietary leucine excess. J Nutr 135(Suppl 6):1571S–5S.
Morris M, Watkins SM. (2005) Focused metabolomic profiling in the drug development process: advances from lipid profiling. Curr Opin Chem Biol 9(4):407–12.
Verpoorte R, Choi YH, Kim HK. (2005) Ethnopharmacology and systems biology: a perfect holistic match. J Ethnopharmacol 100(1–2):53–6.
Wang M, Lamers RJ, Korthout HA, van Nesselrooij JH, Witkamp RF, van der Heijden R, Voshol PJ, Havekes LM, Verpoorte R, van der Greef J. (2005) Metabolomics in the context of systems biology: bridging traditional Chinese medicine and molecular pharmacology. Phytother Res 19(3):173–82.
Zeisel SH, Freake HC, Bauman DE, Bier DM, Burrin DG, German JB, Klein S, Marquis GS, Milner JA, Pelto GH, Rasmussen KM. (2005) The nutritional phenotype in the age of metabolomics. J Nutr 135(7):1613–6.

Contributors

Aniele R Agner – Departamento de Biologia Geral, Universidade Estadual de Londrina, PR, Brazil.

Adaíla Marta Paixão Almeida – Departamento de Química Orgânica e Inorgânica, Centro de Ciências, Universidade Federal do Ceará, P.O. Box 12200, 60021-940, Fortaleza, Ceará, Brazil.

Arjumand Ather – ER-GenTech, Department of Biochemistry and Molecular Biology, University of Ferrara, Via Fosato di Mortara 74, Ferrara 44100, Italy.

J Guillermo Avila – Phytochemistry Lab, UBIPRO-FES-Iztacala, Universidad Nacional Autónoma de México, Coyoacán 04510, México D.F., México.

France Philippe Bernard – Greenpharma S.A. 3, Allée de Titane 45100, Orléans, France.

Fernando Antônio Frota Bezerra – Unidade de Farmacologia Clínica, Departamento de Fisiologia e Farmacologia, Faculdade de Medicina, Universidade Federal do Ceará, P.O. Box 3219, 60.431-970, Fortaleza, Ceará, Brazil.

Nicoletta Bianchi – Department of Biochemistry and Molecular Biology and Biotechnology Center, Ferrara University, Ferrara, Italy.

Monica Borgatti – Department of Biochemistry and Molecular Biology and Biotechnology Center, Ferrara University, Ferrara, Italy.

Janaína Keila P Câmara – Departamento de Química, UFRN, Natal, RN, Brazil.

Cristina MTS Carneiro – Departamento de Farmacologia, UFRJ, Brazil.

Carlos L Céspedes – Instituto de Química, Universidad Nacional Autónoma de México, Coyoacán 04510, México D.F., México.

Ilce MS Cólus – Departamento de Biologia Geral, Universidade Estadual de Londrina, PR, Brazil.

Letícia Veras Costa-Lotufo – Departamento de Fisiologia e Farmacologia, Faculdade de Medicina, Universidade Federal do Ceara, P.O. Box-3157, 60430-270 Fortaleza, Ceara, Brazil.

Tereza Neuma C Dantas – Departamento de Química, UFRN, Natal, RN, Brazil.

Noungoue Tchamo Diderot – Department of Organic Chemistry, Faculty of Science, University of Yaounde I, P.O. Box 812 Yaounde, Cameroon.

José-Antonio Fernández – School of Agronomy (ETSIA) and Head of the Biotechnology Division, Institute for Regional Development (IDR), University of Castilla-La Mancha, Campus Universitario s/n, 02071 Albacete, Spain.

Quoc-Tuan Do – Greenpharma S.A. 3, Allee de Titane 45100, Orleans, France.

Mariana Domínguez – Instituto de Química, Universidad Nacional Autónoma de México, Coyoacán 04510, México D.F., México.

Aurea Echevarria – Departamento de Química, ICE/UFRRJ, Rio Janeiro, RJ, Brazil.

Juliana Echevarria-Lima – Laboratório de Imunonolgia Tumoral, Departamento de Bioquímica Médica, ICB, CSS, UFRJ, Rio de Janeiro, RJ, Brazil.

Andressa Esteves-Souza – Departamento de Química, ICE/UFRRJ, Rio Janeiro, RJ, Brazil.

Tsamo Etienne – Department of Organic Chemistry, Faculty of Science, University of Yaounde I, P.O. Box 812 Yaounde, Cameroon.

Giordana Feriotto – Department of Biochemistry and Molecular Biology and Biotechnology Center, Ferrara University, Ferrara, Italy.

Sona Franova – Department of Pharmacology, Jessenius Faculty of Medicine, Comenius University, Sklabinska 26 037 53 Martin, Slovak Republic.

Ken-ichi Fujita – Department of Environmental Science, Policy and Management, University of California, Berkeley, CA 94720-3112, USA.

Hala Gali-Muhtasib – Department of Biology, American University of Beirut, Beirut, Lebanon.

Roberto Gambari – Department of Biochemistry and Molecular Biology, University of Ferrara, Ferrara, Italy.

Noema F Grynberg – Departamento de Química, ICE/UFRRJ, Rio Janeiro, RJ, Brazil.

Carlos R Kaiser – Instituto de Química, UFRJ, Centro de Tecnologia, Rio de Janeiro, RJ, Brazil.

Mahmud Tareq Hassan Khan – Pharmacology Research Lab, Faculty of Pharmaceutical Sciences, University of Science and Technology Chittagong, Chittagong, Bangladesh.

Isao Kubo – Department of Environmental Science, Policy and Management, University of California, Berkeley, CA 94720-3112, USA.

Elisabetta Lambertini – Department of Biochemistry and Molecular Biology and Biotechnology Center, Ferrara University, Ferrara, Italy.

Ilaria Lampronti – Department of Biochemistry and Molecular Biology, University of Ferrara, Ferrara, Italy.

Antônio J Lapa – Departamento de Farmacologia, Setor de Produtos Naturais, EPM-UNIFESP, São Paulo, SP, Brazil.

Albert Leyva – Department of Pharmacology, University of Missouri-Kansas City, 2411 Holmes, Kansas City, MO 64108, USA.

Tito Monteiro da Cruz Lotufo – Departamento de Engenharia de Pesca, Centro de Ciências Agrárias, Universidade Federal do Ceará, P.O. Box 12168, 60455-760, Fortaleza, Ceará, Brasil.

Aya Kubo – Department of Environmental Science, Policy and Management, University of California, Berkeley, CA 94720-3112, USA.

Maria Aparecida M Maciel – Departamento de Quimica, Universidade Federal do Rio Grande do Norte, Campus Universitario, 59078-970, Natal, RN, Brazil.

Juan C Marín – Instituto de Química, Universidad Nacional Autónoma de México, Coyoacán 04510, México D.F., México.

Toshiro Matsui – Division of Bioscience and Biotechnology, Faculty of Agriculture, Graduate School of Kyushu University, 6-10-1 Hakozaki, Higashi-ku, Fukuoka 812-8581, Japan.

Kiyoshi Matsumoto – Division of Bioscience and Biotechnology, Faculty of Agriculture, Graduate School of Kyushu University, 6-10-1 Hakozaki, Higashi-ku, Fukuoka 812-8581, Japan.

Carlo Mischiati – Department of Biochemistry and Molecular Biology, University of Ferrara, Ferrara, Italy.

Juraj Mokry – Department of Pharmacology, Jessenius Faculty of Medicine, Comenius University, Skalabinska 26, 037 53, Martin, Slovak Republic.

Manoel Odorico de Moraes – Departamento de Fisiologia e Farmacologia, Faculdade de Medicina, Universidade Federal do Ceará, P.O. Box-3157, 60430-270 Fortaleza, Ceará, Brazil.

Maria Elisabete Amaral de Moraes – Unidade de Farmacologia Clinica, Departamento de Fisiologia e Farmacologia, Faculdade de Medicina, Universidade Federal do Ceara, P.O. Box 3219, 60.431-970, Fortaleza, Ceara, Brazil.

Ken-ichi Nihei – Department of Environmental Science, Policy and Management, University of California, Berkeley, CA 94720-3112, USA.

Gabriela Nosalova – Department of Pharmacology, Jessenius Faculty of Medicine, Comenius University, Sklabinska 26 036 01 Martin, Slovak Republic.

İlkay Orhan – Faculty of Pharmacy, Department of Pharmacognosy, Gazi University, 06330 Ankara, Turkey.

Nuno A Pereira – Departamento de Farmacologia, UFRJ, Brazil.

Cláudia Pessoa – Departamento de Fisiologia e Farmacologia, Faculdade de Medicina, Universidade Federal do Ceara, P.O. Box 3157, 60430-270, Fortaleza, Ceara, Brazil.

Angelo C Pinto – Instituto de Química, UFRJ, Centro de Tecnologia, Rio de Janeiro, RJ, Brazil.

Kenia Pissinate – Departamento de Química, ICE/UFRRJ, Rio Janeiro, RJ, Brazil.

Roberta Piva – Department of Biochemistry and Molecular Biology and Biotechnology Center, Ferrara University, Ferrara, Italy.

Regine Schneider-Stock – Department of Pathology, Otto-von-Guericke University of Magdeburg, Germany.

Bilge Şener – Department of Pharmacognosy, Faculty of Pharmacy, Gazi University, P.O. Box 143, 06572 Maltepe – Ankara, Turkey.

Alessia Sereni – Department of Biochemistry and Molecular Biology, University of Ferrara, Ferrara, Italy.

Blanca Serrato – Cenapros INIFAP/Morelia, Alvaro Obregón, Michoacán, México.

Ngouela Silvere – Department of Organic Chemistry, Faculty of Science, University of Yaounde I, P.O. Box 812 Yaounde, Cameroon.

Kenneth D Thompson – MC 0001, Clinical Microbiology, The University of Chicago, 5841 South Maryland Avenue Chicago, IL 60637, USA.

Frederico A Vanderlinde – Departamento de Ciências Fisiológicas, IB, UFRRJ, Brazil.

Valdir F Veiga Jr. – Instituto de Química, UFRJ, Centro de Tecnologia, Rio de Janeiro, RJ, Brazil.

Subject Index

Printed in the United Kingdom
by Lightning Source UK Ltd.
134108UK00002B/34-240/A

9 780444 516190